FATIGUE UNDER THERMAL AND MECHANICAL LOADING:

MECHANISMS, MECHANICS AND MODELLING

Fatigue under Thermal and Mechanical Loading:

Mechanisms, Mechanics and Modelling

Proceedings of the Symposium held at
Petten, The Netherlands, 22–24 May 1995

Edited by

J. BRESSERS
Institute for Advanced Materials,
Petten, The Netherlands

and

L. RÉMY
Centre des Matériaux,
Ecole des Mines de Paris,
Evry, France

M. STEEN and J. L. VALLÉS (Co-Editors)
Institute for Advanced Materials,
Petten, The Netherlands

KLUWER ACADEMIC PUBLISHERS
DORDRECHT / BOSTON / LONDON

A C.I.P. Catalogue record for this book is available from the Library of Congress.

ISBN 0-7923-3993-2
EUR 16353 EN

Published by Kluwer Academic Publishers,
P.O. Box 17, 3300 AA Dordrecht, The Netherlands.

Kluwer Academic Publishers incorporates
the publishing programmes of
D. Reidel, Martinus Nijhoff, Dr W. Junk and MTP Press.

Sold and distributed in the U.S.A. and Canada
by Kluwer Academic Publishers,
101 Philip Drive, Norwell, MA 02061, U.S.A.

In all other countries, sold and distributed
by Kluwer Academic Publishers Group,
P.O. Box 322, 3300 AH Dordrecht, The Netherlands.

Printed on acid-free paper

Printed in the Netherlands

CONTENTS

DAMAGE ACCUMULATION AND CRACK GROWTH

PREFACE

The International Symposium "Fatigue under Thermal and Mechanical Loading", held at Petten (The Netherlands) on May 22-24, 1995, was jointly organized by the Institute for Advanced Materials of The Joint Research Centre, E.C., and by the Société Française de Métallurgie et de Matériaux.

The fast heating and cooling cycles experienced by many high temperature components cause thermally induced stresses, which often operate in combination with mechanical loads. The resulting thermal / mechanical fatigue cycle leads to material degradation mechanisms and failure modes typical of service cycles. The growing awareness that the synergism between the combined thermal and mechanical loads can not be reproduced by means of isothermal tests, has resulted in an increasing interest in thermal and thermo-mechanical fatigue testing. This trend has been reinforced by the constant pull by industry for more performant, yet safer high temperature systems, pushing the materials to the limit of their properties. Dedicated ASTM meetings in particular have set the scene for this area of research. The proceedings of the symposium organized by D.A. Spera and D.F.Mowbray in 1975 provided a reference book on thermal fatigue which reflects the knowledge and experimental capabilities of the mid-seventies. The advent of microcomputers and their increasing availability in the following fifteen years have resulted in a rapid expansion of this field of research by enabling the application of complex strain-temperature histories to testing specimens with a view to simulating the behaviour of critical volume elements of components. The proceedings of the ASTM symposium on thermo-mechanical fatigue organized by H.Sehitoglu in 1991 illustrate the progress achieved in the development of experimental control and in life modelling.

Combined thermal and mechanical loading is a critical issue in components for gas turbines. In Europe, research in this area has been conducted in the collaborative programmes on Materials for Gas Turbines and other High Temperature Applications within the framework of COST 50 and COST 501, which has stimulated cross-border partnership and cooperation. The present symposium is the result of long standing contacts and friendship which developed within the COST framework between the conference organisers, who felt in 1993 that this field of research in Europe had matured to the point that it warranted the organizing of a conference in the old continent. The venue of the Joint Research Centre of the European Commission at Petten was actually a symbol per se.

The symposium addressed the mechanisms, mechanics and modelling of thermal and thermal-mechanical load effects on conventional and advanced structural materials. It was intended as a forum for disseminating recent research results and for discussing and exchanging views amongst material scientists, mechanical engineering experts and design engineers from industry. It brought together some 100 participants from 21 countries, representing the broad range of backgrounds relevant to the topic of the meeting. More than 80 abstracts were received of which 55 were accepted by the Programme Committee. The proceedings contain 50 papers. Much to the regret of the Programme Committee, the final manuscripts of some of the papers were not submitted. Contributed papers were delivered as oral presentations and in two poster sessions. Special thanks are due to Drs. Rezaï-Aria and P. Skelton for managing the difficult task of introducing the poster presentations to the audience. Prof. H. Sehitoglu closed the symposium with an outline of today's main achievements in terms of life prediction of thermally loaded components and with an overview of the perspectives for thermo-mechanical fatigue research in the USA.

The editors wish to thank all the authors and delegates for their contributions. The input of a number of persons has been instrumental in turning the Symposium into an interesting and succesfull event. In particular we wish to thank the members of the International Committee, of the Programme and Organizing Committee, and the session chairmen. Thanks are also due to Dr. H.J. Helms, director of programmes of the JRC, for officially opening the Symposium, and in particular to Mrs. Y. Den Biesen-Hey, Mrs. C. Schauwers and Mrs. G. Schilder for their invaluable assistance during the symposium and with the editing of the proceedings.

J.Bressers, L.Rémy

International Symposium "Fatigue under Thermal and Mechanical Loading : Mechanisms, Mechanics and Modelling"

International Committee

J.L. Chaboche	ONERA, Chatillon, France
A. Del Puglia	University of Firenze, Italy
E.G. Ellison	University of Bristol, Great Britain
P. J. Henderson	SIMR, Stockholm, Sweden
J.W. Holmes	University of Michigan, U.S.A.
J. King	Rolls-Royce plc, Derby, Great Britain
H.P. Lieurade	CETIM, Senlis, France
D. Löhe	Universität Karlsruhe, Germany
H. Mughrabi	Universität Erlangen-Nurnberg, Germany
R. Ohtani	Kyoto University, Japan
V. Panasyuk	Karpenko Phys.-Mech. Inst., Lviv, Ukraine
F. Rezai-Aria	Ecole Polytechnique, Lausanne, Switzerland
H. Sehitoglu	University of Illinois, U.S.A.
P. Skelton	HTMTC, London, Great Britain
M. Steen	IAM-JRC, E.C., Petten, The Netherlands
V.T. Troshchenko	Academy of Sience, Kiev, Ukraine
J.L. Vallés	IAM-JRC, E.C., Petten, The Netherlands

Programme and Organizing Committee

J. Bressers	IAM-JRC, E.C., Petten, The Netherlands
M. De Groot	IAM-JRC, E.C., Petten, The Netherlands
E. Fleury	Ecole des Mines, Paris, France
L. Rémy	Ecole des Mines, Paris, France
B. Seysener	IAM-JRC, E.C., Petten, The Netherlands
M. Steen	IAM-JRC, E.C., Petten, The Netherlands
J.L. Vallés	IAM-JRC, E.C., Petten, The Netherlands

Session chairmen and poster reporters

D. Löhe	Universität Karlsruhe, Germany
H. Mughrabi	Universität Erlangen-Nurnberg, Germany
D. Smith	University of Bristol, Great Britain
A. Bennett	Rolls-Royce plc, Great Britain
R. Ohtani	Kyoto University, Japan
P. Skelton	HTMTC, London, Great Britain
P. J. Henderson	SIMR, Stockholm, Sweden
H. Sehitoglu	University of Illinois, U.S.A.
V.T. Troshchenko	National Academy of Sience, Lviv, Ukraine

Exhibitors

ERA Technology Ltd (UK), SFL (UK), HTLMTC (UK), DARTEC (UK), MTS Systems Corporation (USA)

THE USE OF PLASTIC STRAIN CONTROL IN THERMOMECHANICAL FATIGUE TESTING

H.-J. CHRIST[1], H. MUGHRABI, S. KRAFT, F. PETRY[2], R. ZAUTER[3] and K. ECKERT

Institut für Werkstoffwissenschaften, Lehrstuhl I, Universität Erlangen-Nürnberg, D-91058 Erlangen, Fed. Rep. Germany

[1] *Now at: Institut für Werkstofftechnik, Universität-GH-Siegen D-57068 Siegen, Fed. Rep. Germany*

[2] *Now at: Siemens AG, DIWS-BTQ1, D-86159 Augsburg, Fed. Rep. Germany*

[3] *Now at: Noell-KRC Umwelttechnik GmbH, D-97080 Würzburg, Fed. Rep. Germany*

1. Introduction

Among the variables that govern the mechanical behaviour of materials, the stress σ and the plastic (inelastic) strain rate $\dot{\varepsilon}_{pl}$ play a distinguished role as so-called state variables in mechanical equations of state, cf. Hart [1]. The strain rate plays also an important role in cyclic deformation [2,3], in particular at high temperatures. As is well known from the literature, the effect of (plastic) strain rate on the cyclic stress-strain behaviour as well as on cyclic lifetime gains importance with increasing temperature. Furthermore, many alloys which are used for intermediate and high

Figure 1. Schematic representation of the courses of stress σ, total strain ε and plastic strain ε_{pl} during (a) total strain controlled and (b) plastic strain controlled cyclic deformation.

J. Bressers and L. Rémy (eds.), Fatique under Thermal and Mechanical Loading, 1–14.

temperature applications exhibit dynamic strain ageing effects. In this case, as a consequence of the interaction of dislocations with point defects, which are considered to be responsible for the occurrence of dynamic strain ageing, the sign of the strain rate sensitivity changes with temperature. Whereas at low as well as at high temperatures an increase of the strain rate leads to an enhanced stress amplitude in isothermal fatigue tests at a given constant strain amplitude, in an intermediate temperature interval the opposite effect can be observed.

Fatigue tests on specimens are often carried out under total strain control, mostly by using a triangular command signal which leads to a constant total strain rate $\dot{\varepsilon}$ (in absolute values) during the test as shown schematically in Fig. 1a. Under these conditions, the plastic strain rate changes within each cycle between a minimum value of zero, if a purely elastic deformation takes place immediately after the load reversal point, and a maximum value which approaches the total strain rate for sufficiently high amplitudes at which the total strain approximates to the plastic strain. This behaviour in a total strain controlled test is reflected by the following relationship [3]

$$\dot{\varepsilon}_{pl} = \frac{E \cdot \dot{\varepsilon}}{\frac{d\sigma}{d\varepsilon_{pl}} + E}, \tag{1}$$

where E denotes Young's modulus. It is easily seen that in the limit of vanishing plastic strain ε_{pl} (elastic strain $\varepsilon_{el}>>\varepsilon_{pl}$), $\dot{\varepsilon}_{pl}$ approaches zero, since the slope $d\sigma/d\varepsilon_{pl}$ of the stress-strain curve becomes infinitely steep as the elastic limit is approached. For large plastic strains ($\varepsilon_{pl}>>\varepsilon_{el}$), the slope $d\sigma/d\varepsilon_{pl}$ is much smaller than E and hence $\dot{\varepsilon}_{pl} \approx \dot{\varepsilon}$.

It has been shown in earlier studies for isothermal conditions (e.g. [2,3]) that the continuously changing plastic strain rate can strongly affect the cyclic stress-strain behaviour and can lead to "deformed" hysteresis loops which complicate an interpretation of the loop shape on a micromechanistic basis. The use of true plastic strain control in connection with a triangular command signal as depicted in Fig. 1b overcomes this problem, since the (absolute value of) plastic strain rate is constant during each test and can be held at an identical level for tests at different plastic strain amplitudes by simply adjusting the frequency.

The situation becomes even more complex in typical thermomechanical fatigue (TMF) testing, since the temperature is usually varied simultaneously (e.g. in-phase or out-of-phase) with strain. Therefore, the change of the plastic strain rate within the hysteresis loop is connected with a temperature change, which itself may give rise to a pronounced alteration in the strain rate sensitivity, as described above. Taking these arguments into account, it becomes obvious that TMF hysteresis loops are hard to interpret regarding the dominating microstructural deformation processes. Therefore, a reasonable comparison of isothermal with anisothermal cyclic stress-strain behaviour, which would be an extremely helpful ingredient in the understanding of the basic deformation mechanisms under TMF conditions, seems to be possible only, if a constant plastic strain rate is applied.

In this contribution, a servohydraulic material test system is introduced which permits isothermal and thermomechanical fatigue testing at high temperatures in high vacuum and defined gaseous environments, respectively. Next, it is shown that the plastic strain control mode can be implemented quite easily by a combination of computer calculation and an analogue electronic device. The capability of the test system and the advantages of plastic strain control are demonstrated by means of selected results obtained in the framework of an extensive study of the isothermal and thermomechanical fatigue behaviour of the austenitic stainless steel AISI 304L [4-6]. Finally, these results obtained on a rather ductile material are contrasted against those typical of a high-strength material, namely the monocrystalline nickel-base superalloy CMSX-6, subjected to isothermal and anisothermal fatigue [7-9]. The latter results demonstrate the limitations of plastic strain control.

2. The Material Test System

2.1. GENERAL DESCRIPTION

4	Computer and Peripheral Instruments	3	Instrument Board	2	Control Board	1	Load Frame
		3.1	Oscilloscope	2.1	Data Display	1.1	Upper Crosshead
		3.2	Recorder	2.2	Function Generator	1.2	High Temperature Grips
		3.3	Temperature Controller and Display	2.3	Test Controller	1.3	Vacuum Chamber
		3.4	AD-DA-Converter	2.4	Master Control		
		3.5	Peak/Valley Detector				
		3.6	Vacuum System Control Unit				
		3.7	AD-DA-Converter				

Figure 2. The material test rig for thermomechanical fatigue tests in vacuum or gaseous environment.

The test rig which has been built around a closed-loop servo-controlled electro-hydraulic fatigue machine (MTS 880) is shown in Fig. 2. This system allows uniaxial push-pull loading with maximum/minimum loads of ±100 kN. In order to characterize the cyclic stress-strain response in connection with the microstructural evolution, a uniform stress and strain distribution across the specimen section was preferred compared to the somewhat simpler bending technique [10].

The main components of the test rig are marked and labelled in Fig. 2. The system is equipped with a high-vacuum chamber which surrounds the specimen, high-temperature hydraulic grips and the induction coil of the high frequency (HF) system used for sample heating. The HF generator is not shown in Fig. 2.

Figure 3. A view into the vacuum chamber. (1) specimen, (2) grips, (3) induction coil, (4) extensometer, (5) water cooling jackets, (6) flexible gas inlet and outlet tubes.

The chamber is double-walled and cooled by flowing cold water. Since all flanges are of type conflat and sealed by copper gaskets, the pressures can be reduced down to ultra-high vacuum by means of the vacuum pump system installed which consists of a turbomolecular pump in combination with a backing pump and various devices to measure, control and display the pressure. However, ultra-high vacuum was used only in few cases, because the results obtained on the austenitic stainless steel showed that the transition from high to ultra-high vacuum does not cause an appreciable extension of cyclic lifetime. Therefore, for the sake of easier handling and reduced costs per test, the large front door of the chamber which provides access for sample exchange and mounting, was preferrentially sealed by a viton O-ring. A metal-bonnet-seal valve installed between the chamber and the turbomolecular pump allows a disconnection of the pump system from the interior of the chamber and permits, in combination with a gas supply unit, tests in defined gases.

Fig. 3 shows a view into the vacuum chamber and reveals the parts which surround the (glowing) sample (1). Vacuum-tight high-temperature hydraulic grips guarantee an easy mounting and removal of the fatigue sample (2). The induction coil (3) is specifically designed in such a way that the quartz rods of a high temperature extensometer (4), which is cooled by thermostatically controlled distilled water by means of an almost vibrationless sucking pump, can touch the sample surface with sufficient distance to the windings. The flexible inert gas inlet and outlet tubes (6), which are used in combination with hollow samples, and the water cooling jackets (5) are aids to reach a higher cooling rate under thermomechanical loading conditions.

2.2. SPECIMEN GEOMETRY

The use of plastic strain control requires a highly accurate axial strain measurement in connection with a cyclindrical gauge length of the specimen. Fig. 4 displays the standard geometry of the samples used.

In order to obtain higher cooling rates which determine the upper limit of the frequency of cycling under thermomechanical fatigue test conditions hollow specimens with identical outer dimensions as shown in Fig. 4 were used. For this purpose, firstly a centre hole with a diameter of 5 *mm* was drilled through the solid sample. Next, a perpendicular hole and a sealing face were machined into the button heads. Subsequently, the inner surfaces were electropolished. Finally the openings at the top and the bottom of the sample were closed and sealed by electron beam welding.

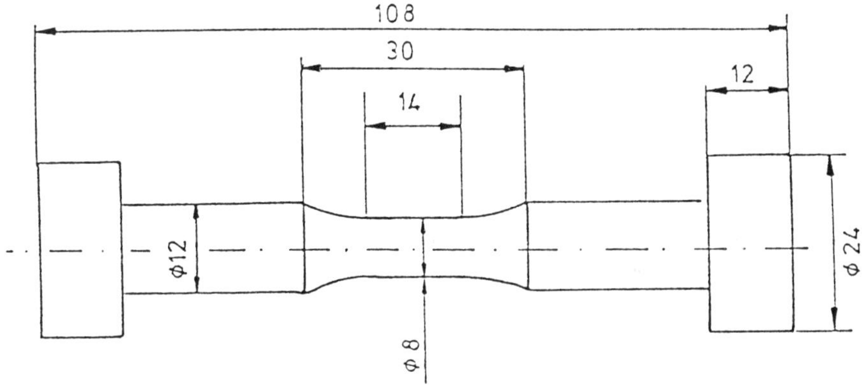

Figure 4. Geometry of the standard specimen.

During the thermomechanical fatigue test, a stream of flowing pre-cooled inert gas with high thermal conductivity (He) is fed through the centre hole in the half cycle of decreasing temperature. The cooling gas is supplied from the side ((6) in Fig. 3) through the perpendicular hole by means of a flexible tube. This tube ends in a hollow screw which is turned in the corresponding thread of the grip head till a small copper gasket presses against the sealing face of the specimen and seals the coolant stream against the vacuum surrounding the outer surface of the sample.

2.3. HEATING AND TEMPERATURE CONTROL

In order to attain a high reproducibility in thermomechanical fatigue testing, basic requirements to the heating system are a fast response, which is necessary to follow the changes in the temperature command signal, and a homogeneous temperature profile along the gauge length of the sample. Both conditions can be met by high frequency induction heating. The test system described in this chapter is equipped with a HF generator supplying a frequency of 200 kHz and a maximum power of 6 kW.

By means of a careful adaption of the geometry of the coil (Fig. 3) the temperature deviation from the cyclically varying command temperature signal could be reduced to less than ±3°C within the gauge length of the solid specimen (Fig. 4). Fig. 5 compares the function generator signal, which is provided from a digital wave table stored in the memory of a digital/analogue converter, with the temperature measured at the sample surface.

As already stated above, the frequency of thermomechanical cycling is limited by the obtainable cooling rate. In order to overcome this problem, hollow specimens were used, and the cooling was improved by using pre-cooled helium as coolant passing through a centre hole. As a disadvantage of this technique a more pronounced temperature gradient results within the gauge length of the sample which can lead to barrelling or ratchetting of the sample.

The temperature measurement was carried out using 0.3 *mm* diameter NiCr-Ni type thermocouples which were attached to the specimen in two different ways. One way was to wind a previously rolled, thin and flat thermocouple around the middle of the gauge length. The part of the thermocouple, where the wires had been connected by spot welding, was pulled against the sample surface using small springs as suspension. This technique is similar to that used in the study of Beauchamp and

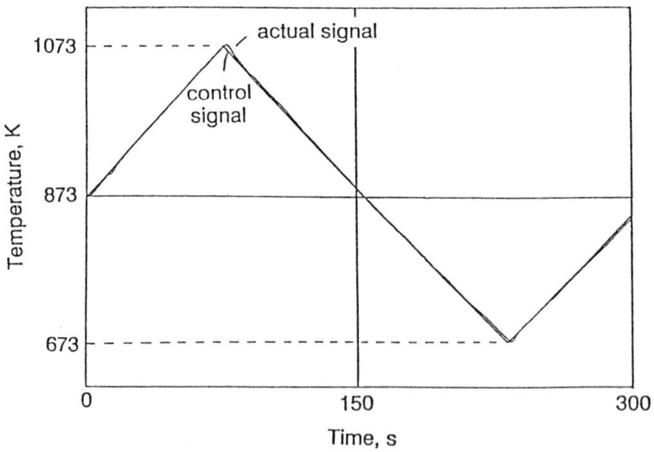

Figure 5. Comparison of command signal and measured temperature during a thermomechanical test.

Ellison [11]. In the case of long-term experiments at high temperatures this arrangement of the thermocouple tended to fracture. Therefore, spot welding of a thermocouple directly at the surface of the sample in the middle of the gauge length was preferred. Premature crack initiation connected with reduced lifetime did not occur during these tests. In tests, in which this type of temperature measurement was applied, damage in the interior of the material due to creep-fatigue interaction prevailed in the case of the stainless steel AISI 304L, whereas the monocrystalline CMSX-6 specimens exhibited predominantly typical fatigue failures.

3. The Plastic Strain Control

It has been shown in earlier studies (e.g. [2,3]) that under isothermal conditions plastic strain control works properly and allows to carry out experiments at identical plastic strain rate for different plastic strain amplitudes applied in the various tests. The experimental method is schematically illustrated in Fig. 6. By means of a minor modification of the analogue electronic circuit of a so-called "analogue strain computer" board of the commercial control electronics, the signals F and ε from the conditioners of the load cell and the extensometer were combined according to:

$$y = \varepsilon - xF \tag{2}$$

Following Hooke's law, the quantity y in equation (2) is equal to the plastic strain, if the value of x, which can be adjusted by means of a potentiometer, is set to the reciprocal product of cross section A and Young's modulus E. This adjustment is simply done by cycling under load control at small load amplitudes (purely elastic deformation). Starting with a value of zero for x, the representation of stress or load vs. y yields a straight line which is turned counterclockwise with increasing x. The correct value of x $(=1/AE)$ is reached, when the line coincides with the ordinate (i.e.

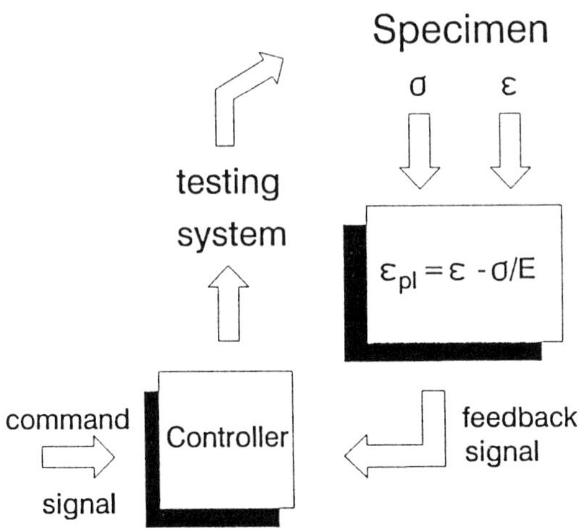

Figure 6. Basic concept of plastic strain control (isothermal case).

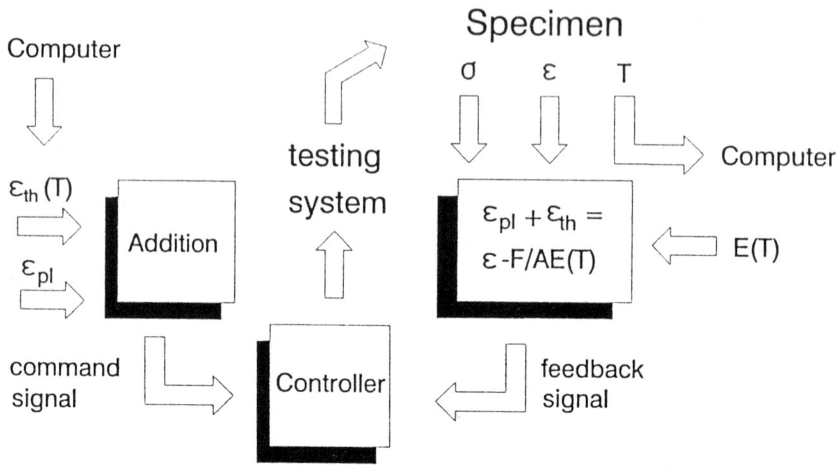

Figure 7. Basic concept of plastic strain control in TMF testing.

no plastic strain). In this position of the potentiometer, y equals the plastic strain. A switch allows to apply y, i.e. the plastic strain, instead of the total strain as feedback signal in the closed control loop of the test system during the actual fatigue test.

In practice, plastic strain control works reliably, although the closed loop control circuit is operated at the borderline of instability due to the fact that, in a stress-plastic strain plot, the Hooke elastic line degenerates to a line of infinite slope. As a consequence, an infinitely small deviation of the plastic strain signal should provoke an infinitely large stress jump and, hence, an interlock of the system should be expected, if the system responded instantly. Fortunately, it is found that, in practice, the response is not fast enough to cause an interlock [3]. The precarious mode of control is merely reflected by a slight oscillatory wiggle of the stress signal.

It should be noted that microstructural changes during cyclic loading may lead to a slight change (reduction) in E with the number of cycles. Therefore, x must be controlled and readjusted if necessary. For this purpose a continuous recording of the stress-plastic strain hysteresis loop by means of a digital storage oscilloscope is very helpful, since changes in E lead to a deviation from infinite slope in the hysteresis loop after the load reversal points in tension and compression, respectively. Furthermore, especially high-strength alloys exhibit an appreciable stress dependency of the Young's modulus [12]. In this case, plastic strain control using a constant E-value is inappropriate and unsatisfactory. In addition, high-strength materials pose other problems (Sect. 4.2).

Under thermomechanical fatigue conditions, the situation becomes more complicated than described so far, since the temperature T varies within the cycle. The extensometer determines a total strain which is the sum of the mechanical strain (elastic and plastic) and the cyclically varying thermal strain ε_{th}. Moreover, the Young's modulus is a function of temperature and, hence, changes also during each cycle. Fig. 7 shows schematically in which way plastic strain control is achieved under TMF conditions.

The subtraction of the elastic strain from the total strain is carried out in an analogue electronic circuit. Therefore, a continuous signal without phase-shift results which is equal to the sum of plastic and thermal strain, if the instantaneously correct value of E is used. The function $E(T)$ had been determined independently by a resonance method and is supplied during the TMF test from a computer via a digital/analogue converter. For the sake of a better stability of the control loop, the thermal strain is not substracted from the total strain. Rather, the thermal expansion is calculated by the computer from the instantaneous temperature of the sample and the thermal expansion coefficient $\alpha(T)$ which has been measured prior to the mechanical tests by means of dilatometry. It could be shown in experiments with zero load and periodically changing temperature that this calculation agrees with the thermal expansion experimentally determined. The sum of this ε_{th} signal and a triangular function generator signal is used as command signal in the control of plastic plus thermal strain.

This control mode commands a very careful and cautious optimization of the control parameters. The best results were obtained by using plain proportional control without any differential or integral contributions. Furthermore, the temperature interval of the TMF test should be limited in such a way that the mechanical properties of the material studied do not change too drastically, since otherwise the control parameters would have to be changed within the cycle. Moreover, a prerequisite of a well-defined stable control is a sufficiently high plastic strain range in order to allow a reasonable control error. Therefore, this technique is well suited for ductile materials such as stainless steels but unfortunately not for high-strength low-ductility materials such as (monocrystalline) nickel-base superalloys subjected to cyclic deformation under conditions of interest with respect to amplitude and temperature. This will be demonstrated in the next section.

4. Selected Experimental Test Results

In order to demonstrate the potential of plastic strain control in TMF tests, some selected results from an extensive study on the thermomechanical fatigue behaviour of AISI 304L stainless steel are reported (the interested reader is kindly referred to refs. [4-6] for more details) first. Then, some work on the monocrystalline nickel-base superalloy CMSX-6 [7-9] will be presented which reveals the limitations of plastic strain control for high-strength low-ductility materials.

4.1. THERMOMECHANICAL FATIGUE OF AISI 304L STAINLESS STEEL

In the case of themomechanical fatigue of the austenitic stainless steel with plastic strain amplitudes of $\Delta\varepsilon_{pl}/2 = 0.5\%$, corresponding to fatigue lives of typically $N_f \approx 1000$ to 3000 cycles to failure, plastic strain control posed no problem and was hence used routinely [4-6]. Using solid specimens, the plastic strain rate was limited by the rather slow cooling rate and was selected as $\dot{\varepsilon}_{pl} = 6.7 \cdot 10^{-5}\ s^{-1}$ in most cases. A a

b

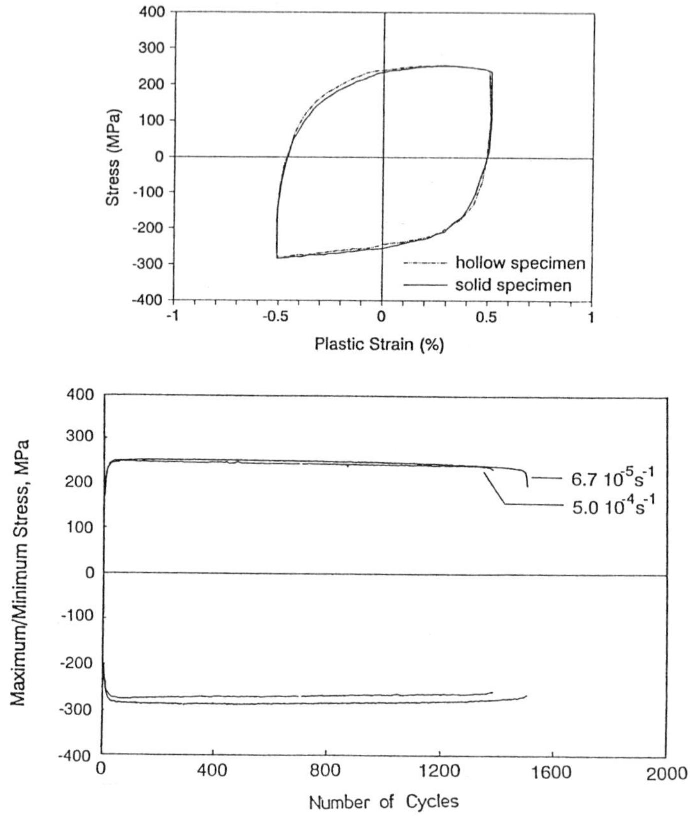

Figure 8. TMF of the stainless steel AISI 304L
a) Hysteresis loops of in-phase TMF loading for solid and hollow specimens
b) Cyclic deformation curves of in-phase TMF loading with two different plastic strain rates.

higher plastic strain rate ($\dot{\varepsilon}_{pl}$ = $5 \cdot 10^{-4}\ s^{-1}$) could be attained only by using hollow, internally cooled samples, cf. Sect. 2.2.

In the case of thermomechanical fatigue of the austenitic stainless steel the fatigue life was found to be shortest in in-phase tests [4-6]. Hence, TMF was studied in most detail under in-phase conditions.

Fig. 8a shows the hysteresis curves obtained for a solid and a hollow specimen in in-phase TMF tests in which the plastic strain and the temperature were cycled synchronously using a triangular command signal, with a plastic strain amplitude $\Delta\varepsilon_{pl}/2$ = 0.5%, a plastic strain rate of $\dot{\varepsilon}_{pl}$ = $6.7 \cdot 10^{-5}\ s^{-1}$ and a temperature range of 400°C to 650°C. It will be noted that the two hysteresis curves are almost identical. This indicates that the test conditions were equally well defined for the solid and the hollow specimen.

The cyclic deformation curves for two in-phase TMF tests performed on a solid specimen ($\dot{\varepsilon}_{pl}$ = $6.7 \cdot 10^{-5}\ s^{-1}$) and a hollow specimen ($\dot{\varepsilon}_{pl}$ = $5 \cdot 10^{-4}\ s^{-1}$) are shown in Fig. 8b. The cyclic hardening/softening curves document that the lower strain rate leads to slightly increased maximum and minimum stress values. This negative strain rate sensitivity has been reported earlier for isothermal tests at intermediate temperatures and is attributed to dynamic strain ageing (DSA) processes. Fig. 8 illustrates that also the TMF behaviour is affected by DSA, even if the maximum temperature of the temperature interval used lies well above the range in which marked DSA occurs in the isothermal case.

4.2. FATIGUE OF THE MONOCRYSTALLINE NICKEL-BASE SUPERALLOY CMSX-6

In Fig. 9a, the hysteresis curves of a monocrystalline CMSX-6 specimen, measured in an isothermal fatigue test at 1100°C under total strain control, using a total strain amplitude of $\Delta\varepsilon/2 = 0.5\%$ and a total strain rate of $\dot{\varepsilon} = 5 \cdot 10^{-3}\ s^{-1}$, are shown in the σ vs. ε- and in the σ vs. ε_{pl}- presentation, respectively. The corresponding dependencies of σ, ε and ε_{pl} on time t are shown in Fig. 9b. Under these conditions,

a

b

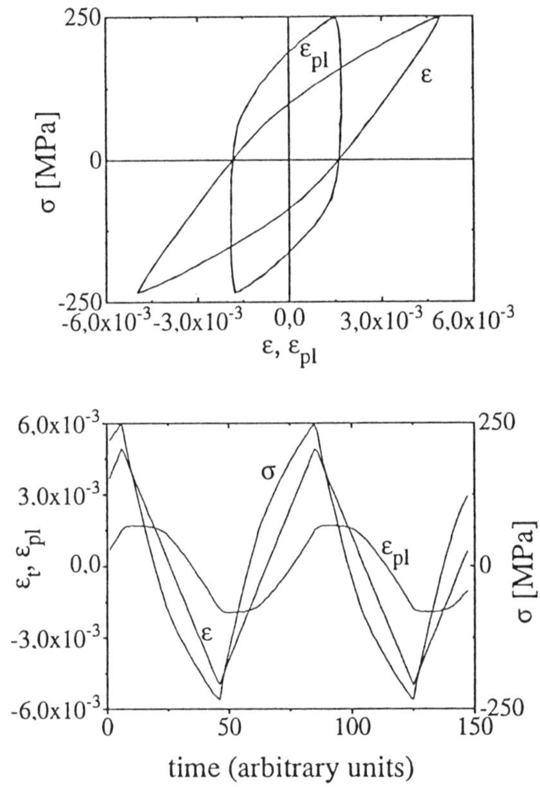

Figure 9. Isothermal fatigue of monocrystalline CMSX-6 at 1100°C under total strain control ($\Delta\varepsilon/2 = 0.5\%$, $\dot{\varepsilon} = 5 \cdot 10^{-3}\ s^{-1}$). From [9].
a) Hysteresis loops in σ vs. ε- and σ vs. ε_{pl}-plot.
b) Dependency of stress σ, total strain ε and plastic strain ε_{pl} on time t.

a

b

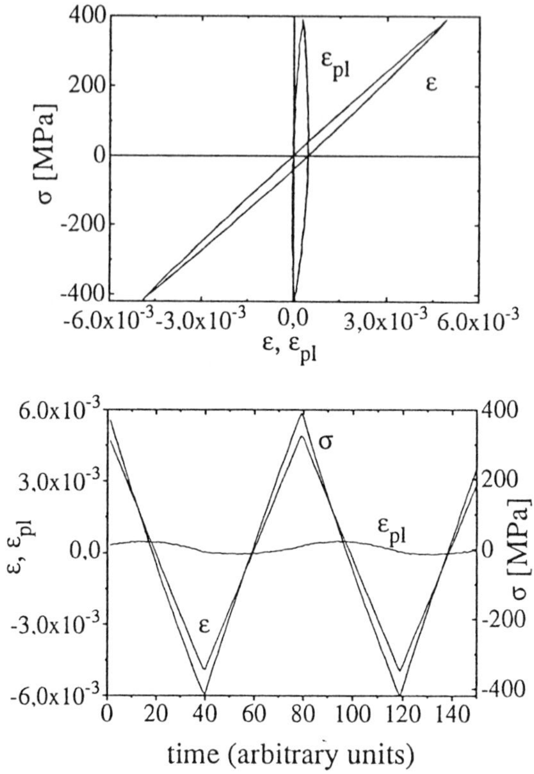

Figure 10. Isothermal fatigue of monocrystalline CMSX-6 at 950 °C under total strain control ($\Delta\varepsilon/2$ = 0.5%, $\dot{\varepsilon} = 5 \cdot 10^{-3}$ s^{-1}). From [9].
a) Hysteresis loops in σ vs. ε- and σ vs. ε_{pl}-plot.
b) Dependency of stress σ, total strain ε and plastic strain ε_{pl} on time t.

i.e. at a higher temperature at which the alloy CMSX-6 is rather ductile and for sufficiently large plastic strain amplitudes ($\Delta\varepsilon_{pl}/2 \approx 1.5 \cdot 10^{-3}$, fatigue life $N_f \approx 1200$), plastic strain control could have been applied equally well but would have had no severe effect because of the rather small difference between $\dot{\varepsilon}$ and $\dot{\varepsilon}_{pl}$.

However, isothermal fatigue at 1100°C does not represent a situation of practical interest. Fig. 10a and b refer to an isothermal test at 950°C with $\Delta\varepsilon/2 = 0.5\%$ and $\dot{\varepsilon} = 5 \cdot 10^{-3}\ s^{-1}$ ($N_f \approx 3000$). In this case, at a lower temperature, the stress is much larger and, as a consequence, the plastic strain amplitude is much smaller ($\Delta\varepsilon_{pl}/2 \approx 1.2 \cdot 10^{-4}$) than in the previously discussed case (Fig. 9). The hysteresis loop is narrow and pointed (Fig. 10b) and, in contrast to Fig 9b, the plastic strain rate at the full amplitude now is about a whole order of magnitude smaller than the total strain rate ($\dot{\varepsilon}_{pl} \approx 0.1\dot{\varepsilon}$). Hence, inspite of the same total strain rates imposed in the two examples discussed, the cyclic deformation behaviour cannot be compared properly because of the very different plastic strain rates. Thus, in the second case (Fig. 10), it would indeed have been advisable to perform the test under plastic strain control. Unfortunately, this was not possible because of the rather non-linear shape of the

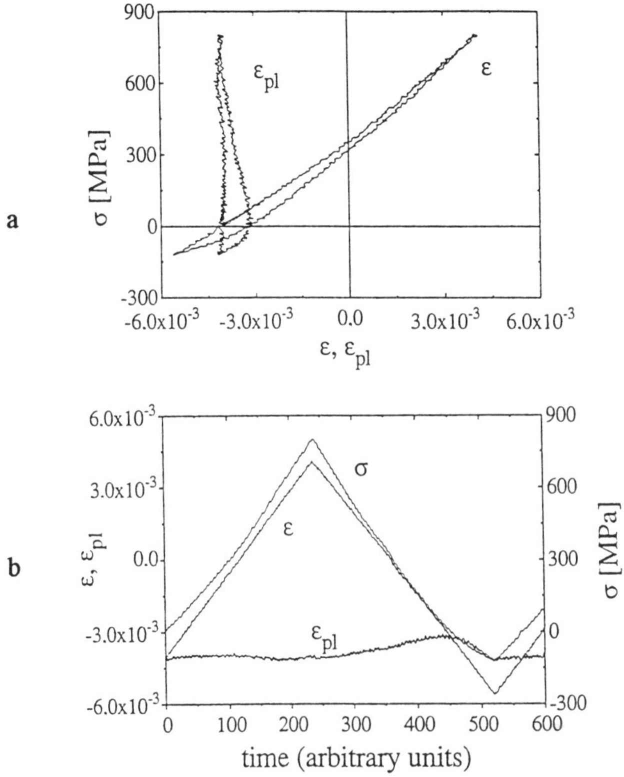

Figure 11. **Out-of-phase thermonechanical fatigue of monocrystalline CMSX-6 in temperature range $700°C<T<1100°C$ under total strain control ($\Delta\varepsilon/2 = 0.5\%$, $\dot{\varepsilon} = 6.7\cdot10^{-5}\ s^{-1}$). From [9].**
a) Hysteresis loops in σ vs. ε– and σ vs. ε_{pl}-plot.
b) Dependency of stress σ, total strain ε and plastic strain ε_{pl}- on time *t*.

elastic loading/unloading curve and because of the rather small plastic strain amplitude of only $\approx 10^{-4}$. This example demonstrates the limitations of the application of plastic strain control.

As a final example, a TMF test on the monocrystalline superalloy CMSX-6 is considered. In the case of this high-strength material, fatigue life was shortest for out-of-phase TMF in the regime of interest [7,8]. Fig. 11 shows the hysteresis loops (Fig. 11a) and the time dependencies $\sigma(t)$, $\varepsilon(t)$, $\varepsilon_{pl}(t)$ for an out-of-phase TMF test on a monocrystalline CMSX-6 specimen. The TMF test was performed under total strain and not plastic strain control because of the difficulties mentioned above, and a total strain amplitude $\Delta\varepsilon/2 = 0.5\%$ and a total strain rate $\dot{\varepsilon} = 6.7\cdot10^{-5}\ s^{-1}$ were used. The temperature range was 700 - 1100°C. The very complex shape of the hysteresis curve is evident from Fig. 11a. Also, it can be noted from Fig. 11b that the plastic strain rate $\dot{\varepsilon}_{pl}$ varies in a very pronounced manner during the cycle, being close to zero almost everywhere except in the approach to maximum compression (at the upper temperature!), where the specimen undergoes microyielding. This most complex behaviour is difficult to analyze in detail in order to obtain a fundamental understanding of the material response. There can be no doubt that this task would be considerably easier, if it were possible to conduct the test in plastic strain control.

Acknowledgements - The financial support of the "Volkswagenstiftung" (Az I/61021), of the "Deutsche Forschungsgemeinschaft" (Az DFG Mu 502/5-1) and of the Bundesminister für Bildung, Forschung und Technologie (Az 03M3038F0) is gratefully acknowledged.

References

1. Hart, E.W., A phenomenological theory for plastic deformation of polycrystalline materials, *Acta Metall.* **18** (1970), 599-610.
2. Mughrabi, H., Cyclic deformation and fatigue: Some current problems, in J. I. Dickson, J. J. Jones, M. G. Akben, H.-J. McQueen, J.-P. Bailon (eds.), *Proc. of ICSMA7*, Pergamon Press, Oxford (1986), pp. 1917-1942.
3. Mughrabi, H., Neuere Ergebnisse der Ermüdungsforschung mit praktischer Relevanz-Grundlagenaspekte der Wechselverformung, in E. Macherauch (ed.), *Werkstoffverhalten und Bauteilbemessung*, DGM Informationsgesellschaft, Oberursel (1987), pp. 49-65.
4. Zauter R., Petry F., Christ, H.-J., Mughrabi H., Thermomechanical fatigue of the austenitic stainless steel AISI 304L, H. Sehitoglu (ed.), *Thermomechanical fatigue behavior of materials*, ASTM STP 1186, Philadelphia (1993), pp. 70-90.
5. Zauter, R., Christ, H.-J., Mughrabi, H., Some aspects of thermomechanical fatigue of AISI 304L stainless steel: Part I. Creep-fatigue damage, *Metall. Mater. Trans. A* **25A** (1994), 401-406.
6. Zauter, R., Christ, H.-J., Mughrabi, H., Some aspects of thermomechanical fatigue of AISI 304L stainless steel: Part II. Dislocation Arrangements, *Metall. Mater. Trans. A* **25A** (1994), 407-413.
7. Kraft, S., Zauter, R. Mughrabi, H., Aspects of high-temperature low-cycle thermomechanical fatigue of a single crystal nickel-base superalloy, *Fatigue Fract. Engng. Mater. Struct.* **16** (1993), 237-253.
8. Kraft, S., Altenberger, I. Mughrabi, H., Directional γ-γ' coarsening in a monocrystalline nickel-base superalloy during low-cycle thermomechanical fatigue, *Scripta Metall. Mater.* **32** (1995), 411-416.
9. Kraft, S., unpublished work (1995).
10. Adam, W. R., Stanley, P., A programmable machine for simulated thermal fatigue testing, *J. Phys. E, Scientific Inst.* 7 (1974), 669-673.
11. Beauchamp D. J., Ellison E. G., A rig for controlled cyclic strain and temperature testing, *J. of Strain Analysis* **17** (1982), 45-52.
12. Sommer, C., Christ, H.-J., Mughrabi, H., Non-linear elastic behaviour of the roller bearing steel SAE 52100 during cyclic loading, *Acta metall. mater.* **39** (1991), 1177-1187.

AN EXPERIMENTAL SYSTEM FOR THERMAL MECHANICAL AND BIAXIAL FATIGUE STUDIES

S. Y. ZAMRIK, D. C. DAVIS and G. T. ROGERS
Department of Engineering Science and Mechanics
The Pennsylvania State University
University Park, PA 16802,
U. S. A.

1. Introduction

This paper describes an experimental system developed to conduct thermal mechanical fatigue (TMF) and biaxial fatigue testing of materials. As such the biaxial mechanical loadings may be either pure torsion or combined axial-torsion using thin-wall tubular-type specimens. Additionally, the system is capable of conducting uniaxial TMF, and isothermal uniaxial and biaxial fatigue testing as well. Currently only a small number of researchers [1] besides the authors are known to be pursuing the experimental testing area of combined thermal and biaxial mechanical fatigue testing and analysis of materials. TMF testing and analysis of materials has been an ongoing research area for over twenty years, however, in all of these previous known cases the mechanical loadings have been the uniaxial.

2. Experimental Testing System for Biaxial TMF

The major components of the TMF testing system are shown in Fig. 1. They are: an axial-torsional load frame capable of applying ±248 kN of axial force and ±2500 kN-m of torque either independently or simultaneously, a system digital controller for direct control of the mechanical cycling and an interface to the thermal loop, a 2.5 kW induction heater for specimen heating, a digital temperature controller, x-y and strip chart graphical plotters, and a control center consisting of a personal computer (PC) with a 486 processor, 16 megabytes of RAM and a 540 megabytes hard drive capability.

J. Bressers and L. Rémy (eds.), Fatique under Thermal and Mechanical Loading, 15–23.

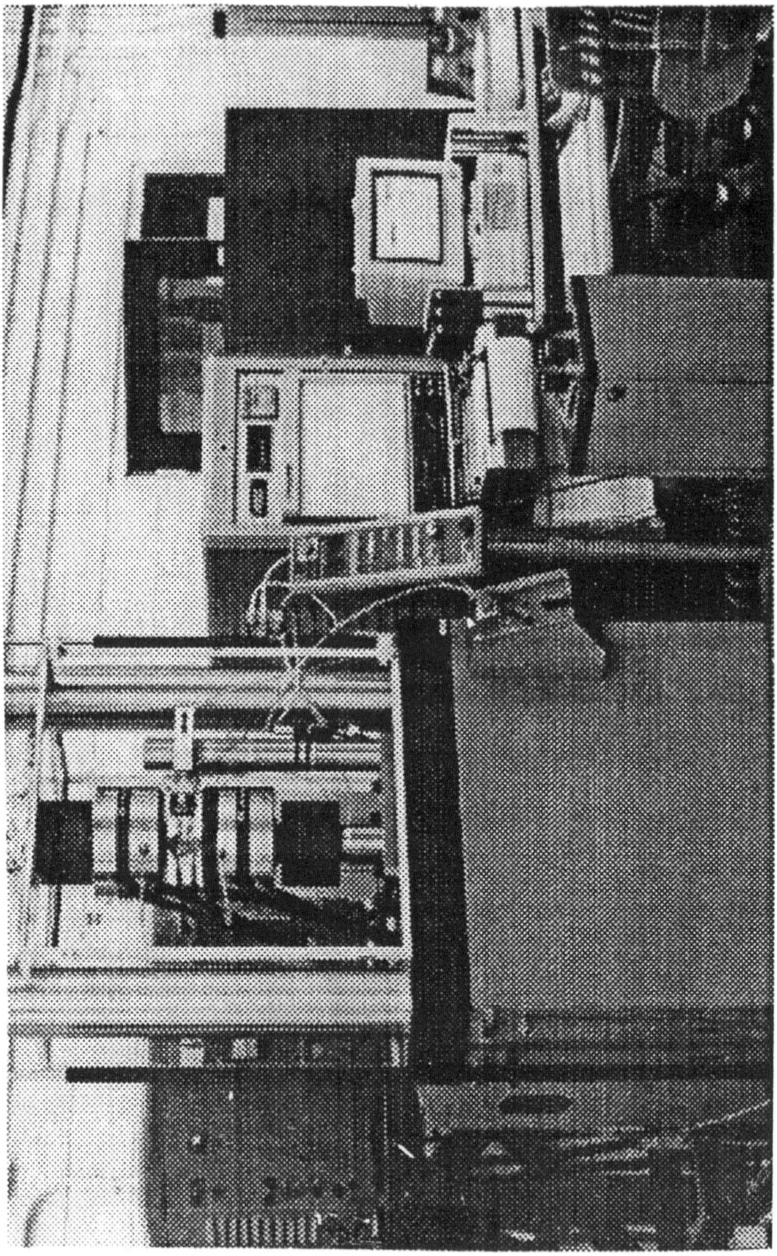

Figure 1. TMF experimental test facility and system.

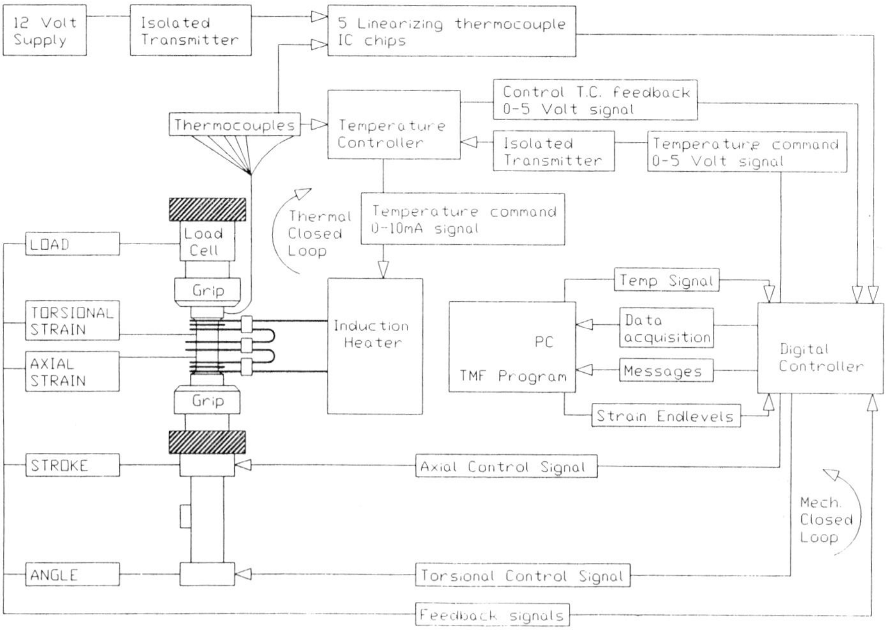

Figure 2. **TMF system component integration schematic.**

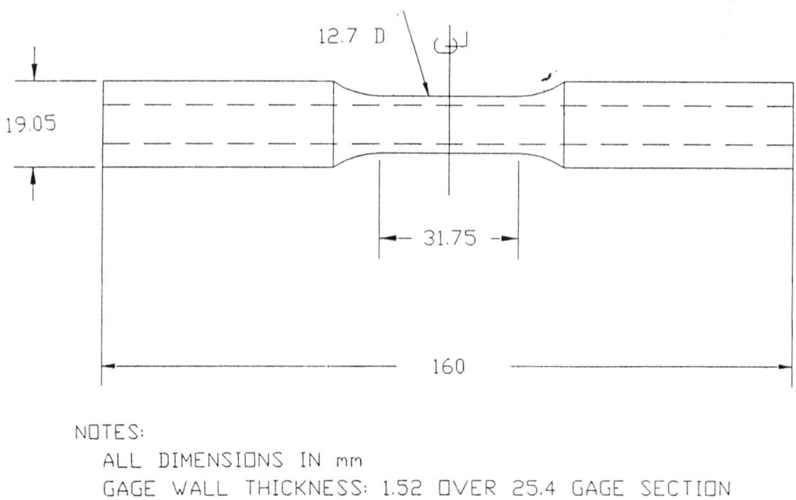

Figure 3. **TMF test specimen.**

2.1. SYSTEM COMPONENT INTEGRATION AND INTERFACES

The schematic in Fig. 2 illustrates the integration and the interfaces between the components of this biaxial thermal mechanical fatigue testing system. There are two closed loops: a mechanical loop and a thermal loop. They are both controlled indirectly through the PC at the control center. Software specifying all operating parameters for uniaxial and biaxial TMF testing, as well as for isothermal uniaxial and biaxial fatigue and creep-fatigue testing, is installed on the PC. The software is written for the workplace shell in OS/2 2.1, a graphical multitasking environment. Through an interface card, the computer commands are communicated to the system digital controller which is the unit responsible for: closed loop control of the axial and torsion channels, load frame actuator and load cell limits, output signals for the graphical plotters, and all analog and digital data acquisition from the thermal and mechanical closed loops. The illustration also shows that the mechanical closed loop control signals are directed from the digital controller to the load frame actuator. The feedback signals from the load frame are stroke and rotation from the actuator, axial and angular strain from the extensometers, and load and torque from the load cell. The load frame is equipped with a moveable hydraulic cross-head and water-cooled hydraulic collets for specimen gripping. The thin-walled tubular type TMF specimen is shown in Fig. 3. Over its 25.4 mm gage section length the wall thickness is 1.54 mm. Using a low stress grind technique and electropolishing, a gage section surface finish (S.F.) of 8-10 can be obtained. Axial and angular strains are measured using a water-cooled biaxial extensometer fitted with conical point ceramic extension probes. An air-cooled axial extensometer fitted with either ceramic or quartz probes is also available for use in pure uniaxial testing.

The thermal loop consists of a digital temperature controller, a 2.5 kW high frequency induction heater, a three-section independently adjustable heating coil, six feedback thermocouples for monitoring the specimen temperature. The temperature controller has pre-established setpoint voltages on which to adjust the output of the induction heater so that the specimen gage temperature follows the desired waveform. One thermocouple is placed slightly above the upper extensometer probe as a feedback to the temperature controller while the five other thermocouples are used to monitor the specimen temperature over the heated gage region. The thermocouples are not welded to the specimen, but the wires are looped around and drawn tight to the specimen surface. The thermocouple wires are then attached to an analog device which linearizes and amplifies their outputs for view at the system digital controller or on the computer (PC) terminal screen, and/or for plotting by the strip chart recorder. The simultaneous thermocouple readings were instrumental in developing a heating coil configuration for maintaining an acceptable dynamic temperature gradient over the specimen gage section.

2.2. BIAXIAL TMF TESTING PROCEDURES

The biaxial TMF testing procedure consists of three main steps. The first step is the system start-up, the second step involves specimen heat-up and thermal cycling at zero load and torque, and the final step is the actual strain-control TMF cycling of the specimen.

The start-up consists of providing electrical power to the system controller, the hydraulic pump, the computer control center, the graphical plotting equipment, the load frame controls and all operating systems. First, hydraulic pressure is provided to the load frame actuator followed by coolant being supplied to the collet grips, extensometer (biaxial), heating coils and the induction heater. The test specimen is then installed through the heating coils, aligned in the collet grips and finally the thermocouples are fixed in place as discussed above.

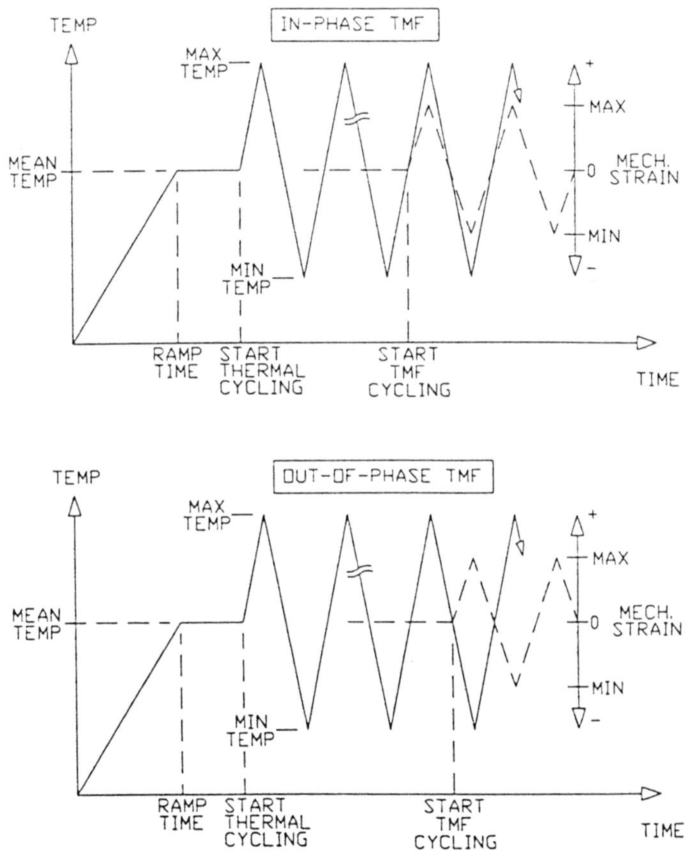

Figure 4. **System start-up and test procedure.**

TABLE 1. System Fatigue Testing Options - Initial Loading Direction Tension or Compression.

ISOTHERMAL CASES	UNIAXIAL TMF CASES	BIAXIAL TMF CASES
Isothermal Continuous Cycling	Uniaxial TMF Continuous Cycling In-phase *	Biaxial (combined Axial-Torsion or Pure Torsion) TMF Continuous Cycling In-Phase *
Isothermal with Tension or Compression or both Hold-Times	Uniaxial TMF Continuous Cycling Out-of-Phase *	Biaxial TMF Continuous Cycling Out-of-Phase *
	Uniaxial TMF In-Phase with Tension or Compression or Both Hold Time	Biaxial TMF In-Phase with Tension or Compression or Both Hold-Time
	Uniaxial TMF Out-of-Phase with Tension or Compression or Both Hold-Times	Biaxial TMF Out-of-Phase with Tension or Compression or Both Hold-Times

* Refers to load-temperature phasing only.

The second step consists of initial specimen heat-up, specimen thermal cycling under zero load and torque, and some specimen setup adjustments as mentioned previously (Fig. 4). This step is mainly controlled by the TMF program which requires the input of several test operating parameters, to include: initial ramp temperature (mean of the maximum and minimum temperature cycle amplitudes), initial ramp time, and maximum and minimum temperature amplitudes. With these parameters input into the TMF program, this thermal cycling and stabilization step can begin. Real time temperature signals from all the thermocouples are displayed at the computer terminal and can be plotted on the strip chart recorder. Some minor adjustments in the temperature controller setpoint voltages, the thermocouple positions, and/or locations of the heating coils may be necessary during this step. These adjustments further ensure that the prescribed maximum and minimum temperature amplitudes of the cycle and a dynamic thermal gradient of no more than 5 - 10°C will be maintained along the specimen gage section during TMF testing. When such a temperature cycle is obtained and is observed to be stable for several cycles, a final temperature cycle is recorded. The thermal strains calculated from this final cycle are used for the thermal compensation in the TMF test. This procedure is necessary to avoid geometric instabilities such as barreling which is sometimes observed in specimens after TMF testing. Even though there will be radial expansion and contraction of the tubular specimen during TMF testing, the angular thermal strain component is considered negligible.

In the final step of the testing procedure the system is switched from a load control mode to a strain control mode to initiate TMF testing. As illustrated in Fig. 4 TMF cycling is started with both the temperature cycle and the mechanical strain (axial and/or torsional) cycles at their mean values. The TMF cycling proceeds in either a pre-selected in-phase or an out-of-phase mode. Both the temperature and the mechanical strain cycles are limited to a constant rate triangular waveform. The convention taken for in-phase biaxial TMF is that at the temperature cycle's maximum amplitude the axial strain amplitude is at the maximum value also and the torsional angle of twist is at a maximum counterclockwise position, and conversely, in the convention taken for out-of-phase biaxial TMF the temperature cycle's amplitude would be at a minimum value when the mechanical strains are in these same positions. For all biaxial fatigue testing the axial and torsional mechanical loadings are currently programmed to be cycled in an in-phase mode, however, the ratio of the applied torsional and axial mechanical strain ranges may have values other than unity. Table 1 summarizes the different types of isothermal fatigue and TMF tests that have been programmed to be conducted by this computer controlled system. The mechanical strain cycling may be either uniaxial, pure torsional or combined axial-torsion. Creep effects can be studied during TMF cycling by imposing a hold in strain and temperature at their maximum and/or minimum amplitudes. Similarly, during isothermal fatigue testing, a hold in strain can be imposed at either the maximum and/or minimum strain amplitudes.

2.3. DATA ACQUISITION AND TEST OUTPUTS

During the load/torque-free thermal cycling and the subsequent TMF testing various types of data acquisition can be performed. The software is written to allow the acquisition of cyclic peak/valley, cyclic mean/span, test maximum/minimum, and timed data from any of the input signals (e.g.: load, torque, strain, displacement, angle of twist, and temperature). One input channel is selected to be monitored in a cyclic data acquisition process and any other desired input signals are recorded when the monitored signal is activated. The TMF cyclic data can be used to generate strain range and stress range life plots. A separate process handles the acquisition of hysteresis loop type (load/torque-strain) data. Hence, for a given cycle, timed data acquisition such as axial load versus axial strain or torque versus angular strain can be obtained. The computer program computes the axial mechanical strain by subtracting the calculated thermal strain component from the total axial strain (input signal). The thermal component of the angular strain is negligible and therefore only the torque versus total (mechanical) angular strain actually is available. All data acquisition may be recorded by either a x-y plotter or strip chart recorder as appropriate, and/or digitally stored in the computer (PC).

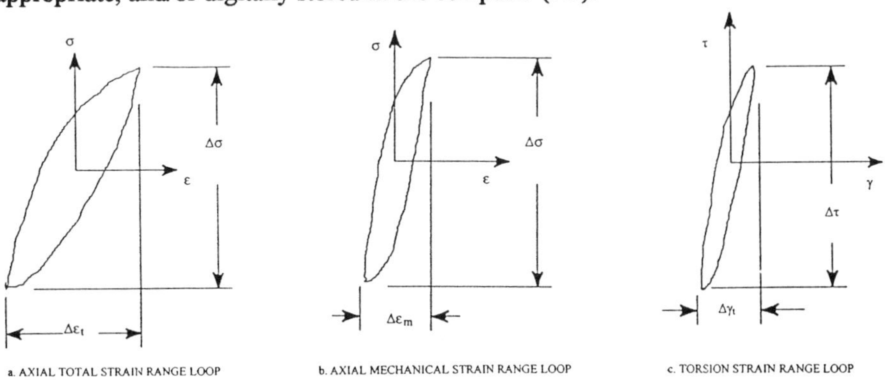

Figure 5. Biaxial TMF hysteresis loops.

A type of data acquisition that is specifically unique to biaxial TMF are the hysteresis loops. In Fig. 5 are shown schematically the three different types of hysteresis loops that result from an in-phase biaxial TMF cycle. In Fig. 5a is the axial stress (σ) versus total axial strain (ε_t), in Fig. 5b is the axial stress versus the mechanical strain (ε_m), and in Fig. 5c is the torsional stress (τ) versus total torsional strain (γ_t). The mechanical strain range ($\Delta\varepsilon_m$) is obtained by subtracting from the total axial strain range ($\Delta\varepsilon_t$) the thermal strains, and as discussed before, because the specimen radial thermal expansion and contraction during TMF cycling is negligible, the total torsional strain range ($\Delta\gamma_t$) is equivalent to the mechanical torsional strain range. Therefore, in biaxial TMF testing two biaxial strain ratios will exist, defined as:

$$\lambda_t = \frac{\Delta\gamma_t}{\Delta\varepsilon_t} \qquad \text{and} \qquad \lambda_m = \frac{\Delta\gamma_t}{\Delta\varepsilon_m} \tag{1,2}$$

Either of these two parameters can be used when comparing biaxial fatigue results. For uniaxial fatigue cycling $\lambda_t = \lambda_m = 0.0$, and for pure torsional fatigue $\lambda_t = \lambda_m = \infty$.

3. Discussion of Some Uniaxial and Biaxial TMF Testing Results

Presented in Table 2 are results from two in-phase TMF uniaxial continuous cycling tests reported previously by the principal authors in [2], and three results from some recent in-phase TMF biaxial (combined axial-torsion) continuous cycling tests of type 316 stainless steel. In all of the TMF tests using type 316 stainless steel, the thermal cycle is between 621°C and 399°C at a frequency of 0.333 cycles per minute. Specimen cooling was by natural convection.

The fatigue life (N_f) from these tabulated test results would show that uniaxial in-phase TMF continuous cycling is more damaging than biaxial (combined axial-torsion) TMF loadings at the same von Mises equivalent mechanical strain ranges. This observation under TMF loadings is consistent with the results discussed in the previous isothermal fatigue studies of this material [3-5]. The most detrimental isothermal fatigue loading was the uniaxial case, followed by biaxial fatigue loadings, and then finally torsional fatigue loadings which had the longest life at the same von Mises equivalent strain (mechanical) ranges. The Z-parameter developed in [5] showed how uniaxial, pure torsion and combined axial-torsion isothermal fatigue data could be correlated onto a single life curve.

4. Summary and Conclusions

A computer controlled experimental test system has been developed to conduct uniaxial and biaxial thermal mechanical fatigue studies of materials. Over fifty different types of TMF and isothermal fatigue tests have been programmed to be conducted. Currently uniaxial and biaxial TMF studies are being conducted using type 316 stainless steel and Inconel 738LC materials. They will be compared with previous results from isothermal fatigue tests on these materials. One of the objectives of this research is to develop a parameter which will correlate uniaxial, pure torsional and combined axial-torsion TMF life curves onto a single thermal mechanical fatigue life curve as was previously demonstrated by use of the Z-parameter [5] for isothermal loadings.

TABLE 2. Thermomechanical Continuous Cycling In-Phase Fatigue Tests of Type 316 Stainless Steel Between 621°C and 399°C (All Stress and Strain Range Values at Mid-Life).

$\Delta\gamma_t$ (%)	$\Delta\varepsilon_t$ (%)	λ_t ($\Delta\gamma_t/\Delta\varepsilon_t$)	$\Delta\varepsilon_m$ (%)	λ_M ($\Delta\gamma_t/\Delta\varepsilon_t$)	$\Delta\tau$ (MPa)	$\Delta\sigma$ (MPa)	N_f
0.0	1.18	0.0	0.80	0.0	0	567	1,025
0.0	1.00	0.0	0.60	0.0	0	494	2,549
1.45	0.83	1.745	0.40	3.625	426	267	1,253
0.80	0.78	1.026	0.40	2.000	323	334	2,954
0.68	0.90	0.756	0.50	1.360	259	411	2,144

Acknowledgements - The authors acknowledge the financial support of the National Science Foundation (Grant Nos. MSS-9215694 and MSS-8906376), and the Pressure Vessel Research Council (Elevated Temperature Design Committee). The assistance of L. Firth and N. M. Renauld in developing this test system is also acknowledged.

References

1. Meersman, J., Ziebs, J., Kuhn, H.-J., Sievert, R., Olchewski, J., Frenz, H., The stress-strain behaviour of IN 738LC under thermomechanical uni- and multiaxial fatigue loading, International Symposium Fatigue Under Thermal and Mechanical Loading, May 22-24, 1995, Petten (N.H.), The Netherlands.
2. Zamrik, S. Y., Davis, D. C. and Firth L. C., Isothermal and thermomechanical fatigue of type 316 stainless steel, To be published in M. J. Verrilli and M. G. Castelli (eds.), *Thermo Mechanical Fatigue Behavior of Materials, ASTM STP*, American Society for Testing and Materials, Philadelphia, 1995.
3. Zamrik, S. Y., Davis, D. C. and Kulowitch, P. J., Failure modes in type 316 stainless steel using biaxial strain cycling, in M. R. Mitchell and R. W. Landgraf (eds.), *Advances in Lifetime Predictive Techniques, ASTM STP 1122*, American Society for Testing and Materials, Philadelphia, 1991, pp. 299-318.
4. Zamrik, S. Y. and Davis, D. C., A ductility exhaustion approach for axial fatigue-creep damage assessment using type 316 stainless steel, *Journal of Pressure Vessel Technology, Transactions of ASME*, **113**, (1991), 180-186.
5. Zamrik, S. Y. , Mirdamadi, M. and Davis, D. C., A proposed model for biaxial fatigue analysis using the triaxiality factor concept, in D. L. McDowell and J. R. Ellis (eds.), *Advances in Multiaxial Fatigue, ASTM STP 1191*, American Society for Testing and Materials, Philadelphia, 1993, pp. 85-106.

ANALYSIS OF THERMAL FATIGUE TESTS FOR SUPERALLOY COMPONENTS

A. KÖSTER, G. LAURENT, G. CAILLETAUD and L. RÉMY
Ecole Nationale Supérieure des Mines de Paris
Centre des Matériaux
URA CNRS 866
B.P. 87 - 91003 EVRY Cedex - France

1. Introduction

Critical components in gas turbines as used in rocket engines are subject to rapid temperature changes during operation and thus experience thermal fatigue (TF) loading. In gas turbine industry, TF is usually investigated using component-like specimens such as Glenny's tapered discs or wedge specimens tested in fluidized beds [1], or rigs based on heating by the flame of a burner and cooling by compressed air [2-3]. However difficulties arose with such facilities e.g. when one wants to investigate high maximum temperatures in excess of 1100°C with fluidized beds, or in terms of reproducibility problems with burner rigs. The purpose of this paper is to give an overview of a thermal fatigue facility that we developed in our laboratory to circumvent such difficulties, and to present a methodology used to calculate the temperature-stress-strain history and the lifetime of specimens with a damage model.

2. Thermal Fatigue Test

The material used in this investigation is a wrought nickel base superalloy which could be used for machining rocket engine blades or discs.

The TF tests are carried out on single-edge wedge specimens with an edge radius of 0.25 mm and a maximum thickness of 6.7 mm. Two geometries were tested as shown in Fig. 1: the one with a total width of 27.6 mm (FT-S) is a standard single-edge wedge specimen [3] and the other is 50 mm wide (FT-L). The larger geometry was employed to increase the thermal gradient during TF testing, and consequently to reduce the lifetime of the specimen by increasing stresses and strains.

The edge area of the specimens was mechanically polished parallel to their longitudinal axis using diamond paste down to 3 μm.

The TF test consists of alternately heating and cooling the leading edge of the specimen. During thermal cycling, the specimen is totally free to expand or to contract so that only mechanical strains resulting from the thermal gradients are seen by the specimen.

J. Bressers and L. Rémy (eds.), Fatique under Thermal and Mechanical Loading, 25–35.

Closed -loop temperature cycling was achieved using a thermocouple spot welded at the mid point of the leading edge at 1.5 mm from the radius. The temperature cycle was obtained by programming a temperature controller which allowed a great versatility in defining thermal cycles (T_{max}, T_{min}, temperature rate, number of cycles,...).

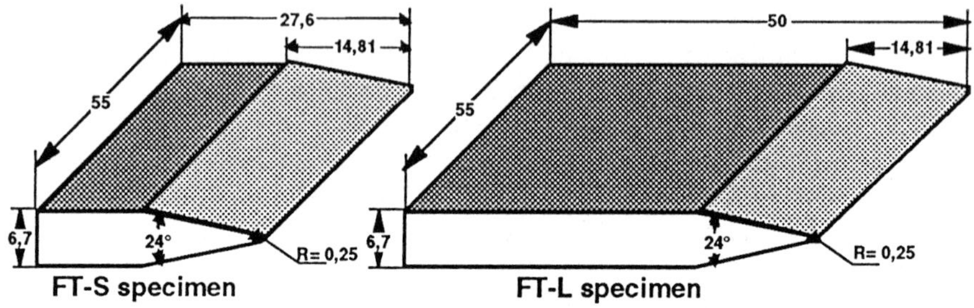

Figure 1. Geometry of Thermal Fatigue specimens.

Heating was provided by a radiation furnace with six 1500-W light bulbs (Fig. 2). It has no thermal inertia. The specimen leading edge is placed in the centre of the furnace on an adjustable holder that kept the tested structure in position without mechanical loading. The furnace was designed to focus lamp radiation on a small area of about 6 mm in diameter.

The cooling system consists of a removable nozzle which moves during the thermal cycle. When the furnace is heating, the nozzle is in a backward position, and when the cooling sequence begins, a pneumatic actuator moves the cooling system near to the specimen (≈3 mm). The cooling was ensured by forced blowing of cold air at 8 bar.

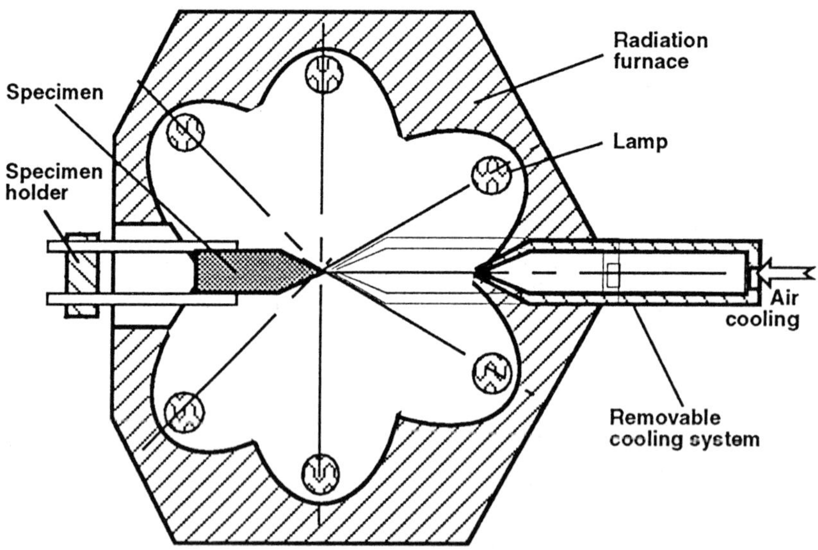

Figure 2. Schematic drawing of the thermal fatigue rig used in testing the wedge-shaped specimens.

3. TF Results

To investigate damage evolution in a wrought superalloy used to make aerospace turbine blades, a thermal cycle was defined between 25°C and 750°C to represent the thermal loading of a blade under operating conditions. Heating and cooling periods were respectively fixed at 60 s and 30 s. The thermal cycle (Figs. 5 and 6, temperature controlled with Th7), is the same for FT-S and FT-L specimens. Tests were regularly interrupted and specimens were examined by scanning electron and optical microscopy to investigate the crack initiation and propagation mechanisms.

As shown in Fig. 3 for the longest crack of each specimen, the crack initiation life is 2.5 times shorter for the FT-L specimen than for the FT-S specimen. Lives for initiation are 800 cycles and 2100 cycles respectively, using a criterion for initiation of 0.2 mm crack length at the wedge tip. The propagation rate increases quickly for the FT-L specimen, and is always higher than the one of FT-S specimen. The propagation rate for FT-S decreases when the crack has reached a length of 3 mm whilst it is constant for FT-L from 0.4 to 4 mm.

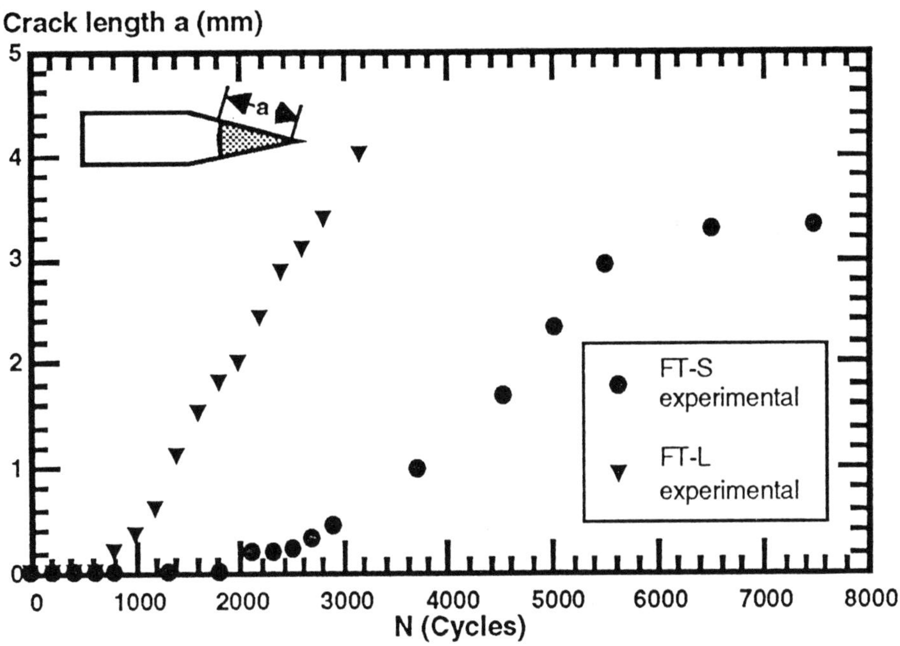

Figure 3. Evolution of the longest crack (a) vs. cycles under thermal fatigue between 25°C and 750°C.

The observation of fracture surfaces has revealed that all cracks initiated from grain boundaries at the wedge tip of the TF specimens. The cracks were oriented perpendicular to the principal loading direction and the crack front was semi-circular (Fig. 4).

Figure 4. Fracture surface of the FT-S specimen after 7500 cycles of thermal fatigue (25°C-750°C).

The crack propagation is mainly intergranular for both geometries, but transgranular propagation was observed for FT-S after 2.5 mm of crack length. In both specimen types, grain boundaries are oxidised.

4. Thermal Analysis

In order to perform a heat transfer analysis, 14 thermocouples (K type) were spot-welded along the mid-section of both specimen geometries. The temperatures from these thermocouples were recorded during a stabilised thermal cycle using a micro-computer. Examples of surface temperatures obtained from both specimens are shown in Figs. 5 (FT-S) and 6 (FT-L).

The transient temperature map of the structures was obtained using a 2-D analysis with the software ZeBuLoN [4] (finite element code developed at ENSMP) in combination with SiDoLo [5], a commercial software for identifying parameters. For reasons of symmetry, only half the middle section was modelled. Specific heat (C_p), thermal conductivity (k) and coefficient of thermal expansion are used in calculations as functions of temperature. The thermal loading of the specimens was described with convective film cooling (h_i) and radiation heat flow ($Ø_i$) which vary with the location (i) in the specimen. The FT-S specimen was virtually divided in 3 zones and the FT-L specimen in 4.

With a set of coefficients (h_i, $Ø_i$), the temperatures are calculated in the whole structure by ZeBuLoN and compared to experimental temperatures with SiDoLo which optimises the coefficients by minimising an error function. The best results for these computations are shown in Fig. 5 for the FT-S specimen and in Fig. 6 for FT-L.

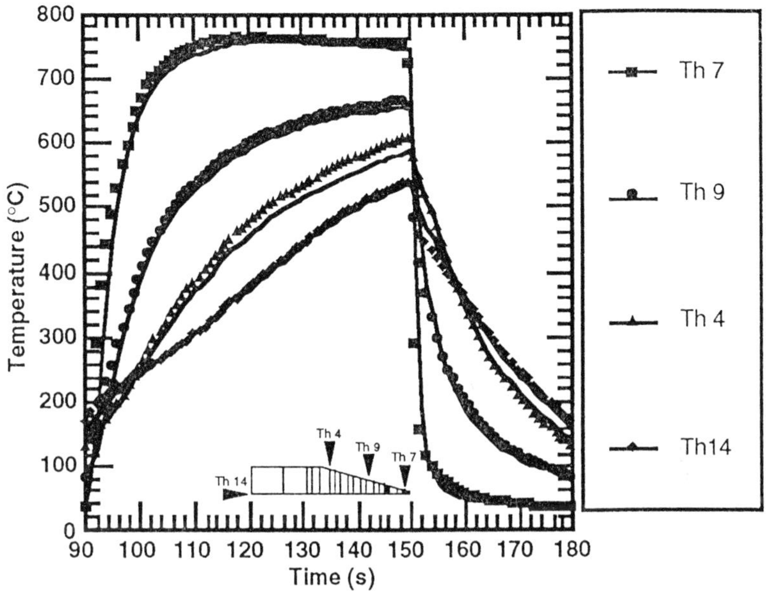

Fig. 5. Experimental and computed temperature profiles for FT-S geometry.

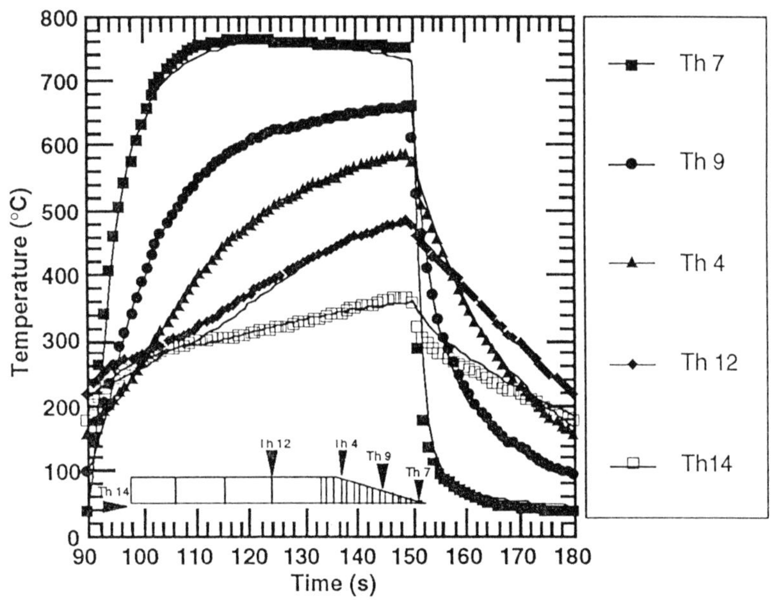

Fig. 6. Experimental and computed temperature profiles for FT-L geometry.

The comparison between the calculated and the experimentally measured temperatures along the section of specimens shows for the most part of the thermal cycle a good agreement, with a maximum difference of 20°C. The results from both thermal analyses were used as input for the mechanical stress-strain analysis.

5. Mechanical Analysis

The software SiDoLo was used to model the mechanical behaviour of the material under non isothermal loading, using isothermal low cycle fatigue (LCF) and thermal mechanical fatigue (TMF) tests results.

The software ZeBuLoN was used to calculate the strains and stresses generated during the thermal cycling, using the same mesh as for thermal analysis.

A viscoplastic Chaboche model with internal variables was used to simulate the stress-strain behaviour of the superalloy [6, 10].

The strain is partitioned into an elastic and a viscoplastic part:

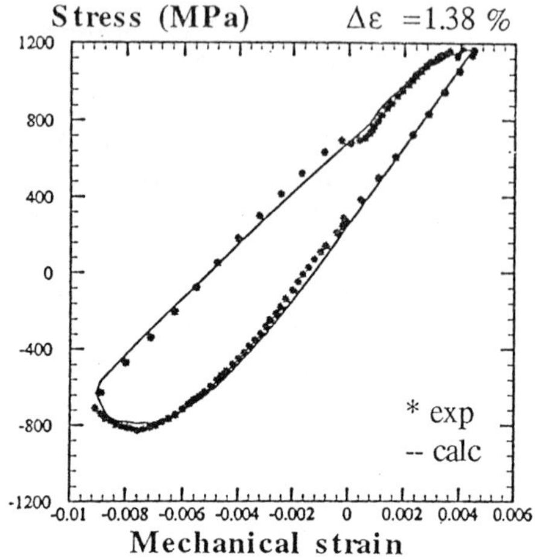

Figure 7. Comparison between experimental and computed stress-strain loops from TMF test.

$$\varepsilon = \varepsilon_e + \varepsilon_v \tag{1}$$

Under uniaxial loading, the viscoplastic constitutive equations are written as:

$$\dot{\varepsilon}_v = \left\langle \frac{|\sigma - X| - R_v}{K} \right\rangle^n \text{Sign}(\sigma - X_v) \quad \text{with } \langle a \rangle = \max(0,a) \tag{2}$$

$$X = X_1 + X_2 \tag{3}$$

$$X_1 = C_1.\alpha_{1v} \quad \text{where} \quad \dot{\alpha}_{1v} = \dot{\varepsilon}_v \tag{4}$$

$$X_2 = C_2\ \alpha_{2v} \text{ where } \dot{\alpha}_{2v} = \dot{\varepsilon}_v - D_2\ \alpha_{2v}\ \dot{v} \quad \text{and } \dot{v} = \left|\dot{\varepsilon}_v\right| \tag{5}$$

$$R_v = R_0 \tag{6}$$

where ε_V is viscoplastic strain, $\dot{\varepsilon}_V$ the rate of viscoplastic strain and v the cumulated viscoplastic strain. Kinematic hardening is described by means of two kinematic variables (X_1 and X_2). R_V represents the elastic domain radius. Young's modulus (E(T)) and Poisson's coefficient (ν) were measured by means of tensile tests at several temperatures. The material parameters K, n, C_1, C_2, D_2, and R_0 were identified in the temperature range 20°C-750°C from LCF and TMF test results [7,8] with SiDoLo which includes constitutive equations in special subroutines. An example of the simulated stress-strain loops is shown in Fig. 7 for a TMF test.

The identified constitutive equations were used in the finite element code ZeBuLoN to calculate strains and stresses using previous thermal analysis for the two specimen geometries. Examples of the results obtained for a mechanical stabilised cycle (20th) are presented in Figs. 8 and 9 for FT-S and FT-L specimens respectively. The stresses generated by thermal cycling are higher at the thin edge of the FT-L specimen relative to the FT-S specimen, and the mean stresses are tensile for both geometries. The mechanical and inelastic strains on the one hand, and the mean strain which is compressive, on the other hand, are largest in the FT-L and FT-S specimen geometries respectively. Inelastic strains are observed up to 5.6 mm and 11 mm from the leading edge of the FT-S geometry and the FT-L geometry, respectively.

6. Life Prediction

Usually, TF is used to compare the thermal cycling endurance of different materials. With the knowledge of the specimen thermal-strain-stress history, the prediction of the TF lifetime is possible using a damage model.

TF crack initiation life which corresponds to the number of thermal cycles required to reach a 0.2 mm crack length, is reported in a Manson-Coffin diagram (using the inelastic strain range, Fig. 10) and compared with fatigue life curves obtained from volume element tests: LCF at 200°C and 750°C and TMF (temperature range between 100°C and 750°C) [7,8]. These tests were performed on the same material as used in the TF tests. The Manson-Coffin diagram shows that the lifetimes obtained with the two TF geometries are in close agreement with the TMF results. TMF and TF lifetimes are 10 times shorter than LCF at 200°C and 8 times longer than LCF at 750°C. Consequently, the Manson-Coffin relation for TMF loading gives a good prediction of the TF specimen life.

The lifetime of TF specimens was also calculated using a fatigue-creep damage model. In this model [9, 10 and 11], the total damage rate results from the cumulation of creep (D_C) and fatigue (D_F) damage coupling is ensured by the total damage dependence of each damage rate :

$$dD = dD_C + dD_F = f_C(\sigma, D, T)\ dt + f_F(\sigma_M, \bar{\sigma}, D, T)\ dN \tag{7}$$

where σ, σ_M, $\bar{\sigma}$ are the current stress, maximum stress and mean stress respectively.

This model had to be defined for the whole temperature range of the TF tests. This was done by identifying model parameters with SiDoLo from LCF test results at high frequency and from creep results. Lifetime prediction for volume elements, tested under LCF at low frequency and TMF conditions was in good agreement with the experiment as shown in Fig. 11. Predictions are unconservative for LCF experiments.

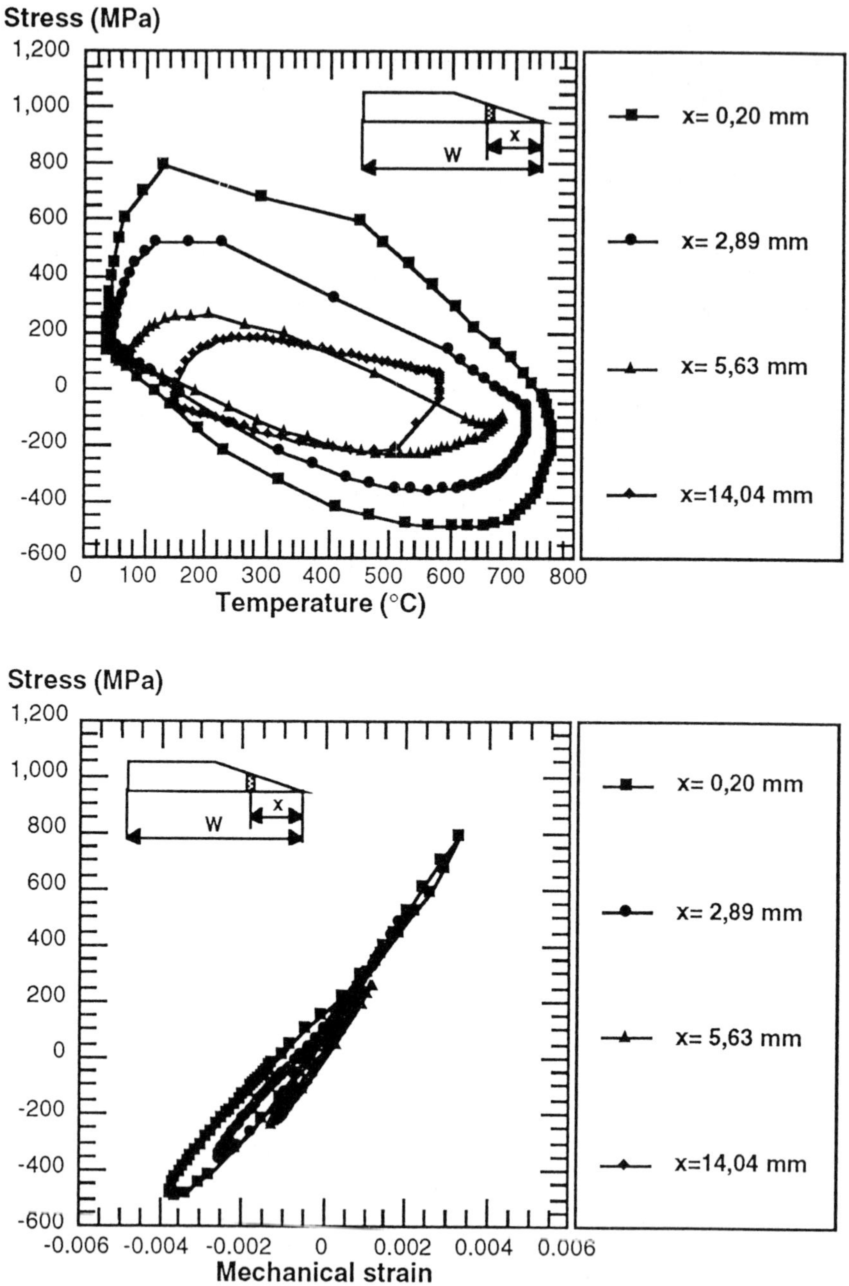

Figure 8. Calculated stress-temperature loops and stress-mechanical strain loops for FT-S geometry.

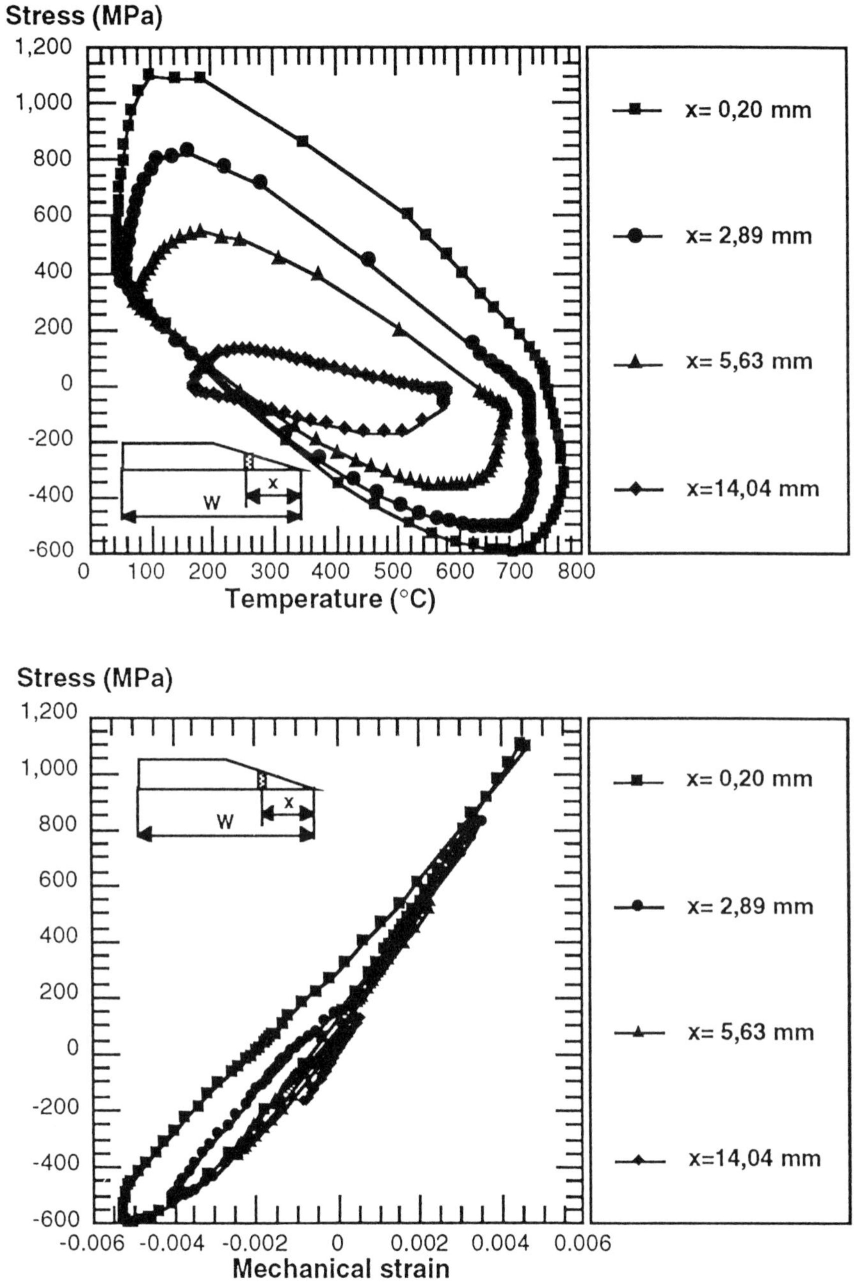

Figure 9. Calculated stress-temperature loops and stress-mechanical strain loops for FT-L geometry.

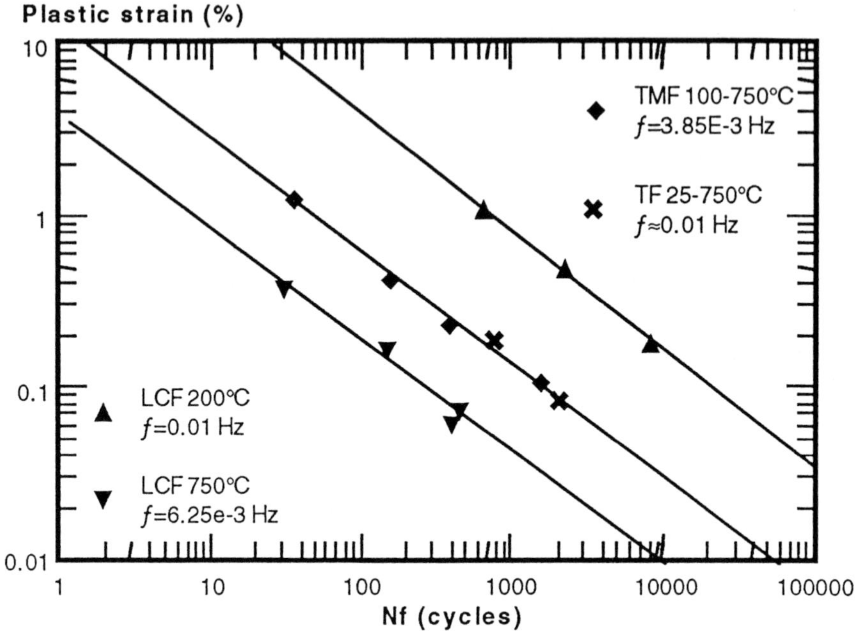

Figure 10. Life time comparison between LCF (200°C & 750°C), TMF and TF tests.

Eq (8) was used to compute the life of the first element at the leading edge of the TF specimen, which is compared to experimental data to 0.2 mm crack depth in Fig. 11. Predictions tend to be slightly conservative, but are within a factor of two of the experimental results.

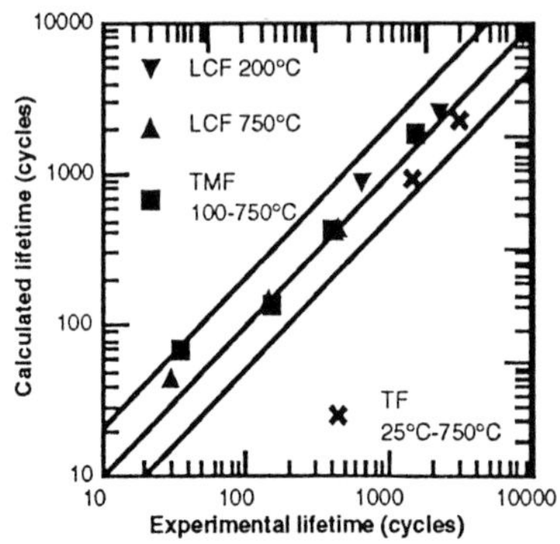

Figure 11. Comparion between calculated life of LCF, TMF, TF specimens and experimental data.

7. Conclusions

A thermal fatigue facility using a radiation furnace and forced air cooling system was developed. The versatility of the TF rig permits various thermal cycles by controlling the temperature history on specimens.

The TF tests were performed using two specimen geometries under a thermal cycle representative of actual blade loading conditions. The effect of the geometry on the crack initiation and propagation was shown, and the damaging mechanisms were briefly described.

A finite element analysis was made to compute the temperature-strain-stress history for the two kinds of specimens. For the thermal analysis, only experimental surface temperature profiles and material thermal properties were used. The mechanical analysis was performed using results from thermal computations and volume element test results to identify the material mechanical behaviour.

Two models were proposed to compute the lifetime to crack initiation under TF loading. The one based on TMF results in combination with a Manson-Coffin relationship was shown to give good predictions of TF data. The other one, based on fatigue-creep damage cumulation, was shown to give good agreement with experiment.

Acknowledgements - Financial support for this study by SEP (Société Européenne de Propulsion) as a part of a research program is gratefully acknowledged.

References

1. Glenny, E., A study of the thermal-fatigue behaviour metals, Paper No 2010, *Journal of the Institute of Metals*, 1970, pp. 449-461.
2. Spera, D.A., Calfo, F.D. and Bizon P.T., Thermal fatigue of simulated turbine blades, *NASA TM X-67820*, National Aeronautic Space Administration, May 1971.
3. Rezaï-Aria, F., François, M. and Rémy, L., Thermal fatigue of MAR-M509 superalloy, *Fatigue Fract. Eng. Eater. Struc.* (1988), 277—289.
4. Burlet, H. and Cailletaud, G., Numerical techniques for cyclic plasticity at variable temperature, *Eng. Comput.* **3**, (1986), 143-153.
5. Pilvin, P., Identification des paramètres de modèles de comportement, *Proc. Mécamat*, Besançon (1988), 155-164.
6. Chaboche, J.-L., Viscoplastic constitutive equations for the description of cyclic and anisotropic behaviour of metals., *Bulletin de l'Académie Polonaise des Sciences*, Série Sc. et Tech. **25**, n° 1 (1977), 33-42.
7. Köster, A., Fatigue of a nickel base superalloy for aerospace turbine blades: application to lifetime prediction, *Ph. D. thesis*, ENSMP, 1995, to be published.
8. Köster, A., Fleury, E., Vasseur, E. and Rémy, L., Thermal-Mechanical Fatigue Testing, *Automation in Fatigue and Fracture : Testing and Analysis, ASTM STP 1231*, C. Amzallag (ed.), American Society for Testing and Materials, Philadelphia (1994), 563-580.
9. Lemaitre, J. and Chaboche, J.-L., *Mécanique des Matériaux solides*, Dunod, 1988.
10. Chaboche, J.-L., Low cycle fatigue and life predictions, C. Amzallag, B.N. Leis and P. Rabbe (eds.), in *ASTM STP 770*, Philadelphia (1982), pp. 81-104.
11. Cailletaud, G. and Chaboche J.-L., Lifetime predictions in 304 stainless steel by damage approach, *ASME*, Orlando(USA), 27 June - 1st July 1982.

THE USE OF AN ADVANCED AUTOMATED CRACK MONITORING SYSTEM TO INVESTIGATE HIGH TEMPERATURE FATIGUE CRACK GROWTH IN A NICKEL-BASED SUPERALLOY

R.A. SMITH[1], Y. LIU[1] and L. GRABOWSKI[2]
[1] *Department of Mechanical and Process Engineering, The University of Sheffield, Mappin Street, Sheffield, S1 4DU, UK.*
[2] *Rolls-Royce plc., P.O.Box 31, Derby DE2 8BJ, UK.*

1. Introduction

This paper describes a new technique designed for testing surface short crack growth behaviour of metallic materials at elevated temperatures using a fully automated optical system. Short cracks as small as 10μm initiated on smooth specimens have been detected and recorded both at room and elevated temperatures up to 700°C. The application of this computer controlled system offers a reliable, time-saving and cost-effective method for the *in situ* study of the initiation and propagation of short cracks in materials subjected to both baseline and dwell fatigue loading sequences at high temperatures. Short crack growth behaviour has been studied in a Nickel-based superalloy. The crystallographic features of short crack growth and the effect of holding time at maximum load ('dwell' loading sequence) on cracking mode change from transgranular to intergranular at high temperature were investigated.

2. Experimental Technique

An automated short crack monitoring and testing system for high temperature use has been developed with the support of Rolls-Royce. This system uses a personal computer to control the optics, the testing machine and data recording units through purpose written software. The monitoring of surface short crack growth was made possible by using a microscope with high magnification lens which accessed a split type furnace. Since the system is fully automated and can carry out the test continuously, it overcomes the difficulties associated with recording high temperature short fatigue crack growth by techniques such as replication. The system is capable of testing and monitoring the growth of freely initiated short cracks as small as 10μm on a smooth specimen surface, and operating under baseline and complex (dwell) loading sequences at temperatures up to 700°C. Higher testing temperature is possible by using a larger furnace.

J. Bressers and L. Rémy (eds.), Fatique under Thermal and Mechanical Loading, 37–46.

This high temperature short crack testing system was developed at Sheffield University based on an original room temperature testing system, introduced by Rolls-Royce [1]. This new system combines the Rolls-Royce design with newly developed system software, hardware and high temperature facilities. There have been many new refinements since the first development phase was reported [2]. The most significant ones are the testing of freely initiated surface short cracks at 700°C and digital image storage, which has been made possible by the arrival of a 3.5 GB mass storage device. The basic working principle is to automatically and systematically scan the surface of a specimen using an optical microscope and record all the scanned information for later processing. The system was fully automated controlled by purpose written software running on a personal computer. This includes the control of a hydraulic testing machine using various loading sequences, the microscope head on stepping motor slides to scan the specimen's surface and the operation of recording units (videos and disks). Since the system is fully automated, it is possible to carry out high temperature crack growth testing continuously without having to interrupt the test and cool the specimen during a data collection, as is necessary if the replica method is used.

The system hardware is shown schematically in Fig. 1. It consists of a PC, an image analyser, a long working distance optical microscope mounted on stepping motor slides, a load sequence generator and two video cassette recorders. There is a large hard disk (3.5 GB) in the PC for storing data collected during a test. The large storage capacity of the disk makes it possible to store as many as 13 000 image frames during a test. The furnace is a split type, controlled by a separate console. A cooling jacket with a front silicon-quartz window is used to protect the optical head. During a test, the temperature in the middle of the specimen is closely monitored by a thermocouple mounted on the specimen, and temperatures are held constant to within ±10°C of the set point.

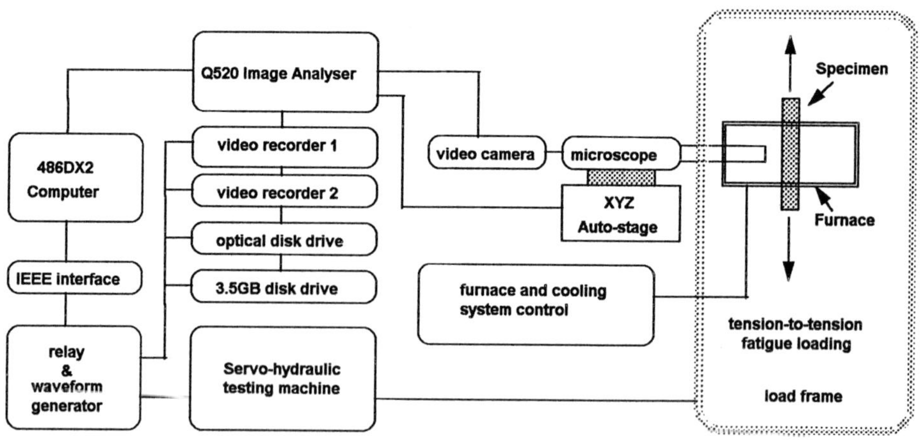

Figure 1. Schematic diagram of the short crack testing system.

The system software was developed using the image analyser's (Q520) high level programming language, QBASIC. The main functions of the software are summarised as follows:

1) Generating load sequences for the tests. This includes baseline fatigue sequences, i.e. sinusoidal and trapezoidal waveforms, and 'dwell' cyclic sequence which introduces hold periods on maximum/minimum loads for fatigue-creep interaction study.
2) Setting a scan pattern on the specimen's surface. The data will be collected frame by frame defined by the scan pattern usually on an area of 5 × 3 mm for a freely initiated short crack growth test.
3) Recording all the image frames collected during the test on either the video tapes and the hard disk. The image frames can be later retrieved and processed to trace back frame by frame to find the crack origin and to measure the length of a crack.
4) The system has been successfully applied to testing the growth of freely initiated short cracks at temperatures up to 700°C. Fig. 2 shows the images extracted from a high temperature short crack test. The crack was initiated from an inclusion and propagated transgranularly on the specimen's surface. Fig. 3 shows the images from a high temperature test using the 'dwell' load sequence. The crack was initiated from a sharp corner. Due to the effect of the hold time at the maximum load, the crack path changed from transgranular to intergranular, suggesting the creep-fatigue interaction.

3. Elevated Temperature Fatigue Crack Growth Testing

3.1. MATERIAL

The material used was Waspaloy, a nickel-based superalloy. Its microstructure comprises an equiaxial γ matrix containing a bi-modal distribution of γ' precipitates. The average grain size was 70μm. This material is being used extensively in civil gas turbine engine applications as a disc material in the hot section. This material has minimum yield and tensile strength of 860 and 1 240 MPa at room temperature, with corresponding maximum values of 778 and 1 089 MPa at 700°C.

3.2. FATIGUE CRACK GROWTH AT ELEVATED TEMPERATURES

Fatigue crack growth tests were carried out under tension-to-tension loading on square section specimens at both 20°C and 700°C. Baseline load sequence of sinusoidal waveform (5Hz) was adopted. Long crack data was obtained using sharply corner notched specimens at 20, 250, 500 and 700°C. Short surface crack data was obtained using smooth plain specimens with a pair of parallel shallow side notches of K_t=1.05. (K_t is the stress concentration factor.) Specimen surfaces were polished and etched before the tests. Test conditions are: σ_{max}=880 MPa at 20, 250, 500°C and σ_{max}=792 MPa at 700°C., R=0.1.

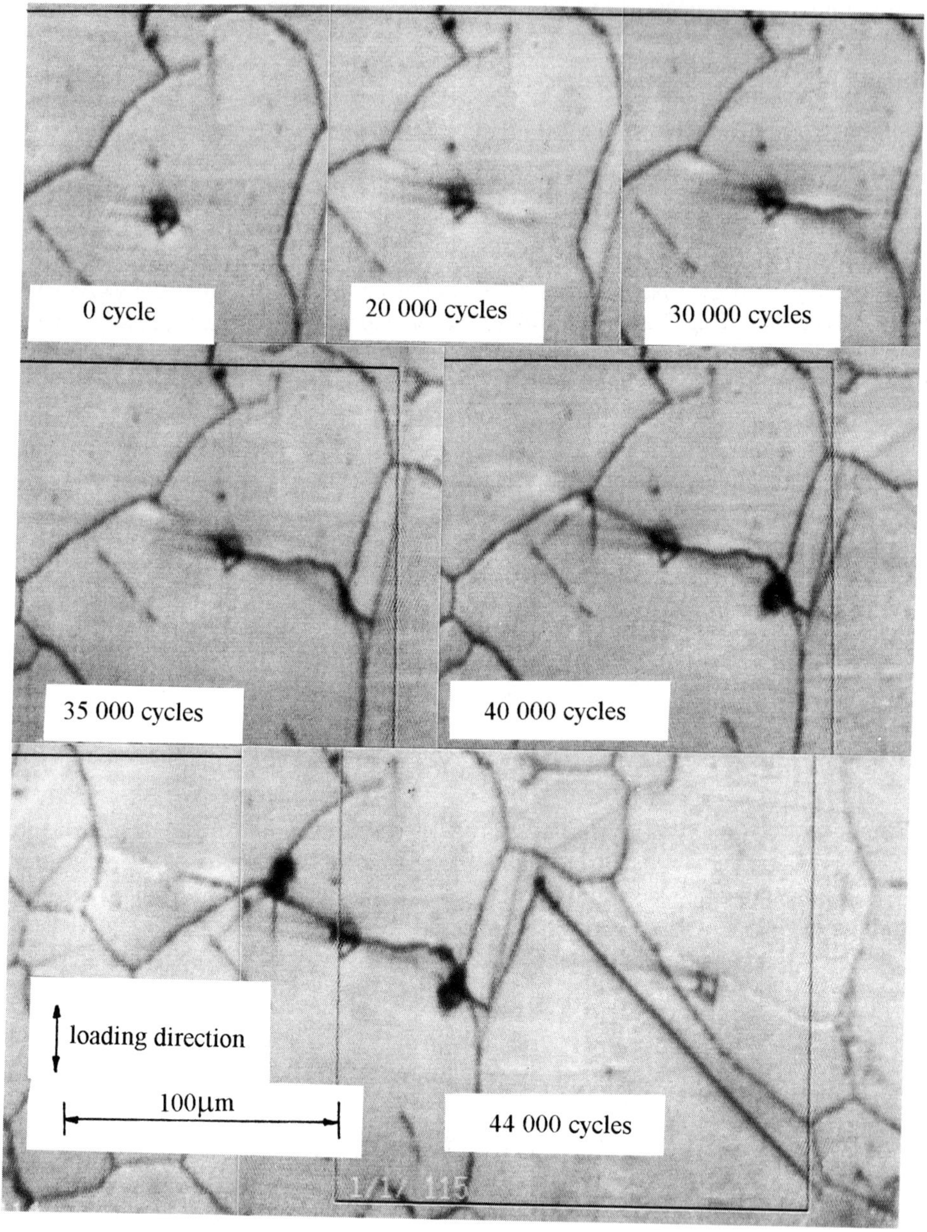

Figure 2. Freely initiated short fatigue crack growth at 700°C.

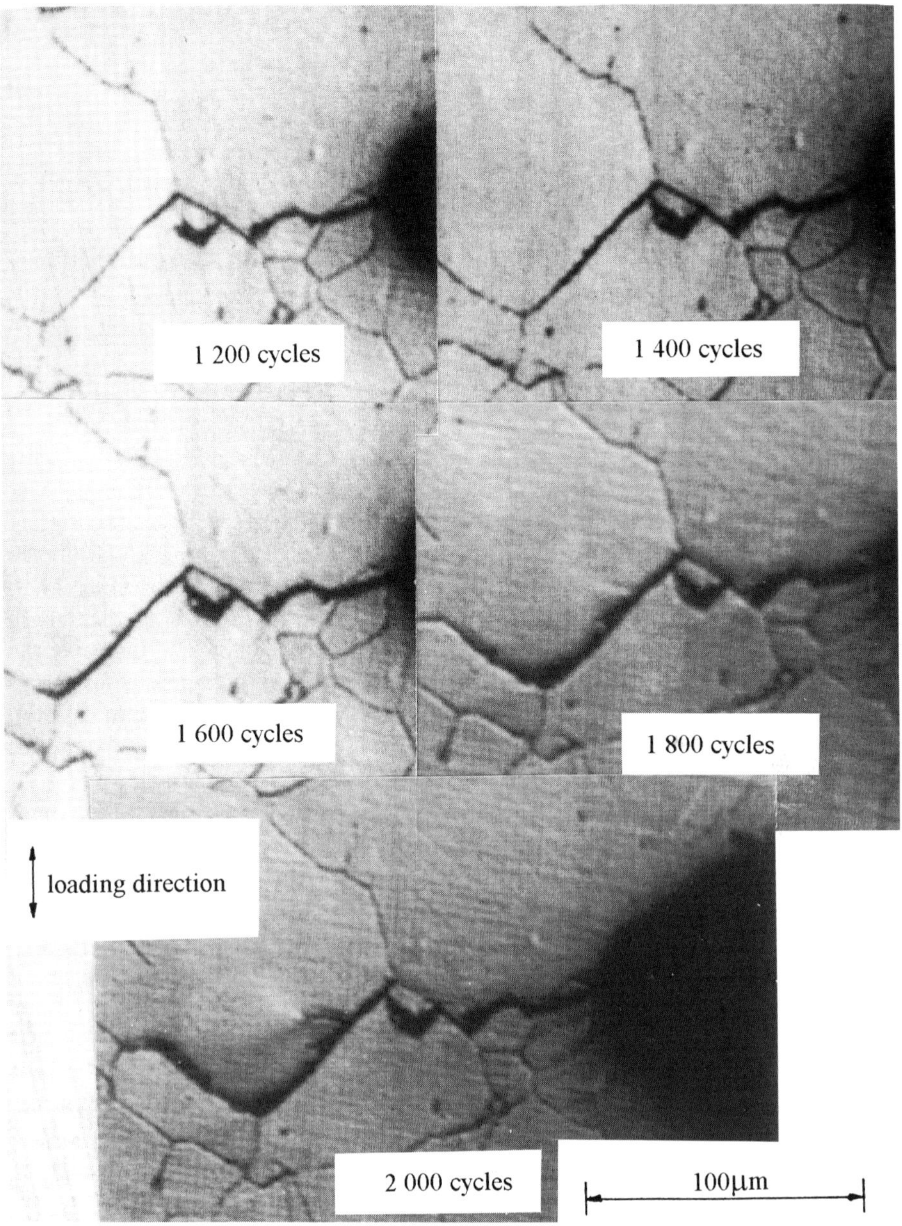

Figure 3. Fatigue crack growth under dwell loading (20s hold time at maximum load) at 700°C; intergranular cracking mode.

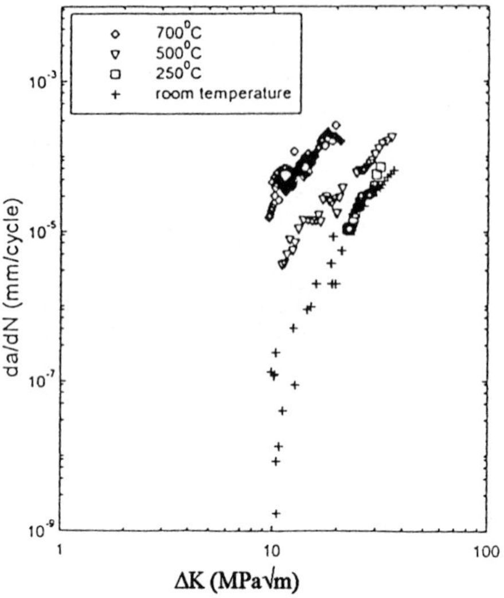

Figure 4. da/dN vs. ΔK at high temperatures.

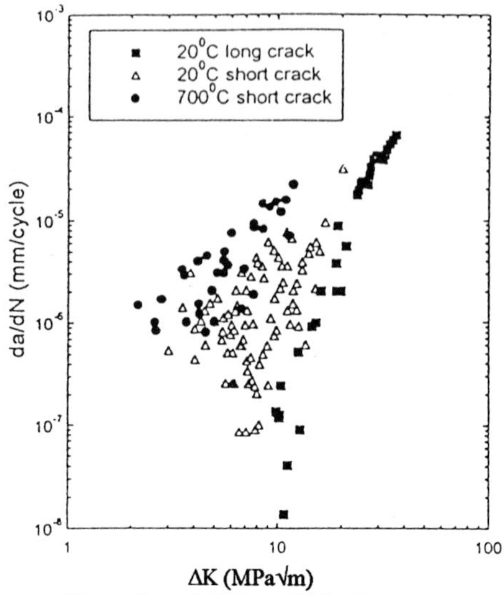

Figure 5. da/dN vs. ΔK for short cracks.

The crack growth rates obtained by sharply corner notched specimens at different temperatures were plotted against the LEFM parameter ΔK in Fig. 4. The short crack growth rates were plotted against the ΔK again in Fig. 5 and will be discussed later. In this graph and the following, all crack lengths are measured as the length projected onto a perpendicular to the loading direction. In the figures of this paper the loading direction is the vertical axis. The stress intensity factor calibrations for corner cracks in this section were taken from the finite element analysis by Pickard [3]. Fig. 4 shows the da/dN vs. ΔK for Waspaloy at different temperatures. At temperatures below 250°C, the temperature effect on the crack growth was small. At above 500°C, the environmental effect is remarkable. At 700°C, growth rates some 10-20 times faster than the corresponding room temperature rates were observed. At elevated temperatures material properties such as modulus and yield stress change and creep deformation and environmental interactions become more important. The test results in Fig. 4 are for relatively high frequency (5Hz) fatigue tests, without consideration of creep and hold time effects. However the introduction of maximum load hold periods did affect crack propagation rates and it will be discussed later in this paper.

The crystallographic features of short crack growth were observed and explained as for example in Fig. 6 which shows that a notch-initiated crack is growing in a transgranular manner at 700°C. The growing crack suffered from resistance from the material's local barriers, such as grain boundaries and slip bands. An extraordinary phenomenon was observed when the crack tip was in grain A in Fig. 6. As explained in Fig. 6, there were two slip systems within grain A: I and II. At first the crack tip tried to extend along the slip system I before it was blocked by the upper grain boundary. Then the crack tip changed its direction to extend along the slip system II before it was blocked by the lower grain boundary. This pattern repeated within the grain to form a zigzag crack path.

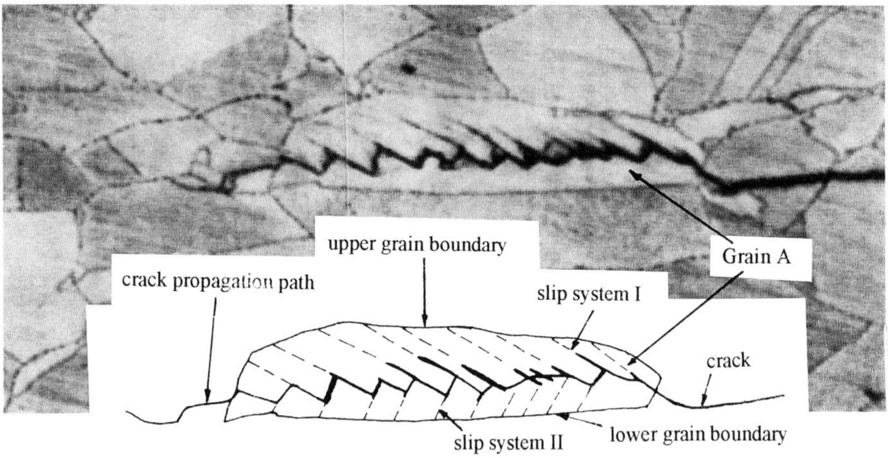

Figure 6. **Microstructural feature of short crack.**

3.3. SHORT CRACK GROWTH AT ELEVATED TEMPERATURE

Due to the considerable difficulties involved in short crack monitoring techniques at high temperatures, very little data for freely initiated surface(thumbnail) crack growth exists, and the problem has been discussed in relatively few publications, as cited in [4] and [5].

By using the automated short crack testing system developed by the authors, high temperature surface short crack growth data can be obtained without interrupting the heating cycle during the tests. Fig. 2 shows some of the images from a freely initiated surface short crack test at 700°C. The crack was initiated from an inclusion. It is worth noting that the inclusion was not an artificial one and other results at high temperatures also suggest that a short crack tends to initiate at inclusions or other types of microstructure discontinuities at high temperature in this material. As a contrast, short cracks initiated at room temperature tend to be stage I shear cracks. The short crack growth rates were plotted against the ΔK again in Fig. 5. The use of ΔK enables a comparison to be made, in the absence of a better alternative, although concerns have been expressed on the inappropriateness of ΔK to characterise short crack growth as cited in [6] and [7]. The short crack growth data shows an apparent acceleration feature comparing with the long cracks at the same nominal stress intensity factor range. The behaviour of the short cracks is anomalous and there is no apparent threshold. Modelling of short crack growth is a far more difficult task than modelling the long crack growth. The problem is that due to the irregularities and discontinuities caused by microstructure interaction, the crack growth rate for short cracks is not a single function of a and σ, as is the case for long cracks which can more readily be characterised by ΔK and the well-known Paris growth law.

3.4. FATIGUE CRACK GROWTH UNDER 'DWELL' LOAD SEQUENCE

Gas turbine discs are subjected to high centrifugal and thermal stresses particularly during take off and climb. Some disc parts may reach temperatures in the creep range and are therefore subjected to possible creep-fatigue interaction. Previous studies have shown faster crack growth rates when tensile dwell periods were included in the loading signal [8].

A typical 'dwell' load sequence, with different hold time at maximum and minimum loads, is shown in Fig. 7. The holding time t_{D1} was 5,10 and 20s in our tests. Before introducing the dwell sequence, a fatigue crack was initiated under the baseline fatigue load sequence, thus providing a sharp crack tip. This type of test takes a considerable time to finish since the effective cycling frequency is low (<0.05Hz for t_{D1}=20s). The test results for different holding times at maximum load are shown in Fig. 8. As shown in Fig. 8, crack growth rates showed no significant change when $t_{D1} \leq 10$s while the crack path was still transgranular. However, when the t_{D1} increased to 20s, the crack growth rates changed rapidly (10 times faster) and the crack growth path changed from transgranular to intergranular as shown in Fig.3.

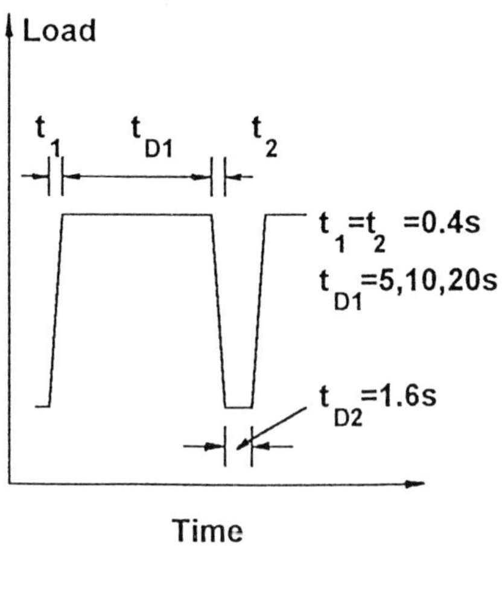

Figure 7. Dwell loading sequence.

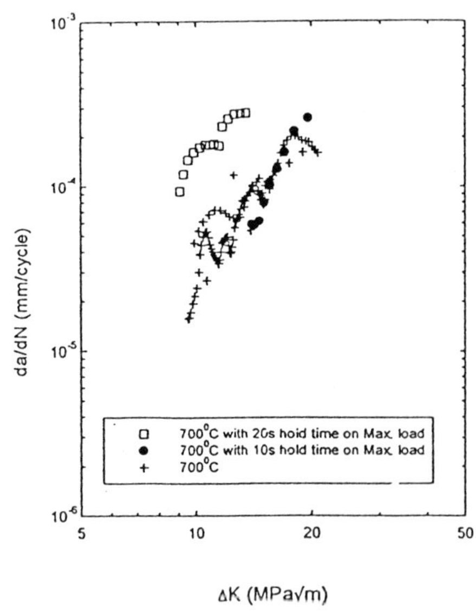

Figure 8. da/dN vs. ΔK under dwell loading.

4. Discussions

The short crack growth behaviour, shown in Fig. 6, is faster than that of long cracks. The long crack growth rates follow the Paris law and show a threshold below which the crack does not grow. The short crack growth data at both room and elevated temperature show anomalous behaviour and no threshold. It can be seen from Figs. 2 and 6 that the short crack growth is highly crystallographic and microstructure related. However the short crack data tends to join the long crack data at larger ΔK values and becomes less scattered. This usually happens when the crack has overcome the stage I shear growth and becomes a stage II tensile crack. In general, stage I crack initiation was most prevalent in Waspaloy [9].

The scatter in short crack growth data and its behaviour compared to long cracks can be explained by the microstructural aspects of the problem:

1) For a given volume of material under fatigue loading, cracks will initiate from the most favourable sites, which may be geometrical imperfections, favourably oriented grains or brittle precipitates. In all cases the local environment of the initiated crack has more influence on short crack development than the bulk of the material. As a result, in the early stage the short cracks will grow more rapidly until some type of microstructural barrier is encountered. This is why the short crack growth rates usually exhibit initial acceleration followed by deceleration.
2) As a short crack propagates into 'normal' material where conditions are near the material's average property and therefore less favourable for crack growth than the initiation site, the increased crack size tends to sample the material more uniformly. It thus becomes a normal long crack, whose behaviour is representative of the bulk properties of the material.
3) Any individual short crack will face unique local conditions due to the various microstructural features distributed in the polycrystalline material. For instance, a crack growing in one grain will have a growth rate that is influenced by lattice orientation and the presence of grain boundaries and precipitates. As a result, each individual short crack behaves uniquely, giving rise to the scatter shown in Fig. 5. This individual behaviour of short cracks imposes great difficulties in modelling the growth behaviour using a unified formula such as the Paris law. This implies that any realistic modelling of short cracks should include some description of the random grain distribution. It is also important to note that the crack is influenced by conditions in the interior of the material as well; our direct observations are confined to the surface.
4) The minima in crack growth rates are associated with the most unfavourable conditions for crack growth. At the early stage of crack growth, the cracks are small and the local driving forces are small. However the cracks are long enough to begin to encounter microstructural barriers. Some short cracks become non-propagating since they have smaller driving forces than the crack growth resistance while others keep advancing across the barriers despite suffering retardation at these barriers.

Acknowledgements - This work is supported by a research contract from Rolls-Royce plc., Derby, Contract Brochure No. PVJ2-126D, to whom grateful thanks are recorded.

References

1. Hussey, I.W., Roche, J., Grabowski, L., A fully automated system for monitoring short crack growth, *Fatigue Fracture Engng Mater Structures,* **14**, (1991), 309-315.
2. Liu, Y., Smith, R.A., Grabowski, L., Automated Short Fatigue Crack Measurements at High Temperatures, in R.A.Ainsworth and R.P.Shelton (eds.), *Behaviour of Defects at High Temperatures,* ESIS 15, Mechanical Engineering Publications, London, (1993), pp. 99-109.
3. Pickard, A.C., The application of 3-dimensional finite element methods to fracture mechanics and fatigue life prediction, EMAS, Warley, West Midlands (1986).
4. Healy, J.C., Grabowski, L, Beevers, C.J., Short fatigue crack growth in a nickel-based superalloy at room and elevated temperature, *Int.. J. Fatigue,* **13**, (1991) ,133-138.
5. Hudak, S.J., Davidson, D.L., Chan, K.S., Howland, A.C., Walsh, M.J., Growth of small cracks in aeroengine disc materials, *Technical Report AFWAL-TR-88-4090,* Southwest Research Institute, USA (1988).
6. Lankford, J., Davidson, D.L., in R.O.Ritchie and J.Lankford (eds.), *Small Fatigue Cracks,* Metal Soc. of AIME, Warrendale, Pa., (1984), pp. 51-72.
7. Leis, B.N., Hopper, T., Broek, D., Kannien, M.F., Critical review of fatigue crack growth of short cracks, *Engng Fract Mech.,* **23**, (1986), 883-898.
8. Byrne, J. Hall, R., Alabraba, M., Grabowski, L., Hussey, I.W., Elevated temperature fatigue crack growth in a nickel base superalloy under dwell conditions, *Proc. Sixth Int. Conf. On Mechanical Behaviour of Materials,* Kyoto, Japan, **4**, (1991), pp. 361-366.
9. Lerch, B.A., Jayarman, N.J., Antolovich, S.D., *Mater. Sci. Engng.,* **66,** (1984), 151-166.

BITHERMAL FATIGUE TESTING: EXPERIMENTAL AND PREDICTION PROBLEMS

A. DEL PUGLIA, F. PRATESI, and G. ZONFRILLO
Dipartimento Meccanica e Tecnologie Industriali,
Università degli Studi di Firenze.
Via S.Marta 3, I-50139 Firenze, Italy.

1. Introduction

The need of thermomechanical fatigue testing is strongly indicated by the actual working conditions of various machine components used at high temperatures. In many cases, temperature is not uniformly distributed on a component and it also changes over the course of time. In these conditions, isothermal testing has been shown to be unsuitable and life prediction must be based on more complex testing procedures.

To this aim, various techniques have been proposed. Among these, widely used is the one known as thermomechanical fatigue (TMF), in which variations of strain and temperature are both linear with time. For the two cases in which the two variations are in phase, IP, or out-of-phase, OP, the definition of a standard procedure is now under way.

In DMTI, another technique [1], known as bithermal testing (BT) has been applied until now. For some characteristics, it may be considered as intermediate between isothermal (LCF) and TMF tests. In BT tests (which, like TMF, may be carried out in both IP and OP conditions) the tension and compression half-cycles are conducted isothermally at two different temperatures, while during the changes of temperature the applied stress is null.

BT testing has several useful advantages. It is easier to implement and to carry out and the long experience acquired with isothermal testing can be more directly applied. Probably, the most significant characteristic of BT testing is that it allows more straightforward correlations among temperature, life and microscopic phenomena. Therefore, by suitable selection of the two temperatures it is possible to investigate the presence of microstructural changes leading to transitions in mechanical behaviour. Moreover, it has been shown that BT tests are the most damaging among the LCF and TMF tests in similar conditions and can thus provide conservative indications for design.

In literature, two main applications of BT testing can be found. In NASA labs, it is used for obtaining input data for the application of the "Total Strain version of Strain Range Partitioning" technique (TS-SRP) to TMF tests [2]. At MPI Stuttgart,

J. Bressers and L. Rémy (eds.), Fatique under Thermal and Mechanical Loading, 47–53.

many experimental procedures have been developed and compared. A particular type of BT procedure gives the shortest lives observed [3].

2. Experimental Problems in Bithermal Testing

On FSX 414 superalloy, the following tests have been carried out with strain rate $\dot{\varepsilon} = 0.001\ s^{-1}$: LCF at 450°C and 950°C, BT at the same temperatures. Additional LCF tests have been performed at 950°C with $\dot{\varepsilon} = 0.03\ s^{-1}$.

The testing procedure and the main experimental results have been reported in [4]. A first group of problems, connected with local ratchetting, determination of elastic moduli and of the plastic strain range, was encountered [5, 6], as summarised here. With the specimen geometry used in LCF tests, local ratchetting occurred in BT tests, with macroscopic changes in the profile of the specimen. This behaviour can be interpreted by a progressive cumulation of plastic strain in limited zones due to the different temperature distributions in the two half-cycles. This problem was overcome by designing a new specimen profile [5].

The apparent elastic moduli not only differ, as expected, between half-cycles (owing to the two temperatures), but also if evaluated in the loading or in the unloading stages of the half-cycle. Due to the uncertainty in the evaluation of the elastic moduli, some difficulties arise for deducing the corresponding plastic strains in the two half-cycles. Moreover, the plastic strain range, $\Delta\varepsilon_{pl}$, shows a change, sometimes remarkable, due to hardening during the whole test. Some procedures for evaluating the elastic moduli and the plastic strain characteristic of the cycle are discussed in [6].

2.1. CYCLIC BEHAVIOUR

While studying in some detail the TS-SRP method and its underlying logic, an interesting field of investigation has been found in the relations controlling the strain behaviour of the material.

The representation of such behaviour is generally provided by the relationship between $\Delta\varepsilon_{pl}$ and the elastic strain range, $\Delta\varepsilon_{el}$, or stress range, $\Delta\sigma$, but it can be based on other quantities. In this investigation, for each half-cycle the maximum (or minimum) elastic strain, ε_{el}, the maximum (or minimum) stress, σ_m, and the hysteresis area, A, have been used. The reference values for each test, carried out until rupture of the specimen occurs, are taken in correspondence to its half-life cycle.

The cyclic behaviour is different in the tension and compression half-cycles, due to the different temperatures. Representations of the flow behaviour as functions of $\Delta\varepsilon_{pl}$ are shown in Fig. 1. Using A (evaluated in MJ/m^3), the sets of data for the various types of test have about the same slope (Fig. 1a) and the behaviour can be represented by regrouping the points in relation to the two temperatures (and strain rate, dashed line). Practically the same behaviour is shown using σ_m or ε_{el} (Fig. 1b), even if more extensive scattering exists. This scattering is enhanced by the very slight slope of the straight lines which fit the experimental points (low value of the exponent in the relation between $\Delta\varepsilon_{el}$ and $\Delta\varepsilon_{pl}$). For the data available, the cyclic relation may therefore already be predicted starting from knowledge of the results of LCF tests only.

2.2 EVOLUTION DURING THE TEST

In order to decrease the observed scattering of the experimental values (particularly evident in the case of ε_{el}) further tests would be useful at lower $\Delta\varepsilon_{pl}$ values. In these conditions, however, life becomes very long and testing is expensive and time-consuming; therefore, it would be interesting to have the required information without having to wait for rupture. In the particular case of the alloy under investigation, the hysteresis cycle does not reach a stable condition and the specimen hardens, especially at 450°C, until a propagating crack develops. The changes in the quantities determining the cyclic behaviour of the material occurring during the test have therefore been investigated.

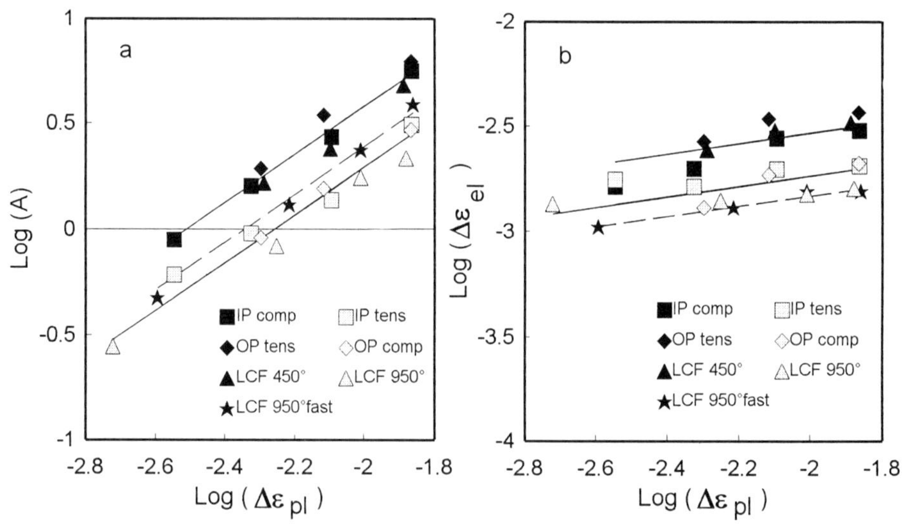

Figure 1. Cyclic behaviour expressed with A and ε_{pl}.

With reference to the half-life values (derived from the relations for the cyclic behaviour), these changes as a function of the life fraction have been approximated with a straight line (considering the logarithms of the quantities involved). Firstly, four different lines - obtained by regression from IP and OP tension and compression half-cycles - have been introduced. They give a very good representation of all the tests. Then the data have been regrouped in two sets related to temperature: the first group includes IP tension and OP compression data, the second OP tension and IP compression data. A regression in these two sets gives a good fit, as shown for example for σ_m in Fig. 2 (where, for greater clarity, only a fraction of the experimental points, taken at regular intervals, has been plotted). Note that the hardening calculated on the basis of the last regression (depending only on temperature) gives at 450°C values of σ_m higher than the experimental ones for compression and lower for tension phases.

3. Problems of Prediction in Bithermal Testing

For the purposes of design, extensive characterisation of the mechanical behaviour of materials is necessary; moreover a life prediction method (LPM) is required for reliable evaluation of strength. For isothermal conditions, several LPMs are available, often providing good agreement with experimental results. For TMF, only recently have LPMs been suggested, which will profit from further applications to experimental results. No model has yet been explicitly developed for BT tests.

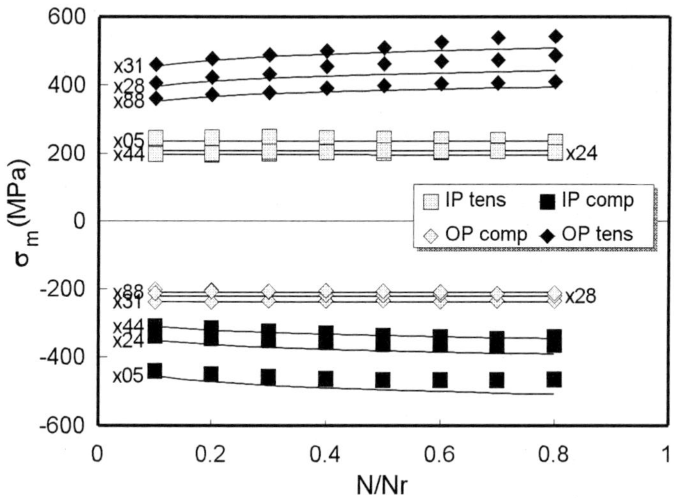

Figure 2. Experimental (symbols) and calculated (lines) values of σ_m (xnn identifies the specimen).

Some isothermal LPMs have been examined and critically evaluated for their potential application to BT (and TMF) testing, either directly or with simple modifications. In previous work on isothermal fatigue, a Continuum Damage Mechanics LPM [7] had been used. This method requires major modifications in order to be applied to BT or TMF, because the temperature variable is not included in the expression of fatigue damage. Therefore, different laws for damage evaluation and for its evolution must be found, for instance connecting the damage to fractions of the cycle and taking the temperature into account. Another LPM considered is the TS-SRP method [1]. It is already used in NASA labs for TMF and it should also work for BT tests. Some difficulties are encountered in evaluating the strength in elementary cycles. For its application to the specific material, some experimental data are needed that are not yet available from the tests carried out. At present, simpler relations are being used, such as Manson-Coffin and Ostergren laws. In these cases, "prediction" has the limited meaning of an interpolation (and some extrapolation) among experimental results and it is valid only for specific test conditions.

3.1. MANSON-COFFIN LAW

For most alloys, the slopes of the straight lines in the Manson-Coffin diagram ($\Delta\varepsilon_{pl} = K\ N_r^{\alpha}$) remain practically constant when temperature changes or hold times are introduced. A regression is performed on the experimental results corresponding to various testing conditions, assigning the same angular coefficient α to all of the lines. The resulting diagram is shown in Fig. 3. The agreement with experimental results is evident (Fig. 5a). This confirms that the exponent depends essentially on the material and is the same for both LCF and BT tests.

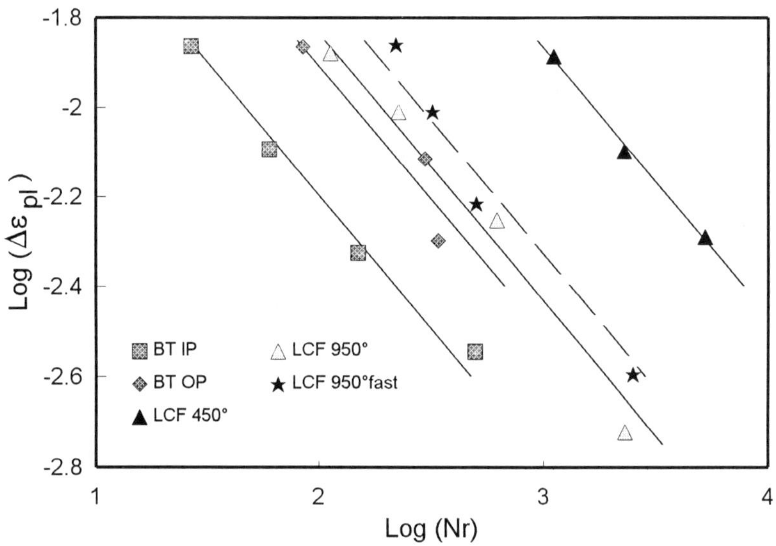

Figure 3. Manson-Coffin law with constant exponent.

3.2. OSTERGREN METHOD

In the Ostergren method, the basis for predicting fatigue life is the energy dissipated by hysteresis during the tensile half-cycle. Although it was proposed for LCF, it can be directly applied to BT tests.

When introducing the BT test, a more complete comparison between LCF and BT tests may be made on the basis of the hysteresis area of both half-cycles, rather than the tension half-cycle alone. The Ostergren method ($A = K'\ N_r^{\beta}$) is found to be fully suitable for representing the experimental results obtained. The slopes β of the straight lines do not change remarkably in the various kinds of test.

As was done for the Manson-Coffin law, regression can be performed on the

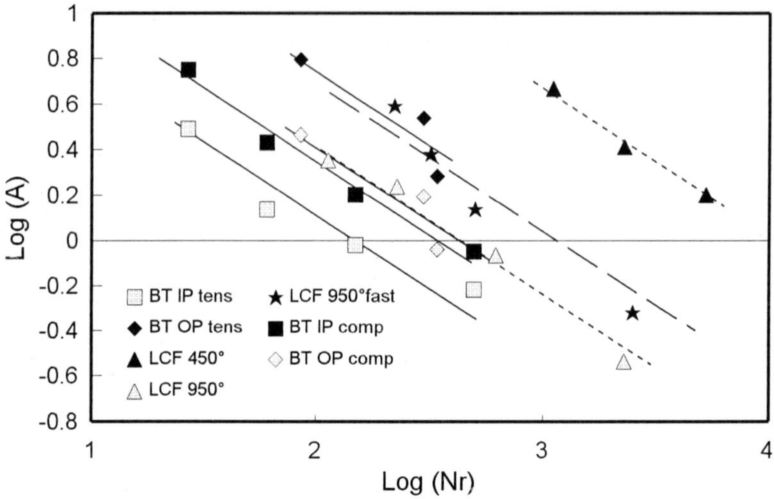

Figure 4. Ostergren law with constant exponent.

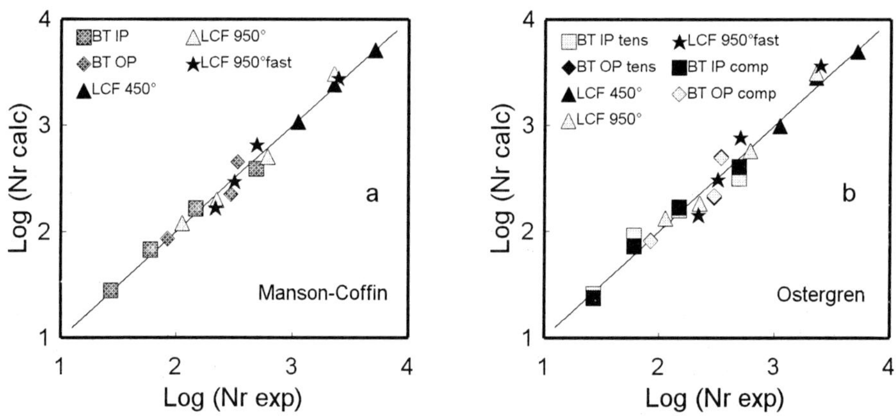

Figure 5. Comparison between experimental and calculated life.

whole set of experimental data, by letting all the straight lines be parallel to each other, as shown in the diagram of Fig. 4. The comparison between predicted and actual life shows good agreement (Fig. 5b).

4. Conclusions

BT tests have been selected and carried out for improving knowledge on the mechanical behaviour of FSX 414 at high temperature. Some experimental problems have been discussed, with particular reference to cyclic behaviour. It has been found that cyclic behaviour is basically a function of temperature.

The application of some LPMs has been investigated. In particular, Manson-Coffin and Ostergren laws have been fitted to the experimental results obtained.

BT testing may be very useful in helping to achieve an understanding of the thermomechanical behaviour of materials; among other things, the experimental problems are limited and the interpretation of results can be straightforward.

References

1. Halford, G.R., Mc Gaw, M.A., Bill, R.C., Fanti, P.D., Bithermal Fatigue: a Link Between Isothermal and Thermomechanical Fatigue, in H.D.Solomon and al. (eds.), *Low Cycle Fatigue*, ASTM STP 942. 1988, pp.625-637.
2. Halford, G.R., Saltsman, J.F., Verrilli M.J., Arya, V.K., Application of thermal life prediction model to high-temperature aerospace alloys B1900 + Hf and Haynes 188, in M.R.Mitchell and R.W.Landgraf (eds.), *Advances in Fatigue Lifetime Predictive Techniques*, ASTM STP 1122, 1992, pp.107-119.
3. Baumgärtner, T., Bothe, K., Hurta, S., Gerold, V., Thermomechanical Fatigue of Nimonic 80A and Alloy 800H. In E.Bachelet et al. (eds.), *High Temperature Materials for Power Engineering 1990*, Kluwer Academic Publishers, Dordrecht, 1990, pp.1087-1096.
4. Del Puglia, A., Betti, F, Giorni E., Pratesi, F., Zonfrillo, G., Low-cycle and bithermal fatigue of a cobalt-base superalloy. In D.Coutsouradis et al. (eds.), *Materials for Advanced Power Engineering 1994*, Kluwer Academic Publishers, Dordrecht, 1994, pp.919-928.
5. Zonfrillo, G., Pratesi, F., Del Puglia, A., Instabilità in fatica ad alta temperatura. Proc. of 25th Conv. AIM, Milan, 1994, pp.62-64.
6. Del Puglia, A., Zonfrillo, G., Pratesi, F., Experimental problems in bithermal fatigue. Proc. of 11th Danubria-Adria Symp. on *Experimental Methods in Solid Mechanics*, Baden (Vienna), 1994.
7. Lesne, P.M., Savalle, S., A differential damage rule with microinitiation and micropropagation. *Rech. mAérosp.* **2** (1987), 33-47.

STRUCTURAL INTEGRITY ASSESSMENT OF THE FUSION REACTOR FIRST WALL USING THE ASME CODE

M. MEROLA[1] and R. MATERA[2]
[1] *Politecnico di Torino, Dipartimento di Energetica Corso Duca degli Abruzzi, 24, I-10129 Torino, Italy.*
[2] *Joint Research Centre, Institute for Advanced Materials. Present address: ITER Joint Central Team, Garching Working Site, Germany.*

1. Introduction

The thermonuclear fusion experimental reactor will be an intrinsically pulsed machine. Therefore the plasma facing components will be subjected to cyclic heat loads thus resulting in a problem of thermal fatigue. The first wall (FW) is the structure that covers about 80% of the total plasma-facing surface and will receive a cyclic heat flux ranging from a zero value up to some hundred kW/m^2.

The Institute for Advanced Materials (Joint Research Centre of the European Union, Ispra site) is carrying out experiments on both simple components and FW prototypes. A Thermal Fatigue Test Facility has been designed and built to make a proper simulation of the working environment of the experimental reactor FW as far as the thermal loads are concerned. A description of the facility can be found in [1]. This apparatus consists of a cylindrical vacuum chamber with diameter 1m and length 1.2m; the thermal heat flux is provided by a series of infrared lamps with 240 mm long filament and a maximum power of 4 600 W each. The maximum heat flux compatible with a long lamp lifetime is 0.8 MW/m^2. This value is sufficient to simulate the thermal loads, including peak loads, to which the FW will be subjected in real operating conditions. Higher fluxes can be obtained at the expense of a shorter lamp life.

The aims of the experimental campaign can be summarised as follows. (1) To evaluate the possible extension of the presently existing nuclear design standards to the fusion environment as far as thermal loads are concerned; (2) to compare the thermomechanical performances of prototypical first wall mock-ups; (3) to study the influence of different manufacturing technologies on thermal fatigue failure mode; (4) to validate computational methods for cycling thermal stress and strain evaluation and thermal fatigue lifetime prediction. All the components tested are made of austenitic stainless steel and are actively cooled by deionised water.

As regards FW prototypes, the testing of one medium-scale prototypical component was recently completed after 50 000 thermal fatigue cycles. It was

J. Bressers and L. Rémy (eds.), Fatique under Thermal and Mechanical Loading, 55–65.

manufactured by Framatome on behalf of The NET Team. The fabrication process used in the corresponding FW concept was adopted. The technology developed envisages the manufacturing of four half-plates, milled to obtain four cooling semichannels. Each pair of half-plates is welded by means of transparency electron beam welding; one then obtains plates with four cooling channels which are welded together, again with an electron beam. The second phase consists of the insertion of tubes without welding into the channels and in their junction to the plate by means of vacuum brazing. Further information concerning the fabrication of the component can be found in the progress reports edited by the manufacturer [2]. Fig. 1 shows a photo of the mock-up.

2. The Thermal Fatigue Test

The thermal fatigue test was carried out in two phases. The first phase lasted about 5 000 cycles and was characterised by a peak heat flux of 590 kW/m^2. The second phase lasted 45 000 cycles, the applied thermal cycle is shown in Fig. 2. The peak heat flux was 660 kW/m^2. The water coolant had a temperature of 15°C and a pressure of 0.15 MPa. The numerical analysis was performed by means of the ABAQUS finite element code (vers. 5.3-1, 1994). The thermal transient analysis was carried out to evaluate the temperature evolution during the thermal cycles. The elastic stress analysis was performed at the end of the heating period and at the end of cycle. The generalised plane strain model (bending allowed) was used to compute stress. Fig. 3 gives the experimental layout and the location of points A and B. The former point was chosen because in that location a long fatigue crack nucleated and propagated as is discussed later. At point B the highest temperature and stress was computed. During phase 2 of the test a maximum temperature of 20 and 386°C was

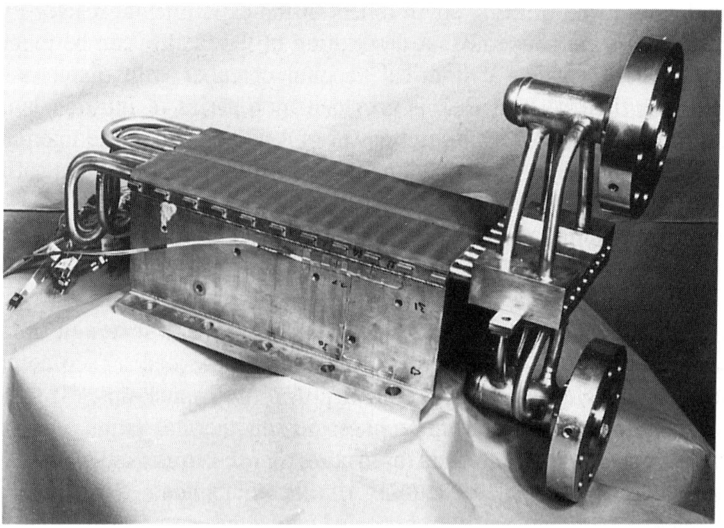

Figure 1. The prototypical first wall mock-up.

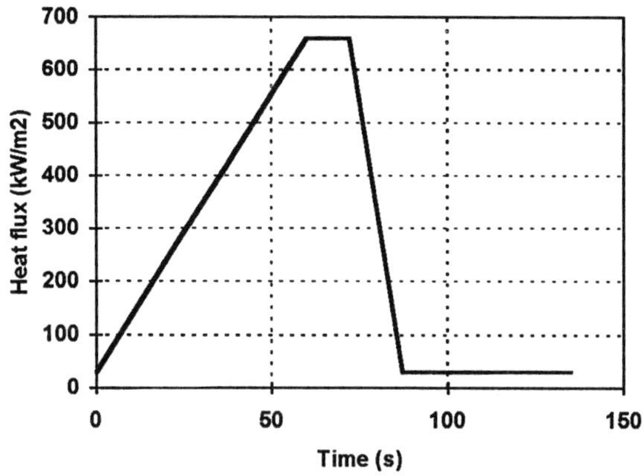

Figure 2. Applied thermal cycle during phase 2 of the test.

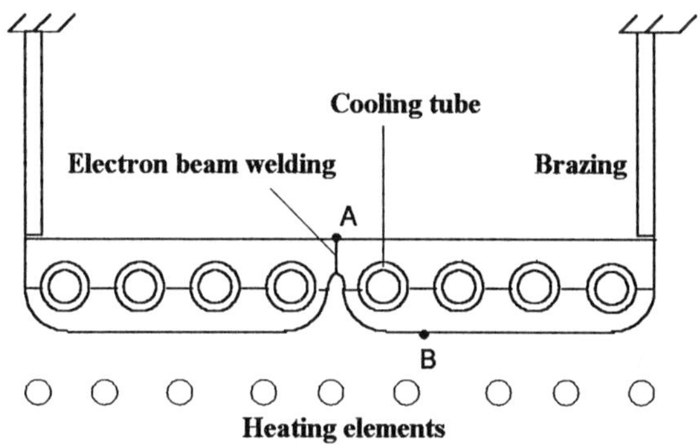

Figure 3. Experimental layout of the thermal fatigue test.

evaluated at points A and B, respectively. The elastic calculation gave a Von Mises equivalent stress of 282 and 835 MPa, respectively.

3. Experimental Results

Non-destructive testing was carried out by Framatome by means of ultrasonics before the fatigue test, after the first phase and at the end of the test. Two parallel cracks were detected after the first 5 000 cycles. They ran along both sides of the central electron beam welding (point A, Fig. 3). After the completion of the 50 000 cycles, these two cracks did not seem to have propagated further.

The component was instrumented by a number of strain gauges (one of them located at point A) and by two displacement gauges. An analysis of the measurements showed that:

- no ratchetting occurred;
- the two above-mentioned cracks probably nucleated after a few tens of cycles and stopped propagating after some hundred cycles.

Further details on the experimental results can be found in [3] and [4].

4. Application of the ASME-III Procedure

The American ASME code [5] deals with fatigue verification in Section III, Division I, Subsection NB-3000 as far as class I components are concerned. The concept of *Alternating Stress Intensity* is introduced and the maximum number of cycles allowed is given by design S/N fatigue curves. Triaxiality is managed using the Tresca criterion. An elastic numerical analysis is required.

Table 1 summarises the evaluation of the maximum allowable number of cycles according to the ASME code. When the figure exceeds 10^6, it is put at infinity. The following definitions were used:

S_{alt} = Alternating Stress Intensity $= \frac{1}{2}\frac{E_{curve}}{E_{ref}}\Delta\sigma_T \times K_e$

E_{curve} = Young Modulus of the design curve = 28 300 ksi (195 300 MPa)
E_{ref} = Young Modulus at temperature T_{ref}
T_{ref} = $(T_{max} + T_{min}) / 2$
K_e = Numerical factor as defined by ASME III, NB-3228.5
S_m = Design Stress Intensity value (used to compute K_e)
$\Delta\sigma_T$ = Tresca equivalent stress range

Two values are given for the design-allowable number of cycles: the former is obtained using the above-mentioned definition of S_{alt}, whereas the latter does not include the factor K_e in the calculation of S_{alt}. In the authors' opinion it is meaningful because K_e is introduced to avoid the $3S_m$ limit on the range of primary plus secondary stress intensity. The $3S_m$ limit is posed to prevent ratchetting but, since it was experimentally proved that the component shakes down, there is no reason to require that the $3S_m$ limit be satisfied and consequently to introduce the K_e factor.

TABLE 1. Design allowable fatigue cycles, ASME procedure.

	Point A	Point B
	Phase 1 of the test	
N incl. K_e	∞	150
N excl. K_e	∞	7 900
	Phase 2 of the test	
N incl. Ke	∞	130
N excl. Ke	∞	6 600

As far as the fatigue usage factor is concerned, one can easily verify that it was zero for point A and above 1 for point B. Therefore this latter point proved to be the most critical. In spite of this no apparent damage was observed on the heated surface. On the other hand a long and deep crack started from point A. This failure did not cause the separation of the two halves of the component.

5. Application of the ASME XI Procedure

5.1. DESCRIPTION OF THE DETECTED FLAWS

In this section the procedure described in the design standard ASME XI will be applied in that location where fatigue cracks were observed. It is worth remembering that two parallel flaws were detected after the first phase. They ran along both sides of the central electron beam welding (point A) at a distance of 5 mm apart. The lengths, measured by means of ultrasonic testing, were 197 and 22 mm and the depths were 2 and 3 mm, respectively. These figures refer to the inspectable region that was about one half of the component; therefore the real length can be guessed to be roughly 400 and 40 mm, respectively. The use of the ASME-XI procedure will be discussed in some detail since its application to the fusion reactor FW is not straightforward.

5.2. CHOICE OF THE ASME-XI DIVISION

The ASME-XI code is organised into the following 3 divisions:

- Div. 1 - Rules for Inspection and Testing of Components of Light-Water Cooled Plants
- Div. 2 - Rules for Inspection and Testing of Components of Gas-Cooled Plants
- Div. 3 - Rules for Inspection and Testing of Components of Liquid-Metal Cooled Plants.

Division 3 appears to be the most relevant as far as the fusion reactor first wall is concerned. In fact liquid metal fast-breeder reactors operate at a temperature level, in a radiation environment and with a choice of materials which are the most similar to those of the plasma facing components. Unfortunately, in the ASME 1992 edition (July) the acceptance standards were still in course of preparation and therefore *Rebus Sic Stantibus* we were forced to use Division 1.

5.3. CLASSIFICATION OF THE FLAWS

The two flaws can be classified as *Parallel Planar Flaws* (ASME-XI, IWA-3350). Since they are located less than ½ in. = 12.7 mm apart, according to the ASME they can be recharacterized as a single flaw with an overall length l of about 400 mm and with a depth a of 3 mm.

5.4. ACCEPTANCE OF THE FLAWS

The in-service acceptance procedures are described in ASME-XI, IWB-3130. They foresee four possibilities:

1. acceptance by volumetric or surface examination
2. acceptance by repair
3. acceptance by replacement
4. acceptance by analytical evaluation.

The first and the last criteria will be discussed hereinafter.

5.4.1. *Acceptance by Volumetric or Surface Examination*

Most of the acceptance criteria concern ferritic materials. Austenitic steels are considered only in the following two cases:

1. austenitic piping (ASME-XI, IWB-3514.3)
2. pressure retaining welds in pump casing and valve bodies (ASME-XI, IWB-3518).

The former procedure was chosen even if we are dealing with a first wall geometry and not with a pipe. This is somewhat arbitrary but it is the only possibility to be able to use the ASME code. However this assumption seems reasonable. The latter procedure requires a nominal wall thickness of at least 2 in. (50.8 mm) to be applied whereas the component thickness t in correspondence of detected flaw is 10.6 mm, i.e. 0.4 in. (this figure was rounded according to what is stated in ASME-XI, IWA-3200(c)). To obtain the maximum allowable crack depth one must refer to table IWB-3514-2 (ASME-XI).

The aspect ratio a/l is $3/400 = 0.0075$, therefore the requirement to be satisfied is: $a/t < 11.745\%$. Therefore the maximum allowable flaw depth is 1.24 mm. Since the real depth a is 3 mm this acceptance criterion is not satisfied. The acceptance by analytical evaluation must be carried out.

5.4.2. *Acceptance by Analytical Evaluation*

The acceptance by analytical evaluation is described in ASME-XI IWB-3600. The following possibilities are foreseen.

1. Acceptance criteria for ferritic steel components 4 in. (i.e. 101.6 mm) and greater in thickness.
2. Acceptance criteria for ferritic components less than 4 in. in thickness.
3. Acceptance criteria for steam generator tubing.
4. Evaluation procedures and acceptance criteria for austenitic piping.
5. Evaluation procedures and acceptance criteria for flaws in ferritic piping.

As far as the material is concerned, the most suitable procedure that can be adopted is the fourth. It is described in ASME-XI IWB-3640. Unfortunately it can not be extended to the first wall geometry since it requires to compute the ratio of flaw length l to pipe circumference. Of course, this ratio can not be defined in the present case.

To be able to apply the ASME code we are therefore forced to use the second above-listed procedure (acceptance criteria for ferritic components less than 4 in. in thickness) which is suitable as regards the geometry but not as regards the material. This procedure is described in ASME-XI IWB-3620. Here is it said that *"These criteria are in the course of preparation. In the interim, the criteria of IWB-3610 may be applied"*. The criteria IWB-3610 refer to the above-listed acceptance criteria for ferritic steel components 4 in. and greater in thickness. This procedure foresees two possibilities to assess the structural integrity:

- Acceptance Criteria Based on Flaw Size (IWB-3611);
- Acceptance Criteria Based on Applied Stress Intensity Factor (IWB-3612).

The latter criteria will be applied as far as the "Normal Conditions" are concerned. The following must be satisfied:

$$K_I < K_{Ia} / \sqrt{10} \qquad (1)$$

where:

K_I is the maximum applied stress intensity factor for the flaw size a_f, i.e. the maximum size for which the detected flaw is calculated to grow in a specified time period, which can be the next scheduled inspection of the component.

K_{Ia} is the available fracture toughness based on crack arrest for the corresponding crack tip temperature.

In rapidly propagating cracks, K_{Ia} is generally lower than K_{IC} for two reasons: (1) because plastic deformation at the moving crack tip is suppressed by the high local strain rate and therefore all the driving force is available for propagation; (2) because the structure contains kinetic energy that can be converted to fracture energy [6]. In case of thermal fatigue cycling no rapidly propagating cracks are expected and therefore we can say $K_{Ia} \cong K_{IC}$. As will be seen later a value of 150 MPa $\sqrt{m}$ can be assumed for K_{IC}, therefore Eq. (1) becomes:

$$K_I < 47 \text{ MPa} \sqrt{m} \qquad (2)$$

The flaw size a_f must be such that the corresponding K_I value satisfies Eq. (2). To compute a_f ASME-XI-appendix A procedure was adopted.

Fig. 4 shows the computed sigma-x distribution in the component thickness in correspondence with point A (y=0) where the flaw nucleated. The x-stress component was the driving force for the detected flaw.

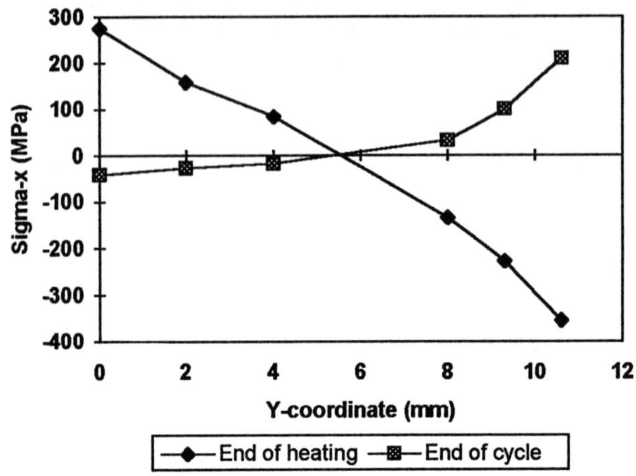

Figure 4. Sigma-x distribution along the component thickness in correspondence with the detected flaw (elasto-plastic analysis).

The above-shown non-linear stress distribution through the wall was conservatively approximated using the linearization technique illustrated in ASME-XI Fig. A-3200-1. Table 2 gives the resulting membrane and bending components at point A.

TABLE 2. Membrane (σ_m) and bending (σ_b) sigma-x components at point A.

	End of heating	End of cycle
σ_m (MPa)	-34	-3.4
σ_b (MPa)	310	-37.6

These values were used to compute the stress intensity factor according to ASME-XI A-3300 procedure. The following equation is recommended:

$$K_I = \sigma_m M_m \sqrt{\pi}\sqrt{a/Q} + \sigma_b M_b \sqrt{\pi}\sqrt{a/Q} \tag{3}$$

where:

M_m is the correction factor for membrane stress (given in ASME-XI Fig. A-3300-3).
M_b is the correction factor for bending stress (given in ASME-XI Fig. A-3200).
Q is the flaw shape parameter as determined from ASME-XI Fig. A-3300-1 using a value of 484 MPa for the yield stress (hardened material).

By the application of the above formula, one can demonstrate that Eq. (2) is satisfied if $a_f < 4.8$ mm. To compute the flaw growth during phase 2 of the test, the following relation was used [7,8].

$$\frac{da}{dN} = C\frac{\lambda^m\left[\Delta K - \Delta K_o\right]^n}{K_{IC} - \lambda \cdot \Delta K} \tag{4}$$

where:

C = 3.122×10^{-6} mm/cycle
λ = 1/(1-R)
R = K_{Imin}/K_{Imax}
m = 1.88 R < 0
ΔK_o = 5.4 (1-0.2R) R< 0
n = 2.95
K_{IC} = 150 MPa √m

Fig. 5 shows the crack propagation after the first 5 000 thermal fatigue cycles as computed by means of the above-mentioned procedure. To obtain a_f=4.8 mm 11,870 thermal fatigue cycles are required. This means that the next non-destructive testing should have been performed before the above-mentioned number of cycles. In spite of this, after 50 000 cycles no crack propagation seems to have occurred, therefore we can conclude that the ASME procedure to assess the structural integrity of damaged component could be extended with safety margins to the present case.

It is worth noting that as concerns a real FW in a fusion reactor environment one should also remember the criteria foreseen for emergency and faulty conditions; i.e. the following relation must be also satisfied:

$$K_I < K_{IC} / \sqrt{2} \tag{5}$$

In this case K_I must include the stress caused by the plasma disruptions events. This latter stress field could make Eq. 5 much more limiting than Eq. 1 and therefore the possible extension of the ASME-XI procedure to the fusion environment can not be simply stated by the above considerations.

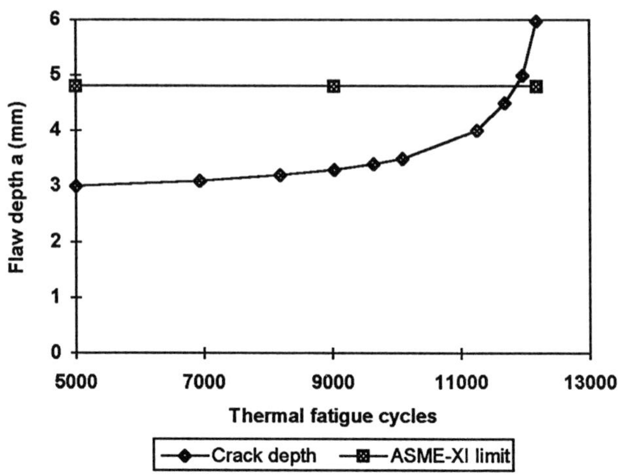

Figure 5. Crack growth according to ASME-XI procedure.

6. Discussion of the Results and Conclusions

The experimental result showed that crack nucleation and propagation did not occur in the most stressed location but where the ASME-III code foresees an infinite design allowable number of cycles. In this respect it is worth briefly recalling the main conclusions of a Shared Cost Action of the European Communities on Liquid Metal Fast Breeder Reactor safety [9]. It was observed that the vacuum electron beam welding (VEB) material is appreciably less subjected to plastic accommodation than the parent material. Mechanical fatigue tests showed that, in low cycle fatigue, the endurance data of VEB fall within the overall scatter of the parent material. However, the fatigue curve of VEB has a steeper slope than the parent material and consequently the high cycle fatigue life is lower. If VEB experimental fatigue curves are used, the lifetime of the welded joint at the computed strain level of point A of the FW component should have been about two orders of magnitude higher. Therefore there seems to be no correspondence between mechanical and thermal fatigue with respect to the endurance prediction. Probably under thermal loading, the different plastic accommodation of the VEB acted as a stress riser generating peak stress where cracks can easily nucleate and propagate. This effect is particularly appreciable under thermal loads. In fact since thermal stress is a self-induced stress, it is very sensitive to the spatial variation of the material properties. On the other hand, in case of mechanical fatigue tests, the stress field in the specimen mainly results from the applied external forces rather than in the spatial variation of the plastic behaviour of the material, therefore the generation of self-induced peak stress is usually not appreciable.

It was experimentally proved that no ratchetting occurred in spite of the very high stress level. In particular the range of the Tresca elastic stress was 600 MPa (point B) and the maximum temperature at this point was 386°C. These figures lead to a S_m value of 110 MPa which is more than 8 times less than the Tresca value. In particular the $3S_m$ limit is greatly exceeded. Therefore it can be concluded that, as far as thermal fatigue on first wall geometries is concerned, this limit as well as the introduction of the K_e factor in the fatigue design procedure appear to be unduly conservative.

The ASME-XI procedure was applied to the detected flaw to evaluate its acceptability. According to the inspection programme scheduled by the ASME-XI IWB-2400, the first in-service examination is carried out after about three years of plant service. One should recall that the thermal fatigue experiment has demonstrated that the crack nucleated in the very first stage of the operating life and that it stopped after some propagation. Taking this into account, the three-year interval might prove to be too long to verify safely the fitness for purpose of a plasma facing component. On the other hand the successive inspection periods seem to be adequate. A procedure for austenitic components other than tubes is lacking in the ASME-XI as far as the analytical evaluation of flaws is concerned. Therefore it should be developed for a proper extension of the design standards to fusion reactor relevant components.

Acknowledgements - The contribution of the Polytechnic of Turin was within the collaboration with the Joint Research Centre of the European Union, Institute for Advanced Materials (collaboration No. 10193-94-05 S0ED ISP I) and was financially supported by The NET Team (contract No. 94-354).

References

1. Matera, R., Merola, M., Antidormi R., Sevini, F., *Experimental Results of the IAEA Benchmark on Lifetime Behaviour of the First Wall of Fusion Machines*, Comm. Europ. Commun., Joint Research Centre Ispra, EUR Report 14891, 1992.
2. *NET First Wall Test Section*, NET contract No. 87.147, FRAMATOME, progress reports No. 1-5, Chalon, France, 1988-1990.
3. Merola, M., *Evaluation of First Wall Thermal Fatigue Testing*, NET contract No. 94.354, Intermediate Report, Part 1, February 1995.
4. Merola, M., Matera, R., Vieider, G., *Study on the Thermal Fatigue Test of a Prototypical First Wall Component*, submitted for publication in Fusion Engineering and Design (1995).
5. ASME III, *Rules for Construction of Nuclear Power Plant Components*, Division 1, Subsection NB, Class 1 Components, 1992 Edition.
6. Anderson, T.L., *Fracture Mechanics*, CRC Press (Boca Raton, Ann Arbon, Boston), 1991.
7. Speidel, M.O., *Fatigue Crack Growth at High Temperatures*, Proceedings of the Symposium on High Temperature Materials in Gas Turbines, Baden, Switzerland, 1974, pp. 207-251.
8. Watson, R.D., Peterson, R.R., Wolfer, W.G., *Lifetime Analysis of Fusion Reactor First Wall Components*, Transaction of ASME 105 (1983) 144-152.
9. *Mechanical Testing of Austenitic Steel Welded Joints*, Boerman D.J., and Krischer, W., (eds.), Comm. Europ. Commun., Joint Research Centre Ispra, EUR Report 13030, 1990.

INITIATION AND GROWTH OF CRACKS UNDER THERMAL FATIGUE LOADING FOR A 316 L TYPE STEEL

A. FISSOLO, B. MARINI, A. BERRADA, G. NAIS and P. WIDENT
*EM/ DECM/ Service des Recherches Métallurgiques Appliquées-
Commissariat à l'Energie Atomique, Centre d'Etude de Saclay
91191-Gif sur Yvette, France*

1. Introduction

In order to assess the behaviour of components submitted to very severe temperature fluctuations, the conditions corresponding to initiation and propagation of cracks under thermal loading must be known. These conditions were determined for the 316 L steel, which is a material widely used in the nuclear industry. In the fast breeder reactors, the mixing of sodium flows at different temperatures (thermal striping) can lead to the development of a crack network on some components [1, 2]. In fusion reactors, the first wall of the tokamak vacuum vessel will suffer from thermal shock even in normal operating conditions due to a periodic ignition of the plasma about every 100 seconds. However, thermal fatigue will also result from successive disruptions of the plasma.

In our laboratory, two types of facility have been developed to study crack initiation and crack growth respectively. This paper gathers all the results obtained in the last ten years. Crack initiation results will be analyzed using the methodology proposed by the French RCC-MR code procedure [3]. Propagation results will be analyzed using Haigh and Skelton's model [4], which is based on the determination of an effective stress intensity. A first analysis was presented in [5].

2. Material

The chemical composition of the material investigated is given in Table 1 :

TABLE 1. Chemical compositions (wt %) of the investigated material (316 L steel).

C	Ni	Cr	Mn	Cu	Mo	Si	Co	S	P	Ta	N	B	Fe
0.024	12.33	17.44	1.82	0.20	2.30	0.46	0.17	0.001	0.027	0.01	0.06	0.0008	Bal.

J. Bressers and L. Rémy (eds.), Fatique under Thermal and Mechanical Loading, 67–77.

Metallographic examinations performed have shown that the average grain size is between 3 and 4 (ASTM-E112-74), the residual ferrite δ concentration is less than 0.5% (Fisher ferritscope) and the inclusion content is less than 3 (ASTM-E45-76).

3. Experimental Procedures

3.1. DETERMINATION OF CRACK INITIATION

Crack initiation was investigated by optical microscopy and was correlated to the detection of at least one crack of a length ranging from 50 to 150 μm. This definition includes some propagation since it corresponds to a crack of several grain sizes.

Thermal fatigue tests are performed with the SPLASH facility (Fig. 1). The specimen is heated by the passage of an electrical current. The two opposite faces are simultaneously cooled by spraying water over a central area of 9 mm wide and 30 mm long. Fig. 2 illustrates the temperature variation at the surface.

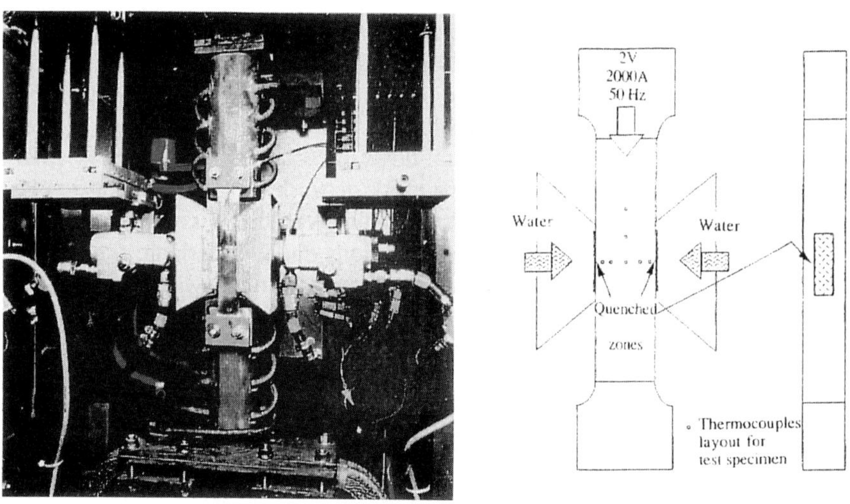

Figure 1. Thermal fatigue test facility SPLASH.

Cyclic stresses result from cyclic thermal gradients between the core and the surface. The specimen has an uniform temperature at the end of the heating phase. Temperature measurements and calculations show that the temperature remains always constant at the core. Thus, quenching leads to a temperature decrease on the surface equal to the temperature gradient between the surface and the core. The basic concept of this facility is similar to that used by D.J. Marsh [6].

Specimen roughness is accurately controlled (Ra ≤ 0.8 μm) before testing to avoid a premature initiation. Such a roughness is in agreement with values used for mechanical fatigue specimens (Afnor NF A-03-401). Temperatures are measured by thermocouples (± 5°C) and continually controlled during testing. Recording of the temperatures at the surface during thermal cycling is required to ensure accurate application of the chosen thermal loading. Such a calibration is obtained with a specific specimen removed before performing the test itself.

Conditions of the thermal fatigue tests are given in Table 2.

TABLE 2. Thermal loading conditions used for crack initiation tests.

ΔT (°C)	100	125	150	180	200	230	250	300	300
T max (°C)	550	550	550	260	550	320	550	450	550
Frequency (Hz)	0.25	0.25	0.25	0.15	0.25	0.15	0.025	0.027	0.25
Cooling time (s)	0.25	0.75	0.25	0.25	0.25	1.3	0.4	12.5	0.25
Hold time (s)	~0	~0	~0	~0	~0	~0	30	20	~0

The hold time is defined as the time during which the temperature is within 5°C of the maximum.

The frequency of tests performed with non significant hold time may be considered as constant since it ranges only between 0.15 and 0.25 Hz.

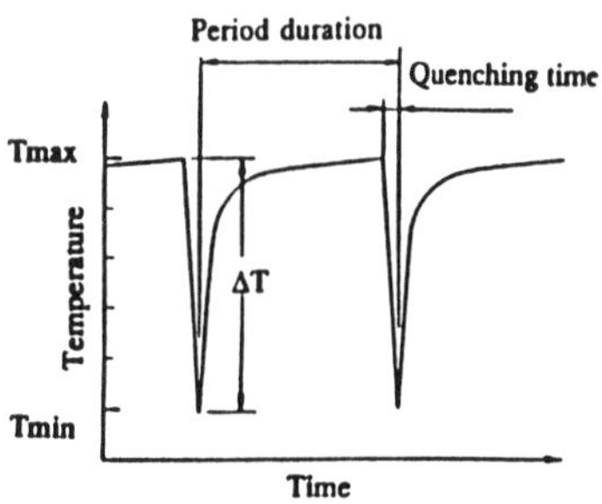

Figure 2. Typical variation of the temperature at the surface of the SPLASH specimen.

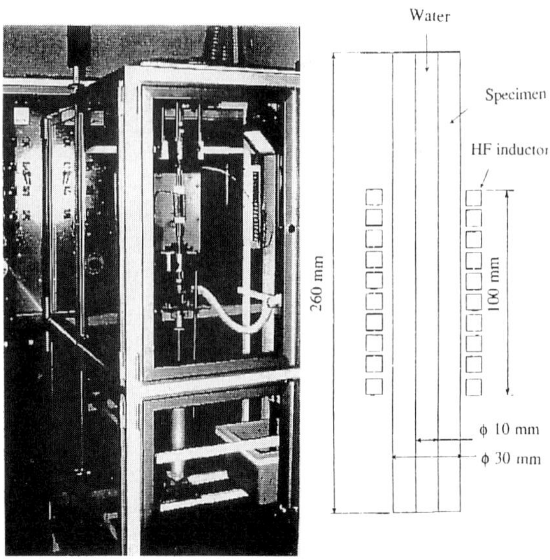

Figure 3. Thermal fatigue test facility CYTHIA.

3.2. DETERMINATION OF CRACK GROWTH FOR A SINGLE CRACK

These tests were performed on the CYTHIA (CYclage THermique par Induction des Aciers) facility (Fig. 3). The specimen is a thick tube periodically heated at the outside by induction using a current of 1 MHz frequency, and permanently cooled by water flow at the inside. To insure the propagation of a single crack, a circular groove of 1 mm depth is machined in the heated length. The determination and calibration of the thermal loading were done using thermocouple measurements. Crack growth was directly measured on the fracture surface. To determine crack propagation during thermal fatigue, specimens were broken after two hours of air oxidation at 450°C.

Furthermore, for the last tested specimen, crack growth as a function of the number of cycles was estimated many times during temporary test interruptions by ultrasonic testing.

Test conditions are presented in the Table 3.

TABLE 3. Thermal loading conditions used for crack growth tests.

ΔT (°C)	T max (°C)	Frequency (Hz)	Cooling time (s)	Hold time (s)
300	380	0.023	30	~ 0

4. Experimental Results

Fig. 4 shows the dependence on the temperature range of the number of cycles to initiate cracking for different maximum cycle temperatures Tmax. Open symbols distinguish tests made with significant hold times from the others.

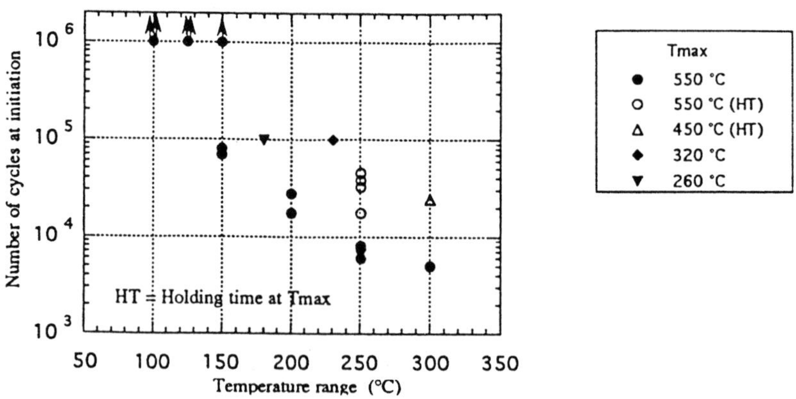

Figure 4. Number of cycles to crack initiation as a function of the temperature range.

Clearly, the number of cycles decreases whith increasing temperature range. Initiation appears only after 6,000 cycles for a temperature range equal to 300°C whereas no initiation is detected after 1,000,000 cycles when it is 100°C.

As for isothermal mechanical fatigue resistance, a limit of endurance can be defined. In our case, it is taken as the maximum temperature range for which no initiation occurs after 1,000,000 cycles. Thus, for a cycle with maximum temperature equal to 550°C, the endurance limit corresponds to a temperature range of 125°C. Furthermore, initiation is delayed when hold times are significant.
Concerning the crack growth tests which are performed on the second facility, Fig. 5 shows a specimen after final rupture. Let us note that the shape of the crack front can be effectively considered as circular.

In Fig. 6, one can observe that estimations obtained from ultrasonic testing (full symbols) are in good agreement with crack propagation measurements performed after rupture (cross symbols). The meaning of the full line curves and of the corresponding q factors will be given in the next section. It has to be pointed out that the development of such a non destructive methodology is very interesting for monitoring crack growth since it allows to obtain more valuable results with fewer specimens.

5. Predictions of Crack Initiation and Propagation under a Thermal Fatigue Loading

5.1. ESTIMATION OF THE NUMBER OF CYCLES TO INITIATION

The RCC-MR (*) code is used in French Fast Breeder Reactors to assess the absence of damage in the components including thermal fatigue loadings [3]. The present results are used to validate this approach. In this code, the temperature range is converted into a total strain range, $\Delta\varepsilon$. Then, estimations can be compared with strain controlled isothermal fatigue.

In our case, the stress state at the surface can be considered as equibiaxial. If one considers that $\Delta\sigma_{eq} = \Delta\sigma_{yy} = \Delta\sigma_{zz}$, strains are given by the relation : (1)

$$\Delta\varepsilon = \begin{pmatrix} -2\nu_s\frac{\Delta\sigma_{eq}}{E_s} & 0 & 0 \\ 0 & \frac{\Delta\sigma_{eq}(1-\nu_s)}{E_s} & 0 \\ 0 & 0 & \frac{\Delta\sigma_{eq}(1-\nu_s)}{E_s} \end{pmatrix} = \begin{pmatrix} -2\nu\frac{\Delta\sigma_{eq}}{E} & 0 & 0 \\ 0 & \frac{\Delta\sigma_{eq}(1-\nu)}{E} & 0 \\ 0 & 0 & \frac{\Delta\sigma_{eq}(1-\nu)}{E} \end{pmatrix} + \begin{pmatrix} -(\frac{\Delta\sigma_{eq}}{K})^n & 0 & 0 \\ 0 & \frac{1}{2}(\frac{\Delta\sigma_{eq}}{K})^n & 0 \\ 0 & 0 & \frac{1}{2}(\frac{\Delta\sigma_{eq}}{K})^n \end{pmatrix}$$

where E_s, υ_s are the secant modulus and Poisson's coefficient corresponding to the point ($\Delta\sigma_{eq}$, $\Delta\varepsilon$) on the cyclic strain hardening curve. The plastic strain is given by:

$$\varepsilon^{pl}_{eq} = \left(\frac{\Delta\sigma_{eq}}{K}\right)^n \qquad (2)$$

One considers that $\Delta\varepsilon_{eq} = K_\nu \; \Delta\varepsilon_1$ where $\Delta\varepsilon_1$ is determined with a pure elastic analysis, and K_ν is a coefficient derived from the cyclic strain hardening curve. This was previously determined from strain-controlled fatigue tests performed on 316L.

(*) *"Règles de Conception et de Construction des Matériels Mécaniques des îlots nucléaires des réacteurs à neutrons Rapides".*

$$\Delta\varepsilon_1 = \frac{2}{3}(1+\nu)\frac{\Delta\sigma_{eq}}{E} \qquad (3)$$

All coefficients used for calculation are taken from the appendix A3-1S of the RCC-MR code [7].

Finite element calculations carried out under thermo-elasticity conditions with generalized plane strain show that the variation of the equivalent stress $\Delta\sigma_{eq}$ can be estimated with a pure biaxial thermoelastic calculation. This is due to the distribution of temperature through the thickness of the specimen. Thermal calculations have shown that the temperature is roughly uniform in the central section whereas it sharply drops near the surface during the quenching time. Assuming that the homogeneous thermal state of the specimen is obtained at maximum temperature, it is finally derived:

$$\Delta\varepsilon = \frac{2}{3}\frac{E_{T\,min}}{E_{T\,max}}\frac{1+\nu}{1-\nu}K\nu\ (\Delta\sigma_{eq}, T_{max})\ \alpha(T_{min})\ \Delta T \qquad (4)$$

where α is the coefficient of linear thermal expansion.

Fig. 7 compares estimations thus deduced with strains obtained from isothermal fatigue curves. The isothermal fatigue data corresponds to failure. However, it was shown that the difference between the number of cycles corresponding to initiation and failure is very small when the strain range is less than 1% for 316 L [8]. The comparison is only made for tests performed with a maximum temperature of 550°C where many results are available. It shows that values ($\Delta\varepsilon$, N) obtained without hold time are always close to the isothermal low cycle fatigue curves established at 450 and 550°C by the French working group CEA - EDF. However, calculated strains are nearer to the 450°C curve which corresponds rather to an intermediate temperature of the cycle. Besides, results are also close to a curve (full line curve) extrapolated from the strain controlled curve proposed by D.S. Wood [9]. Let us note that in case where thermal fatigue comprises significant hold times, the numbers of cycles to crack initiation are larger than those expected from isothermal fatigue data. As a result, such calculations never lead to non conservative estimations. Furthermore, in all the cases, the isothermal strain controlled fatigue RCC-MR design curves underestimate the number of cycles to initiate cracking under thermal fatigue. As observed in the previous part, Fig. 7 exhibits also an endurance limit which corresponds to absence of initiation over 1,000,000 cycles.

5.2. ESTIMATION OF THE CRACK GROWTH

5.2.1. *Modelling methodology*

In order to estimate the propagation under thermal fatigue loading, the model proposed by Skelton is used [4]. As for initiation, thermal loadings are directly converted to mechanical stresses. The application of such a methodology gave good estimations for similar geometry specimens in many cases [10 - 13]. Despite the fact that its basis is deduced from a linear elastic fracture mechanical (LEFM) analysis, it allows to give good predictions even when non negligible plasticity occurs. Indeed,

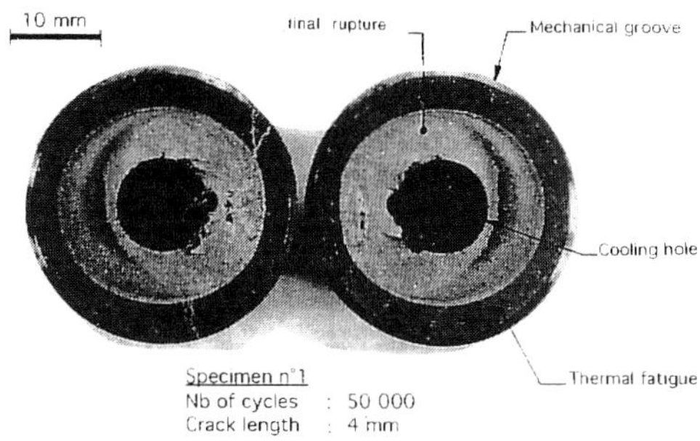

Figure 5. Crack growth tests. Fracture surface of a CYTHIA specimen.

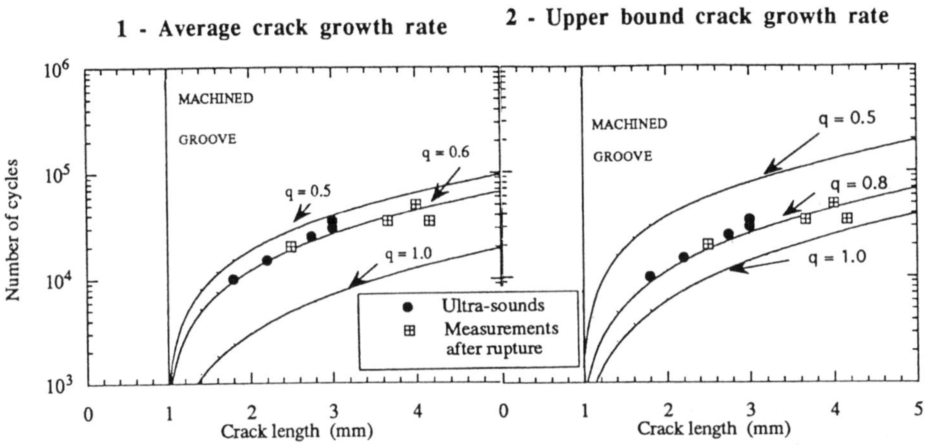

Figure 6. Crack growth tests. Number of cycles as a function of the crack length and predictions deduced from modelling.

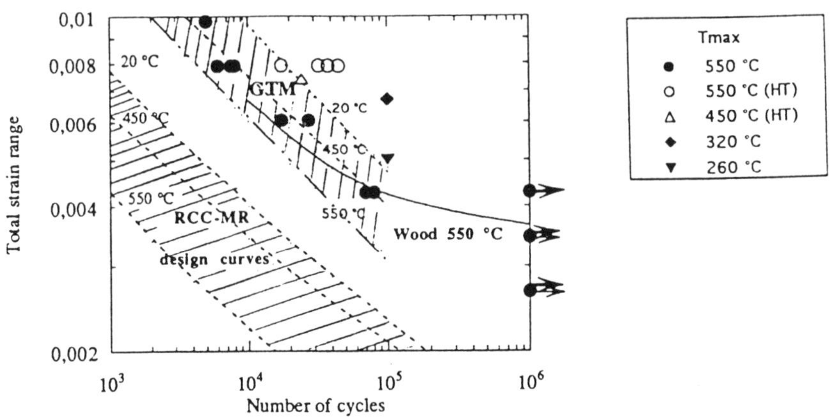

Figure 7. Crack initiation tests. Comparison between estimations deduced from the results (RCC-MR code method) and isothermal strain controlled curves (*).

(*) Isothermal fatigue data corresponds to failure.

Skelton's method has an important advantage compared to approaches based on the J integral [14] since accurate determination of J under these conditions requires complex finite element calculations, whereas effective stresses are more easily deduced.

A first step is devoted to the determination of an effective stress σ_{eff}, which is perpendicular to the crack plane. As shown in Fig. 8, the effective stress is the sum of two terms. The first corresponds to the stress range multiplied by the factor measuring the crack closure. The second is a hypothetical stress normal to the crack face which would produce an elastic strain equal to the actual plastic strain range. In such calculations, stress and strain distributions must be estimated for regions far from the damaged zone.

An effective stress intensity factor can then be deduced as a function of the propagation.

The number of cycles to reach a crack depth is finally estimated assuming that the relation between crack growth rate and stress intensity factor is similar to the one resulting from isothermal fatigue.

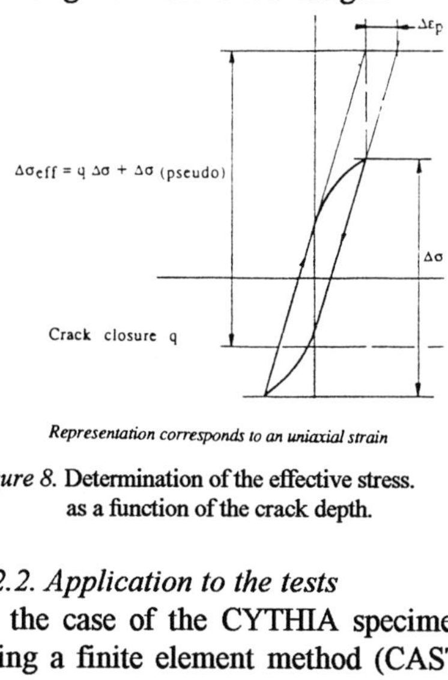

Representation corresponds to an uniaxial strain

Figure 8. Determination of the effective stress. as a function of the crack depth.

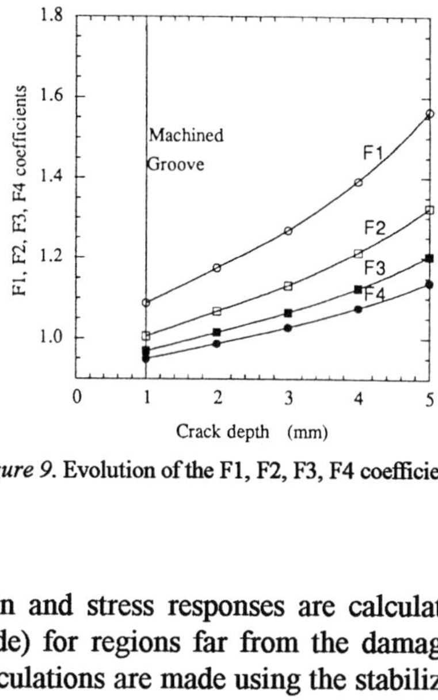

Figure 9. Evolution of the F1, F2, F3, F4 coefficients.

5.2.2. Application to the tests

In the case of the CYTHIA specimens, strain and stress responses are calculated using a finite element method (CASTEM code) for regions far from the damaged zone. A symmetrical stress range is taken. Calculations are made using the stabilized cyclic relation established at the average temperature (300°C). The effective stress range is given as a function of the crack depth x by the following relations, in the plastic domain (Eq. 5) and elastic domain (Eq. 6) respectively :

$$\Delta\sigma_{eff}(x) = q\Delta\sigma_{zz}(x) + \frac{E}{(1+\nu)(1-2\nu)}[(1-\nu)\Delta\sigma_{zz} + \nu\Delta\sigma_{rr} + \nu\Delta\sigma_{\theta\theta}] \quad (5)$$

$$\Delta\sigma_{eff}(x) = \Delta\sigma_{zz}(x) \quad (6)$$

It is assumed that plasticity is negligible when the calculated plastic strain is less than or equal to 10^{-4}. q is the factor measuring the crack closure.

The effective stress is fitted by a third degree polynomial form :

$$\Delta\sigma_{eff}(x) = A_0 + A_1 x + A_2 x^2 + A_3 x^3 \qquad (7)$$

with coefficients depending on the value of the q factor. The effective stress intensity factor is detemined using the superposition method [15], it is given by :

$$\Delta K_{eff} = 1.12(\pi a)^{0.5}[A_0 F_1 + \frac{2a}{\pi} A_1 F_2 + \frac{a^2}{2} A_2 F_3 + \frac{4}{3\pi} a^3 A_3 F_4] \qquad (8)$$

The F_1, F_2, F_3, and F_4 coefficients correspond to uniform, linear, quadratic, and cubic stress profiles respectively. Fig. 9 shows their evolutions with the crack length a. These were calculated in the case of the CYTHIA geometry using finite elements. A refined mesh was used around the tips of the crack since the length was only 50 μm. The stress intensity factors are deduced from the variation of potential energy with crack growth G_θ.

Using a Paris relation, the number of cycles to reach a crack depth a is given by the equation:

$$N(a) = \int_{a_0}^{a} \left(\frac{1}{C\ Keff\ (x)}\right)^{m} dx \qquad (9)$$

if crack growth rate is :

$$\frac{da}{dN} = C\Delta K^{m} \qquad (10)$$

The comparison between measurements and predictions is presented in Fig. 6. Curves were determined using average and upper bound crack growth rates [16] and with closure factors q ranging between 0.5 and 1. Values of q of 0.6 - 0.8 give a good prediction of the experimental results up to a crack length of 4 mm: q = 0.6 corresponds to the average, and q = 0.8 to the upper bound crack growth rate. The same values were already obtained in [10]. A value of q = 1.0 gives conservative predictions.

6. Conclusions

1. The number of cycles corresponding to initiation under thermal loading was effectively determined on the Splash Test Facility. Tests were performed with a temperature evolution during cycling characterized by a maximum between 260 and 550°C and a range between 100 and 300°C. However, as for other test methodologies, definition of initiation is not easy and includes also some propagation.
2. Mechanical calculations use the RCC-MR code procedure. They show that initiation results corresponding to a maximum temperature of 550°C are close to the predictions of rupture obtained under an isothermal uniaxial cyclic loading at an intermediate temperature of the cycle. Such a comparison with the number of

cycles to rupture could seem to be not adapted but for standard fatigue specimens both numbers are very close in the investigated low cycle fatigue domain.

3. These initiation results plainly confirm the conservatism of the French RCC-MR methodology to insure integrity of the structure under thermal fatigue loading.
4. As for mechanical fatigue, an endurance limit is evidenced. Thus, for a temperature range less than 125°C, no crack appears up to 1,000,000 cycles when the maximum temperature is fixed at 550°C. Evidently, such a limit must be at least higher when the maximum temperature becomes lower.
5. Furthermore, the beneficial effect of a significant hold at the maximum temperature was proved. Such a behaviour shows that a direct conversion of a thermal fatigue loading to a mechanical fatigue response is not sufficient.
6. Tests performed on the CYTHIA facility effectively allowed to obtain the number of cycles to reach a given crack depth. The maximum temperature and temperature range are 380 and 300°C respectively. Measurements on the fracture surface are in reasonable agreement with estimations deduced from detection. As a result, many crack growth values can be obtained from one specimen.
7. Concerning the growth of a single crack, estimations deduced from application of the model proposed by Skelton are in good agreement with all the depth measurements in spite of simplifying assumptions. So, application of such a model to a crack network should have to confirm that interaction between the closest cracks leads to very strong rate decreases. Thus, in all the cases, estimation of the growth rate for a single crack should give an upper limit with a given thermal loading.
8 Although the crack closure factor is not determined in experiments, predictions lead to factors about 0.6 - 0.8, which were generally found in similar conditions. Furthermore, it is emphasized that a value of 1.0 gives always a conservative prediction of the number of cycles for a given crack depth.

Acknowledgements - This study was partly supported by the European Technology Fusion Programme and the IPSN safety authority, which are greatly acknowledged.

References

1. Pradel, P., The main objectives of thermal striping studies in progress for French LMFBR Thermal hydraulic and design aspects, *Pressure Vessel Piping Conference*, **98 - 8**, (1985) 143 - 146.
2. Lemoine, P., Marini, B.,Meny, L. Etude de la fatigue thermique superficielle d'un acier inoxydable austénitique, Journée internationale de printemps, *Fatigue à haute température*, (1986), 331 - 344.
3. Code RCC-MR - Filière Rapide - Tome 1 - Volume B - Recueil des règles applicables pour les matériaux de Niveau 1, (1985) . 89 - 95, AFCEN.
4. Haigh, J.R., Skelton, R.P., A strain Intensity Approach To High Temperature Fatigue Crack Growth and Failure, *Materials science and Engineering* (1978) **36**, 133 - 137.
5. Fissolo, A., Marini, B.,Wident, P., Nais, G., Biaxial thermal fatigue on 316L, *Proceedings of the Fourth International Conference on Biaxial/ Multiaxial Fatigue,* Société Française de métallurgie et de matériaux and European Structural Integrity Society, St Germain en Laye, 1994, 513-526.
6. Marsh, D.J., A Thermal Shock Fatigue Study of Type 304 and 316 Stainless Steel. *Fatigue and Engineering Materials and Structures* **4-2** (1981), 179-195.
7. Code RCC - MR - Tome I - Volume Z : Annexe Technique A3 - 1S, AFCEN, Edition of June 1985.

8. Levaillant, C., Thèse d'état de l'Université de Technologie de Compiègne, (1984).
9. Wood, D.S., Proposals for design against thermal striping, *Nuclear Energy*, **19 - 6**, (1980) pp. 433 - 437.
10. Skelton R.P., Crack Initiation and Growth in Simple Metal Components During Thermal Cycling, in R.P. Skelton (ed.), *Fatigue at High Temperature*, Applied Science Publishers London and New York LTD, (1983), pp. 1 - 61.
11. Burlet H., Vasseur S., Cailletaud G., Pineau A., Fatigue Crack Growth under Thermomechanical Loading, Application to life prediction of centrifugal casting equipment, *International Seminar on High Temperature Fracture Mechanisms and Mechanics*, (1987).
12. Burlet, H., Vasseur, S., Besson, J., Pineau, A., *Fatigue of Engineering Materials Stucture*, **12-2**, (1989), pp. 123-133.
13. Green, D., Munz, D., Thermal fatigue crack growth on austenitic steels plates, Private communication June 1991.
14. Dowling, N.E., Crack growth during low cycle fatigue of smooth axial specimens. *Cyclic Stress - Strain and Plastic Deformation Aspects of Fatigue Crack Growth*, ASTM STP 637, American Society for Testing and Materials, (1977) pp. 97 - 121.
15. Buchalet, C.B., Bamford, W.H., Stress Intensity Factor Solutions for Continuous Surface Flaws in Reactor Pressure Vessels, *Mechanics of Crack Growth*, ASTM STP 590, American Society for Testing and Materials, (1976), pp. 385 - 402.
16. Tavassoli, A.A. , Assessment of austenitic stainless steels, *Fusion Engineering and Design* **29** (1995), 371-390.

CRACK GROWTH FROM DEFECTS IN 316L STAINLESS STEEL COMPONENTS UNDER THERMAL FATIGUE/CREEP LOADING

M.P. O'DONNELL[1], R.C. HURST[2], L. LAMAIN[2] and D. TAYLOR[1]
[1] *Dept. Mechanical and Manufacturing Engineering, Trinity College Dublin, Dublin 2. Ireland.*
[2] *JRC/IAM, P.O. Box 2, 1755 ZG Petten, The Netherlands.*

1. Introduction

Type 316L stainless steel has gained considerable importance as a candidate material for the construction of heat exchanger structures in a wide range of engineering applications, both nuclear and non-nuclear. Indeed, the present work results from an earlier investigation of the thermal fatigue behaviour of first wall material for a thermonuclear fusion reactor and its subsequent extension to higher temperatures to support design and operation of components in fast breeder reactors. Under such applications, where complex operational temperatures and stresses are manifested, it is essential to assess component integrity for service in both the ideal material condition and where flaws are introduced either through manufacture or excessive loading.

In order to achieve the required fluctuating thermal gradients consistent with the applications, thick cylinders of 316L stainless steel were induction heated up to 600°C and internally cooled using flowing water. In this way, a repeated thermal shock could be applied to the component and a hold time introduced at elevated temperature to impart a steady state creep loading to the component surface. Spark eroded longitudinal starter notches have been implemented in the specimen surface, and continuous *in situ* monitoring of the crack growth rate from these artificial defects has been achieved using the direct current potential drop (DC/PD) technique, specially adapted for the component type test piece.

The complexity of the stress and temperature fields developed, necessitates the employment of finite element techniques to evaluate the elastic and plastic stress fields across the component wall. For the interpretation of crack growth, these stresses must be converted to stress intensities covering the peak transient conditions applied to an edge crack for the appropriate geometry. This leads to the determination of a suitable thermal fatigue crack growth law. Finally, results from the component experiments were compared with data from mechanical fatigue experiments from the literature.

J. Bressers and L. Rémy (eds.), Fatique under Thermal and Mechanical Loading, 79–88.

2. Experimental Methods

Thermal shock experiments were carried out on an annealed ICL 167 SPH stainless steel which is similar in composition to the low carbon 316L stainless steel. The test material has an austenitic structure, with less than 1% ferrite. Its chemical composition is shown in Table 1 in comparison to that of 316L stainless steel, where values are given in wt% and the balance is Fe [1].

TABLE 1. Chemical composition of ICL 167 SPH and 316L stainless steel.

Material	C	Mn	Si	S	P	Cr	Ni	Mo	N	Co
ICL 167 SPH	≤0.03	1.6→ 2.0	≤0.5	≤0.01	≤0.035	17 → 18	12 → 12.5	2.3 → 2.7	.06 → .08	≤0.25
316L	≤0.03	≤2.0	≤1.0	≤0.03	≤0.045	16 → 18	10 → 14	2.0 → 3.0	-	-

The components tested were in the form of thick cylinders with an outside diameter of 48mm, a 14mm wall thickness, and a length of 120mm (Fig. 1). The thermal shock was generated with a 50kW capacity induction heating system, capable of achieving a surface temperature of 600°C in 7 s. The component was cooled internally by a water flow of 1 L/s.

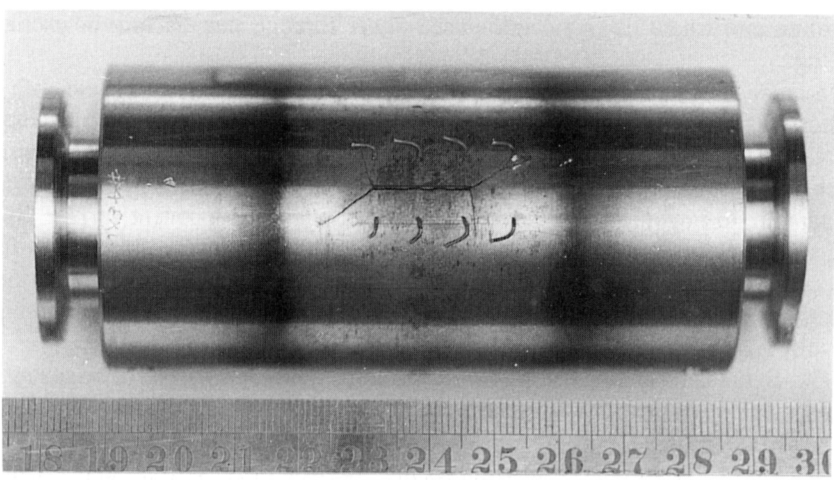

Figure 1. Photograph of a post test specimen showing a longitudinal notch and the position of the PD probes.

Uniform temperature gradients around the uncracked component test section, over a length of 35mm, were achieved through accurate positioning of the induction coil and alteration of its pitch. Axial constraint on the test specimen was avoided by the use of a free expansion unit in the cooling system.

Fig. 1 shows the type of longitudinal notch employed in the present series of tests and the position of the potential drop (PD) probes across the notch. Fig. 2 demonstrates the type of thermal loading experienced by the component, with a thermal cycle of 80°C-600°C at the external surface corresponding to a thermal cycle of 25°C-70°C at the internal surface and hold times of up to 1 hour at maximum temperature.

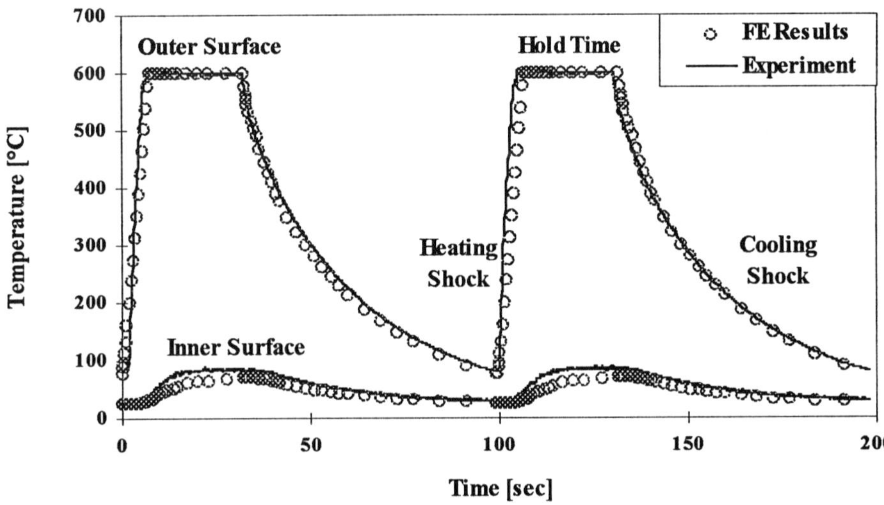

Figure 2. Temperature history imposed on the component - comparison of experimental and finite element results.

The experimental thermal cycling was computer controlled via thermocouples spot welded on the component surface. Continuous monitoring of crack length was achieved with the DC/PD technique [2], where crack length can be determined from the relative PD values, based on calibration curves derived from foil calibration experiments and interrupted test data. Crack growth rates can be found by differentiating the experimental crack length values with respect to the thermal cycles.

3. Finite Element Analysis

Finite element techniques (ABAQUS) were used to determine the temperature history and gradients through the wall thickness over two consecutive thermal cycles, which was sufficient for a stabilised temperature solution. The thermal results were necessary for the subsequent stress analysis, where the thermally induced stresses and strains were calculated for the uncracked component. The stress solutions for the second thermal cycle were used for analysis purposes.

The finite element analysis used a 2-D axisymmetric model with 228 8-noded iso-parametric elements and mesh reduction towards the region of interest across the component thickness. The variation of material properties with temperature and the hardening effects due to fatigue cycling observed in this material were taken into account [1]. In the same way, a 3-D model with 1820 8-noded iso-parametric block elements was used to find the effect of a longitudinal notch on the stress strain response under the same thermal loading conditions. The notch was modelled by element removal giving rise to a worst case solution for the stress concentration.

The externally imposed thermal loading was treated as a nodal boundary condition, where external nodes were constrained to follow the experimental surface temperature during the heating shock and the hold time condition, illustrated in Fig. 2. This was appropriate for this material under the high induction frequency used, where the heating effect was concentrated at the external surface. During cooling the external boundary condition was simply removed. Internally, a heat transfer calculation based on the water/metal interface [3] was used to estimate a surface cooling flux and implemented as an element cooling flux. As the component was continuously cooled internally, this boundary condition was always in operation.

4. Fatigue Analysis

In fatigue analysis, the crack growth rate (da/dN) is normally related to the crack tip stress intensity factor (K). The stress intensity factor embodies the effects of the stress field, the crack size and shape, and the local structural geometry. The stress intensity factor was calculated via the weight function method using a two parameter approximation of the flank displacements, Petroski and Achenbach [4]. The advantage of this method is in the evaluation of stress intensity values from the stress distribution in the uncracked body. Furthermore, the geometry dependent weight functions can be derived from relatively simple loading conditions and then applied to complex stress fields.

As a result of the thermal loading, equal hoop and axial stresses were developed in the component, with negligible stress developed in the radial direction. As crack growth from a longitudinal notch is controlled by the circumferential stress, the hoop stress finite element solutions were used for the evaluation of the stress intensity factor in the current analysis. The maximum stress intensity factor (K_{max}) is derived from the application of the weight function method to the maximum transient hoop stress occurring during the cooling shock.

$$K_{max}(a) = \sqrt{2/\pi a}.\int_0^a \sigma_{hoop}(x).h(a,x).dx \tag{1}$$

where h(a,x) is the weight function calculated for an infinite longitudinal crack in a cylindrical component ($R_{out}/R_{in} = 2$), and x is the variable for distance along the crack face. The calculation is based on the shape function Y(a).

$$Y(a) = 1.122 + 0.411(a/W) + 2.332(a/W)^2 - 1.986(a/W)^3 + 1.972(a/W)^4 \quad (2)$$

where W is the wall thickness. The procedure presented by Lamain [5] was used to calculate the weight functions.

The use of the stress intensity factor is not theoretically justified where highly plastic conditions prevail. The equi-biaxial stress system developed implies that the von Mises and the hoop stress solutions are of a similar magnitude. Examination of the hoop stress profiles generated from the large thermal cycle, Fig. 3, illustrates that plasticity occurs at the external and internal surfaces. Therefore, the stress intensity factor can only be considered in the intermediate elastic region, from $a/W \geq 0.21$ to $a/W \leq 0.64$ for a yield stress of 200MPa. 200MPa is the appropriate yield stress for the cooling cycle, during which the tensile shock at the external surface occurs. The yield stress at 600°C is 110MPa.

5. Results and Discussion

Thermal transients occurs when temperature fluctuations are imposed on components or structures. Thermal shock loading essentially generates fatigue damage. With the introduction of a dwell at maximum temperature, time dependent effects may enhance crack growth behaviour. Crack growth rates under the interaction of creep and fatigue are generally higher than either mechanism acting alone or sequentially, and must be assessed in relation to the component geometry and the service loading conditions.

The analysis of the thermal fatigue tests required a two stage numerical procedure. The first stage consists of determining the stress-strain fields in the uncracked structure from a knowledge of the thermal profiles. The second stage deals with the problem of crack propagation, determination of the stress intensity factor and its variation across the component thickness.

The temperature gradients developed across the wall thickness during the transient thermal loading give rise to high equi-biaxial tensile stresses at the internal surface during heating, and at the external surface during cooling, Fig. 3, with a corresponding compression at the opposite face as the sum of the stress across the wall must equal zero. It is interesting to note that the surface compressive stress at the end of the heating shock relaxes to a tensile stress during the hold period, Fig. 4, which is due simply to the hold time temperature gradient and residual stress effects.

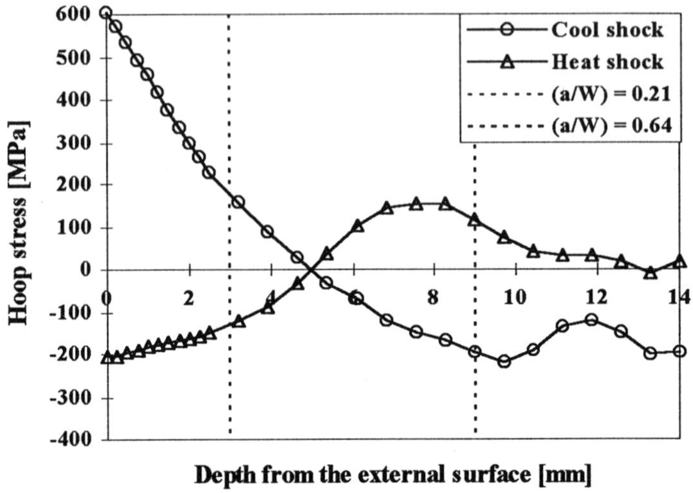

Figure 3. The transient hoop stress distribution.

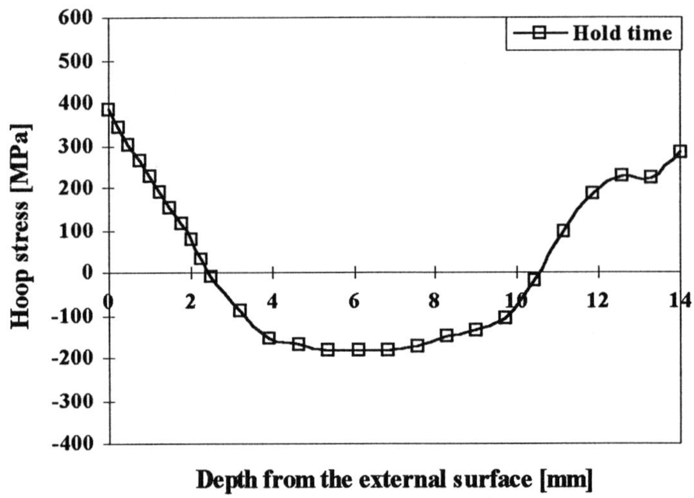

Figure 4. The hold time hoop stress distribution.

The effect of hold time on the crack growth rate resulting from this type of loading is demonstrated in Fig. 5. As the crack grows the effect of hold time appears to reduce, that is, for longer hold time tests initial crack propagation rates are higher with convergence occurring as the crack extends, resulting in the curves becoming more parallel.

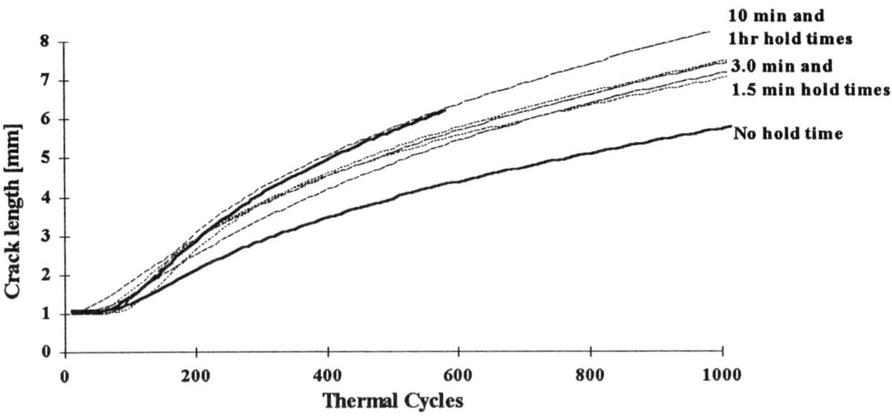

Figure 5. The effect of different hold times on the fracture behaviour.

Complete solutions for both the notched and un-notched maximum stress intensity factor are shown in Fig. 6. The stress intensity values have been corrected for the crack tip plasticity using the plane strain plastic zone size, which effectively increases the crack length by an amount r, where $r = 1/2\pi\ (K/\sigma_y)^2\ (1\text{-}2\nu)^2$. It can be seen that the stress intensity solutions rises steeply initially, followed by a more gradual reduction as the crack moves into the decreasing stress field. This is consistent with the similar decrease in crack growth rate observed experimentally. With the introduction of a defect the initial stress intensity values are greater in the region of stress concentration, with results converging at greater depths. The notched stress intensity values were calculated using the notched finite element stress solutions integrated from the defect root to the crack tip in Eq. (1).

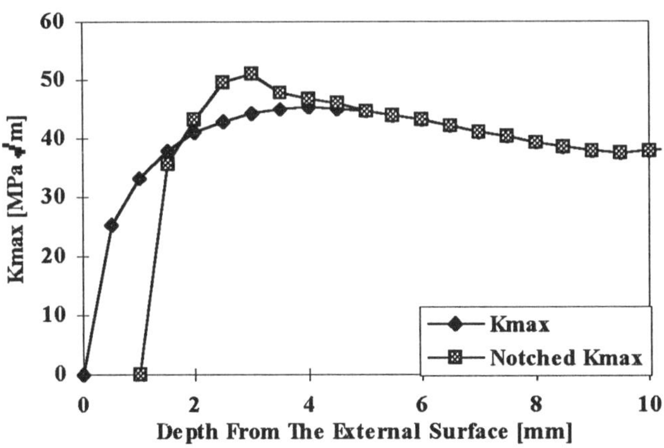

Figure 6. The calculated maximum stress intensity factor.

The variation of R ratio (the ratio of minimum over maximum stress) across the wall thickness is shown in Fig. 7. It can be seen that cyclic behaviour changes from being partially reversed at the external surface, R = -0.35, to a fully reversed cycle, R ≅ -1, over the central region. Towards the inner wall the fatigue cycle becomes more than fully reversed, larger compression than tension, with R ratios reaching values of - 5 at the inside surface.

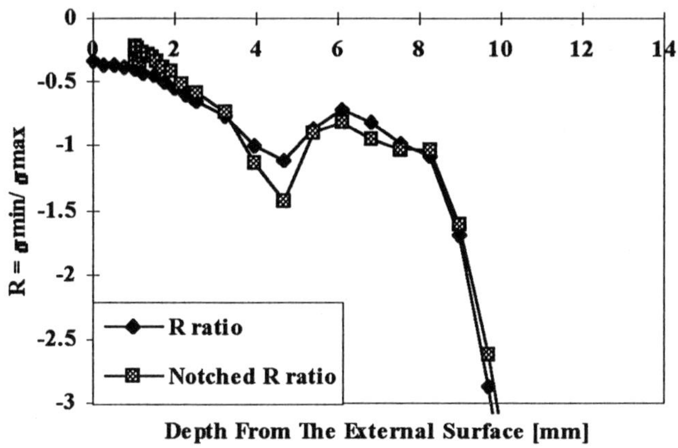

Figure 7. The variation of R ratio across the wall thickness.

To demonstrate the results for thermal fatigue crack growth from 1mm longitudinal notches the calculated stress intensity factor is plotted versus the experimental crack growth rates for tests with different hold times, Fig. 8. The range of the stress intensity factor is derived from Eq. (1), where the notched maximum stress intensity values shown in Fig. 6. Crack growth results are only considered from a/W ≥ 0.21 to a/W ≤ 0.64. The experimental results are presented here with data from mechanical fatigue tests on conventional specimens of 316 and 316L stainless steel tested at different temperatures [6,7].

At the beginning of the experiment the crack growth rate and the range stress intensity are large, with both decreasing as the crack extends due to the reducing stress gradient. This is in contrast to conventional fatigue testing where the stress intensity factor and crack growth rate are normally low with both increasing as the test progresses, traversing the graph in the opposite direction to the experimental data presented here. The results shown in Fig. 8 compare well with the iso-thermal fatigue data, where all the experimental crack growth rates fall between the results at 20°C and 593°C.

In the present analysis the weight functions in the stress intensity factor calculations have been derived for a crack of infinite length. Experimental observations show that the crack develops from the notch, with a length to depth ratio of 15, to a semi-elliptical crack shape, with a length to depth ratio of ≅ 2. An example of a typical surface fracture can be seen in Fig. 1. The effect of the semi-elliptical crack shape is to increase constraint at the crack tip, resulting in lower stress intensity factors. Therefore, as the crack develops, decreasing stress intensity factors and lower crack growth rates would be expected. However, Fig. 8 demonstrates that towards the end of the experiment, where the crack shape becomes semi-elliptical, the experimental crack growth rates still compare well with the stress intensity values based on an infinite crack front. There are three possible reasons for this behaviour.

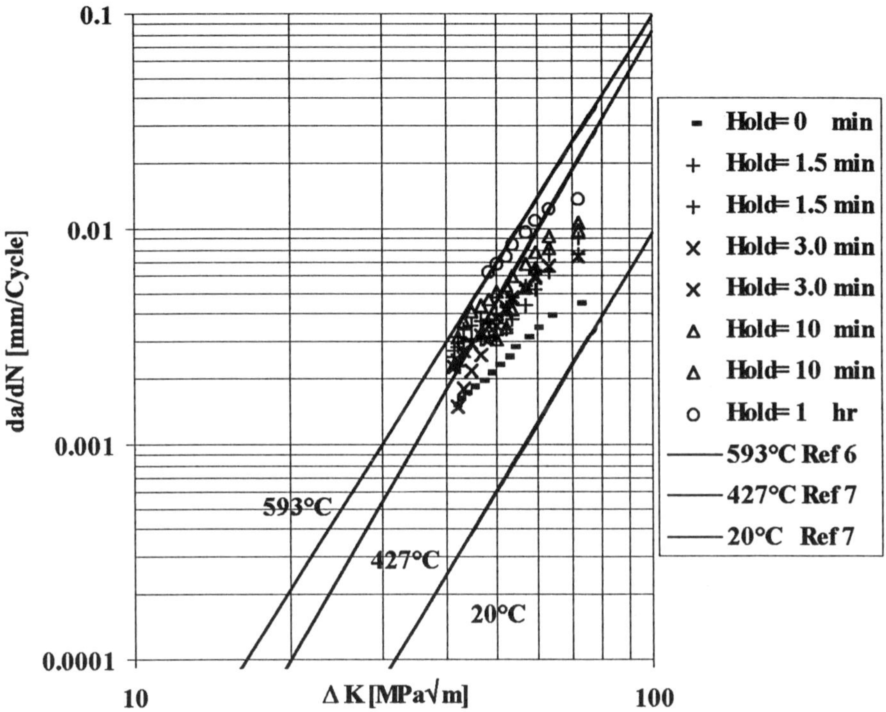

Figure 8. The effect of hold time on the crack growth behaviour.

Firstly, the interaction of crack tip plasticity with section plasticity as the crack approaches the internal wall may further decrease the elastic region over which the stress intensity factor is valid. This results in slightly higher actual stress intensity factors than those elastically predicted, in effect counteracting the constraint imposed by the semi-elliptical crack shape. Secondly, lowering the test frequency can increase crack growth rate especially at elevated temperatures. The maximum test frequency in the current series of tests was ≅0.01Hz calculated without hold time. Finally, the effect of the variation of R ratio across the specimen thickness must also be considered. The importance of R ratio is more evident at low crack growth rates close to the threshold stress intensity factor. Although, some work has been carried out at higher crack propagation rates where it has been reported that crack growth rate increases in general with decreasing R ratio, when crack growth rate is based upon K_{max} [8], (da/dN decreases with decreasing R if ΔK is used as the correlating parameter). This effect was found to be most pronounced under conditions of tension-compression loading or where the maximum stress was varied during the test, both of which occurred in the present experimental programme.

The effect of hold time, previously discussed with regard to Fig. 5, is also apparent in Fig. 8. The increased crack growth rate with longer dwell time is believed to be caused by enhanced crack initiation and early growth resulting from a creep contribution. Due to the hold time stress condition, Fig. 4, and the decrease in temperature with depth, the potential for creep damage is confined to the external few millimetres of the component wall, such that the crack growth rates converge for different hold time tests as the crack extends into a region where thermal fatigue mechanisms act alone. Fig. 8 clearly shows the initial crack growth rates for both the 1hr and 10min hold time tests to be around three times larger than the test with no hold time, reducing to a factor of less than two towards the end of the test. This magnification of thermal fatigue crack growth rate can be thought of as a shift in the Paris law behaviour, which was caused by a weakening in the material resulting from creep damage during the hold time. Future studies are expected to yield an estimation of the hold time contribution to the crack growth rate to be obtained from creep calculations for the temperature and stress conditions during the hold period.

6. Conclusions

Novel crack growth experiments for component type tubular test pieces subjected to thermal fatigue/creep loading have been successfully achieved. The complex thermal stress systems developed were analysed using finite element techniques. The stress intensity factor, calculated by the weight function method was used to correlate thermal fatigue crack growth with isothermal fatigue behaviour. Fatigue/creep interaction apparently occurs during the dwell periods at elevated temperature, increasing the fatigue crack growth rate with longer hold times, but the influence of creep is confined to the external surface for this type of experiment.

References

1. Peckner A.C., Bernstein I.M., *Handbook of Stainless Steels*, McGraw-Hill, Inc., New York, USA, 1977.
2. Hunter C.P., Hurst R.C., Creep crack growth studies on alloy 800H tubes under complex loading conditions, *Materials at High Temperature* **10** (1992), 144-149.
3. Bayazitoglu Y., Ozisik M., *Elements of Heat Transfer*, McGraw-Hill, Inc., New York, USA, 1988.
4. Petroski H.J., Achenbach J.D., *Engineering fracture mechanics* **10** (1978), 257-266.
5. Lamain L.G., *A computer code to analise the fatigue behaviour of tubes: theory and application.* EUR report 14415, May 1992.
6. Michel D.J., Smith H.H., in ASTM-STP **725** (1981), 352.
7. Shahinian P., Watson H.A., Smith H.H., *Fatigue crack growth characteristics of several austenitic stainless steels at high temperature.* ASTM-STP **520** (1973), 387-400.
8. James L.A., The effect of stress ratio on the elevated temperature fatigue-crack propagation of type 304 stainless steel. *Nucl. Tech.* **14** (1972), 163-170.

GROWTH AND COALESCENCE BEHAVIOUR OF SURFACE MULTI-CRACKS IN TYPE 304 STAINLESS STEEL AT ELEVATED TEMPERATURE

C.M. SUH[1], N.S. HWANG[2], S.H. NAHM[3] and H.M. LEE[3]
[1] *Kyungpook National Univ., Dept of Mechanical Engineering*
1370 Sankyuk Dong Pukku, Taegu, 702-701, Korea/S
[2] *Yeungnam Jr. College, Dept of Mechanics,*
1737 Daemyung 7 dong, Namgu, Taegu, 705-037, Korea/S
[3] *KRISS, Materials Evaluation Center, Taedok Science Town,*
P.O. Box 3, 305-606, Korea/S

1. Introduction

Most of the fatigue damage is concerned with fatigue phenomena stemming from a short surface crack [1], which usually starts from small surface flaws [2]. Generally, the surface cracks found in structures and machine elements are not a single crack but multi cracks, and these cracks grow three dimensionally by the procedure of initiation, growth, coalescence and fracture and for this reason, it is very complex to analyse the fatigue crack growth. Studies on the three dimensional character of surface cracks were performed by Kitagawa et al. [1], Suh et al. [3], and Murakami [4] et al. etc.

In this study, the axial loading fatigue tests were carried out to analyse the growth and coalescence of multi-surface cracks at room temperature. The results were analysed by fracture mechanics.

Furthermore, on the basis of these experimental data, a simulation for the growth and coalescence behaviour of multi-surface cracks was developed and compared with the experimental results.

2. Experimental Method and Simulation Programming

2.1. SPECIMEN

In these tests, type 304 stainless steel specimens were used, which were cut off in the perpendicular direction to the rolling direction. There are two types of specimen. One is the specimen with two noches of the same size, with a length of 6mm and a depth of 3mm. The specimen with the two similar notches is called HS2 (see Fig.1a). The other specimen contains one notch of 3mm en length and 1.5mm in depth and two smaller notches of 2.5mm in length and 0.7mm in depth. This latter

J. Bressers and L. Rémy (eds.), Fatique under Thermal and Mechanical Loading, 89–95.

specimen is called HS3A (see Fig 1b). The specimens were polished with till #1200 sand paper on the artificial notched surface, and buffed with chromic oxide.

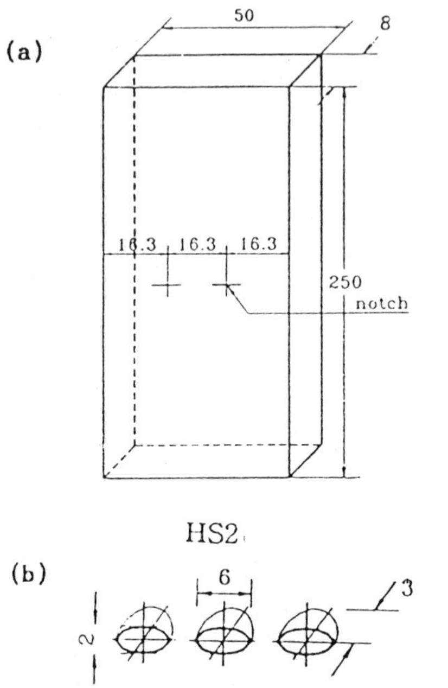

Figure 1.a. Configuration of HS2 specimens with two similar surface notches (the units are mm): (a) geometry (b) details of the notches.

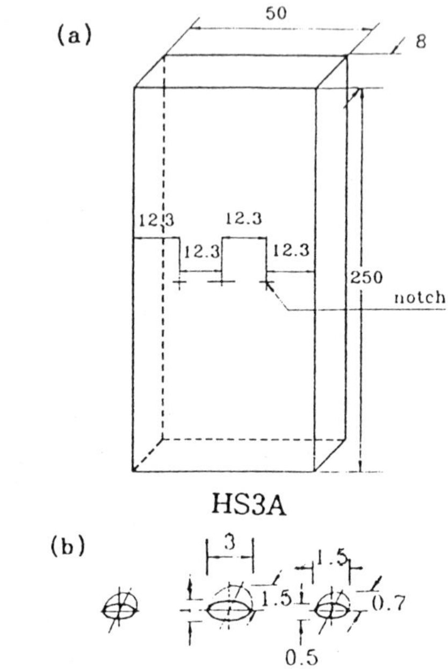

Figure 1.b. Configuration of HS3A specimen with three different surface notches (the units are mm): (a) geometry (b) details of the notches.

2.2. EXPERIMENTAL METHOD

Axial loading control fatigue tests were carried out on a hydraulic testing machine at room temperature with stress ratio 0.1 and a load frequency of 10Hz with sine wave. The surface crack length which initiated at the multi-surface notch was measured by the replica method every 5~7% of the specimen's fatigue life after the specimen was cooled down enough to attach it. The temper color method was applied to measure the depth and observe the shape of crack face. The depth b was determined using the surface crack length 2a.

2.3. SIMULATION COMPOSITION

There are three components in the analysis such as input, analysis and output section. And the programming language used in this simulation was turbo-C which has characteristics of speedy graphic and fast dealing speed. The Newman-Raju formula [5] was adopted to calculate the stress intensity factor for the fracture mechanical analysis in this study. And the interaction condition by Murakami & Nemat-Nasser [4] was mainly adopted as the interaction condition of three dimensional surface cracks. When the adjacent crack distance, δ, (see Fig. 2) attains the reference value δ_o, corresponding to the crack coalescence condition, the cracks coalesce and become a new crack in this simulation. And the coalescence conditions considered are according to the three following cases.

i) surface point connection(SPC) [6] $\delta_o = 0$
ii) ASME Boiler and Pressure Vessel Code Sec. XI [12] $\delta_o = \min(b_1, b_2) \times 2$
iii) BSI PD 6493 [7] $\delta_o = a_1 + a_2$

Here a_1 and a_2 are half the length of the surface cracks, b_1 and b_2 are the depth of the adjacent cracks.

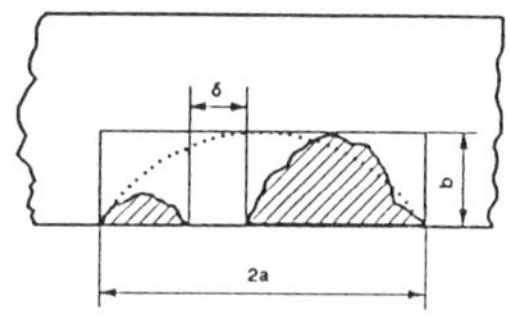

Figure 2. Coalescence condition of adjacent surface cracks.

3. Comparison of experimental and simulation results

3.1. THE FRACTOGRAPHY OF SURFACE CRACKS

Fig. 3a is an example of the fractography of the HS2 specimen. In this photo, the fatigue crack initiated from a semi-circular surface notch and then grew into an ellipse and coalesced. After the coalescence, the stress is concentrated in the coalesced region and it grew rapidly as a semi-ellipse shape and fractured. Fig. 3b shows the fractography of specimen HS3A, which clearly shows that the growth behaviour of the major fatigue crack, initiated from a big notch, governed the fatigue life. As single crack, the major crack grew, penetrated, and then led to fracture of the specimen. This same phenomenon was also seen in the smooth specimen, propagating from a surface fatigue crack.

3.2. S-N CHARACTERISTICS

Fig. 4 shows the relation between maximum tensile stress and fatigue life of experimental data at 538°C by axial loading fatigue tests. In the case of the test data for the same artificial notch size, the data markers O, □, Δ and ◇ represent the experimental results at elevated temperature. In this figure, as the notch number and the stress level increase, the fatigue life decreases constantly.
The data markers, ●, ■, O and ● represent the experimental results at room temperature. In addition, in the case of the test data with different artificial notch size and number, the data markers, Δ and ◇, represent the test results with two and three surface notches at 538°C. The stress levels used are 196.2MPa, 245.3MPa, and 293.2MPa. Here the specimens with two and three notches show a similar fatigue life.

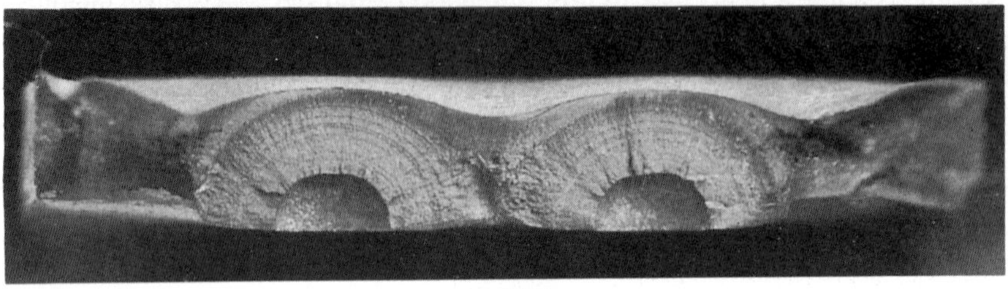

(a) HS2

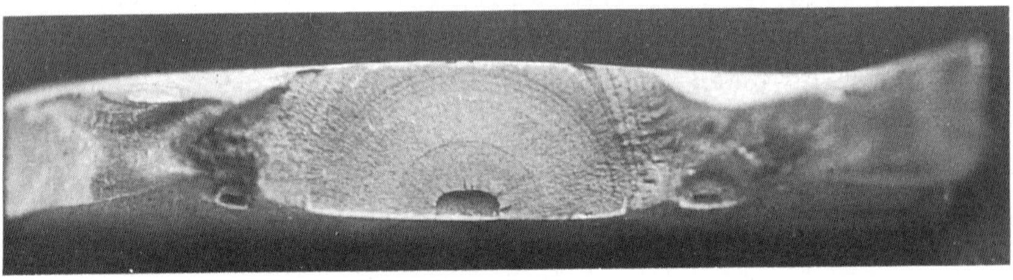

(b) HS3A

Figure 3. Fractography of specimens used in the experiment at 538°C (a) HS2 (b) HS3A.

The increase of the number of small notches had little influence on the fatigue life where the crack might fracture before coalescence. The largest crack is found to mainly govern the fatigue life. In contrast, the cracks which initiated from a surface notch didn't influence fatigue life so much. These results have been reported elsewhere [1,3].

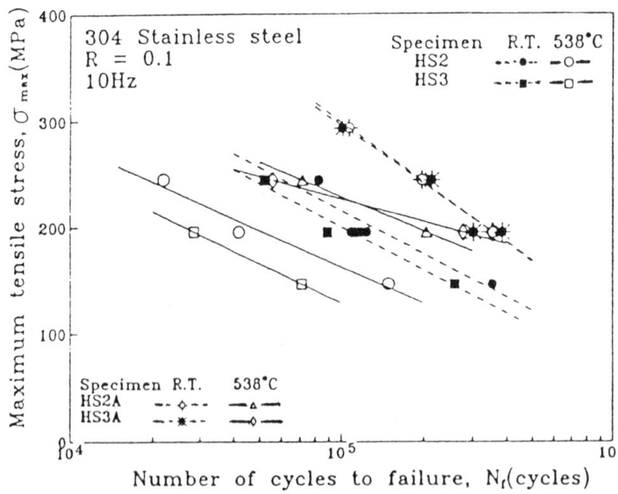

Figure 4. Relation between maximum tensile strength, σ_{max}, and number of cycles to failure, N_f, for all specimens.

3.3. RELATION BETWEEN HALF CRACK LENGTH AND CYCLE RATIO

Fig. 5 shows the relation between the half crack length, a, and the cycle ratio, N/N_f. In this figure, the fatigue behavior of four kinds of specimen is shown. In spite of the difference of specimens, the data of each specimen gathered into a narrow band, and in the region of cycle ratios above 80% the crack growth is similar.

3.4. CHARACTERISTICS OF THE db/dN - ΔK_b PLOT

Fig. 6 shows the relation between crack growth rate and stress intensity factor range of HS2 and HS3 in the depth direction. The solid symbol means the center crack. The crack grows in the perpendicular direction at first and in a certain point it grows slowly. The transient point could be explained by the characteristics at 538°C.

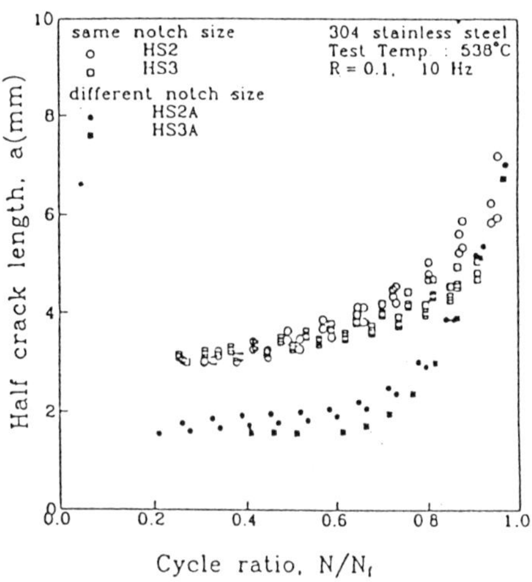

Figure. 5. Relation between half crack length, a, and cycle ratio, N/N_f.

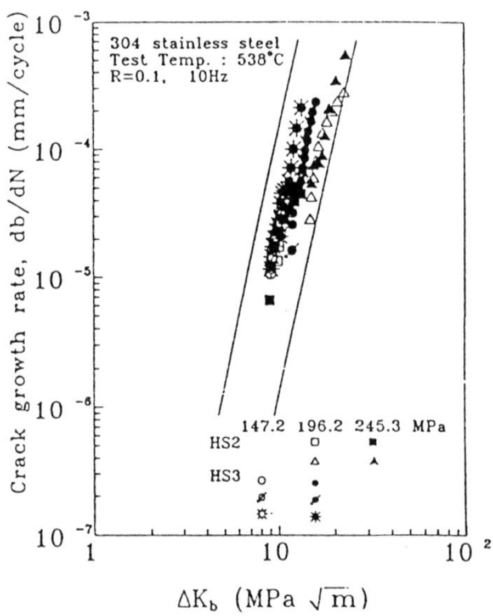

Figure 6. Dependence of crack growth rate upon stress intensity factor range with multi-surface cracks for specimens with the same notch size.

3.5. SIMULATION RESULTS WITH THE MATERIAL CONSTANTS C AND m

The relation between crack growth rate and stress intensity factor is explained by the Paris formula:

$$db/dN = C(\Delta K)^m \quad (1)$$

Here C and m are material constants.

Fig. 7 collects the crack growth coefficient, C and crack growth exponent, m for different specimens and temperature. Those data were assembled in a narrow band regardless of conditions, and the simulation program used these data as input. In this study, mainly the SPC condition was used to simulate and plot because the simulation using the SPC condition gave good agreement when compared with the growth and coalescence behaviour of the experimental results. Also the methods recommended by ASME, and BSI(British Standards Institution) were applied to compare the difference among coalescence conditions.

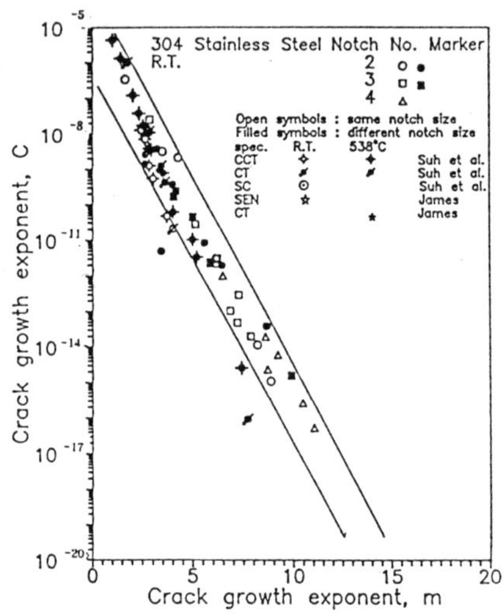

Figure. 7. **Relation between crack growth coefficient, C and crack growth exponent, m compared with the results of James [8].**

Fig. 8 shows the crack growth and coalescence behaviours by simulations at the same crack size, but different crack number, HS3. For this drawing, a Murakami & Nemat-Nasser interaction condition was used for the surface crack initiated and grown from each notch with the same size and the SPC condition for coalescence.

4. Conclusions

Fatigue tests were carried out to study the growth and coalescence behaviour of multi-surface cracks which initiated at the semi-circular surface notch in type 304 stainless steel at elevated temperature, and a simulation program written in C language was developed to predict the growth and coalescence behaviour of the cracks. The important results obtained from this study were:

1. The maximum error between the predicted fatigue lives obtained by the simulation program and those of experimental data was within 10% difference at 538°C, which demonstrates the utility of this simulation.
2. According to the fatigue tests and the simulation results, the fatigue lives of the multi-surface cracks mostly depend on the growth and coalescence behaviour of the major crack rather than on the behaviour of the sub-cracks.

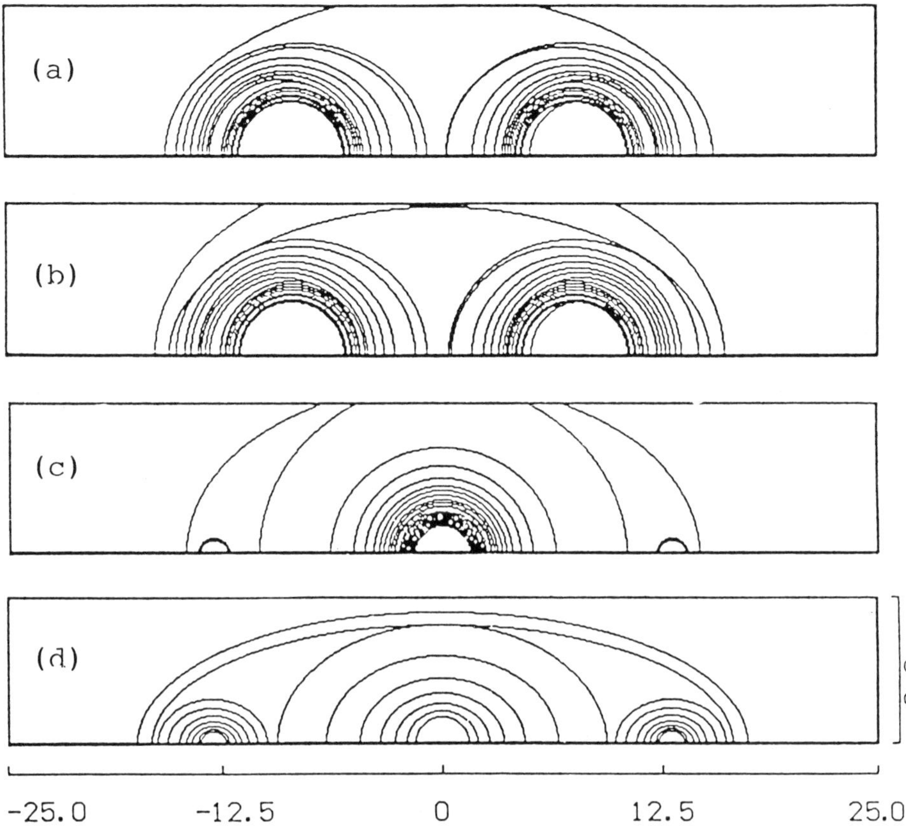

Figure. 8. Simulation results applied to material constants C and m of HS2 HS3 specimen (a) HS2-1 (b) HS2-2 (c) HS3A-1 (d) HS3A-2.

References

1. Kitagawa, H., Takahashi, S., Suh, C.M., and Miyashita, S. Quantitative analysis of fatigue process-Microcracks and slip lines under cyclic strains, ASTM STP 675, 1979, pp.420-449.
2. ASME Boiler and Pressure Vessel Code, Sec. X I Rules for Inservice Inspection of Nuclear Power Plant Components, Appendix A, 1980.
3. Suh, C.M, J.J. Lee, Y.G. Kang and H.J. Ahn, Fatigue Crack Initiation and Growth Life Prediction with Statistical Consideration, KSME & JSME joint Conf, 1990, pp.684-689.
4. Murakami, Y and S. Nemat-Nasser, Interacting Dissimilar Semi-Elliptical Surface Flaws under Tension and Bending, *Engng. Frac. Mech*, **16**, 1982, 373-386.
5. Newman, J.C, Jr. and Raju, I.S. Stress Intensity Factors Equation for Cracks in Three-Dimensional Finite Bodies ASME STP 791, 1983, pp.238-265.
6. Yuuki, R and Yoshida, T. Fatigue Life Prediction System for Structures, *JHPI*, **r24**, No.3, 1986, 115-123.
7. BSI PD 6493 Guidance on Some Methods for the Deviation of Acceptance Levels for Defects in Fusion Welded Joints, 1980.
8. James. L.A., Frequency Effects in the Elevated Temperature Crack Growth Behavior of Austenitic Stainless Steels-A Design Approach, *J. of Pressure Vessel Technology*, **101**, 1979, 171-175.

DAMAGE AND FATIGUE LIFE OF SUPERALLOY IN738LC UNDER THERMO-MECHANICAL AND LOW CYCLE FATIGUE LOADING

W. CHEN[1], A. DUDKA[1], H. CHEN[2], D. MUKHERJI[2], R.P. WAHI[2] and H. WEVER[1]

[1] *Institut für Metallforschung, TU Berlin, D-10623*

[2] *Hahn-Meitner-Institut Berlin GmbH, D-14109 Berlin*

1. Introduction

Contradictory results exist in literature about the influence of various types of thermomechanical fatigue loading (TMF), e.g. in-phase (IP), out-of-phase (OP) and diamond-type cycling on the cyclic life of high temperature alloys, and about their relationships to the cyclic life of isothermal low cycle fatigue (LCF) tests at the maximum temperature of the corresponding TMF tests [e.g. 1 - 4]. This is obviously related to the differences in the basic properties of materials tested as well as to the details of test procedures and parameters employed. In order to be able to understand these complicated and at times controversial relationships it is essential to have an insight into the damage processes taking place in different types of alloys and under various loading conditions.

In the present study the fatigue life and damage of IN738LC, a widely used investment cast nickel base superalloy for application as blade material in land-based gas turbines, have been investigated under both TMF and LCF loading. To minimise the thermal stresses the rate and therefore the range of temperature variation both in IP and OP TMF tests were kept relatively small and the thermal strains were fully compensated. These types of IP and OP TMF tests provide a good basis for comparison of results with each other and with those obtained from LCF tests earlier.

2. Material and Experimental Procedure

The chemical composition of the alloy IN738LC examined in the present study is shown in Table 1. The cast bar stock was subjected to hot isostatic pressing at 1453 K and 100 MPa to minimise casting porosity. After solution treatment and a two stage ageing treatment [5] the specimens contained a unimodal distribution of roughly cuboidal γ' precipitates with an average edge length of about 450 nm and a volume fraction of about 43%. The volume fraction of carbides amounted to about 1%. In the present study the same type of solid specimens were used for both TMF

J. Bressers and L. Rémy (eds.), Fatique under Thermal and Mechanical Loading, 97–102.

and LCF tests, since the hollow specimens showed frequently different test results from those obtained from solid specimens under the same LCF test conditions [6].

TABLE 1. Chemical composition of IN738LC in wt. %.

Al	B	C	Co	Cr	Mo
3.40	0.01	0.11	8.60	15.93	1.75
Nb	**Ta**	**Ti**	**W**	**Zr**	**Ni**
0.82	1.80	3.47	2.61	0.04	balance

Both the TMF and LCF tests were conducted in a servohydraulic MTS testing system under symmetric (R_s = -1) total axial strain control [7, 8]. In TMF tests the specimens were inductively heated to the mean temperature T_m of 1123 K without any load. In order to keep the mechanical strain amplitude ε_m and the mechanical strain rate $\dot{\varepsilon}_m$ constant in both the IP and OP tests, the total strain amplitude and the total strain rate were varied accordingly, using the thermal expansion and contraction data measured on the same specimen by a temperature cycling before starting the test. Both the IP and OP tests were carried out at different constant mechanical strain amplitudes varying between 0.4 to 1.0 % and at a constant mechanical strain rate of 10^{-5} s^{-1}. The maximum cooling rate, with no thermal gradient across the specimen diameter, is limited by the specimen geometry and thermal properties of the material. This limit restricts the maximum rate $\dot{T}$ of temperature variation in the present tests. The temperature range ΔT is determined by ε_m, $\dot{\varepsilon}_m$ and. $\dot{T}$. In the present study a relatively small temperature variation rate of 6 to 15 K/min was used, depending on the mechanical strain amplitude. In all the tests the minimum and maximum temperatures were 1023 and 1223 K. The LCF tests were performed using the same mechanical test parameters at 1223 K and 1023 K.

Crack initiation sites and propagation paths were examined by optical and scanning electron microscopy. The damage parameters, e.g. length and density of cracks, were measured on the longitudinal section of the failed specimens for the test conditions mentioned above.

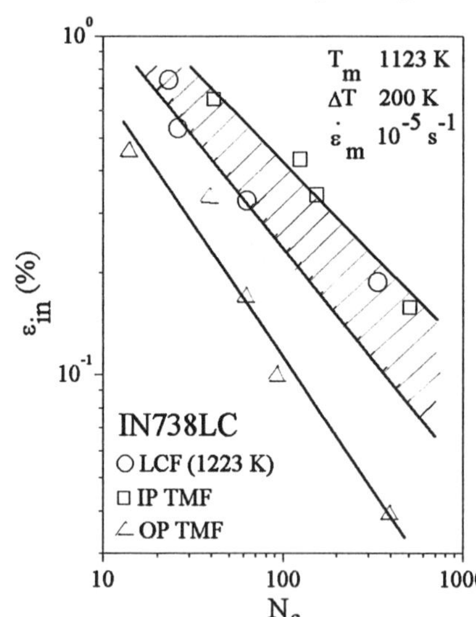

Figure 1. Cyclic life N_f as a function of inelastic strain amplitude ε_{in}.

3. Results and Discussion

The fracture life (N_f) of the alloy under IP, OP as well as under LCF conditions, as a function of inelastic strain amplitude ε_{in} are plotted in Fig. 1. The data points of the IP and the LCF tests at 1223 K are closer to each other. The

OP tests showed a distinctly shorter fatigue life specially at small ε_{in}. A possible reason for the observed shorter life is the higher tensile stress amplitude (> 550 MPa) under OP test condition in comparison to those (< 300 MPa) under IP TMF loading and LCF loading at 1223 K [7].

Microscopic examinations revealed that in both the IP and OP tests as well as the LCF tests at 1223 K crack initiation occurred essentially at the intersections of grain boundaries with the specimen surface, (Fig. 2). The severe oxidation of the crack surface at these sites suggests that the oxidation of grain boundaries within the higher temperature phase of the load cycle, i.e. tensile deformation range in an IP test and compressive deformation range in an OP test, weakens the grain boundaries. They become a preferential site of crack nuclei formation under these test conditions. The same type of crack initiation at grain boundaries was also observed under creep-fatigue loading at 1223 K [9]. In the LCF test at a lower temperature of 1023 K such crack initiation sites could not be detected since the oxidation of grain boundaries was negligibly small at this temperature (Fig. 3) and apart from the main crack leading to fracture, the failed specimen contained almost no other cracks. Based on these observations we conclude that the crack initiation is assisted by the oxidation of grain boundaries at the specimen surface under both the TMF loading and the LCF loading at 1223 K. The density of such crack nuclei at the surface of the specimens tested under IP TMF loading was measured to be about three times higher than that under OP test condition.

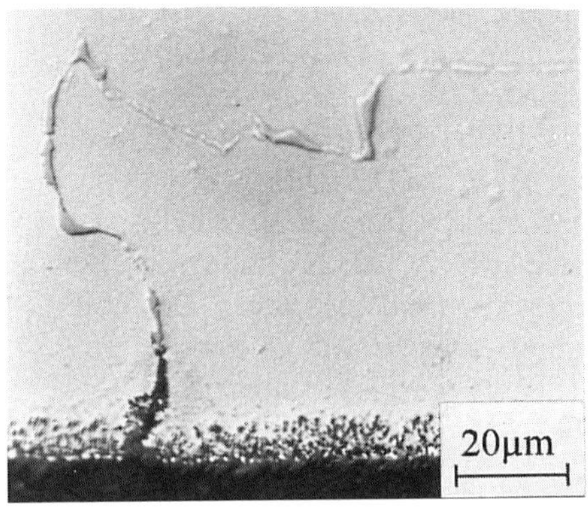

(a)

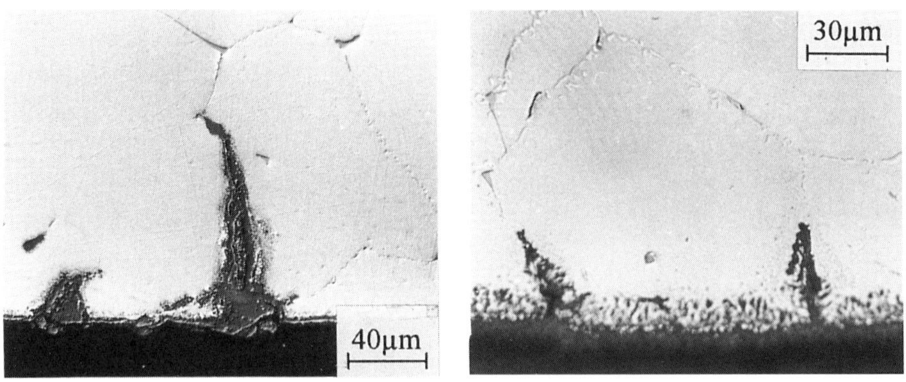

(b) (c)

Figure 2. Micrographs showing crack initiation under a) IP TMF, b) OP TMF and c) isothermal LCF loading.

The crack propagation essentially took place along the grain boundaries (Fig. 4a) and the dendrite boundaries (Fig. 4b) under all loading conditions. The number density of cracks measured in the IP specimens is quite high and is comparable with the value of the LCF tests at 1223 K, indicating the occurrence of a competitive growth of individual cracks in both types of tests, (Fig. 5). In the OP specimens only one crack within the gauge length was found, which led to failure of the specimens. Obviously under OP TMF loading crack growth is a more predominant process compared to that under IP TMF loading due to the influence of the higher tensile stress amplitude of the OP tests (Fig. 6). The same behaviour was also observed in the LCF test at 1023 K.

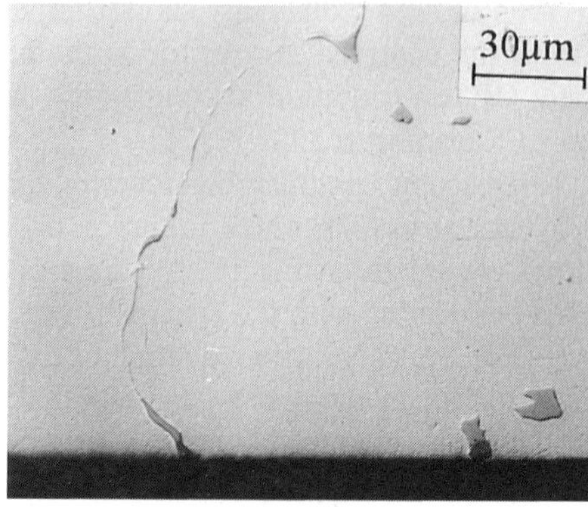

Figure 3. Negligible grain boundary oxidation in LCF test at 1023 K.

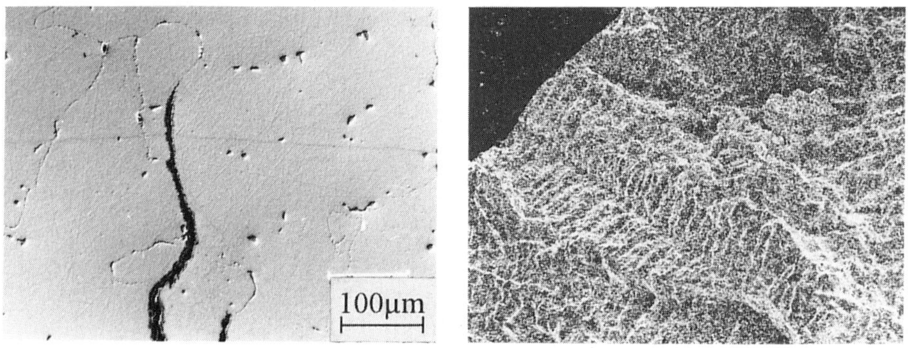

(a) (b)

Figure 4. Micrographs showing crack growth along a) grain boundary and b) dendrite boundary.

Based on the above microscopic examinations the life behaviour under IP, OP and LCF tests can be rationalised as follows:

- Both in the IP and the LCF tests the temperature and the stress are similar in the tensile part of the cycle, but not so in the compressive part. The observed damage is also similar in both these tests. It is therefore concluded that the tensile part of the cycle in the IP test determines the nature and amount of damage. This explains the observed similarity of the dependence of N_f on ε_{in} of the IP and the LCF tests. Thus the LCF test is a reasonable approach to predict the cyclic life of IP test under the present condition of testing.

- Crack initiation has been found to depend on the oxidation of grain boundaries at the specimen surface. Therefore no drastic change in crack initiation behaviour between IP and OP tests in the present study is expected. This is under the assumption that the sign of the load does not influence the process of grain boundary diffusion of oxygen very much. Different crack growth paths might cause a difference in fatigue life [2]. In the present study both the IP and OP tests have the same crack growth paths, i.e. grain and dendrite boundaries. The critical crack length for catastrophic crack growth, however, is much smaller in the case of OP tests because of the higher tensile stress under this test condition (Fig. 6). Additionally, the high tensile stress in an OP test occurs at a temperature where the crack propagation rate has been measured to be higher than that at a higher temperature of 1123 K [10]. This may be due to the improvement in ductility of IN738LC at higher temperatures. Both factors may combine in shortening the life of OP TMF tests in this study.

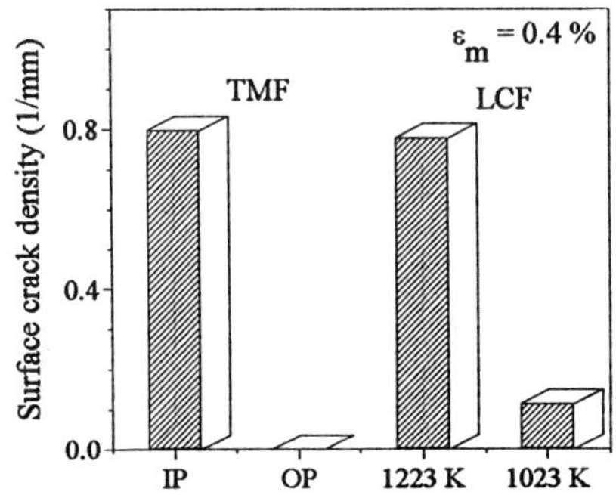

Figure 5. Surface crack density under TMF and isothermal LCF loading.

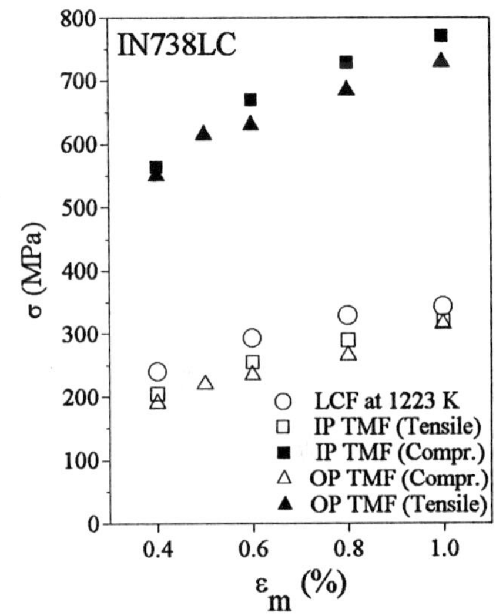

Figure 6. Tension and compression stress amplitude σ in TMF and LCF tests.

4. Conclusions

The present investigation of the TMF and LCF behaviour of the nickel base superalloy IN738LC has shown the following main results:

- The most damaging loading condition is the OP TMF testing, with the shortest fatigue life, whereas the IP TMF tests under the present conditions showed a higher fatigue life which is comparable to that of the LCF tests at 1223 K.
- Crack initiation occurred at intersection sites of grain boundaries with the specimen surface under IP and OP TMF loading and LCF loading at 1223 K.
- Under all test conditions crack propagation took place along the grain and dendrite boundaries. The IP TMF tests showed a similar crack growth behaviour as the LCF tests at 1223 K. The same similarity has been found between the OP tests and the LCF test at 1023 K.

- The shorter fatigue life under the OP TMF loading could be rationalised in terms of a higher tensile stress amplitude and therefore higher stress intensity at the crack tip, in combination with a reduced ductility at the temperature where the maximum in tensile stress occurs.

References

1. Natta, A., Kuwabara, K., Thermal-Mechanical Fatigue Failure and Life Prediction, in R. Ohtani et al. (eds.), *High Temperature Creep-Fatigue*, Elsevier Appl. Sci., London, (1988), pp. 203-222.
2. Henderson, P.J., Linde, L., High Temperature Thermo-mechanical and Low Cycle Fatigue of the Oxide Dispersion Strengthened Alloy MA754, in Y. Hosoi et al. (eds.), *Aspects of High Temperature Deformation and Fracture in Crystalline Materials*, JIM, Sendai, Japan, (1993), pp. 527-534.
3. Russell, E.S., Practical Life Prediction Methods for Thermal-mechanical Fatigue of Gas Turbine Buckets, in *Proc. Conf. on Life Prediction for High Temperature Gas Turbine Materials*, EPRI, Syracuse Uni., N.Y., (1986), pp. 3.1-3.39.
4. Vasseur, E., Remy, L., High Temperature Low Cycle Fatigue and Thermal-mechanical Fatigue Behaviour of an Oxide-Dispersion-Strengthened Nickel-base Superalloy, *Mater. Sci. Eng.*, **A184** (1994), 1-5.
5. Mukherji, D., Jiao, F., Chen, W., Wahi, R.P., Stacking Fault Formation in γ' Phase During Monotonic Deformation of IN738LC at Elevated Temperatures, *Acta metall. mater.*, **39** (1991), 1515-1524.
6. Ziebs, J., private communication.
7. Chen, H., Chen, W., Wahi, R.P., Wever, H., Cyclic Life of Superalloy IN738LC under In-phase and Out-of-phase Thermomechanical Fatigue Loading, *Z. Metallkde.*, **86** (1995), 423-427.
8. Jiao, F., Chen, W., Mukherji, D., Zhu, J., Wahi, R.P., Deformation Behaviour and Microstructural Evolution in IN738LC under LCF Loading, in M. Jono and T. Inoue (eds.), *Mechanical Behaviour of Materials VI*, Pergamon Press, London, (1991), pp. 385-390.
9. Chen. H., Rumi, M., Chen, W., Wever, H., Wahi, R.P., Cyclic Life and Macrocrack Initiation of IN738LC under Creep-Fatigue Loading at 1223 K, in Y. Hosoi et al. (eds.), *Aspects of High Temperature Deformation and Fracture in Crystalline Materials*, JIM, Sendai, Japan, (1993), pp. 513-518.
10. Scarlin, R.B., Fatigue Crack Growth in a Cast Ni-base Alloy, *Mater. Sci. Eng.*, **21** (1975), 139-147.

THE EFFECT OF THERMO-MECHANICAL FATIGUE LOADING ON SURFACE OXIDATION OF IN738LC SUPERALLOY

S. ESMAEILI, C.C. ENGLER-PINTO JR., B. ILSCHNER
and F. RÉZAÏ-ARIA
Laboratory of Mechanical Metallurgy,
Department of Materials,
Swiss Federal Institute of Technology,
CH - 1015, Lausanne, Switzerland

1. Introduction

Turbine blade materials are subjected to severe cyclic thermo-mechanical fatigue (TMF) loading in a highly oxidizing environment. To protect the blades from environmental damage they are coated. Nevertheless, during turbine operation, coating degradation may occur [1–3] and the base metal may be exposed to the environment. Therefore, the study of the effect of thermo-mechanical loading on surface oxidation of superalloys can result in a better understanding of the damage which occurs in real blades. Investigations showing the enhancement of the oxidation process by high temperature isothermal fatigue or temperature cycling have already been reported [4–7].

The objective of this paper is to study the surface oxidation behavior of IN738LC under out-of-phase TMF loading between 400–900°C with and without a hold time. IN738LC is a conventionally cast nickel-base superalloy widely used as a blade material in power generation turbine engines. Although an extensive body of data is available on fatigue, oxidation and creep-fatigue interactions for this alloy, quantitative studies devoted to non-isothermal fatigue modification in the oxidation kinetics are still needed.

2. Material and Experimental Procedures

The chemical composition of IN738LC in weight percent is 0.11C-16Cr-8.5Co-1.75Mo-2.6W-1.75Ta-0.9Nb-3.4Al-3.4Ti-0.05(Zr-B)-balance Ni. The TMF tests were carried out on tubular specimens in laboratory air [8]. The out-of-phase cycle was used with the temperature varying between 400–900°C. To study the effect of constant strain dwell period on the oxidation-TMF interaction, some specimens were tested with a ten minute hold time at 900°C. The period of the thermo-mechanical cycles without hold time was 124 s. The TMF test conditions of the experiments used for the investigation of the oxidation-TMF interaction are summarized in Table 1.

J. Bressers and L. Rémy (eds.), Fatique under Thermal and Mechanical Loading, 103–108.

After completion of the TMF tests the specimens were cut in longitudinal sections. To preserve the surface oxide integrity in metallographic preparations, thin layers of nickel were deposited on the samples. Both optical microscopy and scanning electron microscopy (SEM) were used to examine the oxidation related microstructural features. Quantitative metallography on the specimens was performed using SEM in order to corelate the surface oxide thickness with the number of thermo-mechanical cycles. Surface oxide thickness was taken as the average of 50 measurements where the oxide-scale integrity was preserved.

TABLE 1. TMF test conditions.

mechanical strain range, %	hold time, minutes	number of cycles, N	N / N_f^*, %	total cycling time, hours
0.7	10	41	100	8.2
0.6	10	205	100	41.2
0.5	0	370	50	12.7
		742	100	25.5
	10	200	35	40.2
		570	100	114.6
0.35	0	2060	100	71.2
	10	912	100	183.4

* N_f = Number of cycles to fracture.

3. Results and Discussion

Fig. 1 shows a typical electron micrograph of a TMF cycled specimen demonstrating different microstructural features. The results of the quantitative oxidation analysis are shown in Figs. 2 and 3.

Fig. 2 shows the oxide thickness as a function of the square root of the total cycling time in hours (including the hold time). It can be seen that the tests with and without hold time at 0.5% and 0.35% mechanical strain range reveal a unique parabolic oxidation kinetics. Therefore, the introduction of the ten minute strain hold, although increasing the surface oxide thickness, did not alter the oxidation kinetics and its dependence on the mechanical strain.

The oxidation process of the 0.35% strain range tests and the initial stages of oxidation of the 0.5% strain range tests may also fit a linear law [9], which means that the kinetics of oxidation may be a combination of linear and parabolic. For this kind of behavior, the oxidation reactions can be interface controlled (linear) during initial stages and diffusion limited (parabolic) after extended oxidation [10]. However, considering the scatter of the experimental data in the oxide-scale measurements and based on the results of 0.5% and 0.35% TMF tests, parabolic kinetics appears to be a better approximation to describe the overall oxidation kinetics.

The change in surface oxide-scale thickness with time was therefore fitted by a simple parabolic time dependent function as follows:

$$x = k_p \sqrt{t} \qquad (1)$$

where x is the oxide-scale thickness and k_p is the parabolic rate constant.

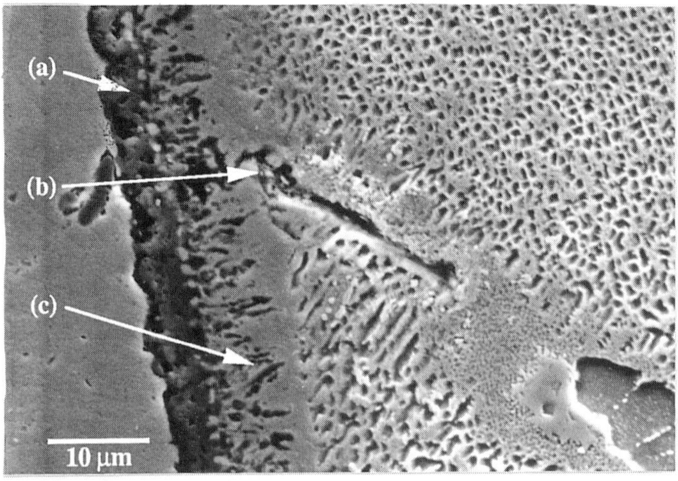

Figure 1. Electron micrograph of a TMF cycled specimen at a mechanical strain range $\Delta\varepsilon_m = 0.5\%$ (10 min. hold time) showing: (a) the surface oxide scale, (b) oxide penetration, and (c) γ'-depleted zone including a discontinuous oxide layer.

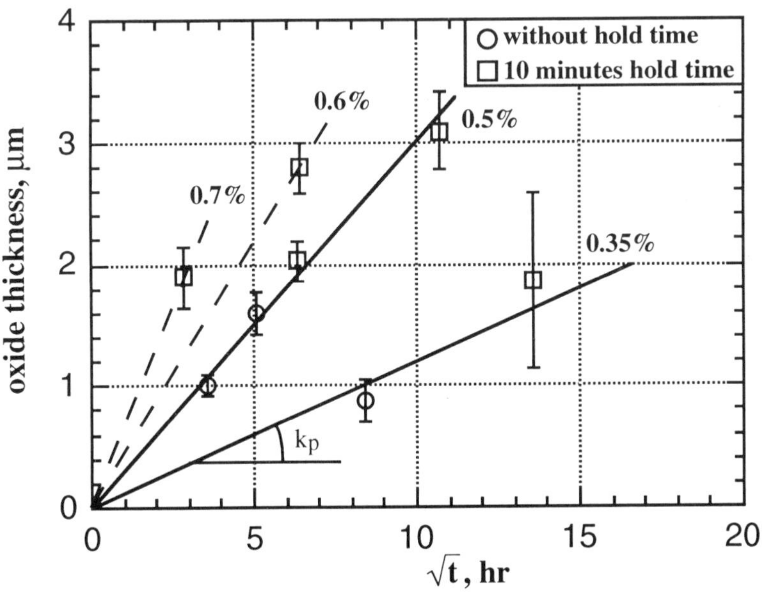

Figure 2. Oxide thickness as a function of the square root of the total cycling time (including the hold time).

Fig. 3 shows the variation of k_p as a function of the mechanical strain range. It is seen that k_p increases by increasing the mechanical strain range, $\Delta\varepsilon_m$, according to a power-law relationship:

$$k_p = A\ (\Delta\varepsilon_m)^n \tag{2}$$

where A and n are the constant coefficient and exponent, respectively.

Therefore, the governing equation for the mechanical strain enhancement is :

$$x = A\ (\Delta\varepsilon_m)^n\ \sqrt{t} \tag{3}$$

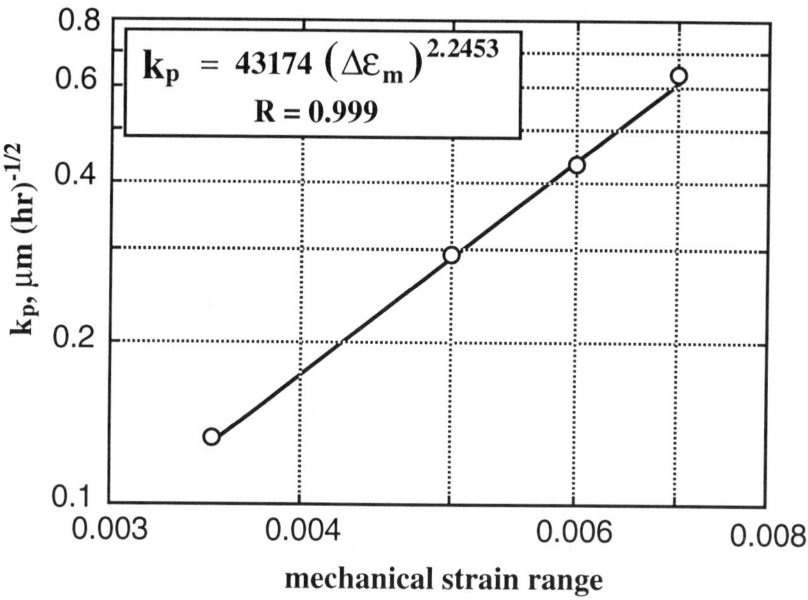

Figure 3. The parabolic rate constant, k_p, as a function of the mechanical strain range (Eq. 2).

The oxidation process under out-of-phase TMF loading is a complex process. The alloy is first subjected to a compressive stress in heating to the maximum temperature followed by a tensile loading in cooling. During compression at high temperatures, a complex interaction of fatigue-creep-oxidation weakens the mechanical properties of the alloy (in particular in the sub-surface zone) which will be later subjected to a tensile loading at lower temperatures, when the oxide is less ductile.

The simple relationship in Eq. (3) describes the synergistic effects of time and mechanical strain on the oxidation process of IN738LC under TMF loading. However, this equation can only be considered as a first approximation to describe the surface oxidation kinetics. It should be noted that it predicts no oxidation ($k_p = 0$) when the alloy is subjected to the same thermal cycling under no strain ($\Delta\varepsilon_m = 0$). This is obviously not possible, since an oxidation rate constant, k_p^o, must exist, to describe the oxidation kinetics when the specimen is cycled under no strain. This k_p^o should be valid up to a minimum mechanical strain range, $\Delta\varepsilon_m^o$ which corresponds to the lower limit of validity of Eq. (2), as illustrated in Fig. 4.

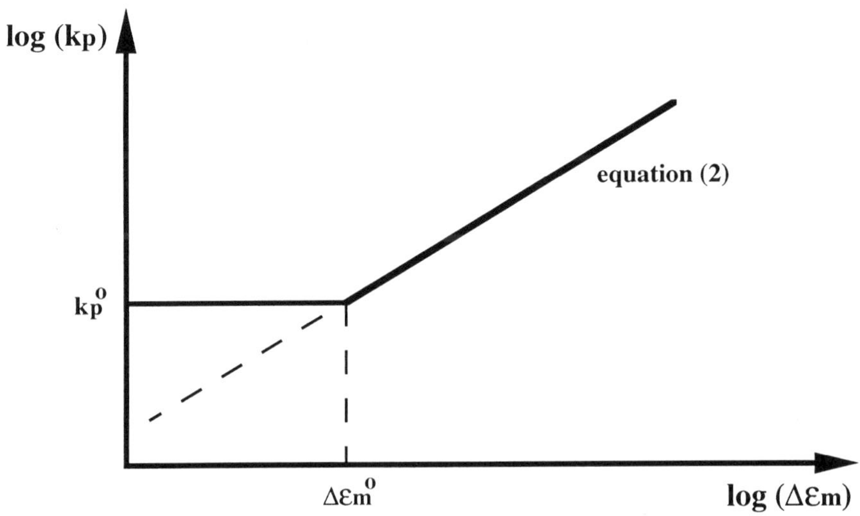

Figure 4. Schematic representation of k_p as a function of $\Delta\varepsilon_m$ in OP-TMF tests.

4. Summary

This study of the effect of out-of-phase TMF loading on the surface oxidation of IN738LC superalloy has shown that:

(1) The kinetics of oxidation can be approximated by a parabolic law.
(2) Mechanical strain enhances the surface oxidation process. The parabolic rate constant increases with the mechanical strain according to a power-law relationship.
(3) While a ten minute hold time at constant strain during TMF cycling contributes to the increase of the surface oxide layer thickness, it does not change the general kinetics law which expresses the effect of the cyclic strain range.

References

1. Nicoll, A.R. and Hilderbrandt, U.V., Thermal Fatigue of Coated Superalloys, in *Environmental Degradation of High Temperature Materials*, Spring Residential Conference, series 3, N13, **2**, The Institute of Metallurgists, March 1980, p. 6.
2. Wood, J.H., and Goldman, E.H., Protective Coatings, in C. T. Sims, N. S. Stoloff and W. C. Hagel (eds.) *Superalloys II*, John Wiley & Sons Inc., 1987, p. 359.
3. Rhys-Jones, T.N., Surface Stability of Materials Used in Aeroengine Applications Surface Stability, in T.N. Rhys-Jones (ed.), *Oxidation-Corrosion-Erosion-Wear*, The Institute of Metals, London, 1989, p. 1.
4. Skelton, R.P. and Bucklow, J.I., Cyclic oxidation and crack growth during high strain fatigue of low alloy steel, *Metal Science*, **12**, 1978, 64–70.
5. Reger, M.and Rémy, L., Fatigue oxidation interaction in IN 100 superalloy, *Metallurgical Transactions*, **19A**, 1988, 2259–2268.
6. Wright, P.K., Oxidation-Fatigue Interactions in a Single-Crystal Superalloy, in *Low Cycle Fatigue*, H.D. Solomon, G.R. Halford, L.R. Kaisand and B.N. Leis (eds.), ASTM STP 942, 1988, pp. 558–575.
7. Rémy, L., Oxidation effects in high temperature creep and fatigue on engineering alloys, in *Corrosion-Deformatoin Interactions*, T. Magnin and J.M. Gras (eds.), Fontainebleau, France, 1992, pp. 425–459.
8. Engler-Pinto Jr., C.C., Härkegård, G., Ilschner, B., Nazmy, M.Y., Noseda, C. and Rézaï-Aria, F., Thermo-Mechanical Fatigue Behavior of IN738LC, *Materials for Advanced Power Engineering*, 1994, D. Coutsouradis et al. (eds.), Part I, 1994, pp. 853–862.
9. Esmaeili, S., Engler-Pinto Jr., C.C., Ilschner, B. and Rézaï-Aria, F., Interaction Between Oxidation and Thermo-Mechanical Fatigue in IN738LC Superalloy — *I, Scripta Metall. Mater.*, **32**, 1995, pp. 1777–1781.
10. Kofstad, P., *High Temperature Corrosion*, Elsevier Applied Science, London , 1988.

PHENOMENOLOGICAL DAMAGE PARAMETERS FOR TMF LIFE PREDICTION OF DS CM 247 LC

M. BAYERLEIN, W. HARTNAGEL and CH. SOMMER
ABB Corporate Research
P.O. Box 10 13 32,
D-69115 Heidelberg, Germany

1. Introduction

Raising the efficiency of gas turbine engines is a main objective of gas turbine manufacturers. A preferred approach is to increase the operating temperature of the engine. Therefore blading materials with higher temperature capability are required. This led from conventionally cast (CC) equiaxed grain components to the introduction of directionally solidified (DS) columnar grain superalloy blades, e.g. in DS CM 247 LC, and single crystal (SC) components. Other approaches to achieve a higher operating temperature are improvement of blade cooling and optimization of design and life prediction methodologies.

Blading materials suffer from thermal-mechanical fatigue (TMF), creep, oxidation and hot corrosion. Most of these degradation mechanisms are more or less well understood and under control. With regard to TMF there is still a number of open questions concerning mechanical behavior and the relevant deformation and failure mechanisms. This holds also for the comparison of life data obtained under isothermal low-cycle-fatigue (LCF) and TMF loading. Since isothermal LCF tests can be performed much easier and cheaper than TMF tests, there is an interest to assess TMF life on the basis of isothermal LCF data. In the literature it is often proposed to use LCF life data obtained at the maximum temperature of the TMF cycle, e.g. [1].

In order to fully exploit the high temperature mechanical capability of the blading materials, advanced damage models which are capable of predicting TMF lives much more accurately than today are required. In case of anisotropic materials the orientation dependence has to be considered.

Within several independent reviews the most promising high temperature fatigue life prediction methods have been covered [2-5]. With regard to description of TMF behavior, best results were obtained with the models successfully applied for isothermal LCF situations. Three broad classes of life models can be distinguished in general [6]:

J. Bressers and L. Rémy (eds.), Fatique under Thermal and Mechanical Loading, 109–118.

- phenomenological models (related to macroscopic life and cycle parameters),
- cumulative damage models (related to damage explicitely caused by creep and plasticity),
- crack growth models (related to local inelastic strain at a void or a crack tip).

The phenomenological models have the advantage of simplicity and a rather direct relationship to data bases. However they are not very amenable to accounting for significant interaction effects when different damage mechanisms (e.g. cyclic creep, fatigue) operate either simultaneously or sequentially. The most important phenomenological models are: Maximum Tensile Stress, Coffin-Manson, Ostergren, Frequency Modified Ostergren, Smith-Watson-Topper (SWT), etc. The current paper is confined to the phenomenological damage models.

The evaluations of phenomenological damage models reported in literature are mainly related to CC materials, especially IN 738 LC, cf. [7-10]. The few publications concerned with anisotropic blading materials refer to coated systems, e.g. [4, 11, 12]. With regard to IN 738 LC, Meersmann et al. [9] claim inelastic strain range, total strain range, maximum stress range and the related energy products to be the most important parameters governing out-of phase (OP), in-phase (IP), clockwise diamond (CWD) and counter-clockwise diamond (CCD) TMF life (450-950°C). Engler-Pinto et al. [10] investigated OP TMF (400-900°C) on IN 738 LC. They found Coffin-Manson and Smith-Watson-Topper to correlate the TMF lives well. However the statements of the two papers are generally based on 3 or 4 data points only and are not supported by statistical evaluations.

Concerning TMF life prediction on the basis of isothermal LCF data, Kuwabara *et al.* [7] reported that fairly good results were achieved on a total strain basis for IN 738 LC, IN 939 and Mar-M 247 for IP and OP TMF loading (300-900°C) using LCF data obtained for the maximum temperature of the TMF cycle. The authors found inelastic strain not to be suitable in most cases. Embley and Russell [8] investigated IN 738 LC and GTD-111 under OP, CWD and CCD TMF loading conditions (427-821°C). Good predictions were obtained for the OP cycle for both materials on equal total strain basis using LCF data measured at the maximum temperature of the TMF cycle. However the data base was rather limited. Inelastic strain did not provide good correlations since inelastic strain values measured under these loading conditions are usually very small. The authors claimed maximum stress and maximum temperature to be the most important parameters controlling TMF life.

The OP TMF experiments of the present work (400-1000°C, $R_\varepsilon=-\infty$) simulate the TMF loading of a volume element of a land-based gas turbine blade close to the surface of the leading edge. The objective is to determine suitable phenomenological damage parameters for the three relevant directions (longitudinal, transverse and diagonal) of DS CM 247 LC on the basis of these data and the macroscopic physical quantities controlling TMF life. Additionally, the possibility of adequately predicting TMF life from isothermal LCF life data obtained at the maximum temperature of the TMF cycle is discussed. Therefore, TMF and LCF life data are correlated with microstructural results obtained by TEM investigations.

2. Experimental Details

2.1. MATERIAL

The composition of DS CM 247 LC in weight % is Ni: bal., Cr: 8.1, Co: 9.2, Mo: 0.5, W: 9.5, Ta: 3.2, Ti: 0.7, Al: 5.6, Zr: 0.01, B: 0.01, C: 0.07, Hf: 1.4. The material was used in the fully heat-treated condition. The microstructure consisted mostly of cuboidal γ' particles embedded in the γ matrix. The volume fraction of γ' was determined to be about 70%. The alloy was provided in form of slabs with the longitudinal direction being parallel to [001]. The maximum angular deviation from [001] was limited to 15°.

2.2. TESTING

Threaded cylindrical specimens with a diameter of 10 mm and a gauge length of 30 mm were machined out of slabs in longitudinal (0°), transverse (90°) and diagonal (45°) direction. Isothermal LCF tests and TMF tests were run on closed-loop servohydraulic testing machines which were equiped with 200 kHz induction heating facilities. Temperature was measured with a thermocouple which had been locally rolled to facilitate fixing around the circumference in the middle of the specimen. For strain measurement a strain gauge system with 15 mm distance between the ceramic tips was applied. The specimens were subjected to controlled temperature-strain-time cycles using commercial software.

Isothermal LCF tests were performed according to ASTM E 606. Testing temperatures were 500°C, 850°C, 950°C and 1000°C. All tests were conducted under total strain control with a constant strain rate of 6%/min. Additional tests with a strain rate of ~0.2%/min, 1%/min and 24%/min, respectively, were carried out to consider the effect of strain rate on mechanical behavior and life.

The TMF tests were performed with temperature cycles between 400°C and 1000°C, a heating/cooling rate of 3.8 K/s, $R_\varepsilon = -\infty$ and a frequency of $3.2 \cdot 10^{-3}$ s^{-1}. Cooling was achieved by heat flux through the grips. No forced cooling was applied in order to achieve a low dynamic axial temperature gradient (maximum 1 K/mm in the gauge length covered by the extensometer). All tests were carried out for various mechanical strain ranges with temperature and mechanical strain being out-of-phase. The number of cycles to crack initiation $N_i(2\%)$ was determined at a decrease of the tensile stress by 2%, the number of cycles to failure at a decrease by 50%.

2.3. TEM INVESTIGATIONS

Thin sections of LCF and TMF specimens were investigated by transmission electron microscopy (TEM) with regard to deformation-induced dislocation structures.

3. Results and Discussion

3.1. EVALUATION OF PHENOMENOLOGICAL DAMAGE PARAMETERS WITH REGARD TO TMF LIFE

Fig. 1 presents the stress-temperature curve of the first three cycles and the stabilized loop. Due to stress relaxation at high temperatures a positive mean stress develops. The magnitude of the mean stress shift depends on the mechanical strain range. The TMF lives for longitudinal, transverse and diagonal directions are plotted in a normalized $\Delta\varepsilon_t$ - N_i (2%) - representation in Fig. 2. For a given total strain range, N_i (2%) is highest for the longitudinal direction followed by the transverse and the diagonal direction. The corresponding stress ranges at $N = N_i(2\%)/2$ are shown in a

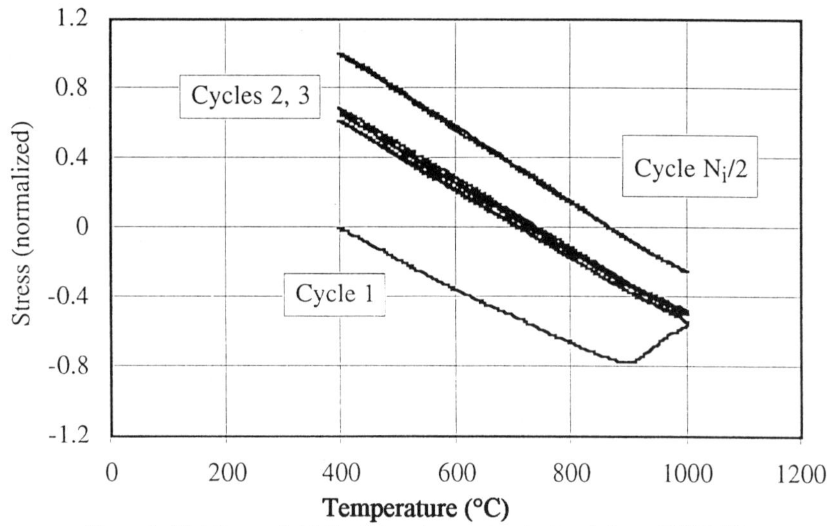

Figure 1. First three and stabilized stress-temperature hysteresis loops (OP TMF).

normalized $\Delta\sigma$–N_i (2%) - representation in Fig. 3. It is apparent that the data points for longitudinal and diagonal directions fall on a common line in the logarithmic representation. On an equal stress range basis, the lives for the transverse direction are shorter by a factor of about 2.5 compared to longitudinal and diagonal direction. Since plastic strain ranges were found to be very small (less than 0.05%), the stress ranges are mainly governed by the magnitude of Young's modulus for a given total mechanical strain range. Young's modulus increases from longitudinal via transverse to the diagonal direction. From the good correlation in Fig. 3, it can be concluded that stress range has a strong effect on life under the given TMF testing conditions. However, especially with regard to the transverse direction there have to be more physical parameters influencing TMF life.

In order to determine these quantities, the various phenomenological damage parameters given below were evaluated with regard to their correlative capability concerning the TMF lives for the three directions. For all parameters linear regressions were performed and correlation coefficients and standard deviations were calculated.

Coffin-Manson	$N_i\,(2\%) = C_o \cdot \Delta\varepsilon_{in}^{C_1}$	(1)
Total strain	$N_i\,(2\%) = C_o \cdot \Delta\varepsilon_t^{C_1}$	(2)
Ostergren	$N_i\,(2\%) = C_o \cdot (\Delta\varepsilon_{in} \cdot \sigma_{max})^{C_1}$	(3)
WT	$N_i\,(2\%) = C_o \cdot (\Delta\varepsilon_t \cdot \sigma_{max})^{C_1}$	(4)
$\Delta\varepsilon_{in}$:	inelastic mechanical strain range	
$\Delta\varepsilon_t$:	total mechanical strain range	
σ_{max}:	maximum tensile stress	
C_o, C_1:	constants	

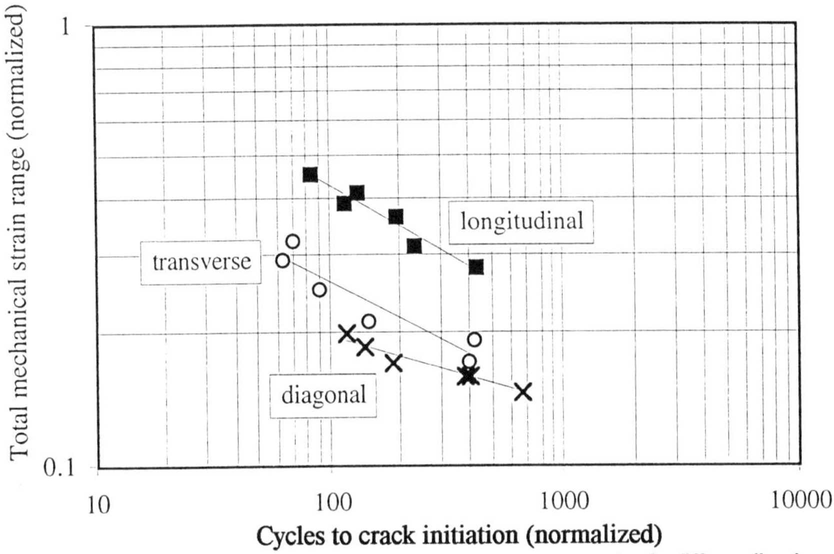

Figure 2. TMF life data in normalized $\Delta\varepsilon_t$ - N_i (2%) representation for different directions.

The evaluation performed with the TMF data for the longitudinal direction revealed unattractive results for the parameters including inelastic mechanical strain range (Inelastic strain, Ostergren). Since inelastic strains are very small for superalloys under the present loading conditions, these data are affected by a disproportionally large error. Consequently these parameters were not considered for the evaluation of transverse and diagonal TMF data.

The correlation coefficients and standard deviations evaluated for the different directions are given in Table 1. Optimum correlation coefficients (>0.98) and standard deviations (<0.04) were found for the SWT parameter for both longitudinal and diagonal direction. Correlation and standard deviation determined for the Total strain parameter are worse but acceptable. For the transverse direction, correlation was in general found to be worse compared to the other two directions, however, still satisfactory. Best results were again obtained for the SWT parameter (corr. coeff: 0.90, stand. dev: 0.13). The SWT parameter was also reported by Engler-Pinto *et al.* [10] to correlate OP TMF lives (400-900°C) well for CC IN 738 LC.

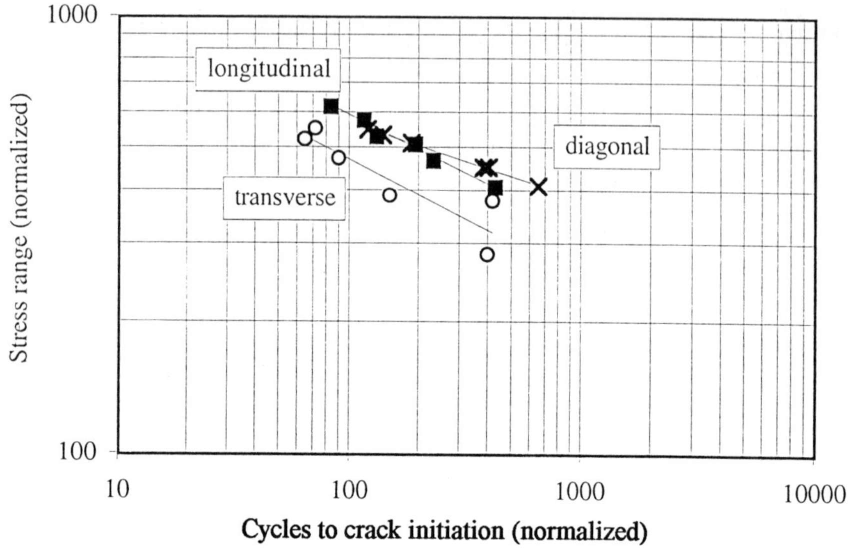

Figure 3. TMF life data in normalized $\Delta\sigma$ - N_i (2%) representation for different directions.

TABLE 1. Correlation coefficients and standard deviations for the various phenomenological damage parameters (TMF data separately fitted for each direction).

TMF Cycle: 400 - 1000°C, OP	corr.coeff. r^2	stand. dev. log N_i(2%)
longitudinal		
Coffin-Manson	0.748	0.094
Total strain	0.945	0.066
Ostergren	0.786	0.087
Smith-Watson-Topper	0.981	0.039
transverse		
Total strain	0.879	0.141
Smith-Watson-Topper	0.896	0.131
diagonal		
Total strain	0.948	0.075
Smith-Watson-Topper	0.987	0.037

The results indicate that the life of DS CM 247 LC investigated under the applied OP TMF loading conditions is governed by the quantities total strain range, stress range and maximum tensile stress.

3.2. COMPARISON OF ISOTHERMAL LCF AND OP TMF LIFE DATA

For the comparison of TMF and LCF life data, the total mechanical strain range - N_i (2%) representation was selected. Total mechanical strain range was proven above to be a suitable parameter to correlate TMF lives. Due to different testing frequencies (factor 7 to 22) TMF and LCF life data can not be directly compared. In order to enable a comparison on an equal frequency basis, the effect of strain rate (frequency) on LCF life was investigated for various temperatures and a medium reference strain range. LCF lives were found to be shorter for the tests performed with the lower TMF testing frequency. The ratio of life reduction was determined for the reference strain range at 850, 950 and 1000°C. In order to put LCF tests on an equal frequency basis compared to TMF tests all LCF life data were corrected using this factor.

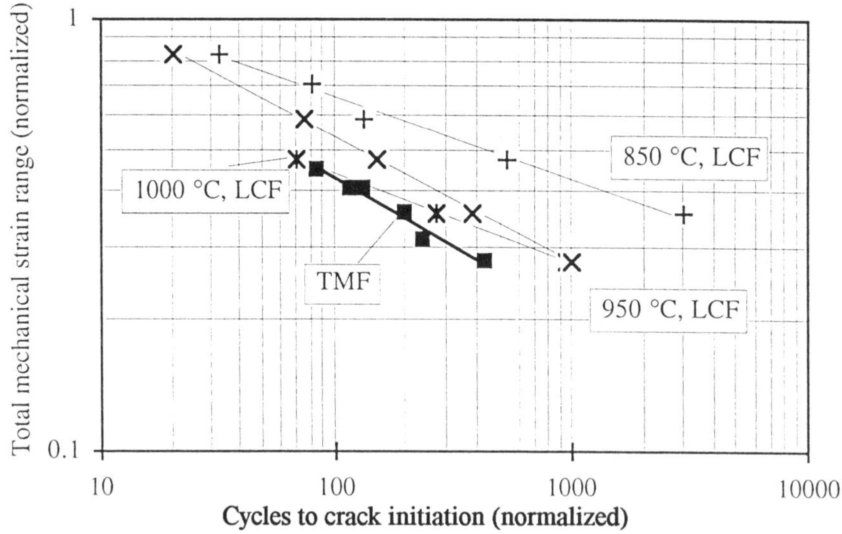

Figure 4. TMF and LCF life data in normalized $\Delta\varepsilon_t$ - N_i (2%) repesentation. Different testing frequencies for TMF and LCF data.

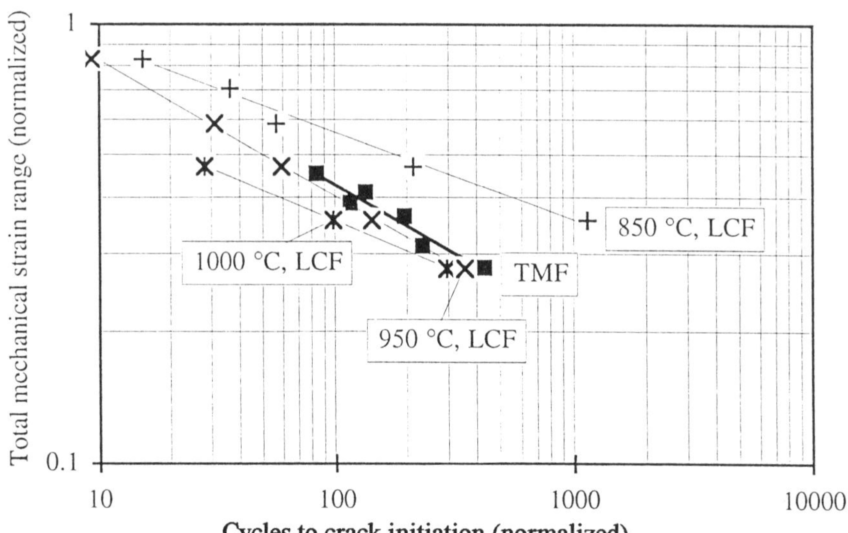

Figure 5. TMF and LCF life data in normalized $\Delta\varepsilon_t$ - N_i (2%) repesentation. LCF lives corrected to frequency of TMF tests.

Subsequently, the variation of testing frequency among the LCF tests due to different strain ranges (all tests performed with constant strain rate) was considered. Differences in frequency with regard to the reference strain range tests were calculated for all LCF tests and the effect on life was assessed and considered using LCF life data measured in dependence on frequency for the reference strain range. It was assumed that the effect of frequency on life is approximately equal for all the strain ranges considered in this investigation.

The effect of frequency on life becomes apparent by comparison of LCF and TMF life data exemplarily given for the longitudinal direction in Figs. 4 and 5. For the LCF life curves in Fig. 5, a correction of lives to the lower TMF testing frequency was performed, which shifts the LCF curves to lower lives (factor 2.5). The comparison of TMF and LCF data on an equal frequency basis shows that TMF lives can be conservatively predicted by the 950°C LCF curve. This holds also for the transverse direction. In the literature it is often proposed to use LCF life data obtained at the maximum temperature of the TMF cycle for TMF life prediction. Here, this would be 1000°C and lead to an even more conservative life prediction.

Conservative life predictions with LCF data obtained at the maximum temperature of an OP TMF cycle were also reported for superalloys with equiaxed grain structure by Kuwabara *et al.* [7] for IN 738 LC (300-900°C) and Embley and Russell [8] for IN 738 LC and GTD-111 (427-871°C).

3.3. MICROSTRUCTURAL INVESTIGATIONS

Fig. 6 presents the dislocation structures in cyclic saturation that develop under LCF loading at 500 and 1000°C and under 400-1000°C OP TMF loading. For the LCF deformation at 500°C a high dislocation density in the matrix channels and pronounced cutting of γ' was observed. The specimens cycled at 1000°C showed a relatively low dislocation density in the matrix. Dislocations were mainly located in the γ-γ'-interface. No traces which indicate cutting of γ' precipitations were found. In the TMF loaded specimens cutting of the γ' precipitations as well as a beginning formation of a dislocation network in the γ-γ'-interface was observed. An initial state of directional coarsening (rafting) of the γ' precipitations parallel to the stress axis was found in the TMF loaded specimens. The directional coarsening is a consequence of asymmetrical loading in the OP TMF cycle which causes compressive stresses to occur at the high temperatures. The orientation parallel to the stress axis under compression stresses is due to the negative γ-γ'-misfit of DS CM 247 LC. The specimens cycled under symmetrical LCF loading with nearly equal stresses in tension and compression did not show any rafting.

The results of the TEM investigation show significant differences between the microstructures formed under LCF and TMF loading. The coincidence of the TMF life data with the high temperature LCF life data found in the present investigation is thus obviously fortuitous, since it can not be correlated with the microstructural findings.

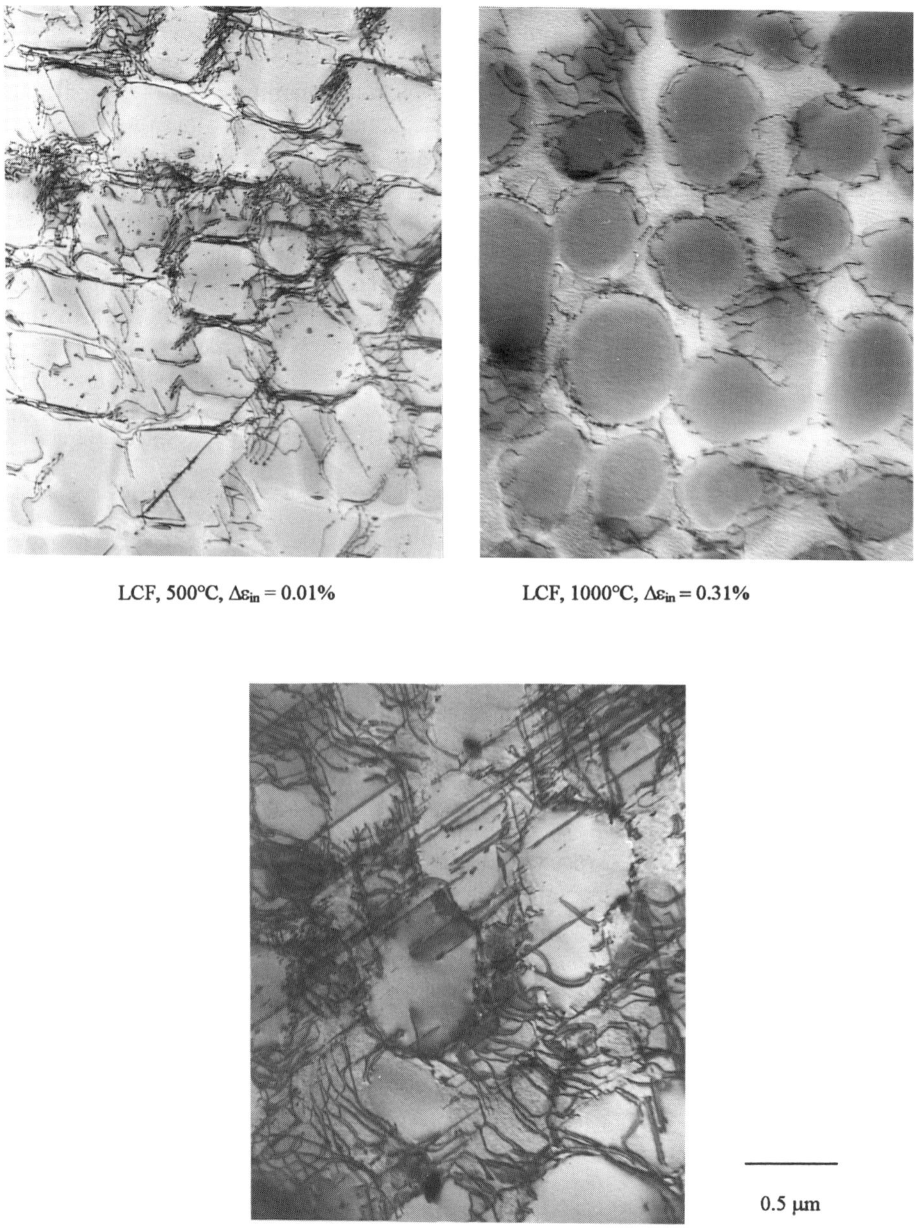

LCF, 500°C, $\Delta\varepsilon_{in} = 0.01\%$

LCF, 1000°C, $\Delta\varepsilon_{in} = 0.31\%$

0.5 μm

TMF, 400-1000°C, OP, $\Delta\varepsilon_{in} = 0.01\%$

Figure 6. Comparison of LCF and TMF dislocation structures of DS CM 247 LC at cyclic saturation.

4. Conclusions

1) TMF life data under the present OP TMF loading conditions can be well correlated with both SWT parameter and Total strain parameter for the various directions of DS CM 247 LC. Parameters related to inelastic strain did not show good correlations.
2) TMF lives for the applied OP TMF loading conditions can be predicted conservatively by isothermal LCF life data gained at the maximum temperature of the TMF cycle on equal testing frequency basis. This could be proven for the longitudinal and the transverse direction.
3) The dislocation structure of OP TMF loaded specimens showed features which are typical for both low and high temperature LCF deformation mechanisms.
4) The coincidence between the LCF life data obtained at the maximum temperature of the TMF cycle and the OP TMF life data can not be correlated with the microstructural findings. Thus the coincidence has to be regarded as fortuitous.

References

1. Viswanathan R., *Damage Mechanisms and Life Assessment of High Temperature Components*, ASM International, Metals Park, Ohio, 1989.
2. Lloyd G.J., Wareing J., Life-Prediction for Combined Creep Fatigue Endurance, *Metals Technology* (1981), 297-305.
3. Priest R.H., Ellison E.G., An Assessment of Life Analysis Techniques for Fatigue Creep Situations, *Res. Mechanica* 4 (1981), 127-150.
4. Swanson G.A., Linask I., Nissley D.M., Norris P.P., Meyer T.G. and Walker K.P., *Life Prediction and Constitutive Models for Engine Hot Section Anisotropic Materials Program*, NASA CR-174952, 1986, pp. C-1 - C-29.
5. Fischmeister H.F., Danzer R., Buchmayr B., Life Time Prediction Models, in W. Betz et al. (eds.), *High Temperature Alloys for Gas Turbines and Other Applications 1986*, D. Reidel Publishing Company, Dordrecht, 1986, pp. 495-549.
6. Swanson G.A., Linask I., Nissley D.M., Norris P.P., Meyer T.G. and Walker K.P., *Life Prediction and Constitutive Models for Engine Hot Section Anisotropic Materials Program, NASA CR-179594,* 1987, p. 18.
7. Kuwabara K., Nitta A. and Kitamura T., Thermal - Mechanical Fatigue Life Prediction in High - Temperature Component Materials for Power Plant, in Conference on *Advances in Life Prediction Methods*, Albany, N.Y., ASME, New York, 1983, pp. 131-141.
8. Embley G.T. and Russell E.S., Thermal-Mechanical Fatigue of Gas Turbine Bucket Alloys, in *First Parsons International Turbine Conference*, Parsons Press, Trinity College, Dublin, 1984, pp. 157-164.
9. Meersmann J., Ziebs J. and Kühn H.-J., The Thermo-Mechanical Behavior of IN 738 LC, in D. Coutsouradis et al. (eds.), *Materials for Advanced Power Engineering 1994*, Kluwer Academic Publishers, Dordrecht, 1994, pp. 841-852.
10. Engler-Pinto C.C Jr., Härkegård G., Ilschner B., Nazmy M.Y., Noseda C. and Rézai-Aria F., Thermo-Mechanical Fatigue Behavior of IN 738 LC, in D. Coutsouradis et al. (eds.), *Materials for Advanced Power Engineering 1994*, Kluwer Academic Publishers, Dordrecht, 1994, pp. 853-862.
11. Bill R.C., *Micromechanisms of Thermomechanical Fatigue - A Comparison with Isothermal Fatigue*, NASA Tech. Memo, NASA TM - 87331, 1986.
12. Guedou J.-Y. and Honnorat Y., Thermomechanical Fatigue of Turbo-Engine Blade Superalloys, in H. Sehitoglu (ed.), *Thermomechanical Fatigue Behavior of Materials*, ASTM STP 1186, Philadelphia, 1993, pp.157-175.

TEMPERATURE DEPENDENCE OF THE INTRINSIC SMALL FATIGUE CRACK GROWTH BEHAVIOUR IN NI-BASE SUPERALLOYS BASED ON MEASUREMENT OF CRACK CLOSURE

M. OKAZAKI[1], H. YAMADA[1] and S. NOHMI[2]
[1] *Department of Mechanical Engineering, Nagaoka University of Technology, Tomioka 1603-1, Nagaoka, Japan 940-21*
[2] *Kawasaki Heavy Industries Co. Ltd., Kawasaki-cho 1-1, Akashi, Japan 673*

1. Introduction

Ni-base superalloys have been used with high reliability for blades and vane applications in gas turbines for the last decades [1-13]. Recently the application of directionally-solidified and single crystal alloys has become common in advanced gas turbines. On these materials, several kinds of studies, such as on the effects of temperature [1-5], loading frequency [1,4,6], stress multiaxiality [9], environment [3], microstructures [13], and crystallographic orientation [8] on the crack growth during fatigue, creep and creep-fatigue, have been performed. On the other hand, experimental evidence has also shown that small crack growth behaviour may not conform to that measured conventionally with physically long cracks [10-19]. However, most of the foregoing studies have dealt with the propagation of physically long cracks by using the compact tension specimen as an example, and there is little quantitative information on the process of small crack initiation and the subsequent propagation in Ni-base superalloys, which generally dominates the major part of the fatigue life [10-13]. Furthermore, the crack closure phenomenon [20], which plays an important role in fatigue crack propagation, has not been taken into account even in the foregoing few studies at all, due to the difficulty of the measurement at high temperatures.

In this work, the effect of temperature on the physically small fatigue crack growth behaviour of a directionally-solidified and a single crystal Ni-base superalloy was investigated at temperatures between 873 and 1223 K. The small crack growth behaviour thus obtained was compared with the physically long crack propagation behaviour. The factors which lead to the lack of similitude law in propagation rate between small and long cracks were also discussed, based on the measurement of crack closure levels and on the chemical analysis near the crack tip, utilizing the electron probe microanalyzer.

J. Bressers and L. Rémy (eds.), Fatique under Thermal and Mechanical Loading, 119–129.

2. Experimental Procedure

Two kinds of Ni-base superalloys were tested in this work: the first one was a single crystal alloy , CMSX-2, and the second one was a directionally-solidified alloy, denoted by CM247LC-DS. The microstructures of these materials consist of a γ matrix strengthened by cubic γ' precipitates. The size of the γ' precipitates centers around about 0.5 μm (approximately 60% by volume). The chemical compositions, the condition of heat treatments and the mechanical properties of these materials are given elsewhere [10-12]. From these materials, the solid cylindrical smooth specimen (7 mm in diameter) and the center notched (CN) specimen (2 mm in thickness and 14 mm in width) were machined so that the axis of all specimens was within 5° from the <001> crystallographic orientation. The plane portions in the CN specimen lay within 5° from a {100} crystallographic plane.

Fully-reversed load-controlled fatigue tests were conducted at a frequency of 5 Hz under a load ratio of -1 in air, by means of an electro-hydraulic fatigue testing machine. The specimen was heated by using induction heating. The effect of temperature on the propagation behaviour of small and long fatigue cracks was investigated at temperatures of 873, 1023, 1123 and 1223K. The physically long crack propagation tests were carried out using the CN specimen, according to the ASTM standard E647-88. The naturally initiated surface small fatigue crack growth behaviour, on the other hand, was investigated by periodically replicating the smooth specimen surface by means of acetyl cellulose film, after cooling off the specimens to room temperature. The small crack length investigated ranged from about 20μm to 500 μm, and the length was measured on the replica using an optical microscope. The small crack propagation tests were carried out by choosing the values of the applied stress amplitude, σ_a, so that the ratio of σ_a to the proof stress, σ_Y, was approximately constant at all test temperatures: σ_a was from 55% to 60% of σ_Y. The crack opening-closure behaviour of small and long cracks was also investigated at high temperatures by utilizing the originally-developed electro-optical displacement detector system, by which the crack opening displacement could be detected with a resolution of about 0.2 μm up to 1223 K. The details of the procedure are given in Ref. 12.

3. Results and Discussion

3.1. PHYSICALLY LONG CRACK GROWTH BEHAVIOUR

The relationship between the physically long crack propagation rate, da/dN, and the stress intensity factor range, ΔK, in CMSX-2 is given as a typical result in Fig. 1a, where the shaded areas depict the scatterbands of long crack growth rates in CM247LC-DS at 873 K reported elsewhere [11]. It can be seen that the resistance to crack propagation, as well as the fatigue threshold, increased with temperature in CMSX-2. It is also found that the crack growth rate in CMSX-2 is comparable to that of CM247LC-DS at a temperature of 873 K. The experimental results of crack closure for the physically long cracks, akin to the behaviour of small cracks described later, indicated that the crack opening ratio, $\Delta K_{eff}/\Delta K$(ΔK_{eff} being the effective stress intensity factor range), decreased with the increase of test temperature. SEM observation of the crack propagation path showed that the crack tip as well as the crack wake were covered with oxidation products at the test temperature of 1123 K (Fig. 2a). This suggests that the oxide-induced crack closure became significant, resulting in an apparent increase of fatigue threshold in Fig. 1a. On the other hand, the elastic modulus can also contribute to the temperature dependence in Fig. 1a, since it is noted from a comprehensive survey of published data that fatigue crack propagation rates vary in proportion to elastic modulus in many engineering materials [24].

Fig. 1b depicts the relationship between the physically long crack growth rates and the effective stress intensity factor range divided by the elastic modulus, $\Delta K_{eff}/E$, where E is the elastic modulus parallel to the loading axis. The shaded band represents the scatterband of the physically long crack growth rates obtained under different R ratios on CM247LC-DS at 873 K [11, 12]. It is found that the crack growth rates are represented by unique curves independent on test temperatures. This experimental result shows that the increase of crack propagation resistance with temperature shown in Fig. 1a mainly results from the temperature dependence of the crack closure and of the elastic constant.

3.2. NATURALLY-INITIATED SMALL CRACK GROWTH BEHAVIOUR

The rates of the naturally-initiated small fatigue crack growth were measured using

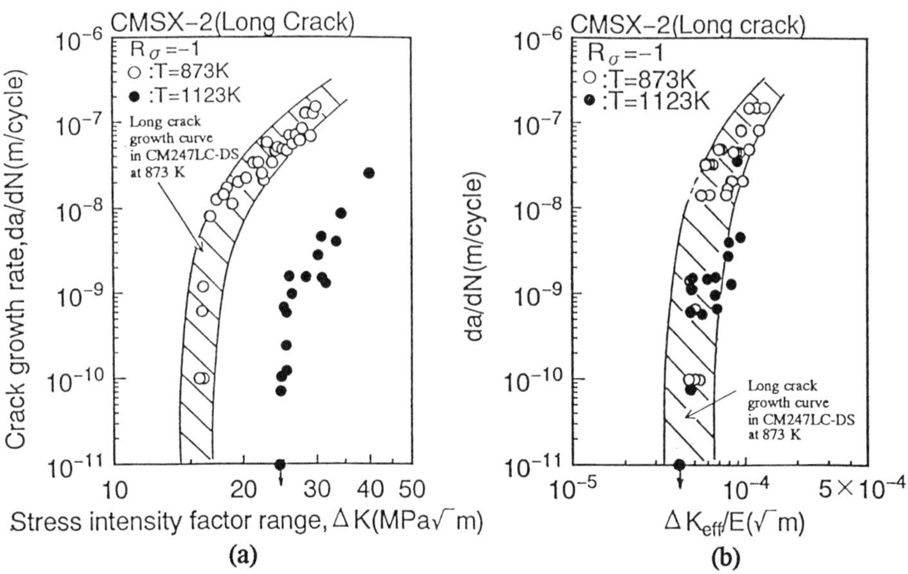

Figure 1. Long crack growth rates, da/dN, in CMSX-2 correlated with a) stress intensity factor range, ΔK (b) $\Delta K_{eff}/E$.

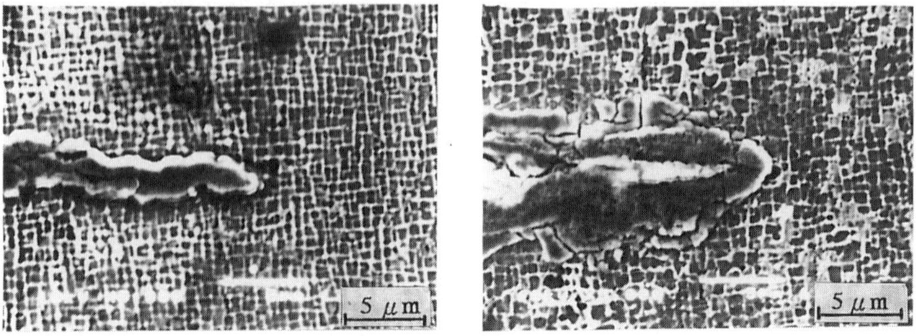

(a) (b)

Figure 2. Morphology of crack tip in CM247LC-DS at 1123 K.
(a) Long crack (ΔK= 25 MPa $\sqrt{m}$, ΔK_{eff} = 3.8 MPa $\sqrt{m}$).
(b) Small crack (ΔK= 12.6 MPa $\sqrt{m}$, ΔK_{eff} = 3.8 MPa $\sqrt{m}$).

smooth specimens. Some of the results are compared with those of the physically long cracks in Fig. 3, as a function of ΔK, where ΔK for the small cracks was evaluated from the Raju-Newman equation by approximating the semi-circular geometry of the surface crack [22]. The data points and the shaded bands in Fig. 3 show the crack growth rates measured for many small cracks in several smooth specimens, and the long crack growth curves at 873 K obtained in the previous section, respectively. It is found that the growth rates of the small cracks are considerably higher than those of long cracks, and the small cracks grow even at ΔK levels lower than the long crack threshold at all test temperatures. It is also found that the rates of small crack propagation increase with the increase of the test temperature. It is reiterated that the physically long cracks displayed decreasing growth rates with increasing test temperature, a behaviour completely contrary to that of the small cracks (Fig.1a). This experimental evidence illustrates the importance of the investigation of small fatigue crack growth rates themselves. The details of the acceleration phenomenon of the small cracks will be discussed in the next section. As given elsewhere [10,11], most of the small cracks initiated from microscopic pores located just near the specimen surface.

An example of the original P-δ curve (where P and δ are the applied load and the crack opening displacement at the center of the crack, respectively) monitored by the electro-optical equipment described in Sec. 2 [12], is presented in Fig.4. The presence of crack closure causes the non-linearity of the P-δ curve during the unloading process [20, 21]. In order to clearly identify the crack closure level, the P-δ curve was converted into a so-called subtracted P-δ curve by the computer in which the reduced displacement, defined as the difference between the actual displacement and a linear line fit through the linear portion of the upper unloading curve, was correlated with P [21]. The crack closure level was identified by magnifying the non-linearity of the subtracted P-δ curve. The subtracted P-δ curve is also given in Fig. 4. While some noise signals (which must result from the fluctuation of air, an unavoidable phenomenon in optical devices so far as used at high temperatures) are superposed on the original P-δ curve, the subtracted P-δ curve enables to identify the crack closure level. The arrows in Fig. 4 mark the crack closing level thus determined from the subtracted P-δ curves. It is also important to note that the crack closure level thus

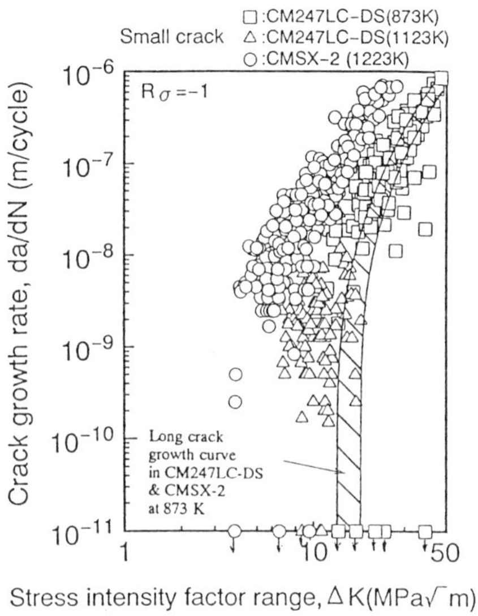

Figure 3. Some examples of small crack growth rates in terms of ΔK.

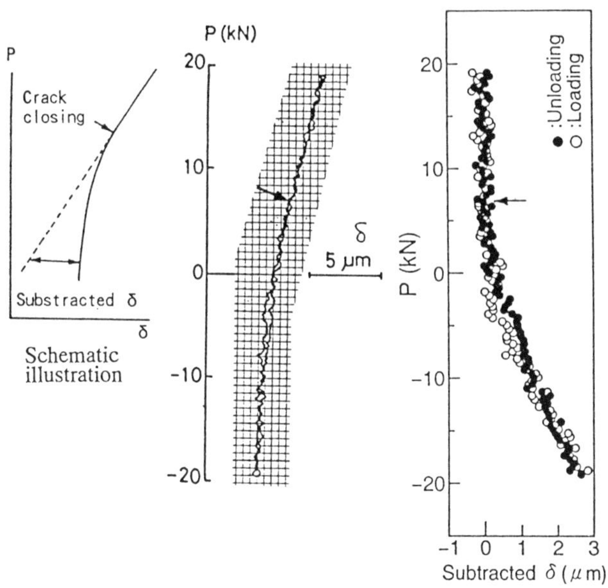

Figure 4. An examples of P-δ curve and the corresponding subtracted P-δ curve (the crack of 58 μm in half length, 1123 K, CM247LC-DS).

determined involves an error in the effective stress intensity factor range, ΔK_{eff}, due to the noise signal, which is estimated at about 20%. The relationship between the crack opening ratio, $\Delta K_{eff} / \Delta K$ and the small crack length is given in Fig. 5. Since the stress amplitude was kept constant during the small crack propagation tests in this work, the increase of crack length corresponds to an increase of ΔK. It can be concluded from Fig. 5 that the crack opening ratio decreases when the temperature is increased from 873 K to 1223 K. At a given test temperature, the ratio is almost independent on the small crack length.

Based on the results obtained, the small crack growth rates given in Fig. 3 are correlated with ΔK_{eff} and with $\Delta K_{eff}/E$ in Figs. 6a and 6b, respectively. It is found from Fig. 6 that the intrinsic small fatigue crack propagation resistance, in which the factor of crack closure is excluded, decreases with increasing temperature. Such a temperature dependence is still found even when the temperature dependence of the elastic modulus is taken into account (Fig. 6b). As mentioned earlier, the value of ΔK_{eff}, evaluated in this work possibly involves an error of about 20% because of the noise on the P-δ curves. However, the increase of small crack growth rate with temperature in Fig. 6 is significantly larger than the 20% error band. It should be noted that the above results are contrary to those of the physically long cracks shown in Fig. 1. Long crack propagation data therefore do not provide adequate information for the purpose of reliability considerations [10-19].

In order to qualify the temperature effect on small fatigue crack propagation, the crack propagation paths at the respective temperatures were observed by SEM. A typical result of CM247LC-DS at 1123 K is presented in Fig. 2b. It is found that the crack plane is extensively covered with oxidation products. SEM observation with high magnification reveals that there was a γ' depleted zone adjacent to the oxide layer near the fatigue cracks (Fig. 8), where the profile of the γ' precipitates is not clearly identified. In order to study the oxidation effect, the chemical analysis near the crack propagation plane was carried out by utilizing an electron probe microanalyzer

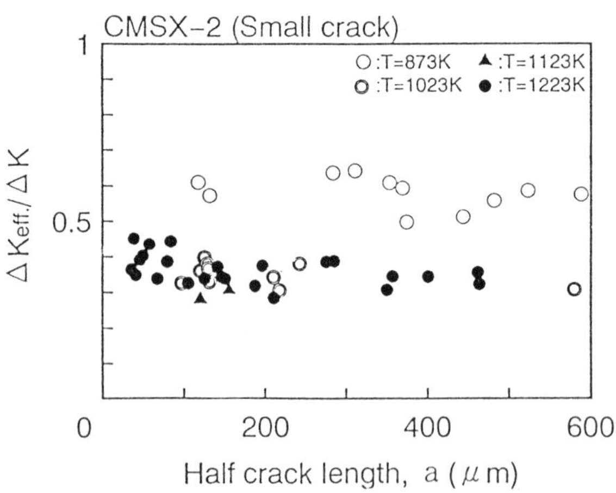

Figure 5. Change of crack opening ratio, $\Delta K_{eff}/\Delta K$, with the small crack length.

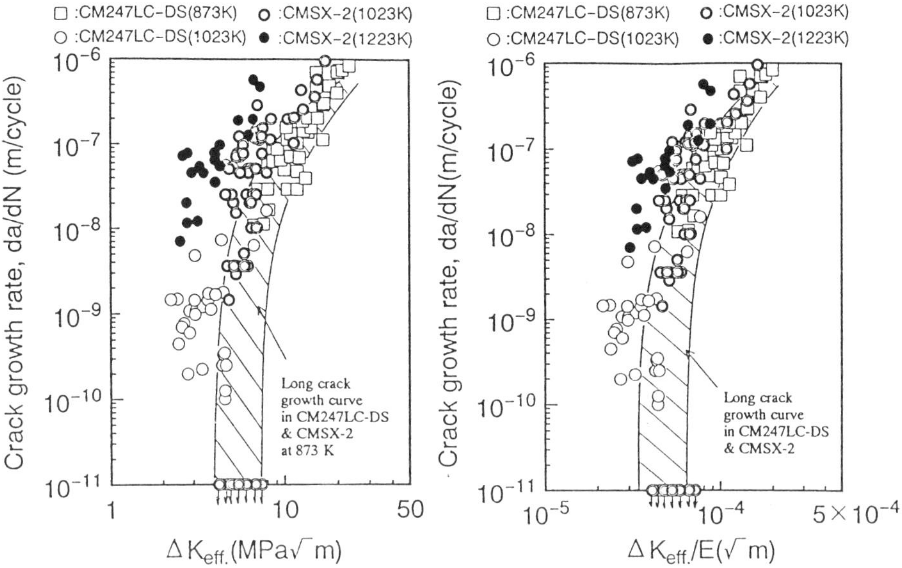

(a) (b)

Figure 6. Temperature dependence of small crack growth rates in terms of (a) ΔK_{eff}, (b) $\Delta K_{eff}/E$.

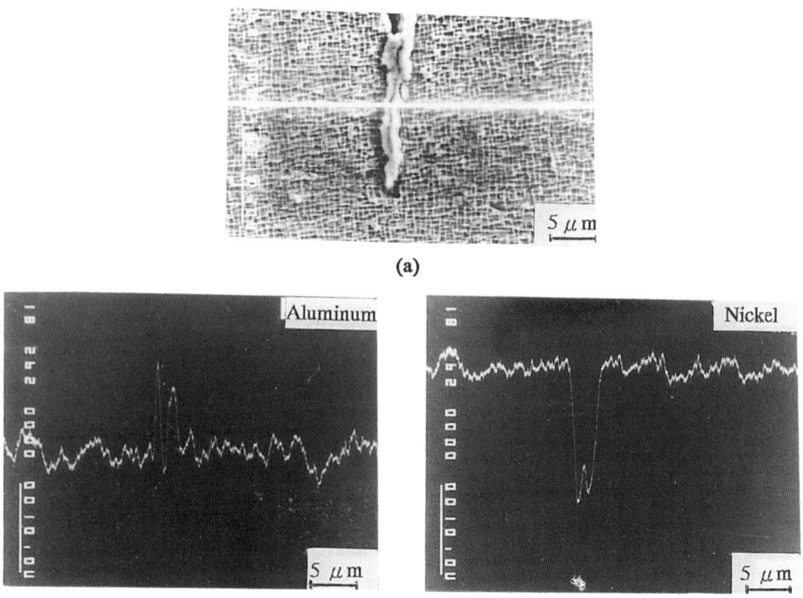

(a)

(b) (c)

Figure 7. EPMA analysis near small crack propagation plane in CM247LC-DS at 1123 K.
(a) SEM image (b) Line analysis of Al (c) Line analysis of Ni

(EMPA), and is presented in Fig. 7. It can be seen from this figure that around the crack propagation plane the content of Ni is depleted and that of Al becomes

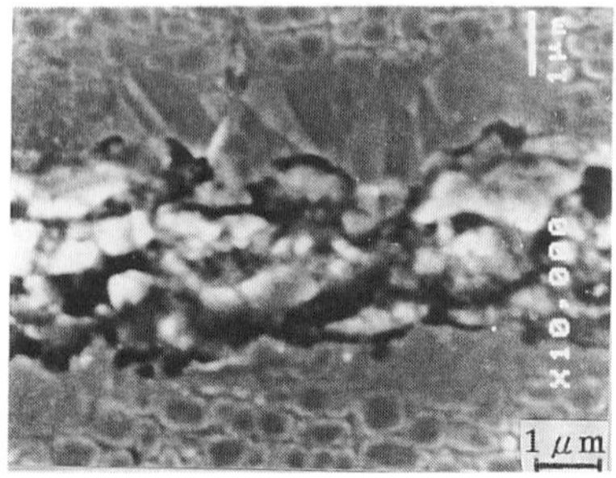

Figure 8. SEM observation near crack propagation plane in CMSX-2 at 1223 K.

enriched, respectively. This analysis confirms that the oxidation at the crack tip not only leads to the formation of oxides but also to the depletion of γ' precipitates [3, 23]. While the γ' depleted zone size was not measured quantitatively in this work, it was observed by SEM that the size increased with the test temperature. It is naturally postulated that the depletion of γ' precipitates may result in a decrease of the high temperature strength. From the foregoing experimental results it can be reasonably concluded that the local degradation of high temperature strength resulting from the depletion of γ' precipitates due to oxidation near the crack tip, as well as the temperature dependence of the crack closure level and of the elastic modulus, lead to the intrinsic temperature dependence of the small crack growth rates shown in Fig. 6.

The comparison of Fig. 2a and Fig. 2b shows that the γ' depleted zone size of the small crack is significantly larger than that of the long crack, although the reason is not clear. Note that both type of cracks were under a comparable ΔK_{eff} level (the level of ΔK in the former is lower than in the latter). This suggests that the degree of oxidation attack near the crack tip is different for small and long cracks, which results in the lack of similitude in crack growth rates between them. It is assumed that the difference of higher order terms, including the T-stress [24] in the crack tip stress field, between small and long cracks (even if under the same ΔK_{eff} level) would be related to the above difference of the γ' depleted zone size. It can also be pointed out that the small crack easily nucleated from the oxidation spike on the smooth specimen.

3.3. LACK OF SIMILITUDE LAW BETWEEN SMALL AND LONG CRACKS.

As shown in Fig. 3, the small cracks exhibit notably higher growth rates than the long cracks in CM247LC-DS and CMSX-2, especially at higher temperature, when the crack closure phenomenon is not taken into consideration. The following factors, which possibly induce such a lack of similitude between small and long cracks, can be pointed out [13-19]:

(i) Difference of crack closure level between them [12,14,18,19],

(ii) Difference of crack growth mechanism between them [12,18,19],
(iii) Difference of chemical influence between them [15,18, 19],
(iv) Application limitation of linear elastic fracture mechanics to small cracks [18,19],
(v) While the fatigue crack growth rate is generally influenced by microstructure and crystallographic orientation, the sensitiveness to these factors is different for small and long cracks [10,12,13,16].

Although it is generally very difficult to separately evaluate the effects of the above factors quantitatively, the dominant ones will be discussed hereafter.

The comparison between Figs. 3 and 6 shows that the difference of growth rates between small and long cracks on the basis of ΔK_{eff} is clearly smaller than that on the basis of ΔK under all test conditions. This indicates that the crack closure phenomenon more or less contributes to the lack of similitude between small and long cracks at all test temperatures.

Next, in order to exclude the factor of crack closure, let us consider the effect of other factors by means of Fig. 6. It is interesting to note that the growth rates of small cracks agree with those of the long cracks within the scatterband at 873 K, (see the square symbols). At this temperature, as reported elsewhere [11,12], both the small and long cracks mainly propagate in the γ matrix by Mode I fracture, and hence the propagation mechanism of both crack types can be considered to be quite similar; in other words, the contribution of factors of (ii) and (v) is negligible. Furthermore, factor (iii) would not be significant at 873K, because the depletion of γ' precipitates resulting from oxidation attack is scarcely found. Thus, it can be concluded that at 873 K where the effect of oxidation is not noticeable, the difference of crack closure level between the long and small cracks is mainly responsible for the lack of similitude between both crack types, and the role of other factors is negligible.

At higher temperature, on the other hand, the accelerated growth rate of small cracks compared with the long cracks becomes remarkable as the test temperature is raised, even when the role of crack closure is excluded (see other symbols than the squares in Fig. 6). This suggests that in addition to crack closure, some other factors must be active in enhancing the growth rate of small cracks at higher temperatures. The small crack problem at higher temperature can be reasonably understood as follows: provided that factor (iv) is dominant, the degree of acceleration of the small cracks compared with the long cracks would be independent on test temperature, because the ratio of the applied stress amplitude to the 0.2% proof stress was selected to be constant at all temperatures (Section 2). However, the experimental results show that the higher the test temperature is, the more notable is the acceleration of the small cracks. This suggests that the contribution of (iv) is not large. Furthermore, it is considered that the contribution of the factors (ii) and (v) is not plausible, because the small and long cracks propagated by Mode I type fracture (Fig. 2). As shown in the previous section, on the other hand, a rising of the test temperature increases the chemical effect (factor (iii)), as suggested by the depletion of γ' precipitates occurring around the crack tip. Additionally, this effect is more remarkable in small cracks than in long cracks, even if they grow under a comparable ΔK_{eff} level (compare Figs. 2a with 2b). From the above discussion, it is reasonably concluded that both crack closure and the environmental attack lead to the lack of similitude between long and small cracks at higher test temperature. However, further research is needed to quantitatively evaluate the environmental factor on small crack growth rates, as well as on the details of the crack propagation mechanism.

4. Conclusions

In this work, the effect of temperature on the propagation behaviours of physically long and small cracks was investigated at temperatures between 873 and 1123 K in a single crystal and a directionally-solidified Ni-base superalloy (denoted by CMSX-2 and CM247LC-DS, respectively). The factors which lead to the lack of similitude in propagation rates between small and long cracks were also discussed, based on the measurement of crack closure levels and the chemical analysis near the crack tip utilizing the electron probe microanalyzer.

The main conclusions are summarized as follows:

(1) The intrinsic small crack propagation resistance, taking crack closure into consideration, decreased with increasing temperature. SEM observation and EPMA chemical analysis revealed that the depletion of γ' precipitates around the crack tip resulting from chemical attack was closely related to the above temperature dependence of the intrinsic small crack growth rates. It was also observed that the γ' depletion in small cracks was more remarkable than that in long cracks, even when both cracks were grown under a comparable ΔK_{eff} level.

(2) In contrast to the small crack growth rates shown in (1), the physically long crack behaved in a simple manner as follows: the growth rates at all test temperatures could be represented approximately by an unique curve, by taking into account the temperature dependencies of the crack closure level and of the elastic modulus.

(3) The typical characteristics, which indicate the lack of similitude between small and long cracks, were found at all test temperatures: the propagation rates of small cracks were significantly higher than those of long cracks. The small cracks grew even at ΔK levels lower than the long crack threshold. At lower temperature the major factor contributing to the lack of similitude in crack growth rates was the difference of closure level between long and small cracks. At higher temperatures, the difference of environmental attack between small and long cracks was also responsible for the above small crack propagation problem, in addition to the crack closure phenomenon.

References

1. Leverant G.R. and Gell M., The influence of temperature and cyclic frequency on the fatigue fracture of cube orientated Ni-base superalloy single crystals, *Metall. Trans. A* **6**, (1975), 368-371.
2. Crompton J.S. and Martin J.W., Crack growth in a single crystal superalloy at elevated temperature, *Metall. Trans. A* **15**, (1984), 1711-1719.
3. Gell M. and Leverant G.R, Mechanics of high temperature fatigue, in Carden A.E. and McEvily, A.J. (eds.), *Fatigue at Elevated Temperatures*,ASTM STP 520, ASTM, Philadelphia, 1972, 37-67.
4. Hoffelner W., Effect of environment on crack growth behaviour of superalloys under various loading conditions, *Mat. Sci. Tech.* **3**, (1987), 765-771.

5. Gell M. and Leverant G.R., The effect of temperature on fatigue fracture in a directionally-solidified nickel-base superalloy, in A.K.Head and N.J. Grant (eds.), *Proceedings of ICF-2*, Brighton, 1969, 565-575.
6. Webster G.A., High temperature fatigue crack growth in superalloy blade materials, *Mat. Sci. Tech.* **3**, (1987), 716-724.
7. Sadananda K. and Shahinian P., Analysis of crystallographic high temperature fatigue crack growth in a nickel base alloy, *Metall. Trans.* A **12**, (1981), 343-351.
8. Chan K.S., Hack J.E. and Leverant G.R., Fatigue crack growth in Mar-M200 single crystals, *Metall. Trans.* A **17**, (1986), 1739-1750.
9. Chan K.S., Hack J.E. and Leverant G.R., Fatigue crack propagation in Ni-base superalloy single crystals under multiaxial cyclic loads, *Metall. Trans.* A **17**, (1986), 1739-1750.
10. Okazaki M., Tabata T. and Nohmi S., Intrinsic stage 1 crack growth of direcionally-solidified Ni-base superalloys during low-cycle fatigue at elevated temperature, *Metall. Trans.* A **21**, (1990), 2201-2208.
11. Okazaki M., Imai T., and Nohmi. S., Naturally-initiated fatigue small crack growth in a Ni-based superalloy at elevated temperature, in K.T. Rie (ed.), *Low Cycle Fatigue and Elasto-Plastic Behavior of Materials* -3, Elsevier Appl. Sci., 1992, 539-545.
12. Okazaki M., Ohshima S. and Nohmi S., Intrinsic small fatigue crack growth based on the crack opening-closing measurement in Ni-base superalloys at high temperature, *Trans. Jap. Soc. Mat. Sci.* **43**, (1994), 860-866.
13. Brown C.W. and King J.E., The relevance of microstructural influences in the short crack regime to overall fatigue resistance, in R.O. Ritchie and J. Lankford (eds.), *Small Fatigue Cracks*, AIME, Warrendale, 1986, 73-95.
14. Haddad M.H. and Topper T.H., Prediction on non-propagating cracks, *Eng. Frac. Mech.* **11**, (1979), 573-584.
15. Gangloff R.P., Crack size effect on the chemical driving force for aqueous corrosion fatigue, *Metall. Trans. A* **16**, (1985), 958-969.
16. Chan K.S., Effect of cross slip on crystallographic cracking in anisotropic single crystals, *Acta. Metall.* **35**, 1987, 981-987.
17. Hudak S.J., Small crack behavior and the prediction of fatigue life, Trans. ASME, *J. Eng. Mat. Tech.* **103**, (1981), 26-35.
18. Ritchie R.O. and Lankford J., in Ritchie R.O. and Lankford J.(eds.), *Small Fatigue Cracks*, AIME, Warrendale, 1986, 1-5.
19. Tanaka K., Mechanisms and mechanics of short fatigue crack propagation, *Int. J. Jap. Soc. Mech. Eng.* **30**, (1987), 1-13.
20. Elber W., The significance of fatigue crack closure, in *Damage Tolerance in Aircraft Structures*, ASTM STP 486, ASTM, Philadelphia, 1971, 230-42.
21. Kikukawa M., Jono M., and Hora H., *Int. J. Frac.* **13**, (1977), 699-705.
22. Raju I.S. and Newman J.S., An empirical stress intensity factor equation for the surface crack, *Eng. Frac. Mech.* **11**, (1979), 817-27.
23. Aghinton E. and Bamberger M., High temperature low cycle fatigue of a nickel-based Mar-M200+Hf in Air and Ar+20% O2 environment, *J. Mat. Sci.* **26**, (1991), 1873-81.
24. Suresh S., in *Fatigue of Materials*, Cambridge University Press, Cambridge, 1991, Chap. 9.

THERMAL FATIGUE BEHAVIOUR OF THE NICKEL-BASED SUPERALLOY CMSX-4

M. BLÜMM, F. MEYER-OLBERSLEBEN and F. RÉZAÏ-ARIA
Laboratory of Mechanical Metallurgy,
Depertment of Materials, Swiss
Federal Institute of Technology,
C.H. - 1015 Lausanne, Switzerland.

1. Introduction

Single crystal (SC) nickel-based superalloys are candidate materials for advanced gas turbine blades. The main goal in using SC blades is to increase the inlet gas temperature, thereby improving the turbine efficiency and also to extend the durability of the blades by increasing there creep-fatigue-oxidation resistance. The elimination of grain boundaries in SC alloys has improved the life of the blades in comparison to conventionally cast components. The maximum load bearing capacity in thermal fatigue of SC blades is expected to coincide with the <001> crystallographic orientation, since the modulus of elasticity has its lowest value in this direction leading to low thermal stresses [1-2].

The design of SC blades presents a great challenge since these alloys are anisotropic in mechanical behaviour. Although SC blades are oriented along the <001> crystallographic direction, different volume regions having different secondary crystallographic orientations are subjected to a complex state of stress [3].

Investigations have already been devoted to the anisotropic isothermal low cycle fatigue [3-7] or thermo-mechanical fatigue [8] behaviour of SC superalloys. To our knowledge no systematic investigation has yet been reported in the open literature on the anisotropic thermal fatigue behaviour of SC superalloys. Thermal fatigue (TF) is one of the life limiting factors of vanes and blades. Thermal strains and stresses arise in particular from temperature gradients in the blades during start-up and shut-down operations.

This contribution deals with an assessment of the anisotropic thermal fatigue behaviour of single crystal CMSX-4. The TF behaviour is reported for <001>, <011> and <111> crystallographic orientations taken as the main load axis of the blade-shaped single wedge specimens. The TF behaviour of the <001> orientations was investigated for different maximum temperatures, T_{max}, of the thermal cycle. The other crystallographic orientations were studied at $T_{max} = 1100°C$.

J. Bressers and L. Rémy (eds.), Fatique under Thermal and Mechanical Loading, 131–139.

2. Material an Experimental Procedures

2.1. MATERIAL

CMSX-4 is a rhenium-containing nickel-based superalloy with the density of 8.7 kg/dm^3. Its chemical composition is given in Table 1. Two batches of the alloy were used. The first batch (I) was cast at Thyssen Feinguss, Germany, in a high gradient furnace and delivered by MTU Munich GmbH, Germany. The second batch (II) was cast at Howmet Exeter Casting, England, using the seed technology and delivered by EGT Limited, GEC Alsthom, England. A comparable interdendritic distance was found in both batches. The batches were received in fully heat treated conditions with regular cubic γ'-precipitates of 380 to 400 nm edge length. The alloy was investigated in the non-hipped and uncoated condition.

TABLE 1. Chemical composition of CMSX-4 in weight %.

Co	Ta	Cr	W	Al	Ti	Mo	Re	Hf	C	Ni
9.5	6.5	6.4	6.3	5.6	1.0	0.6	3.0	0.1	0.01	bal.

2.2. THERMAL FATIGUE SPECIMENS

Thermal fatigue tests were carried out on blade-shaped single wedge specimens, Fig. 1a. The specimen geometry was slightly different in the x-dimension, namely type A (batch I) and type A′ (batch II). The specimens were cut by electroerosion, then ground and finally manually polished parallel to the edge to 1 μm surface finish.

The longitudinal y-axes of the TF specimens were parallel to the disired crystallographic orientations as shown in Fig. 1b. The crystallographic orientation of each specimen was determined by the back reflection Laue technique.

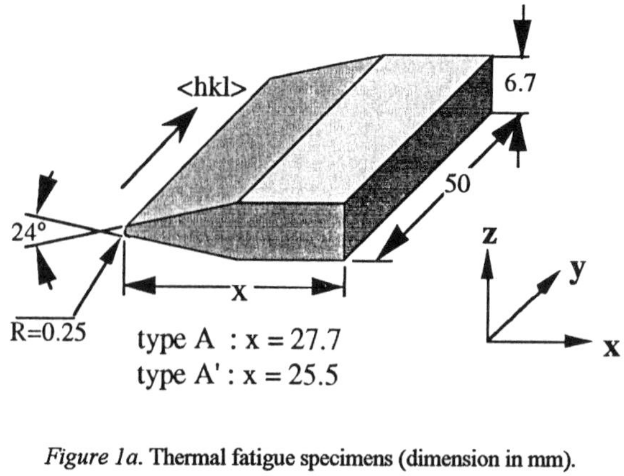

Figure 1a. Thermal fatigue specimens (dimension in mm).

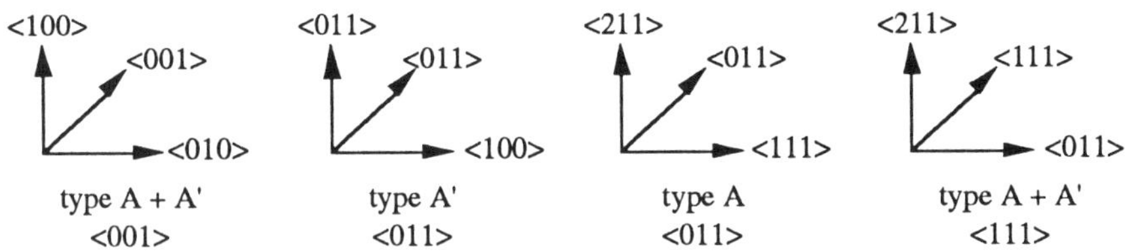

Figure 1b. Orientations of the TF specimens.

2.3. THERMAL FATIGUE TEST PROCEDURE

The TF tests were performed under similar temperature-time cycles of 60 s heating and 20 s cooling periods (Fig. 2). A thermal fatigue test rig using a 6.0 kW Hüttinger solid state induction generator with 3000 kHz frequency for heating and air blast for cooling the wedge tip of the specimens was used [9], (Fig. 3). A Ni/NiCr thermocouple spot-welding at 0.5 mm from the edge monitored the temperature during the test. The specimens were externally unconstrained during thermal cycling. The temperature-mechanical strain hysteresis loop was measured along the edge length over 50 mm by an extensometer [9].

The TF behaviour of the <001> orientation was investigated at different maximum temperatures between 1000°C and 1150°C. The <111< and <011> orientations were examined at T_{max} = 1100°C. The minimum cycle temperature was maintained at 200°C for all experiments. The tests were regularly interrupted to investigate crack initiation and propagation by means of scanning electron microscopy (SEM) and optical microscopy.

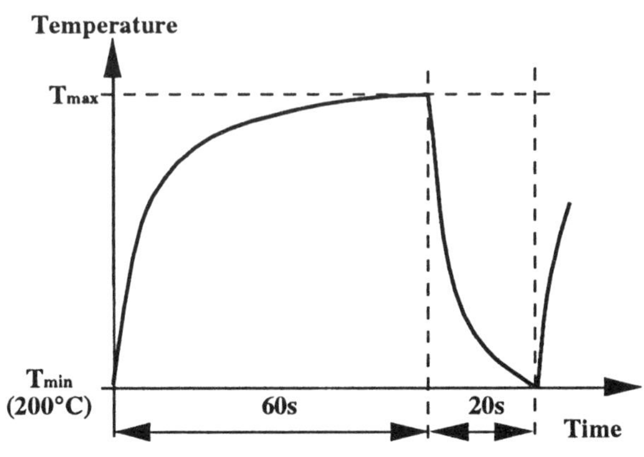

Figure 2. Thermal fatigue cycle.

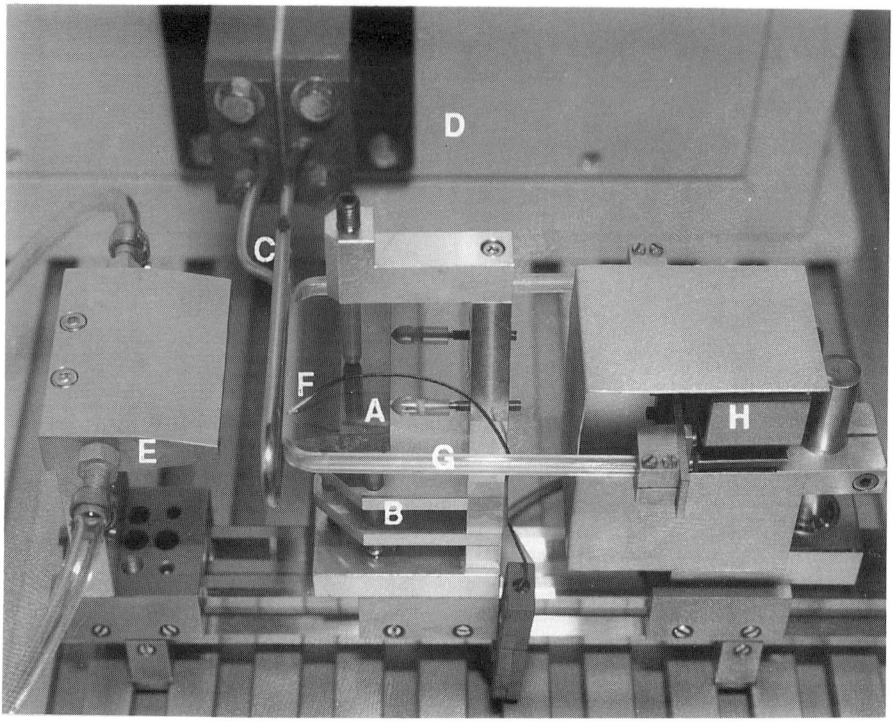

Figure 3. View of the thermal fatigue rig with: (A) specimen, (B) specimen support, (C) induction coil, (D) HF-genarator, (E) air nozzle, (F) thermocouple and (G, H) extensometer device.

3. Results and Discussion

No slip bands were observed either at the wedge tip or on the wedge surfaces of the <001> TF specimens. On the other hand, multiple slip bands were visible in <011> and <111> oriented specimens. Slip bands appeared at the wedge tip and on the two wedge surfaces at the edge region. They propagated towards the bulk (over about 1.5 to 2 mm) with an increasing number of TF cycles. Fewer slip bands appeared in the specimen type A′. Fine slip bands were visibele between the primary large spaced bands. Microscopic observations and laser profilometry measurements along the tip and on the wedge surfaces [10] have revealed that the density of the slip bands decreases form the central zone to the corners of the specimens.

Laser profilometry measurements of the wedge surfaces along the y-axis revealed that the central zone of the <011> and <111> specimens displaced and curved along the z-axis. Fig. 4 shows an example of such displacement for three TF-specimens with different crystallographic orientations. The curvature near the wedge tip was maximum, indicating the localisation of plastic shear deformation in the central zone of the <011> and <111> specimens. In <001> specimen only a slight slope can be noticed in the profile curve, which can be explained with an irregular manual polish of this specimen before the TF-test. The scatter is due to oxidation.

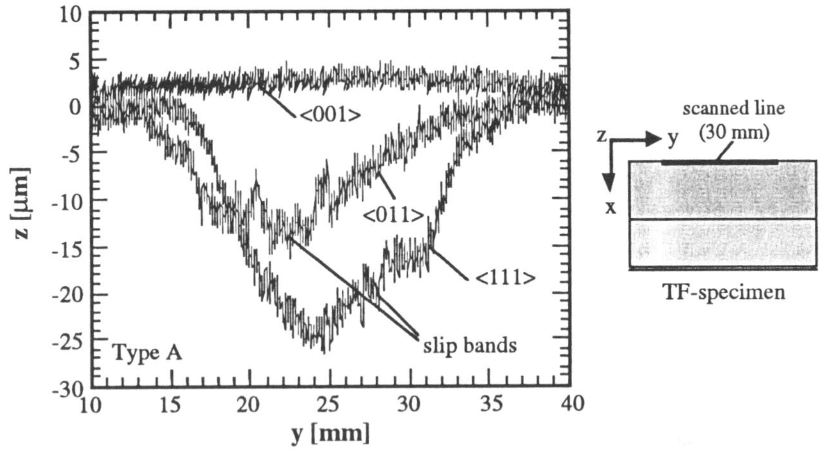

Figure 4. Profilometry measurements near the wedge tip: <111> specimen after 70, <011> specimen after 230 and <001> specimen after 1220 thermal cycles.

Strain measurements show that the mechanical strain-temperature hysteresis loops are smaller in <011> and <111> in comparison to the <001> oriented specimen (Fig. 5). The mechanical strain ε_{mec} is thereby defined as the difference between the total strain measured over 50 mm and the free thermal dilation.

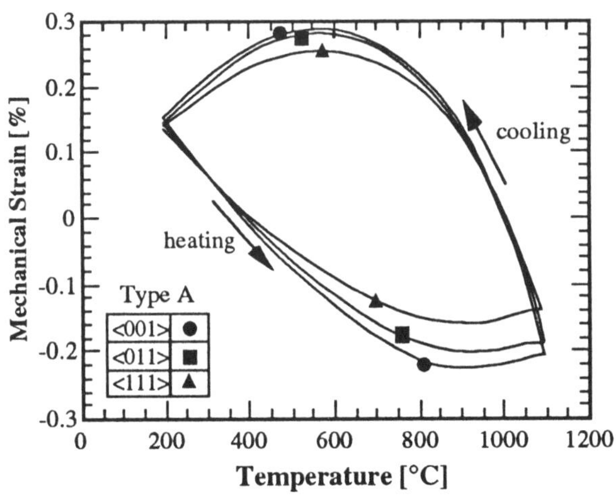

Figure 5. Mechanical strain-temperature loops for specimen type A, all orientations.

3.1. CRACK INITIATION AND PROPAGATION

CMSX-4 has a high oxidation resistance. The wedge tip of the specimens remained essentially unaffected during TF cycling. In the <001> specimens cracks initiated at oxidised cast micropores at the wedge tip, Fig. 6a. In <011> and <111> oriented specimens cracks initiated on slip bands, frequently on the wedge surfaces, Figs. 6b. and 6c. However, in general pores were also located on the crack initiation sites. Some cracks which initiated from the wedge surfaces linked with the small cracks initiated from the wedge tip to form a unique through-thickness crack. Moreover, while in <001> specimens only a limited number of cracks were detected, multiple crack initiation was observed in the two other crystallographic orientations.

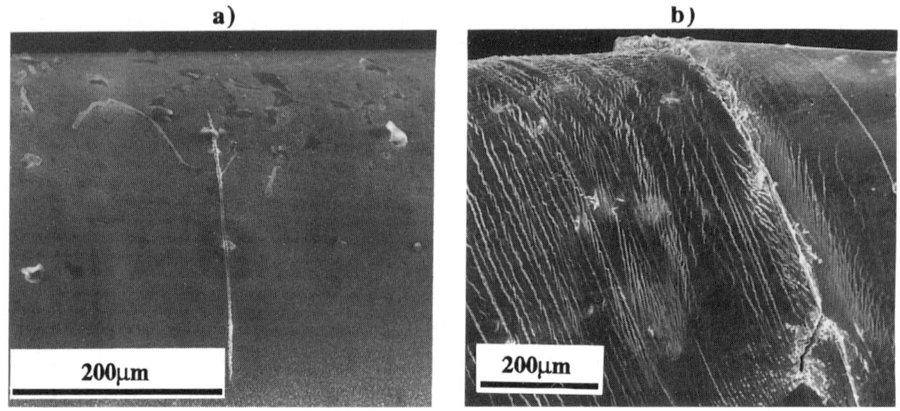

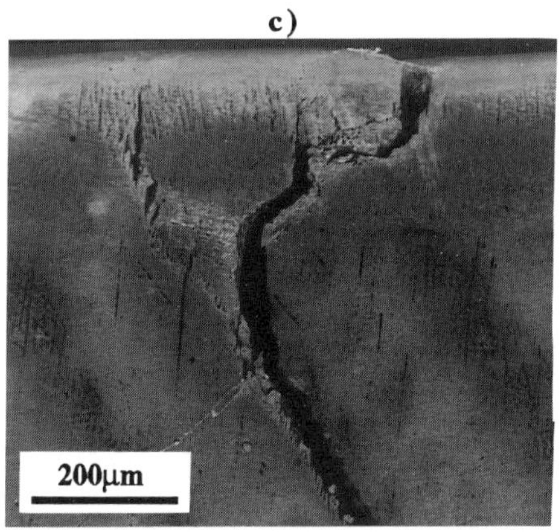

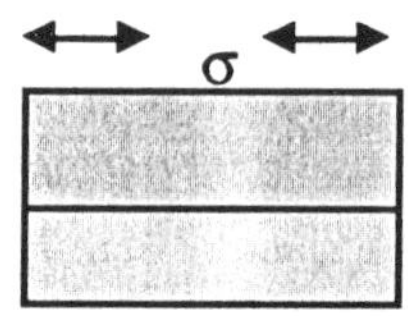

Figure 6. Main cracks in the different oriented specimens of type A
a) <001>, 1700 thermal cycles
b) <011>, 440 thermal cycles
c) <111>, 150 thermal cycles.

In <001>specimens cracks propagated perpendicular to the y-axis. Cracks which were longer than about 2 to 3 mm deviated with a certain angle in respect to the y-axis on crystallographic planes. In <011> and <111> specimens cracks propagated first along the slip bands changing their propagation direction by zigzagging and by passing from one slip band to another at the intersection of two slip bands. They followed then a direction perpendicular to the y-axis. The crack tip was subjected to a complex mixed-mode loading. In addition, fatigue striations were observed on the fracture surfaces.

3.2. CRACK PROPAGATION CURVES AND THERMAL FATIGUE LIFE

Fig. 7 shows the crack propagation behaviour of all three crystallographic orientations as a function of the number of thermal cycles. A higher crack growth rate was observed for <011> and <111> specimens in comparison to the <001> specimens. Intuitively, for the same mechanical strain a higher stress intensity factor is expected for <011> and <111> specimens because of the higher modulus of elasticity in these crystallographic orientations in comparison to the <001> orientation. SEM observations at room temperature showed in addition that due to the activation of multiple slip bands cracks remained open, thereby increasing the crack velocity.

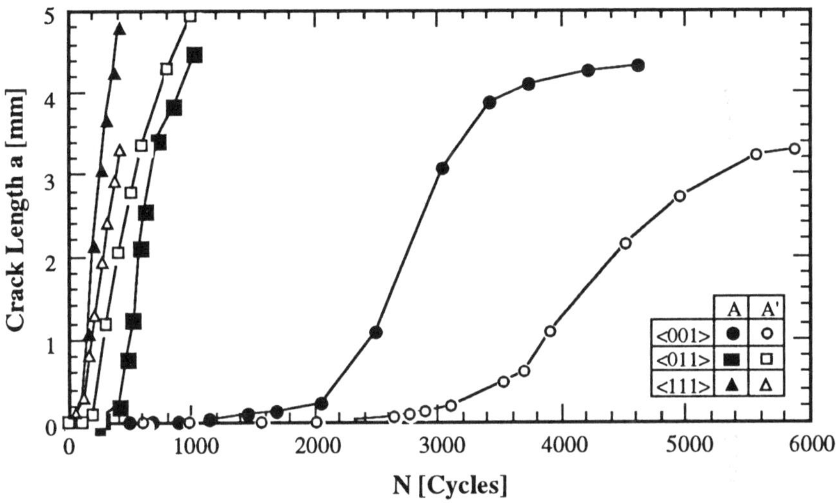

Figure 7. Effect of crystallographic orientation on the thermal fatigue crack propagation, $T_{max} = 1100°C$.

The definition of a crack initiation life is very difficult for <011 and <111> specimens. Therefore, only a TF-life corresponding to an 1 mm crack depth, N_f, is reported here. Fig. 8 depicts the RF-life as a function of T_{max}, and mechanical strain range for all crystallographic orientations and both RF specimen types. These curves show the drastic effect of the crystallographic orientations on the TF-life of CMSX-4. The complex crack initiation mechanisms in <011> and <111> oriented specimens decreased drastically the number of cycles to crack initiation in comparison to the <001> specimens.

To rationalise these experimental results an elastic stress range was defined as $\Delta\sigma = E\Delta\varepsilon_{mec}$. The modulus of elastic E of each orientation was calculated from the elastic constants at 615°C [11] which corresponds to the temperature of maximum tensile stress during the cooling part of the thermal cycle (Fig. 5). The variation of the TF-life as a function of this stress range is reported in Fig. 9. This figure shows that under the test conditions reported here the lower TF-life in <011> and <111> specimens can be explained by a higher thermal stress generated in these orientations as has previously been reported in investigations devoted to the anisotropic isothermal low cycle fatigue and thermo-mechanical fatigue behaviour of SC superalloys [3, 8, 12].

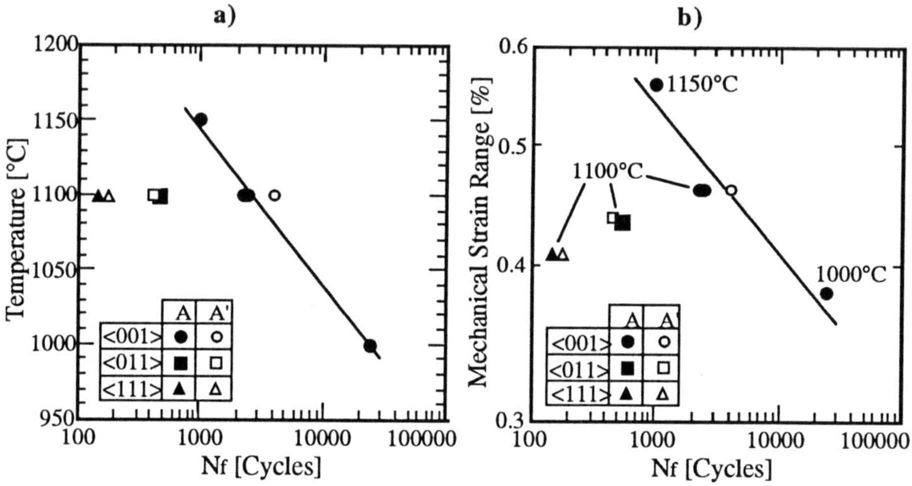

Figure 8. Varation of the thermal fatigue life, N_f, as a function of: a) T_{max}, b) mechanical strain range.

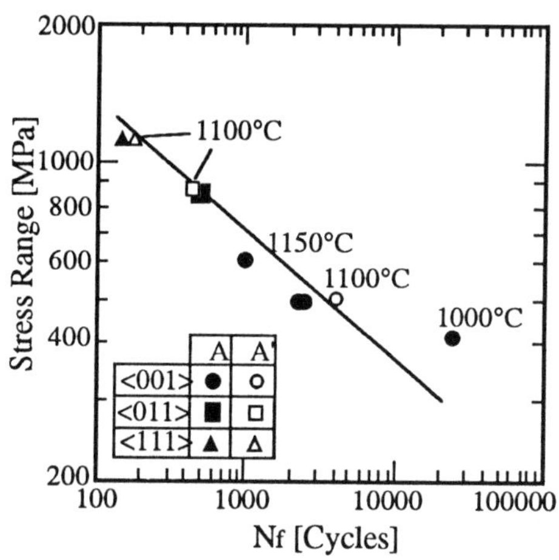

Figure 9. Variation of the thermal fatigue life, N_f, as a function of stress range.

4. Summary

The anisotropic thermal fatigue behaviour of CSMX-4, a rhenium-containing single crystal nickel-based superalloy, was investigated for <001>, <011> and <111> crystallographic orientations. Blade-shaped single wedge specimens were used. It was shown that higher thermal stresses in <011> and <111> orientation decreased drastically the TF-life. Multiple slip bands appeared on <011> and <111> specimens. Higher thermal fatigue crack growth rates were observed for <011> and <111> specimens. In <001> specimens cracks initiated on the cast micropores. In <011> and <111> specimens cracks initiated mainly on the slip bands. Micropores were also present on these crack initiation sites. A multiaxial loading took place in <011> and <111>oriented specimens. The experimental results were explained by a stress-based criterion.

Acknowledgements - The authors would like to acknowledge MTU Munich GmbH, Germany and EGT Limited, GEC Alsthom, England for providing the alloy free of charge as well as the Swiss Priority Program on Materials Research for supporting this investigation. Prof. B.Ilschner is acknowledged for many fruitful discussions and support.

References

1. Versnyder, F.L., Superalloy Technology - Today and Tomorrow, in R. Brunetaud et al. (eds), *High Temperature Alloys for Gas Turbines,* Reidel, Dordrecht, 1982, pp. 1-49.
2. Erickson, G.L. and Harris, K., DS and SX for Industrial Gas Turbines, in D. Coutsouradis et al. (eds.), *Materials for Advanced Power Engineering,* Part II, Kluwer Academic Publishers, Netherlands, 1994, pp. 1055-1074.
3. Dala, R.P., Thomas, C.R. and Dardi, L.E., The effect of Crystallographic Orientation of the Physical and Mechanical Properties of an Investment Cast Single Crystal Nickel-Base Superalloy, in Duhl, D.N. et al. (eds.), *Superalloys 1988,* Seven Springs Mountain Resort, Champion, Pennsylvania, 1988. pp. 185-197.
4. Gabb, T.P., Gayda, J. and Miner, R.V., Orientation and Temperature Dependence of Some Mechanical Properties of the Single-Crystal Nickel-Base Superalloy Réne N4: Part II. Low Cycle Fatigue Behaviour, *Metall. Trans. A* **17** (1986), pp. 497-505.
5. Gabb, T.P., Welsch, G., Miner, R.V. and Gayda, J., The Low Cycle Fatigue Deformation Response of a Single-Crystal Superalloy at 659°C, *Mater. Sci. Eng.* **A198** (1989), pp. 189-202.
6. Defresne, A. and Rémy, L., Fatigue Behaviour of CSMX-4 Superalloy [001] Single Crystals at High Temperature. I: Low cycle Fatigue of Notched Specimens, *Mater. Sci. Eng.* **A129** (1990), pp. 45-53.
7. Fleury, E. and Rémy, L., Low Cycle Fatigue Damage in Nickel-Base Superalloy Single Crystals at Elevated Temperature, *Mater. Sci. Eng.* **A167** (1993), pp. 23-30.
8. Fleury, E. and Rémy, L., Behaviour of Nickel-Base Superalloy Single Crystals under Thermal-Mechanical Fatigue, *Metall. Mater. Trans. A* **25** (1994), pp. 99-109.
9. Meyer-Olbersleben, F., Engler-Pinto Jr., C.C. and Rézaï-Aria, F., On Thermal Fatigue of Nickel-Based Superalloy, M.J. Verrilli and M.G. Castelli (eds.), *Thermo-Mechanical Fatigue Behaviour of Materials,* 2nd Volume, ASTM STP 1263, American Society for Testing and Materials, Philadelphia, 1995.
10. Blümm, M. and Rézaï-Aria, F., unpublished results, Swiss Federal Institute of Technology, Switzerland, 1994.
11. Bayerlein, U., *Zur Ermittlung der Textur und Gefügeabhängigkeit der elastischen Eigenschaften sowie der Einkristallkonstanten von Superlegierungen bei höheren Temperaturen,* Fortschrittsberichte VDI, Reihe 5: Grund- und Werkstoffe, Nr. 236, VDI Verslag, Düsseldorf, 1991.
12. Li, S.X. and Smith, D.J., An Overview of Combined Fatigue and Creep Response of Single Crystal Nickel Base Superalloys, in D. Coutsouradis et al. (eds.), *Materials for Advanced Power Engineering,* Part II, Kluwer Publishers, The Netherlands, 1994, pp. 1055-1074.

COMPARISON OF SINGLE CRYSTAL SUPERALLOYS FOR TURBINE BLADES THROUGH TMF TESTS

J.C. LAUTRIDOU, J.Y. GUEDOU and J. DELAUTRE
Materials and Processes Department, SNECMA
B.P. 81, 91003 EVRY CEDEX - France

1. Introduction

The greatest advance in metal temperature and stress capability for turbine blades in the last thirty years has been the result of the development of directionnally solidified single crystal superalloys. Two groups of alloys, called first and second generation, have been developed. The second generation alloys were essentially developed to improve the creep resistance and the long time phase stability in order to allow increased operating temperatures. Creep resistance is not the only property to be considered for life prediction. In complex shaped highly cooled turbine blades, stresses of thermomechanical origin may become at least as important as the centrifugal stress causing the creep of the material during the stabilized regime of the engine cycle. Those thermomechanical stresses are generated in particular during aircraft take off and landing operations. A limited amount of work has been done to compare the performance of the two generations of single crystal alloys regarding thermomechanical fatigue resistance.

The aim of the present study was to compare the mechanical properties of four single crystal superalloys belonging to the first (AM1, AM3) and second generation (MC2, CMSX4). In order to get a complete comparison, creep rupture, low cycle fatigue and thermal mechanical fatigue (TMF) tests were carried out.

2. Materials and Heat Treatments

AM1 is a French patented [1] first generation single crystal alloy introduced by SNECMA for turbine blading in the military M88 engine. AM3 is also a first generation single crystal [2] which offers very attractive properties due to its low density ($8.25x10^3$ kg/m^3). MC2 is a second generation single crystal, developed by ONERA [3], which has the particularity to be rhenium free. CMSX4, developed by CANNON MUSKEGON corporation [4], is a rhenium containing alloy (3% wt) as the majority of second generation single crystals. The chemical compositions of the four single crystals used in this investigation are shown in Table 1.

J. Bressers and L. Rémy (eds.), Fatique under Thermal and Mechanical Loading, 141–149.

TABLE 1. Nominal compositions (wt %) of the single crystal superalloys.

	Ni	Co	Cr	W	Al	Ta	Mo	Ti	Re
MC2	base	5,08	7,89	7,99	5,05	5,96	2,11	1,49	-
CMSX4	base	9,70	6,47	6,50	5,63	6,50	0,56	1,01	2,9
AM3	base	5,61	8,04	4,91	5,86	3,43	2,20	2,00	-
AM1	base	6,5	7,8	5,7	5,2	7,9	2	1,1	-

The single crystal bars with their axis oriented in the [001] direction were given the following heat treatments :

TABLE 2. Heat treatment conditions.

ALLOY	AM1	AM3	MC2	CMSX4
solution treatment	1300°C/3h	1305°C/3h	1315°C/3h	multi-step from 1277°C to 1321°C (16h)
precipitation treatment	1100°C/5h	1050°C/16h	1100°C/5h	1080°C/4h
aging treatment	87°C/16h	850°C/24h	850°C/24h	870°C/16h

These heat treatments produce a homogeneous distribution of cuboidal γ' precipitates with a mean size of 0.5 μm for AM1, 0.4 μm for AM3 and MC2, and of 0.3 μm for CMSX4.

3. Results and Discussion

3.1. CREEP PROPERTIES

An extensive characterization of the creep-rupture strength of the four alloys was conducted between 760°C and 1150°C employing specimens machined from cylindrical bars of 10 mm in diameter. The creep capability of the alloys is compared in Fig. 1. It appears from this comparison that the two second generation single crystals, CMSX4 and MC2, exhibit the highest resistance especially as soon as the testing temperature is higher than 1000°C. CMSX4 is better than MC2 at temperatures lower than 1000°C but the difference between these alloys is small at higher temperatures, MC2 being better than CMSX4 in the range 1050°C to 1100°C and CMSX4 being better than MC2 at 1150°C. The better behaviour of CMSX4 can be attributed to the beneficial effect of rhenium which is known to reduce rates of diffusion and thus retards coarsening of the γ' strengthening phase [5]. The particular behaviour of MC2, which is rhenium free, can be attributed to γ and γ' strengthening by a particularly high level respectively of molybdenum and titanium as compared to the other alloys [3]. The temperature advantage of CMSX4 over AM1 varies from 25°C around 850°C to 50°C around 1150°C.

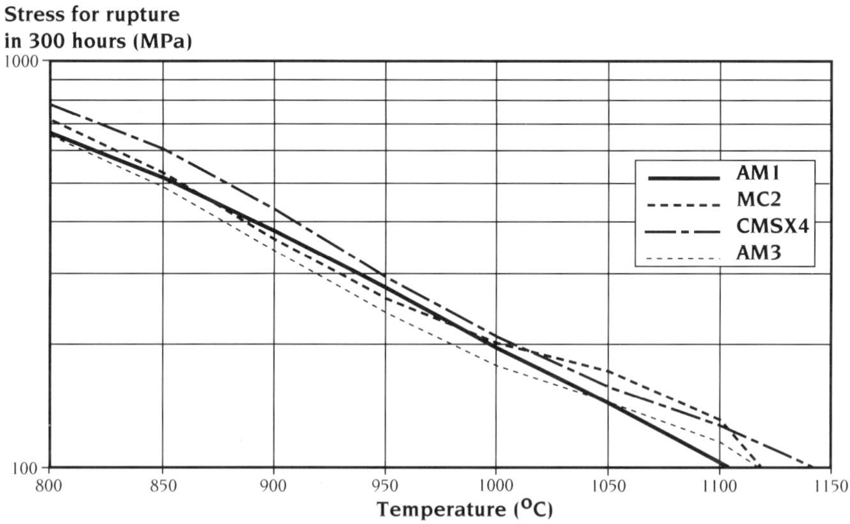

Figure 1. Creep-rupture capability comparison.

3.2. LOW CYCLE FATIGUE PROPERTIES

Strain controlled low cycle fatigue (LCF) tests were performed at 800°C on uncoated specimens using a wave frequency of 0.25 Hz. This temperature was chosen because it nearly corresponds to the maximum tension peak of the thermal-mechanical cycle simulating the operating conditions at the leading edge of a turbine blade. Solid cylindrical specimens of 4.37 mm in diameter and 13 mm in gauge length were used for LCF. The LCF lives are presented in Fig. 2 as a function of stress range at half life. The best alloy appears to be AM1. MC2 and CMSX4 offer a very similar fatigue behaviour, worse than AM1 but better than AM3. A fractographic examination of the rupture surface of each specimen was conducted. For each alloy, cracks initiate in the same way from subsurface micropores. In order to try to explain the differences between the alloys it was decided to compare the mean stress values of the stabilized mid-life hysteresis loops. The mean stress values were found to be higher for CMSX4 and MC2 (320 MPa) as compared to those of AM1 and AM3 (250 MPa) indicating that a higher mean stress relaxation occurs in AM1 and AM3. This can explain the better behaviour of AM1, as compared to CMSX4 and MC2, assuming that the life is mostly spent in the crack propagation phase [6] and thus a higher mean stress should lead to a reduction of the life.

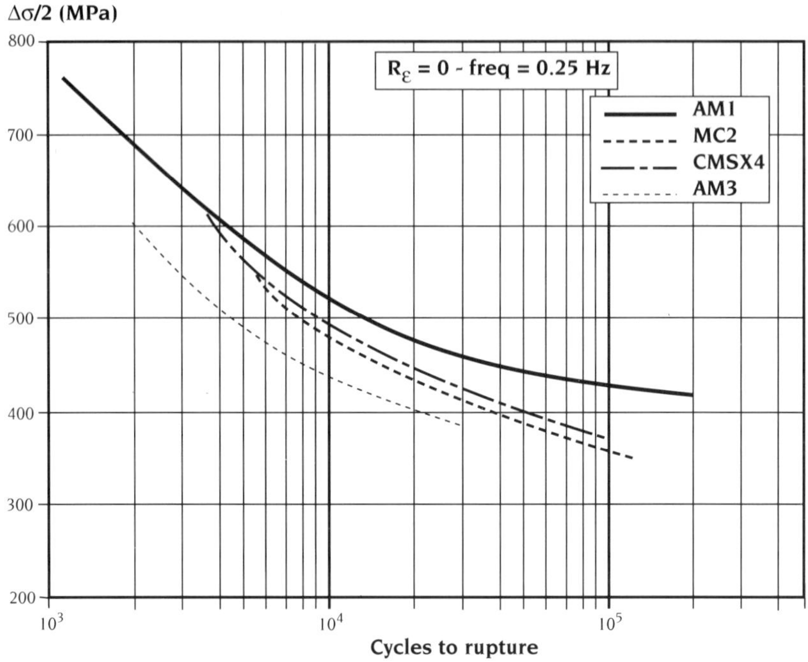

Figure 2. LCF lives comparison at 800°C.

3.3. THERMAL-MECHANICAL FATIGUE PROPERTIES

3.3.1 Uncoated materials

Hollow cylindrical fatigue specimens (9 mm internal diameter) were machined from bars of 20 mm in diameter. All the materials were investigated in the bare condition. The tests were performed within the Centre d'Essais Aéronautiques de Toulouse, and the experimental procedure used was close to that of SNECMA which has been shown elsewhere [7]. The TMF cycle used for this study is presented in Fig. 3.

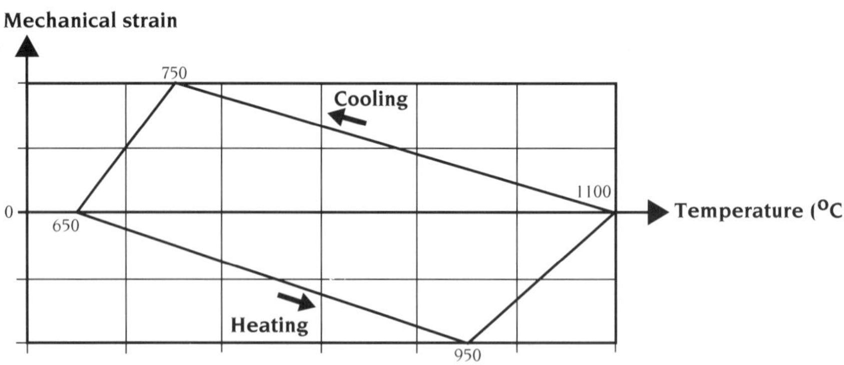

Figure 3. "4 slopes" TMF cycle definition.

It consists of a triangular temperature cycle between 650°C and 1100°C associated with an alternate four slopes strain wave (R_ε = - 1). The frequency of this cycle is 5.6x10^{-3} Hz (180 s per cycle). This cycle was defined to simulate the temperature and strain histories at the leading edge of a turbine blade. The four uncoated materials have been submitted to the same cycle under the same strain range (± 0.5%), up to macroscopic crack initiation that is determined by a 20% maximum tensile load decrease after stabilization. The results of two tests on each alloy are shown in Fig. 4, illustrating a better fatigue resistance of AM1,AM3 and MC2 as compared to CMSX4 which exhibits on average a reduction of life by a factor of 2.

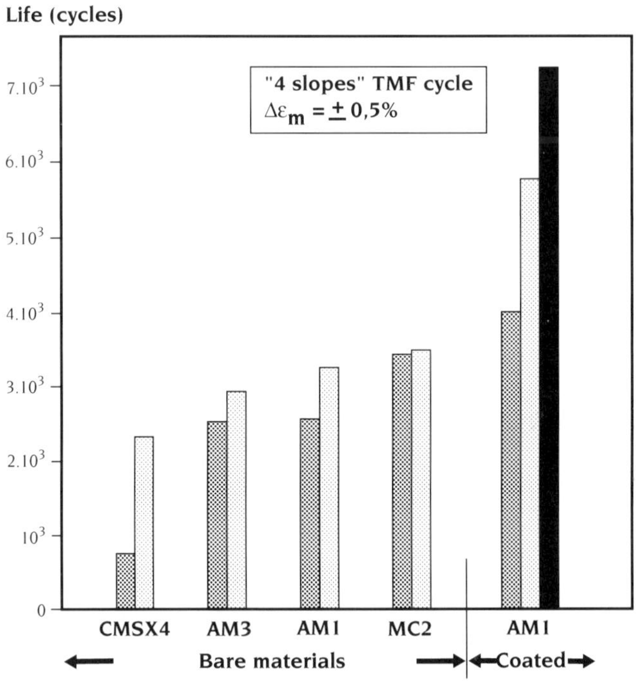

Figure 4. TMF lives comparison.

Two kinds of investigation were made in order to try to explain the particular behaviour of CMSX4. At first an examination of the stress-strain hysteresis loops at mid life was made for the four alloys. No major difference was observed between the alloys concerning stress and plastic strain amplitudes. Secondly, fractographic examinations of failed specimens were performed. These investigations reveal that for each alloy crack initiation sites were located at the oxide scale on the surface of the specimens. This is in agreement with the results obtained by others [6] on Ni-base single crystals tested in TMF at the same strain range. These observations support the view that long lives under TMF or thermal fatigue are mainly controlled by fatigue oxidation interactions and that the TMF resistance cannot be compared with LCF lives obtained at a temperature close to the maximum peak stress of the TMF cycles because the damage mechanisms are not the same. It was indeed shown in section 3.1. that at this temperature (close to 800°C) failure occured by crack initiation from subsurface micropores. As pointed out by Fleury and Remy [6] LCF isothermal tests at high temperature, close to the maximum temperature of the TMF cycle, should be more representative of damage mechanisms operating in TMF tests. Consequently, the TMF resistance of bare materials should be more dependent on the oxidation behaviour of the material rather than on its creep resistance.

3.3.2 Coated materials

Aluminide coating has been shown to improve TMF life on superalloys [8] especially at low strain levels corresponding to longer lives due to their beneficial effect against oxidation. The same kind of TMF tests as mentioned earlier have been performed in the same loading condition on AM1 chromium-aluminium coated specimens (AM1 + CA). The results are compared to the previous ones obtained on bare specimens on Fig. 4. A mean life improvement by a factor of 2 is observed as compared to bare AM1. On these specimens crack initiation sites were observed at the interface between the Cr-Al coating and the single crystal material. An illustration of this phenomenon is shown in Fig. 5.

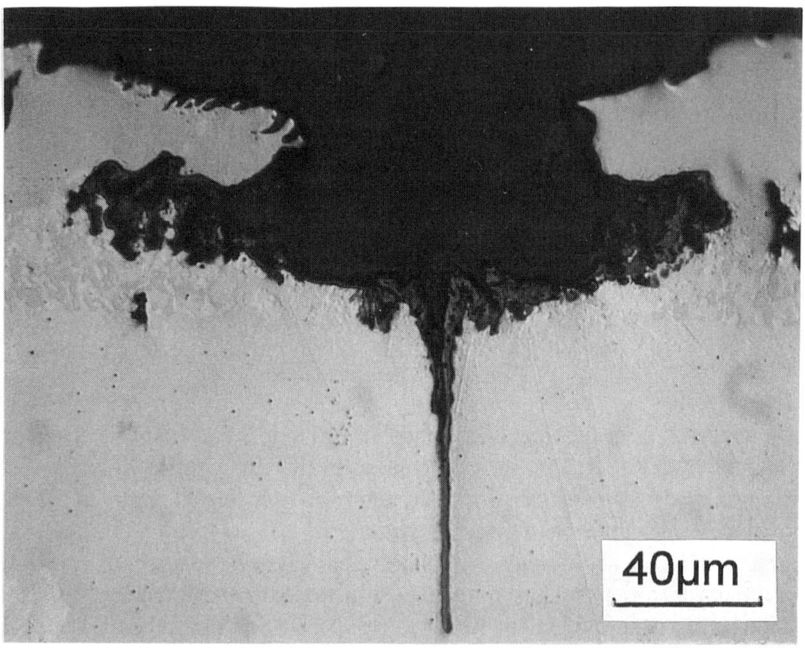

Figure 5. Illustration of fatigue crack initiation on coated AM1 under TMF ("4 slopes" cycle).

First, the coating cracks, then an oxide develops at the interface between the coating and the substrate and finally a crack propagates into the substrate until rupture. The coating will prevent oxidation of the substrate provided it can withstand the mechanical loading without cracking. It is therefore very important to know the ductile-brittle transition temperature of the coating and to make sure that the TMF cycle of actual blades does not enter the brittle range to prevent the coating from cracking at the very first cycle and thus protect the substrate from localised oxidation.To measure the effect of such a situation another TMF cycle was applied. This new TMF cycle (cycle "W") is shown in Fig. 6, where it can be seen that it contains a peak stress at the lowest temperature (600°C) which is within the brittle domain of the coating.

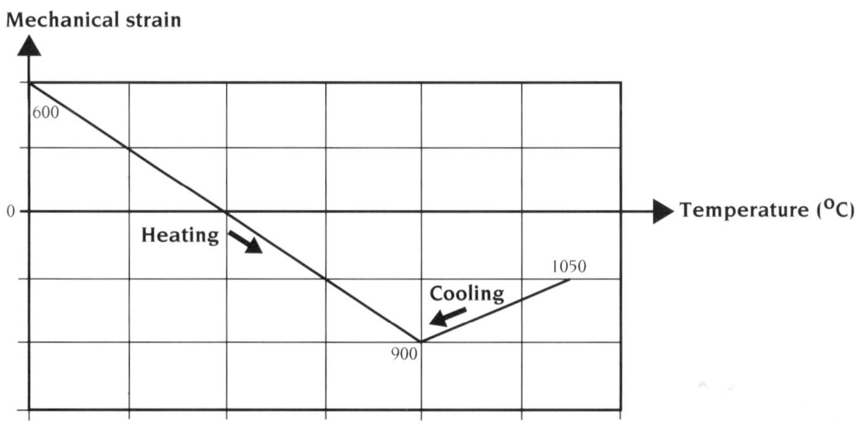

Figure 6. "W" TMF cycle definition.

Several tests were performed on AM1 chromium-aluminium coated specimens. The results are compared in Fig. 7 to those obtained on coated AM1 under comparable strain conditions but with the four slopes TMF cycle previously used. It appears that a life reduction factor of 3, with respect to the coated material using the "4 slopes" TMF cycle, is obtained when the "W" cycle is applied. The metallurgical examination of the specimens (Fig. 8) shows that a great number of straight cracks have crossed the coating and are growing into the substrate perpendicularly to the applied stress. This is attributed to the brittleness of the coating at the lowest cycle temperature which leads to its rupture in the very first cycles. The single crystal is thus no longer protected against oxidation and the fatigue life is then considerably reduced. Moreover, multiple cracking of the coating leads to the development of a circumferential crack at the surface of the specimen which undoubtedly contributes to a further reduction of life due to a geometrical effect. This clearly shows that in some operating conditions the coating cannot prevent the single crystal from oxidation and demonstrates tha necessity of developing single crystals of good intrinsic resistance to the fatigue oxidation damage that operates in TMF.

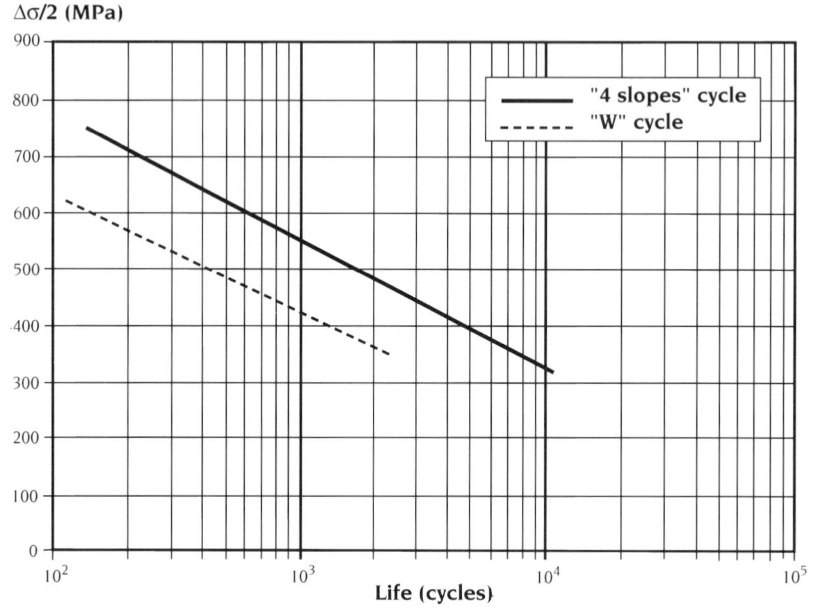

Figure 7. Influence of TMF cycle definition on life for coated AM1.

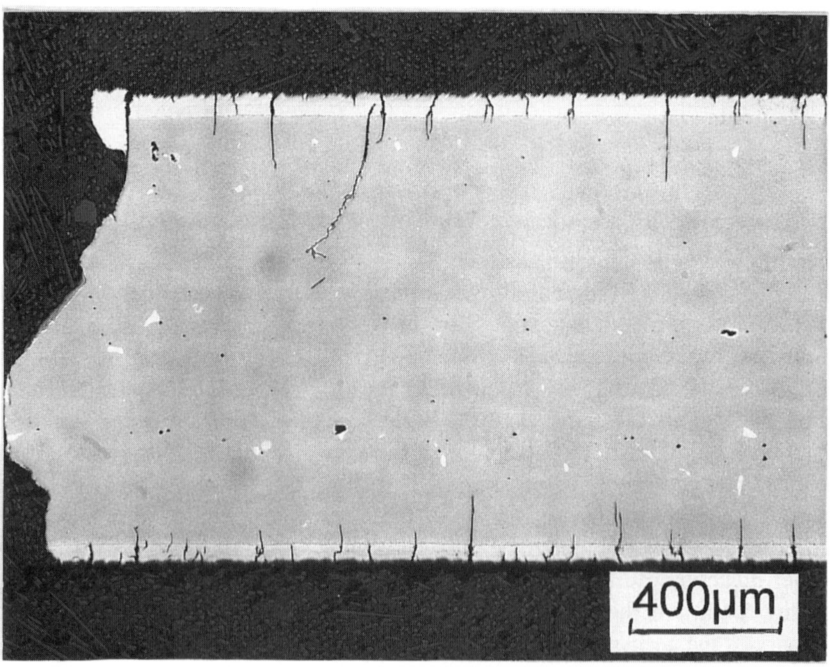

Figure 8. Illustration of fatigue crack initiation on coated AM1 under TMF ("W" cycle).

4. Conclusion

Creep, low cycle fatigue and TMF tests have been performed on four turbine blade single crystal superalloys. The main objective of the study was to compare the mechanical properties of first and second generation single crystals. The superior creep resistance at high temperature of the second generation single crystals (MC2 and CMSX4) over the first generation (AM1 and AM3) was confirmed. The low cycle fatigue resistance at 800°C in strain controlled conditions was found to be very dependent upon the stress-strain behaviour of the alloy, the best results being obtained for the material which displays the lowest mean stress. At this temperature fatigue life was not sensitive to environment and cracks initiate at subsurface micropores. Under TMF cycling, which is more representative of service operating conditions of turbine blades, the damage is predominantly associated with the oxidation-fatigue interaction. In bare condition, the second generation single crystals are not found to show better results than the first generation, CMSX4 presenting the lowest results. The TMF fatigue resistance of aluminide coated AM1 was found to be better than that of bare AM1 when the specimens were not stressed into the brittle domain of the coating, but was substantially reduced in the opposite situation.

It is clear that creep, fatigue and oxidation damage are operating in turbine blades during service and that complex interactions exist between those mechanisms. This indicates that the development of single crystal superalloys cannot only be based on the achievement of a high creep resistance, and that it is also necessary to evaluate the TMF resistance of materials in the bare and coated conditions. The development of prediction models that would explicitly treat interactions between those damage mechanisms would be of great help for the optimisation of future materials.

References

1 Bachelet, E., Lamanthe, G., In *National Symposium Single Crystals Superalloys*, Villard de Lans, France 1986.

2. Khan, T., Recent Developments and Potential of Single Crystal Superalloys for advanced turbine blades in *High Temperature Alloys for Gas Turbines and Other Applications*, Dordrecht, 1986, pp. 21-50.

3. Caron P., Khan T., Development of a new nickel based single crystal turbine blade alloy for very high temperatures, In *European Conference on Advanced Materials and Processes*, Aachen (FRG) november 1989.

4. K. Harris, G.L., Ericksson, R.E., Schwer Development of CMSX-4 for Small Gas turbines, TMS-AIME, *Fall Meeting*, Philadelphia, 1983.

5. Giamei, A.F., Anton D.L., Rhenium additions to a Ni-base superalloy : Effects on microstructure, *Metall. Trans A*, **16**, pp. 1997-2005.

6. Fleury, E., Rémy L., Thermal-Mechanical fatigue behaviour of Nickel base superalloy single crystals *Proceedings of ICF 7*, Houston, 1989, pp. 1133 - 1140.

7. Guedou, J.Y., Honnorat Y., Thermomechanical Fatigue of turbo-engine blade superalloys - *ASTM - STP 1186*, Philadelphia, 1993, pp. 157-175.

8. Bernard, H., Rémy, L., *High Temperature Materials for Power Engineering*, Kluwer Academic Publishers, Liege, 1990, pp. 1185-1194.

THERMO-MECHANICAL FATIGUE BEHAVIOUR OF SRR99

C.C. ENGLER-PINTO JR., F. MEYER-OLBERSLEBEN
and F. RÉZAÏ-ARIA
Laboratory of Mechanical Metallurgy,
Department of Materials,
Swiss Federal Institute of Technology,
CH - 1015, Lausanne, Switzerland

1. Introduction

Thermal gradients arising during transient regimes of start-up and shut-down operations produce a complex thermal and mechanical fatigue loading which limits the life of turbine blades. The thermo-mechanical fatigue (TMF) test consists in applying independent temperature-time and mechanical strain-time cycles to an LCF specimen in order to simulate the thermo-mechanical loading experienced by turbine blades [1].

Due to different service conditions and design philosophies, several strain-temperature cycles have been proposed for TMF tests [1–3]. It has been shown that the TMF behaviour and resistance of superalloys depend upon the cycle history used. In the present investigation a strain-temperature cycle is proposed for TMF investigations, which was deduced from strain measurements on a blade-shaped specimen during thermal fatigue (TF) tests [4]. It is named *thermal-fatigue-based* (TFB) cycle. By the use of this cycle, the life and the behaviour under TF and TMF loading can be compared under similar strain–temperature histories.

This contribution deals with an investigation on the thermo-mechanical fatigue behaviour in the [001] orientation of the single crystal nickel-based superalloy SRR99. TMF tests were performed under out-of-phase (OP) and TFB cycles. The results are compared to the ones obtained under TF experiments on blade-shaped single wedge specimens [4, 5].

2. Material and Experimental Procedures

The nominal composition of SRR99 is given in Table 1. The alloy was received as cast solid bars (Ø20 mm∗150 mm) in the fully heat treated condition with cubic γ'-precipitates of about 360 nm edge length.

TABLE 1. Nominal composition of SRR99 in weight%.

Ni	Co	Al	Ti	Ta	W	Cr	Mo	C
65.4	5.0	5.6	2.2	2.9	9.5	8.5	0.03	0.020

The TMF tests were performed under total strain control on hollow cylindrical specimens (1 mm wall thickness). The specimens were induction heated and the

J. Bressers and L. Rémy (eds.), Fatique under Thermal and Mechanical Loading, 151–157.

temperature was measured by an infrared two-colour pyrometer. A linear heating and cooling rate of 10 K/s was used. A complete description of the test facility and procedures is given in [6]. The temperature ranges were 600–900°C and 600–1100°C for the OP cycle and 600–1100°C for the TFB cycle. The tests performed at T_{max} = 1100°C were regularly interrupted to take replicas of the surface which were used to investigate crack initiation and propagation by means of scanning electron microscopy (SEM).

The TMF results are compared to the TF life previously obtained on self constrained blade shaped specimens cycled between 200–1100°C [4, 5]. The TF tests were performed under temperature-time cycles of 60 s heating and 20 s cooling periods. The tests were regularly interrupted to allow microscopic observations by SEM and optical microscopy directly on the specimen. Fig. 1 shows the thermal cycles used for the TF and TMF experiments, while the mechanical strain-temperature cycles under TMF (600–1100°C and $\Delta\varepsilon_m$ = 0.7%) are compared to TF (200–1100°C and $\Delta\varepsilon_m$ = 0.69%) in Fig. 2. It can be seen that the mechanical strain ratio, $R_\varepsilon = \varepsilon_{min}/\varepsilon_{max}$, was approximately -1 for all tests.

One important difference between TMF and TF tests is the applied strain rate. In fact, the strain rate at the very beginning of heating and cooling under TF ($-2.4 \cdot 10^{-3}$ s^{-1} and 10^{-2} s^{-1} respectively) is much higher than under the TMF experiment ($-2.5 \cdot 10^{-4}$ s^{-1} and $1.5 \cdot 10^{-4}$ s^{-1}). Therefore, if the mechanical behaviour of the alloy is sensitive to the strain rate, the TF and TMF results must be compared with care.

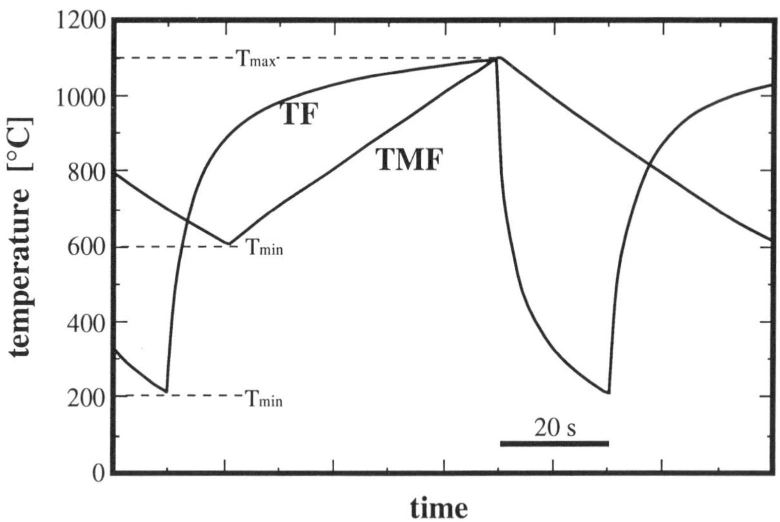

Figure 1. Temperature-time cycles used for TF and TMF tests.

3. Results and Discussion

For both OP and TFB cycles, the specimen is under compression at the highest temperature part of the cycle. Therefore, stress relaxation occurs in compression during the first cycles, a higher relaxation being observed under the TFB cycle. This stress relaxation results in the development of a positive mean stress. The stabilised stress-mechanical strain and stress-temperature hysteresis loops for the OP and TFB cycles ($\Delta\varepsilon_m$=0.7%) are reported in Fig. 3.

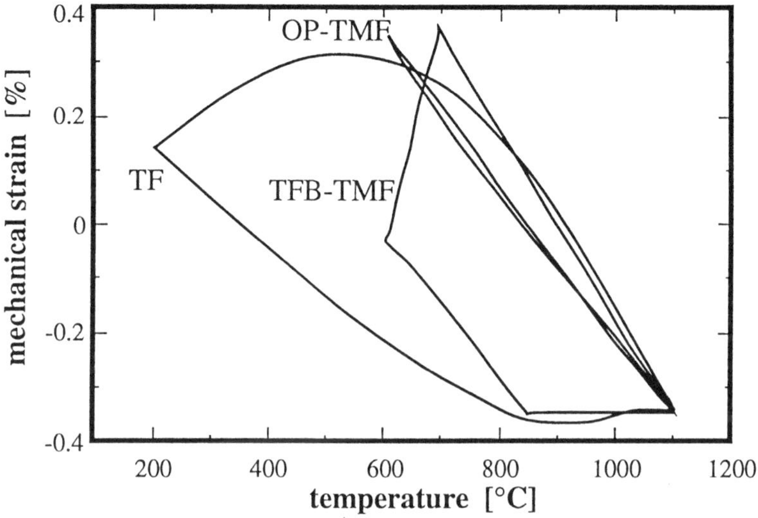

Figure 2. Mechanical strain-temperature cycles used for TF and TMF tests.

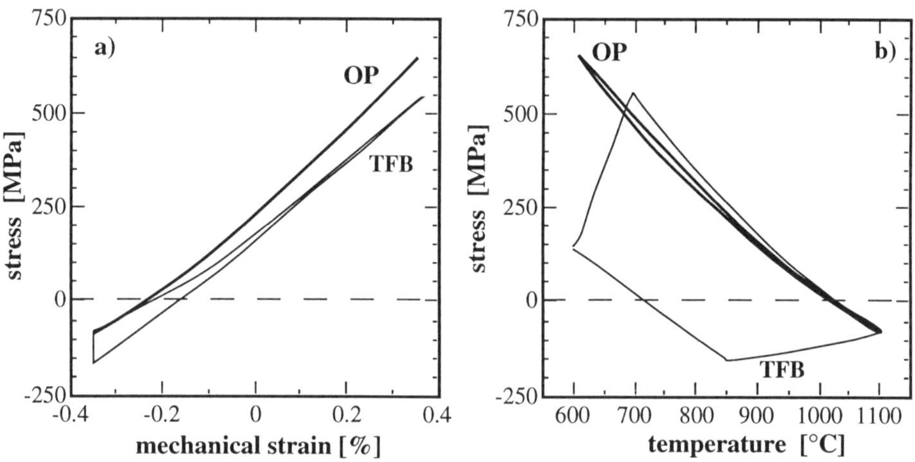

Figure 3. Stabilised hysteresis loops for OP and TFB cycles: a) stress–mechanical strain loop, b) temperature–stress loop.

Similarly to thermal fatigue tests [5], oxide-scale spalling was observed during TMF tests for lower mechanical strain ranges. Under higher strain multiple crack initiation was observed; in this case the main crack is formed by the coalescence of small cracks. More cracks were nucleated under OP cycle. Cracks always initiated on oxidised residual cast pores located in interdendritic zones, Figs. 4a-c.

Two fatigue lives are defined in the present investigation: crack initiation life (N_i), corresponding to the number of cycles to form a 0.1 mm crack depth, and total fatigue life (N_f), defined as the number of cycles where a decrease of 5% in the maximum stress is observed. This 5% reduction in stress corresponds approximately to the formation of a single 1 mm crack depth (= TMF specimen wall thickness).

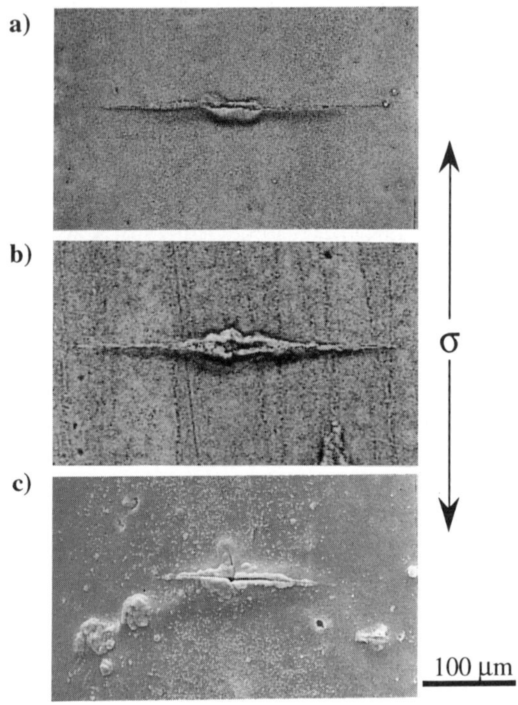

Figure 4. SEM micrographs of surface crack initiation at micropores: a) OP-TMF, 1500 cycles (replica), b) TFB-TMF, 4000 cycles (replica), c) TF (200–1100°C), 1040 cycles.

The TMF and TF lives are reported in Figs. 5a-f. It can be seen that the OP cycle results into a shorter life as compared to the TFB cycle (about a factor two). A decrease of the maximum cycle temperature from 1100°C to 900°C extends significantly the total fatigue life, as is shown for the OP cycle (the crack initiation life was not determined for the tests performed at T_{max} = 900°C).

Under the test conditions applied, TF tests result in a lower life than the TMF tests conducted under TFB cycles (similar strain-temperature history). On the other hand, comparable lives are obtained in TF tests and OP-TMF tests conducted between 600–1100°C.

During creep, high temperature low-cycle fatigue, thermo-mechanical fatigue and thermal fatigue loading, the γ-γ' microstructure of superalloys changes according to the applied strain-temperature-time history [7–9]. Microstructure investigations near the surface (≈50 μm depth) have revealed that, for the same strain range (0.7%), three different γ-γ' rafted structures are developed under the two TMF cycles and the TF loading (Fig. 6).

For the TFB specimen (5400 cycles) the γ' rafted structure is well developed parallel to the stress axis (Fig. 6b), because the alloy is under significant compression between 850–1100°C (Fig. 3b). This type of rafted structure is typically obtained under compressive creep tests. Under the OP test (1600 cycles) the rafted structure is less developed (Fig. 6a), as the time spent in compression at high temperatures is shorter. Under TF (2000 cycles, Fig. 6c) the original cubic γ'-structure is significantly coarsened, but less rafted than after TFB cycling. This leads to the presumption that the alloy is subjected to lower compressive stresses near the surface at high temperature for the thermal fatigue test.

The fatigue life curves and the observations of the microstructure clearly show that different damage accumulation mechanisms can occur under non-isothermal fatigue loading depending upon the strain-temperature-time history applied. For the test condition reported here, it seems that the rafted γ-γ' structure developed parallel to the stress axis improves the TMF resistance of the alloy. Therefore, although thermal fatigue and OP-TMF experiments result into comparable fatigue lives, the damaging mechanisms near the surface are probably different as they present distinct γ-γ' structures.

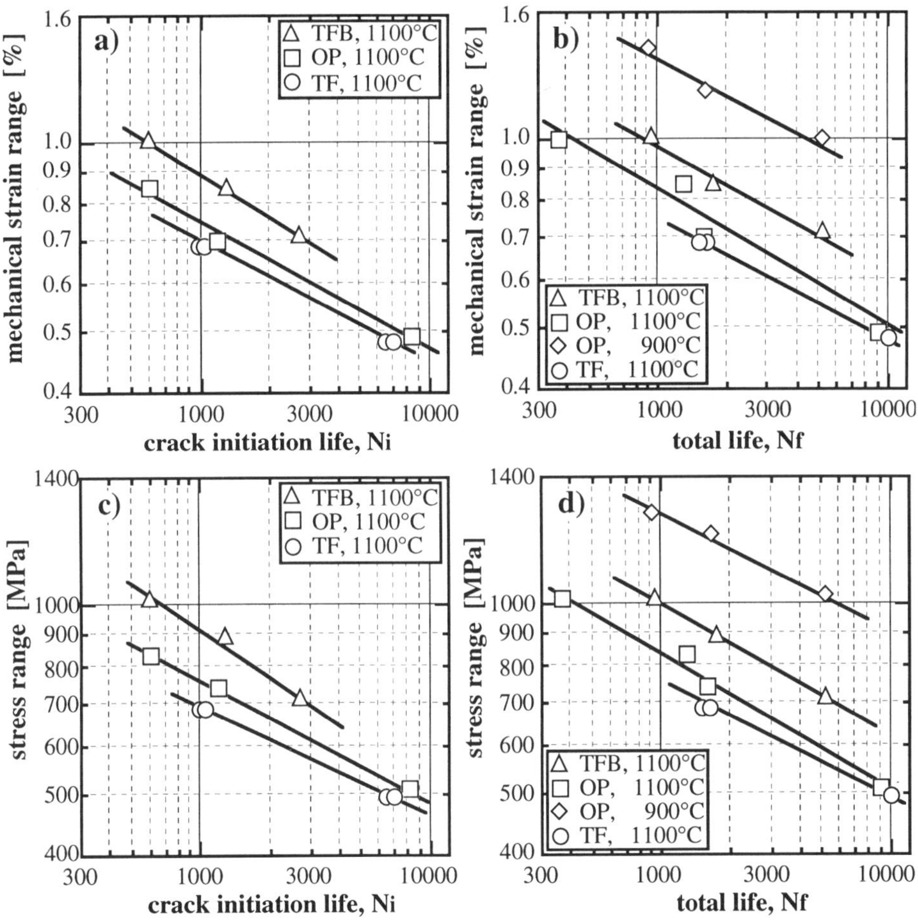

Figure 5. Crack initiation life and total life as a function of: (a), (b) mechanical strain range $\Delta\varepsilon_m$ and (c), (d) stress range. The stress ranges presented for the thermal fatigue (TF) tests were obtained from elastic calculations.

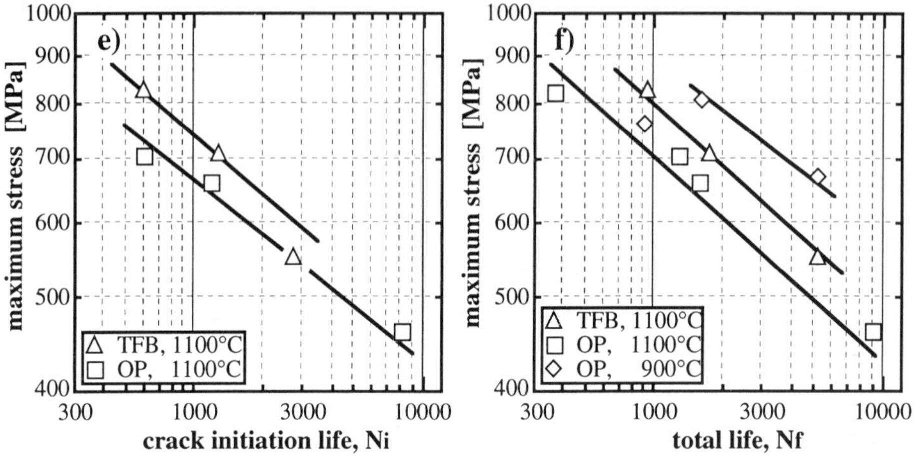

Figure 5 (cont.). **Crack initiation life and total life as a function of: (e),(f) maximum stress.**

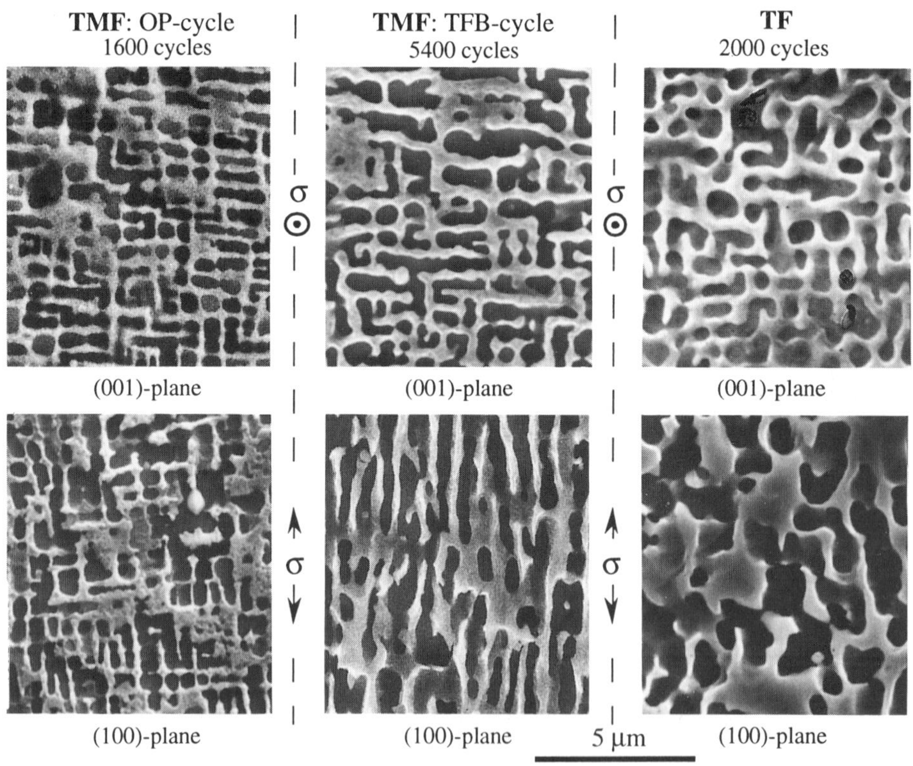

Figure 6. **γ-γ' microstructure after cycling with 1100°C maximum cycle temperature and 0.7% mechanical strain range: a) OP-TMF test, b) TFB-TMF test, c) TF test.**

4. Conclusion

The thermo-mechanical fatigue behaviour of SRR99 was investigated under out-of-phase and *thermal-fatigue-based* cycles. Stress relaxation leading to a positive mean stress occurred under both TMF cycles. For the test conditions investigated, SRR99 presents a shorter TMF life under the OP cycle than under the TFB cycle. Moreover, the life under thermal fatigue is shorter than under the TFB-TMF cycle, which experiences a similar mechanical strain-temperature history. An identical mechanism of crack initiation at the interdendritic oxidised pores is observed for all tests, but the evolution of the microstructure is not the same due to the different strain-temperature-time histories. The development of a rafted γ-γ' structure parallel to the stress axis seems to have a beneficial effect on the TMF life.

Acknowledgements - The authors would like to acknowledge MTU Munich GmbH, Germany for providing the alloy free of charge, the Swiss National Foundation and the Swiss Priority Program on Materials for supporting this investigation. C. Engler-Pinto thanks also the Secretary for Science and Technology of Brazil (CNPq) for granting his PhD scholarship. Prof. B. Ilschner is acknowledged for many fruitful discussions and support.

References

1. Hopkins, S.W., Low-Cycle Thermal Mechanical Fatigue Testing, in D.A. Spera and D. Mowbray (eds.), *ASTM STP 612*, American Society for Testing and Materials, 1976, pp. 157–169.
2. Malpertu, J.L. and Rémy, L., Influence of Test Parameters on the Thermal-Mechanical Fatigue Behaviour of a Superalloy, *Metall. Trans. A* **21**, (1990), 389–399.
3. Kraft, S., Zauter, R. and Mughrabi, H., Aspects of High-Temperature Low Cycle Thermomechanical Fatigue of a Single Crystal Nickel-Base Superalloy, *Fatigue Fract. Eng. Mater. Struct.* **16**, No. 2, (1993), 237–253.
4. Meyer-Olbersleben, F., Engler-Pinto Jr., C.C. and Rézaï-Aria; F., On Thermal Fatigue of Nickel-Based Superalloys, Michael J. Verrilli and Michael G. Castelli, (eds.), *Thermo-Mechanical Fatigue Behavior of Materials: 2nd Volume, ASTM STP 1263*, American Society for Testing and Materials, 1995.
5. Meyer-Olbersleben, F., Goldschmidt, D. and Rézai-Aria, F., Investigation of the Thermal Fatigue Behaviour of Single-Crystal Nickel-Based Superalloys SRR 99 and CMSX-4, *Superalloys 1992*, S.D.Antolovich et al., eds., TMS, Warrendale, 1992, pp. 785–794.
6. Engler-Pinto Jr., C.C., Härkegård, G., Ilschner, B., Nazmy, M.Y., Noseda C. and Rézaï-Aria, F., Thermo-Mechanical Fatigue Behavior of IN738LC, *Proceedings of Materials for Advanced Power Engineering 1994*, Part I, Kluwer Academic Publishers, 1994, pp. 853–862.
7. Pineau, A., Influence of uniaxial stress on the morphology of coherent precipitates during coarsening elastic energy considerations, *Acta Metall.*, **24**, (1976), 559–564.
8. Miyazaki, T., Nakamura, K. and Mori, H., Experimental and theoretical investigations on morphological changes of γ' precipitates in Ni -Al single crystals during uniaxial stress-annealing, *J. Mater. Sci.*, **14**, (1979), 1827–1837.
9. Arrell, D.J. and Vallés, J.L., Interfacial dislocation based criterion for the prediction of rafting behaviour in superalloys, *Scripta Metall. Mater.*, **30**, (1994), 149–153.

IMPORTANCE OF CRACK GROWTH TO DAMAGE UNDER TMF LOADING

E. E. AFFELDT[1], J. TIMM[2] and A. BENNETT[3]
[1] *MTU Motoren- und Turbinen-Union München GmbH*
P. O. Box 500640, D 80976 Munich, Germany
[2] *Institute for Advanced Materials, Joint Research Centre*
P. O. Box 2, 1755 ZG Petten, The Netherlands
[3] *Rolls-Royce plc*
P. O. Box 31, Derby DE24 8BJ, United Kingdom

1. Introduction

Thermomechanical fatigue experiments were conducted under strain-control with two different strain-temperature loops, simulating the loading situation at critical locations of aero-engine turbine blades. These cycles (-135°lag cycles with R_ε-ratios equal to zero and to $-\infty$, T_{min}= 300°C, T_{max}=1050°C; and an in-phase cycle with R_ε=0 and T_{min}= 300°C, T_{max}=850°C) were applied to two different single crystal nickel base alloys (SRR99 and CMSX6). Crack initiation and propagation were monitored in-situ by means of a computer vision system throughout the test at predefined numbers of cycles. It was shown that the specimen life under the influence of the -135°lag cycle is mainly controlled by crack propagation. Scanning electron microscopic investigations of the side and fracture surface of the specimens reveal multiple crack initiation at sites of preferential oxidation. The fracture surface indicates that crack propagation is mainly in a penny-shaped configuration. The in-phase cycle causes crack initiation from internal porosity; therefore an analysis based on surface observations is not possible. A detailed description of the TMF testing is given in [16]. To analyze the crack propagation under TMF loading a number of isothermal crack growth tests were conducted to study the influence of time, temperature and R-ratio on the crack growth rate.

2. Materials and Experimental Procedure

The materials used in this study were the two nickel base superalloys SRR99 and CMSX6. The material was cast in cylindrical form and grown to single crystals with the longitudinal direction within an angular deviation of smaller than 10° of the crystallographic direction [001]. From these cylindrical bars corner crack (CC) specimens (with square cross section of 8x8 mm^2 and threaded ends) were machined, after X-ray diffraction analysis via the Laue technique, to align the side faces of the speci-

J. Bressers and L. Rémy (eds.), Fatique under Thermal and Mechanical Loading, 159–168.

men with crystallographic planes of type {100}. To localise crack initiation a small artificial spot welding defect was produced at the corner, with a depth of 50 - 100 μm. The DC potential drop technique was applied to monitor crack propagation. To obtain a sufficiently high resolution, 50 μm thick platinum leads were spot welded near the edges of the notch. Crack depth was calculated from the potential values measured based on [1] calibrated with a beach mark technique [2] with the assumption that the crack could be handled as semicircular and the notch is only an extension of the crack length. The stress intensity factors were calculated with regard to [3] based on crack extension in the depth direction and the assumption of a quarter-circular crack, because the study concentrated on crack propagation in the [001]-direction. The crack growth rate is calculated between individual measurements if the crack extension exceeds 10 μm and by the difference in crack length divided by the difference in cycle numbers. The corresponding stress intensity factor range is based on the mean in crack length and stress range. For these tests, which were expected to start with low crack propagation rates, blockwise testing was applied. The waveform is defined as a trapezoidal signal abbreviated by the time (in seconds) for up-loading, dwell at maximum load, down-loading and dwell at minimum load; e. g. 1/1/1/1 means 1 s for up-loading, 1 s at maximum load and so on.

3. Results

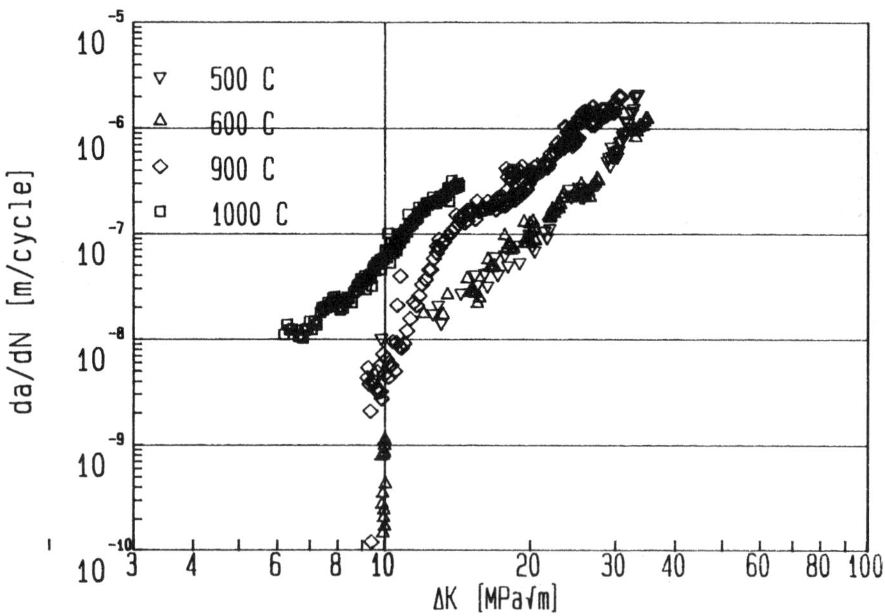

Figure 1. Influence of temperature on crack propagation rate of SRR99 with standard waveform (1/1/1/1).

The influence of temperature was tested with the standard waveform (1/1/1/1). The crack propagation rate of SRR99 is shown in Fig. 1 as a function of the stress intensity factor range. There is an clear acceleration with increasing temperature above 600°C. A comparable result was obtained with CMSX6 (Fig. 2). But some differences are obvious. Whereas the results of CMSX6 indicate no influence of a crack growth threshold for temperatures below 1050°C, that influence is obvious in most of the results for SRR99. Even at 600°C a clear deviation from the linear correlation in the Paris regime is observed. It was made certain that there is no retarding effect of the crack initiation procedure. Crack initiation was done with a sinusoidal signal of 5 Hz. At the end of this procedure the crack propagates in both materials with comparable rate. Changing to the standard signal for the crack propagation rate test at 600°C and 900°C, leads to a considerable deceleration in SRR99, partly necessitating an increase in the stress range, but not in CMSX6. This indicates that at higher temperatures the crack growth threshold is higher for lower frequencies. A comparison of the crack propagation rates of CMSX6 and SRR99 yields one scatterband for all K-values higher than 12 MPa√m at 500°C and 600°C. For lower stress intensity factor ranges the crack propagation rate of SRR99 at 600°C deviates considerably from the Paris line indicating the approach of a threshold. The same observation holds for 900°C if K is less than 15 MPa√m. At higher stress intensity factor ranges the crack propagation rate of both materials is comparable. The tests at higher temperatures show that the threshold now again is lower than at 900°C and the crack propagation rate is higher, but as the temperature deviates by 50°C, a comparison of the materials is impossible.

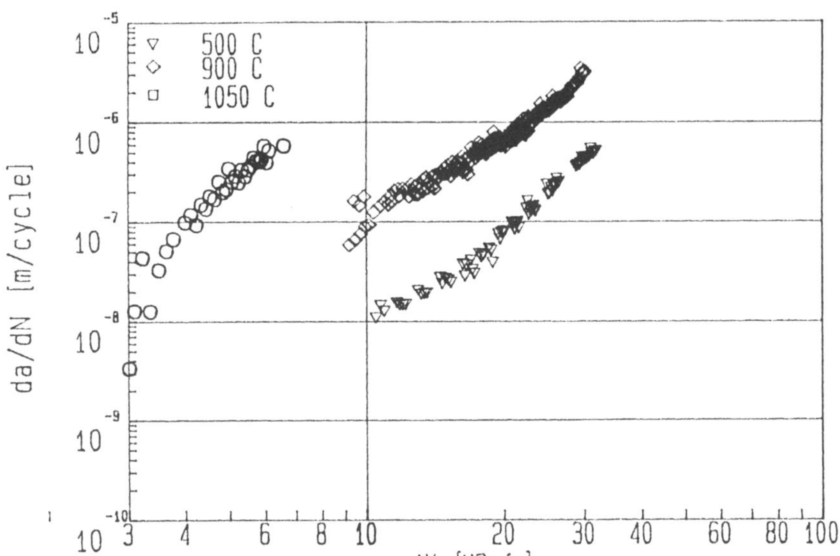

Figure 2. Influence of temperature on crack propagation rate of CMSX6 with standard waveform (1/1/1/1).

The influence of the different loading rate applied during the TMF test (which is 30 s for heating from T_{min} to T_{max} and 60 s for cooling) compared to the crack growth test signal 1/1/1/1 was investigated by testing with a signal of 30/0.5/60/0 at 900°C. This temperature was chosen, because a comparison of the above mentioned results with the crack propagation rates in TMF tests indicates comparable rates for a temperature near to 900°C. The results for both materials are given in Figs. 3 and 4

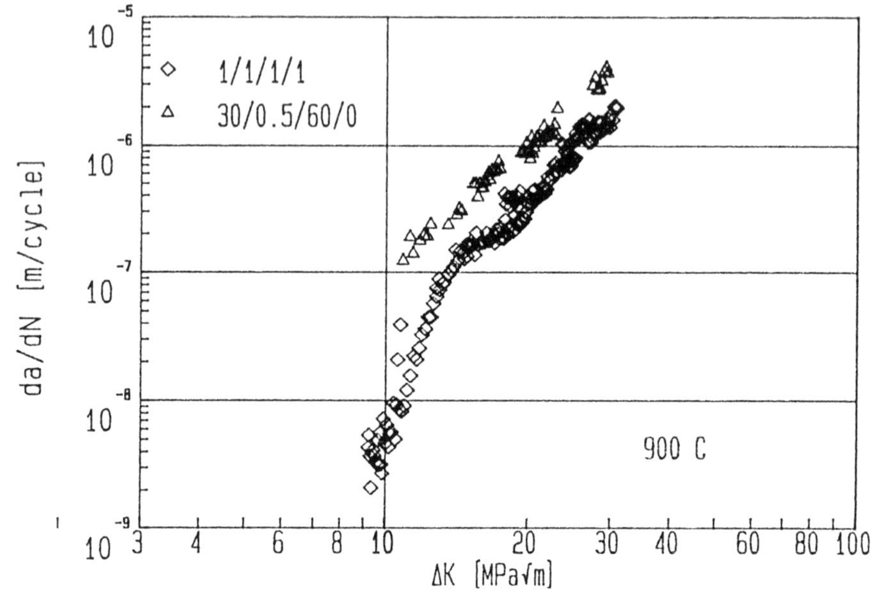

Figure 3. Influence of loading rate on cack propagation rate of SRR99 at a temperature of 900°C.

respectively. Compared to the standard signal the crack growth rate is higher by a factor of about 2.5 for both materials. The influence of the different R-ratio in the TMF test (TMF cycling with a strain-ratio R_ε=0 and -∞ leads after some percentage of the lifetime to a comparable stress-ratio of approximately R_σ=-0.5) and R_σ=0 in the crack growth test was investigated only for CMSX6. Fig. 5 shows a comparison of the crack propagation rate with R_σ=0 to a test with R_σ=-0.5. For the R_σ =-0.5 test the stress intensity factor range was evaluated in two different ways, one accounting for the whole stress range, the other only for the tensile part of it. The R_σ =0 result lies between the two, but nearer to the result based on the whole range. All cracks propagated on planes approximately perpendicular to the stress axis. The fracture surface of tests performed at 500°C and 600°C is smooth, with some small irregular features and a regular pattern of lines perpendicular to the crack growth direction. This pattern is usually described as striations, but a comparison of the spacing at a crack depth of 0.5 mm (ΔK = 12 MPa$\sqrt{m}$, da/dN = 0.02 μm/cycle) and at a depth of 2 mm (ΔK = 30 MPa$\sqrt{m}$, da/dN = 0.3 μm/cycle) shows almost the same spacing of 0.5 μm (Fig. 6). Striations are usually thought to reflect the plasticity controlled opening (blunting) and closing of the crack tip during one cycle, which would contradict the crack growth measurement. This indicates that the pattern in Figs. 6a, b is caused by some other plasticity related effect, e.g. correlated to the periodicity of the microstructure. At higher temperatures the fracture surface is covered with an oxide scale, which determines the appearance; differences in loading rates are not discernible.

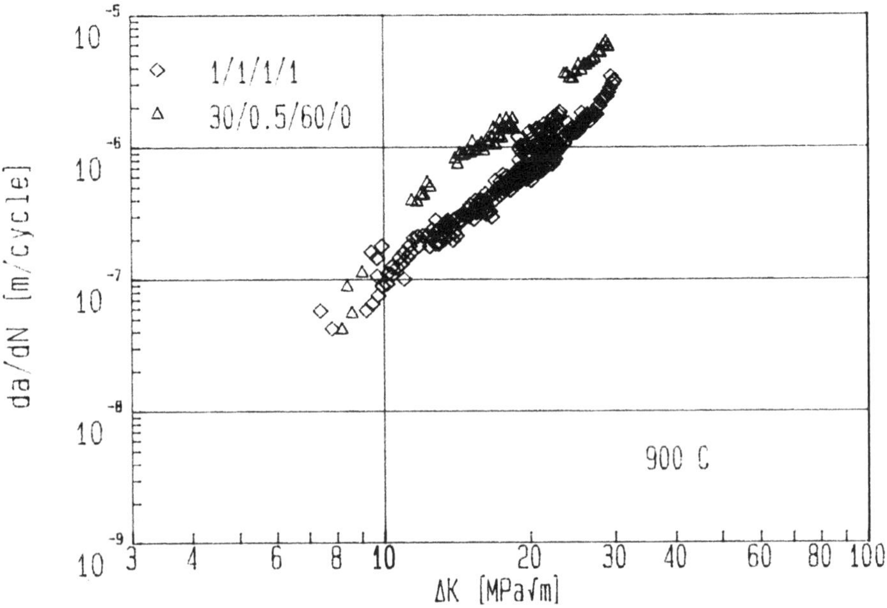

Figure 4. **Influence of loading rate on crack propagation rate of CMSX6 at a temperature of 900°C.**

4. Discussion

Crack propagation in nickel-base superalloys shows only little influence of temperature and frequency at intermediate temperatures [4-6]. Part of that influence can usually be accounted for by the well-known temperature dependence of the elastic constants (Young's modulus) and the (cyclic) yield strength, both controlling the crack tip plasticity and therefore the crack advance per cycle. As temperatures increase above 600°C and for frequencies below 1 Hz, additional temperature dependent processes accelerate the crack propagation rate and the crack path changes from transgranular to partly or totally intergranular. There is some disagreement in the literature [5-7] but most authors attribute that acceleration to an influence of the oxidation, based on several observations, e.g.

- the crack propagation rate in vacuum [8] is not accelerated at temperatures where acceleration under the influence of air/oxygen is observed
- the crack propagation rate in the interior of specimen [4] is not accelerated at temperatures where on acceleration is observed for cracks at the specimen surface
- the crack propagation rate is often orders of magnitude higher than under static load.

In single crystals the crack propagation rate is additionally influenced by the mode (stage I/II) and the crystallographic orientation of the crack plane and crack growth direction [10-13]. To avoid the effect of such variables, this study concentrated on a fixed crystallographic orientation of the specimens. In order to differentiate the different mechanisms it is possible to show the crack propagation rate for a fixed

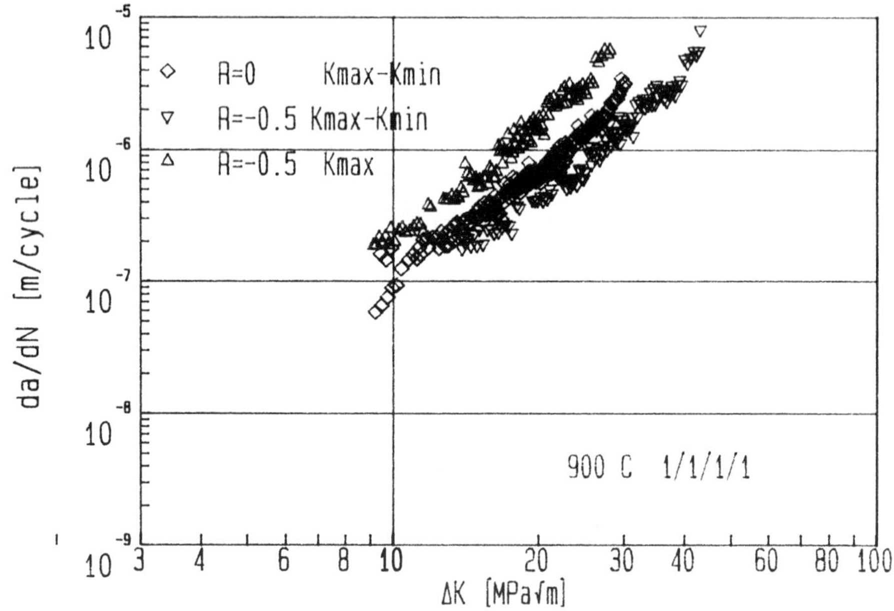

Figure 5. Influence of R-ratio on crack propagation rate of CMSX6 at a temperature of 900°C.

Figure 6. Fracture surface at a crack depth of 0.5mm (left) and 2.0mm (right) showing comparable striation spacing.

crack driving force, e.g. the crack opening displacement as a function of the inverse temperature (Fig. 7) [4]. Data from previous studies on the same alloy were used.

The crack propagation rate is controlled by an almost temperature-independent part (fatigue) and a thermally activated acceleration (cc: corrosion or creep), which is probably an oxidation phenomenon as previously discussed. That can be described by a linear superposition [4]:

$$\frac{da}{dN} = \frac{da}{dN}(fat) + \frac{da}{dN}(cc) \tag{1}$$

The thermally activated part may be written as:

$$\frac{da}{dN}(cc) = C_0 \exp(-\frac{Q}{RT})* f(t)* K^n \tag{2}$$

It is obvious that the influence of a different loading rate or period t can be transferred to another temperature. The above-mentioned description characterizes the behaviour in the Paris regime. Concerning the experimental data, the crack propagation rate of both alloys is independent of thermally activated processes at temperatures below 800°C, which leads to comparable rates and frequency independent behaviour for 500°C and 600°C over most of the stress intensity factor range investigated. At 900°C the behaviour changes drastically. Whereas at the upper end of the ΔK range the crack gowth rates of both materials approach one another, they differ by an order of magnitude at ΔK=10 MPa$\sqrt{m}$. That difference is, however, mainly caused by a difference in threshold. There is some evidence in the literature [14,15], in agreement with the reported results, that in nickel-base alloys the threshold for crack propagation increases with temperature, sometimes resulting in a cross-over of the crack growth curves for different temperatures. A possible explanation for that behaviour is that the same process which accelerates the crack propagation rate at high stress intensity factor ranges caused the higher threshold [10]. The oxide build-up leads to a closure effect. For SRR99 that behaviour is also found in [10]. Comparing the data of this study to [10, 13] shows that the threshold values are less than those reported in [10, 13]. That agrees with the in-house experience that the threshold for CC specimens is lower than those of the usually applied Compact Tension geometry (CT). At higher temperatures two opposing effects now shift the threshold back to lower values. The faster crack growth rate due to faster oxidation rates at higher temperature is countered by the dramatically declining yield strength.

To compare the isothermal crack propagation rate to the rates under TMF loading, one has to account for the different loading rate and the different R-ratio. That is done by shifting the SRR99 curve for 900°C and slow loading rate by a factor of 2 which allows for the change in R-ratio. The result is given in Fig. 8, which shows data points of the crack growth rate for several cracks on the side surface of a TMF specimen. Interference can partly decelerate their growth rates. But the main crack, causing failure, is not influenced. Therefore an upper bound, as shown in the diagram, is thought to best represent the results. Comparison to the isothermal data shows that the crack growth rates approach one another at ΔK=10 MPa$\sqrt{m}$, but the slope of the isothermal data is steeper. The observation that the crack growth rate under TMF loading is comparable to that under isothermal LCF at 900°C (which is accelerated by oxidation, Fig. 7) indicates a thermal activation, but less than for an

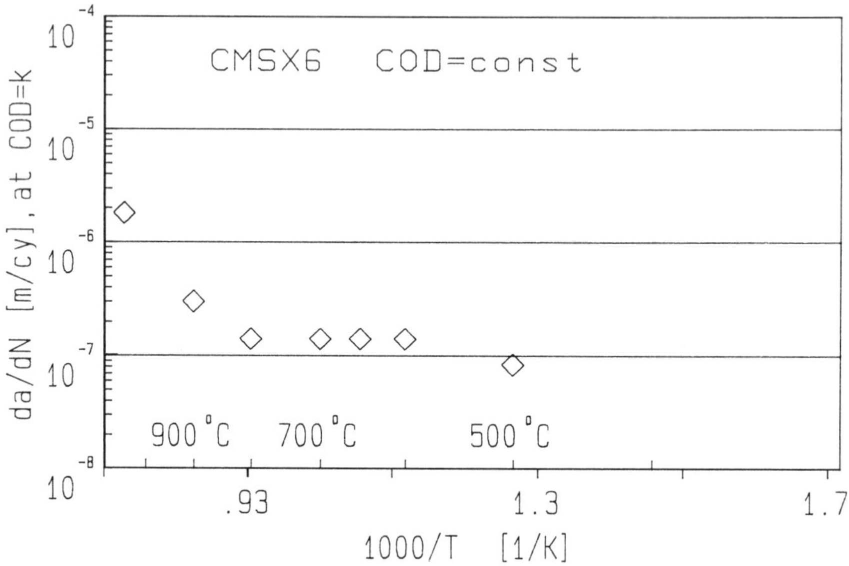

Figure 7. Crack growth rate of a constant crack opening displacement as function of the inverse temperature.

isothermal crack propagation rate at maximum temperature. That is consistent with the interpretation that the acceleration in TMF is based on the build-up of oxide at the crack tip, which should be influenced by the fact that in TMF cycling the crack tip spends only a small part of the test cycle at high temperature. On the other hand the different slopes of the curve and the fact that there are cracks propagating at a stress intensity factor range much smaller than the value estimated from the isothermal tests indicate that the TMF behaviour can hardly be represented by LCF data at some equivalent temperature. A more detailed investigation of the relationship between stress and temperature shows that at high temperatures the stresses are low, causing crack closure. Therefore neither plastic deformation nor oxidation occurs at the crack tip. As a result of that, the crack growth rate is not accelerated by oxidation. Also an increase of the threshold, as one would expect for isothermal crack growth, caused by an oxide induced closure, is suppressed. At low temperatures, stresses reach their maximum but plasticity is small and oxidation has only a small chance to embrittle the crack tip of the now-open crack. This is a plausible explanation of the surprisingly low crack growth rate under TMF loading.

5. Summary and Conclusion

The number of cycles till fracture under TMF loading is mainly controlled by crack propagation [16]. To study the contribution of different parameters to the isothermal crack propagation rate during LCF, the influence of temperature, loading rate and

R-ratio were investigated for the two single crystal nickelbase alloys SRR99 and CMSX6. It was found that the crack propagation for a trapezoidal signal of 0.25 Hz is mainly controlled by fatigue below temperatures of 800°C, but the threshold value in SRR99 is increased with respect to CMSX6 as the result of oxide induced closure. At 900°C the crack propagation rate is accelerated with respect to the lower temperatures for both alloys at high stress intensity factor ranges, but the different behaviour of the threshold causes an order of magnitude difference at a stress intensity factor range of about 10 MPa$\sqrt{m}$. A comparison with the crack propagation rate under TMF loading indicates rates comparable with those for isothermal crack growth at 900°C. A different slope in the Paris regime and a difference in the threshold value shows that a clear understanding necessitates additional investigations focusing on:

- a study of the influence of the specimen form,
- a comparison of the crack growth rate under isothermal and TMF loading tested with the same specimen and crack length measurement method.

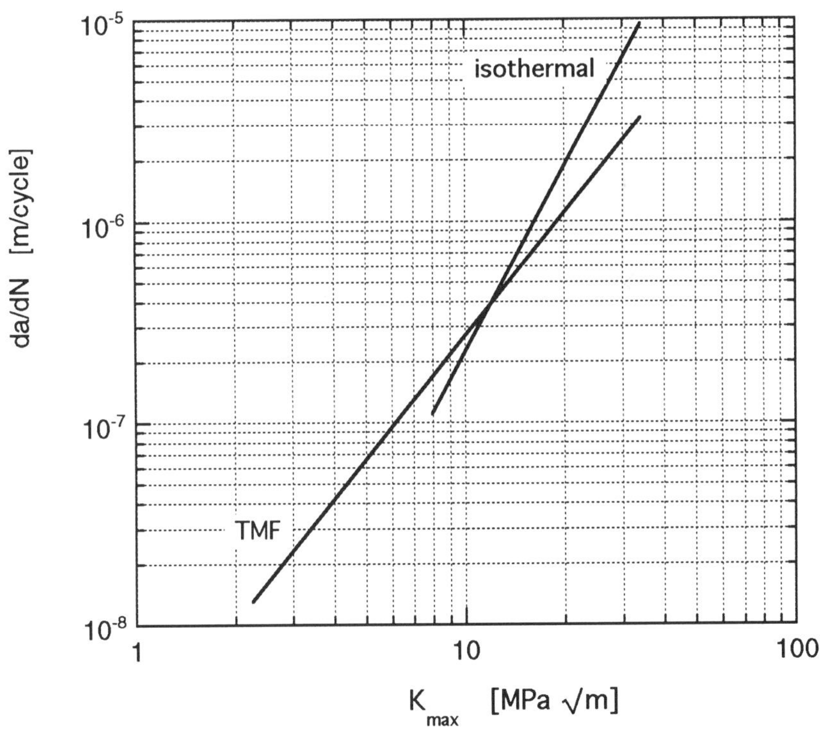

Figure 8. Comparison of crack growth rate for TMF loading with isothermal crack growth rate at 900°C.

Acknowledgements - This work is part of the Brite-Euram project BE3338-89 with financial support from the EEC, coordinated by Rolls-Royce (UK). The authors thank J. Haug for his assistance.

References

1. Roe, G.M. and Coffin L.F., Unpublished Research, General Electric Corporate Research and Development, Schenectady 1978.
2. Wei, W., Flöge, H. and Affeldt, E.E., Investigation of the Low Cycle Fatigue Behaviour of IN100 using a Beachmark Technique, *Scripta Metall.* **25**, (1991) pp. 1757-1761.
3. Newman, J.C., Jr. and Raju, I.S., Stress-Intensity Factor Equations for Cracks in Three-Dimensional Finite Bodies, *Fracture Mechanics: Fourteenth Symposium ASTM STP 791*, J.C. Lewis, G. Sines, American Society for Testing and Materials, Philadelphia, (1983) pp. I-238-I-265.
4. Affeldt, E.E., Flöge, H. and König, G., Influence of Time and Temperature on Crack Growth Rate in the P/M Nickelbase Alloy Udimet 700 with Three Different Microstructures, *Low Cycle Fatigue and Elasto-Plastic Behaviour of Materials - 3*, Elsevier Science Publishers Ltd, Essex, (1992) pp. 166-171
5. Nikbin, K.M., Predictions of the Initial Non-Steady-State Crack Growth Behaviour in the Creep/Fatigue of the Nickel-Base Superalloy AP1, *J. Materials Engineering and Performance* **2**, (1993), pp. 537-544.
6. Ghonem, H., Nicholas, T. and Pineau, A., Elevated Temperature Fatigue Crack Growth in Alloy 718 - Part II: Effects of Environmental and Materials Variables, *Fatigue Fract. Engng Mater. Struct.* **16**, (1993) pp. 577-590.
7. Nicholas, T. and Weerasooriya, T., Hold-Time Effects in Elevated Temperature Fatigue Crack Propagation, *Fracture Mechanics: Seventeenth Symposium. ASTM STP 905*, J.H.Underwood,et al., American Society for Testing and Materials, Philadelphia, (1986) pp.155-168.
8. Pelloux, R.M. and Huang, J.S., Creep-Fatigue-Environment Interactions in Astroloy, *Creep-Fatigue-Environment Interactions*, R.M. Pelloux, N.S. Stoloff, The Metall. Soc. of AIME, Warrendale, (1980) pp.151-164.
9. Arana-Antelo, M., *et.al.*, Initiation and Growth of Thermomechanical Fatigue Cracks in coated and uncoated Turbine Blade Materials, Brite-Euram report MAT48-TM9, 1994.
10. Henderson, M.B. and Martin, J.W., The Influence of Orientation, Temperature and Frequency on Fatigue Crack Growth, In S. D. Antolovich et al., (eds.), *Superalloys 1992*, The Minerals, Metals & Materials Soc. (1992), pp.707-716.
11. Antolovich, B.F., Saxena, A. and Antolovich, S.D., Fatigue Crack Propagation in Single -Crystal CMSX-2 at Elevated Temperature, *J. Materials Engineering and Performance* **2**, (1993), pp. 489-496.
12. Howland, C., The Growth of Fatigue Cracks in a Nickel Base Single Crystal, In K.J. Miller, E.R. de los Rios, (eds.), *The Behaviour of Short Fatigue Cracks*, Mechanical Engineering Publishers, London, (1986) pp.229-239.
13. Yang, R., Creep-Fatigue Crack Growth in a Nickel Base Superalloy, Ph. D. Thesis, Imperial College of Science and Technology, London, 1991.
14. Galatolo, R., Fatigue Crack Growth Threshold at high Temperature on Engine Disc Superalloys, *European Propulsation Forum*, London, (1993) pp. 7.1-7.7.
15. Van Stone, R.H. and Krueger, D.D., Near-Threshold Crack Growth in Nickel-Base Superalloys, T.A. Cruse, (eds.), *Fracture Mechanics: Nineteenth Symposium.* ASTM STP 969, American Society for Testing and Materials, Philadelphia, (1988) pp. 883-906.
16. Bressers, J., Timm, J., Williams, S., Bennett, A. and Affeldt, E., Effects of Cycle Type and Coating on the TMF Lives of a Single Crystal Nickel Based Gas Turbine Blade Alloy, to be published in M. J. Verilli, M. G. Castelli, (eds.), *Thermo-Mechanical Fatigue Behaviour of Materials ASTM STP 1263*, American Society for Testing and Materials, Philadelphia 1995.

INFLUENCE OF CYCLE SHAPE AND SPECIMEN GEOMETRY ON TMF OF AN ODS NICKEL-BASE SUPERALLOY

M. MARCHIONNI, D. RANUCCI and E. PICCO
CNR - ITM
Via Induno 10, 20092 Cinisello B. (Milano), Italy

1. Introduction

The gas turbine components of jet engines and power plants are usually subjected to cyclic stresses and strains during thermal transients induced by start up and shut down. The resulting thermal/mechanical fatigue cycling determines times to failure and fracture mode of the components that are not always well reproduced by laboratory isothermal low cycle fatigue tests. In order to improve the information on high temperature mechanical properties of the alloys used for turbine components like vanes and blades, the thermal mechanical fatigue technique was introduced several years ago [1, 2]. The most important advantage of this technique is the capability to test the material with temperature and strain conditions as close as possible to those of the components in service, in particular for those new, high technology alloys that may exhibit peculiar response to in service loading, due to heterogeneity, anisotropy, etc. [3].

Since TMF testing is a time consuming technique, the main goal of the researchers is to investigate whether low cycle fatigue, sensibly shorter and less expensive, is able to predict the fatigue life also during TMF conditions. The results obtained till now are contradictory as in some cases isothermal LCF and TMF lives are similar [4, 5], while in other cases they are strongly different [6, 7].

This paper describes the results of a study where the thermal mechanical fatigue (TMF) behaviour of ODS MA 6000 superalloy was investigated in the temperature range of 550°C - 1050°C. Different thermal mechanical histories, diamond type (simulating in service conditions), in phase and out of phase were applied to solid and hollow specimens in order to analyse the influence of cycle shape and specimen geometry on TMF life.

2. Material and Experimental Procedures

Inconel MA 6000 alloy is a powder metallurgy nickel-base superalloy produced by the mechanical alloying process and containing a finely distributed dispersoid of yttrium oxide (Y_2O_3). The presence of these fine particles allows to extend the operating

J. Bressers and L. Rémy (eds.), Fatique under Thermal and Mechanical Loading, 169–178.

temperature to about 1100°C [8]. The nominal composition of the alloy is given in Table 1.

The material has been supplied in the form of hot rolled bars after the following heat treatment: 1230°C for ½ hr air cooled + 950°C for 2 hrs air cooled + 850°C for 24 hrs air cooled. The final microstructure presents large elongated grains in the hot rolling direction.

The TMF tests were performed by a servohydraulic closed loop machine on cylindrical specimens with 7 mm diameter and 12 mm gauge length. Some tests were performed on hollow specimens with 7 mm external diameter and 1.5 mm thickness. Hollow was made by spark erosion and no polishing was performed in the internal surface.

The samples were heated by an induction coil and the temperature was controlled by a thermocouple spot-welded onto the specimen.

TABLE 1. Chemical composition of MA 6000 alloy in wt%.

C	Ta	W	Fe	Zr	N	Cr	Ti
0.059	1.97	4.12	0.94	0.13	0.16	14.94	2.23

Al	B(ppm)	Mo	Ni	S	O	Y_2O_3
4.26	110	2.04	bal	0.005	0.58	1.09

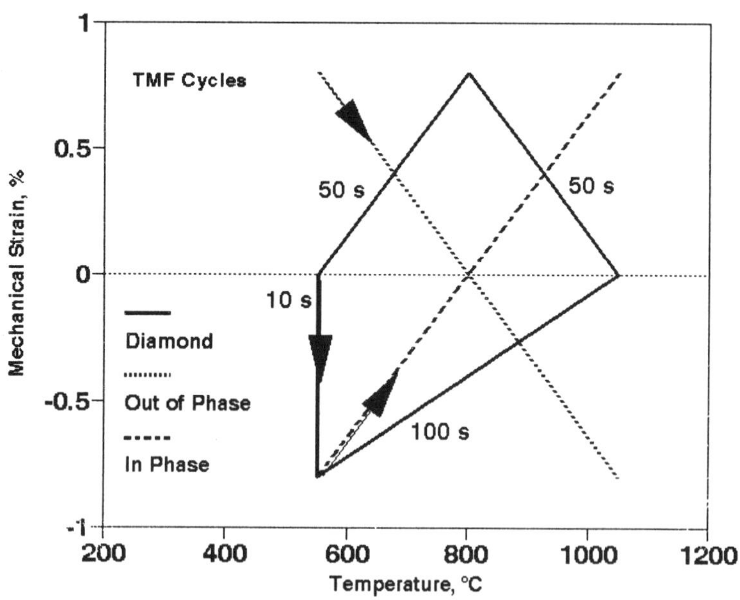

Figure 1. Shape of the TMF cycles.

The TMF cycles adopted are shown in Fig. 1. Arrows indicate the cycling direction and starting point. In the diamond cycle (D cycle) the time is indicated for each segment respectively.

The tests were carried out in the temperature range of 550°C - 1050°C fully reversed in total strain range control. The D cycle models the temperature variations of uncooled vanes during in-service start-up and shut-down [9]; the cycle time of 210 s is the minimum value obtainable without cooling on the specimen. The influence of cycle shape on the TMF results was analysed by performing in phase (IP) and out of phase (OP) cycles with the same temperature range of the diamond cycle (550°C - 1050°C) and cycle time of 210 s. The influence of the specimen geometry was investigated by performing tests on hollow samples using the D cycle, due to the scarce material availability and to the long endurance of TMF experiments. The fatigue damage was analysed by scanning electron microscopy (SEM) observations of fracture surfaces.

3. Experimental Results and Discussions Testing

3.1. TMF RESULTS

The hysteresis loops generated by D, IP and OP cycles are sensibly different, as shown in Figs. 2 - 5, where the stress response as a function of the mechanical strain is plotted. The hysteresis loops of solid and hollow specimens when applying a D cycle are similar in shape, but the stress response is slightly higher and hysteresis loop area is smaller for hollow specimens.

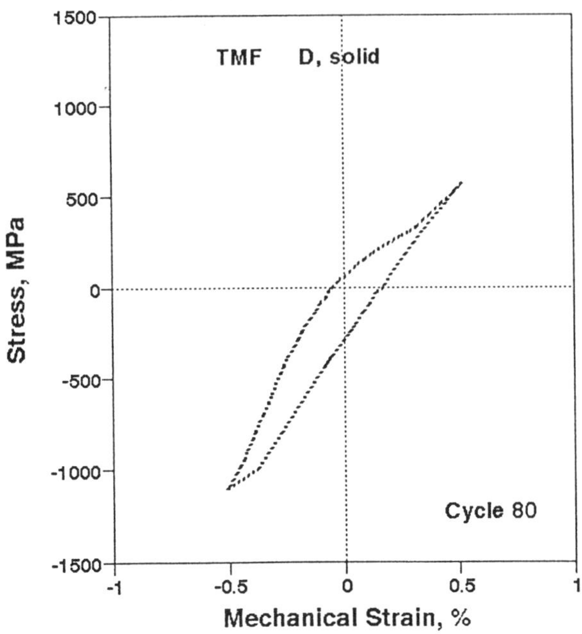

Figure 2. Hysteresis loop generated in a D cycle.

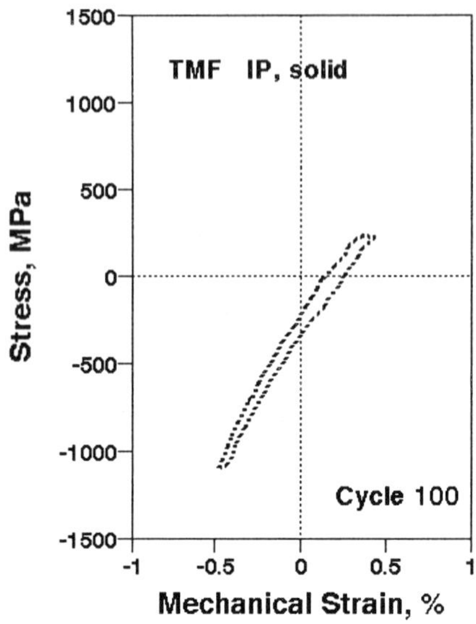

Figure 3. Hysteresis loop generated by a P cycle.

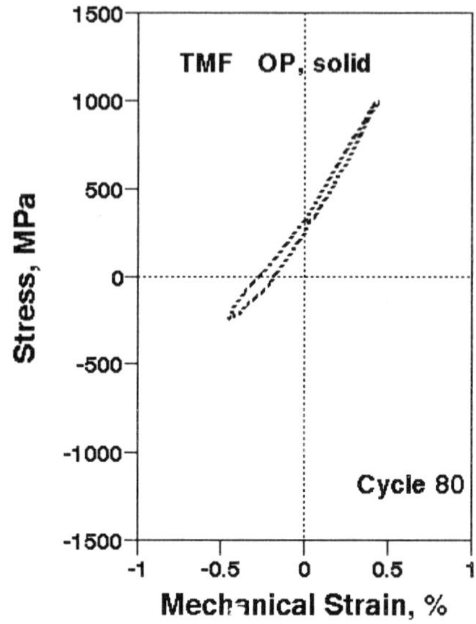

Figure 4. Hysteresis loop generated by an OP cycle.

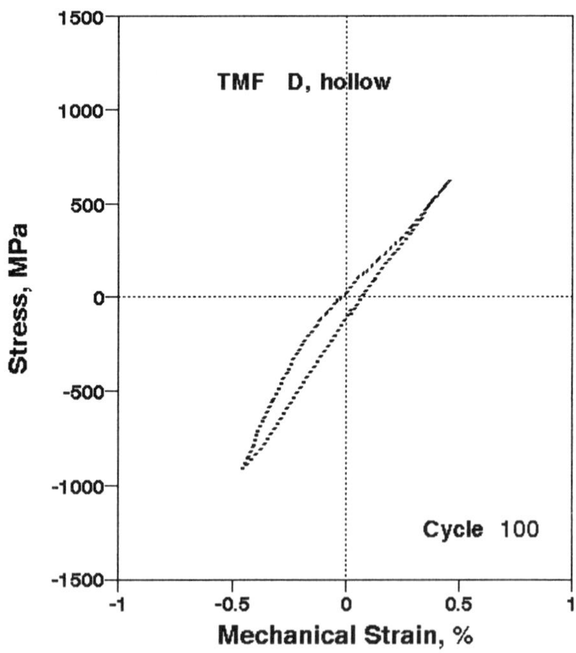

Figure 5. Hysteresis loop generated by a D cycle on a hollow specimen.

Fig. 6 presents the TMF results plotted as mechanical strain range versus number of cycles to failure. Failure was assumed when a 20% stress response was reached. In some test the complete failure occurred before stress drop. The longest fatigue life is obtained on solid specimens tested applying by a D cycle. When IP and OP cycles are used or hollow specimens are tested, a remarkable fatigue life reduction and a wider increased scatterband are observed. However, taking into account the complexity of the TMF testing procedure and the results obtained in a round robin testing performed by European laboratories [10], a maximum scatterband of a factor of 2 to 4 for each type of experiment can be considered satisfactory.

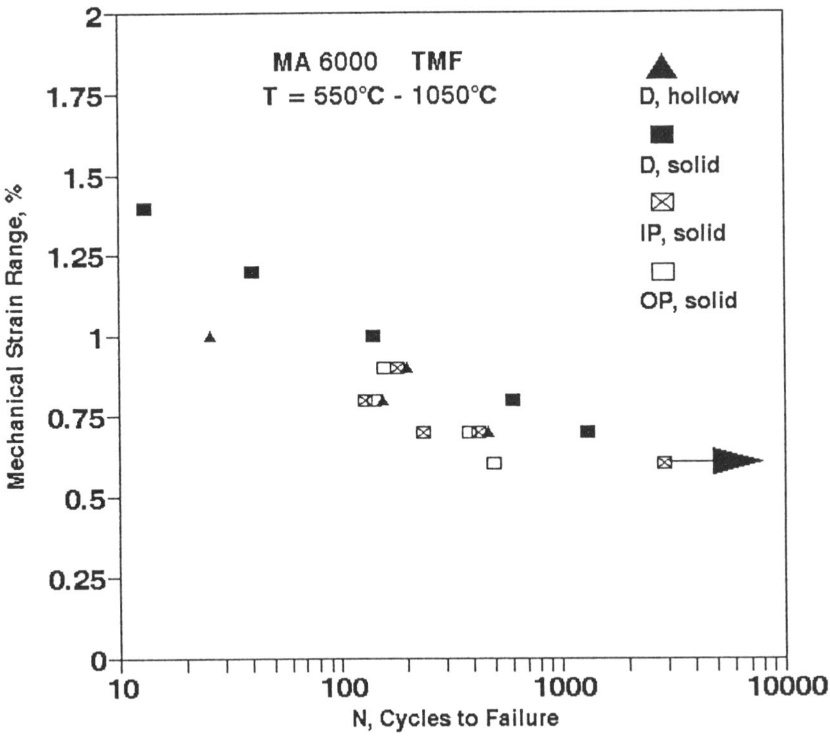

Figure 6. Comparison of TMF results in different testing conditions.

3.2. FRACTURE SURFACE EXAMINATION

Examination of the fracture surface after isothermal fatigue damage revealed crack (or cracks) initiation at the external surface of the specimen and propagation in stage II mode with or without the presence of fatigue striations according to the material ductility or brittleness [10].

This behaviour is also observed in TMF testing, even if the time per cycle is considerably longer. After a relatively short or long time to crack initiation, correlated to the cycle shape and to the imposed strain range, cracks generally start on the external surface and propagate transgranularly in stage II mode.

Fig. 7 shows an example of crack initiation on a specimen tested with a D cycle at a mechanical strain range of 1%. Initiation is located at the surface and cracks propagate inward with the presence of fatigue striations. In this specimen multiple crack initiation has been observed. One of the cracks initiated in the specimen bulk at a casting pore (Fig. 8). However multiple crack initiation is found more frequently at lower mechanical strain ranges.

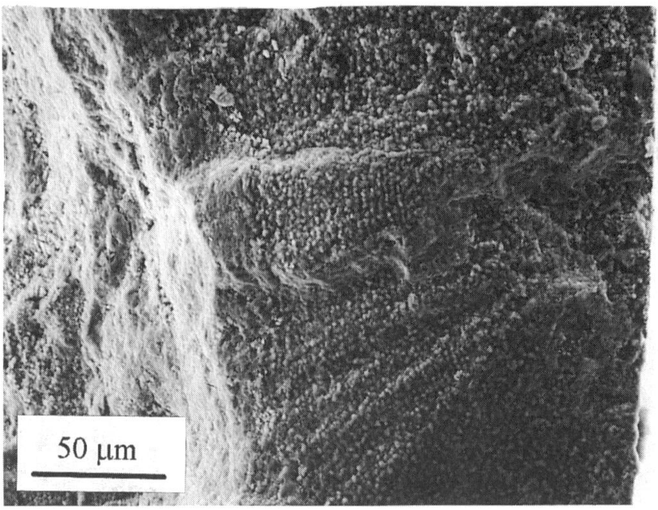

Figure 7. Aspect of the fracture surface: D cycle, $\Delta\varepsilon = 1\%$, N= 140.

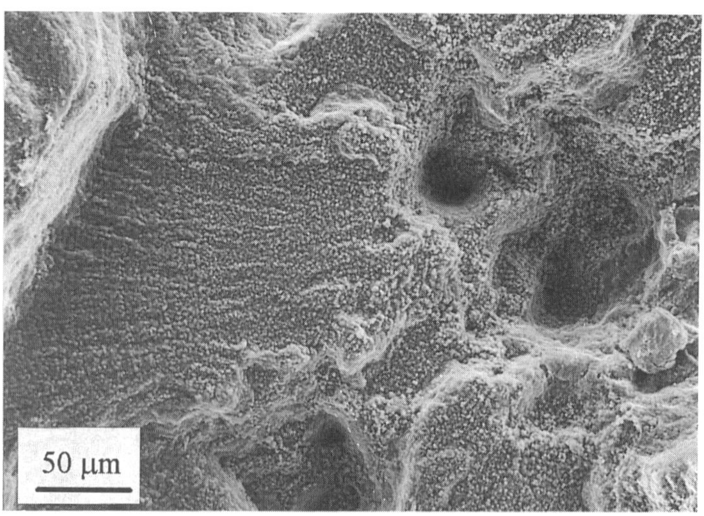

Figure 8. The same specimen of Fig. 7 showing a crack initiation in correspondence of pores.

The appearance of the fracture surface morphology is similar for IP and OP tests. Surface crack initiation and transgranular propagation are observed on the specimens tested in these experimental conditions. The specimens tested with the IP cycle show no fatigue striations (Fig. 9) and a large distribution of secondary cracks (Fig. 10). In the fracture surfaces produced by OP cycling several zones with fatigue striations and no secondary cracks have been observed (Fig. 11).

In hollow specimens tested with a D cycle, cracks frequently initiate at the internal surface (Fig. 12) and fatigue striations are also apparent. In the same tests external surface crack initiation has also been observed.

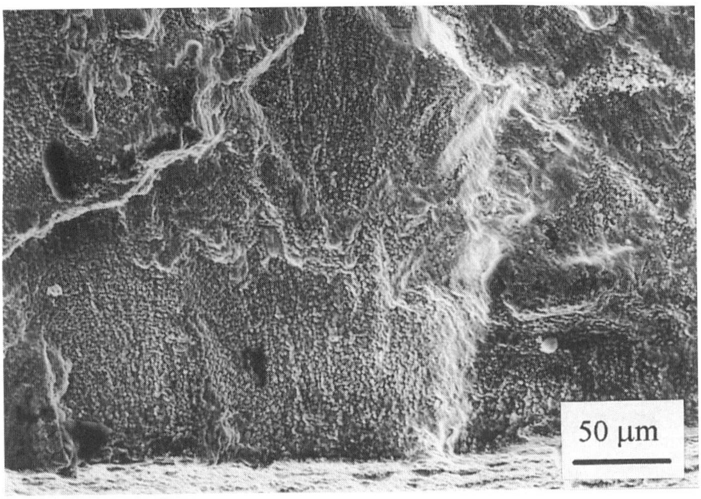

Figure 9. Fracture surface showing crack initiation: IP cycle, $\Delta\varepsilon = 0.9\%$, N= 182.

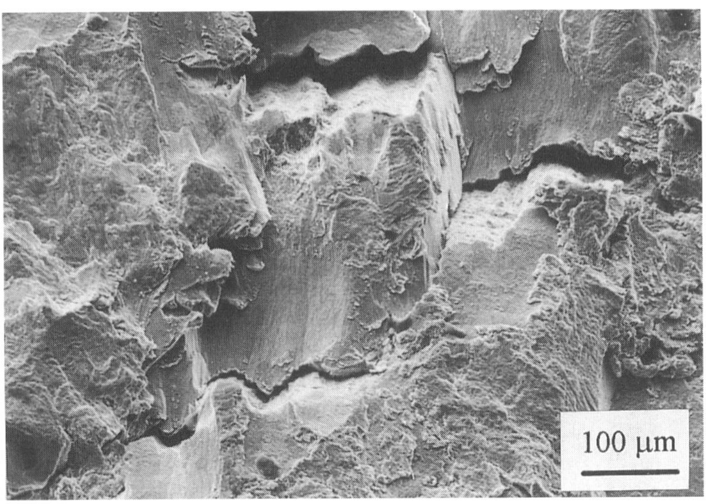

Figure 10. The same specimen of Fig. 9 showing large secondary cracks.

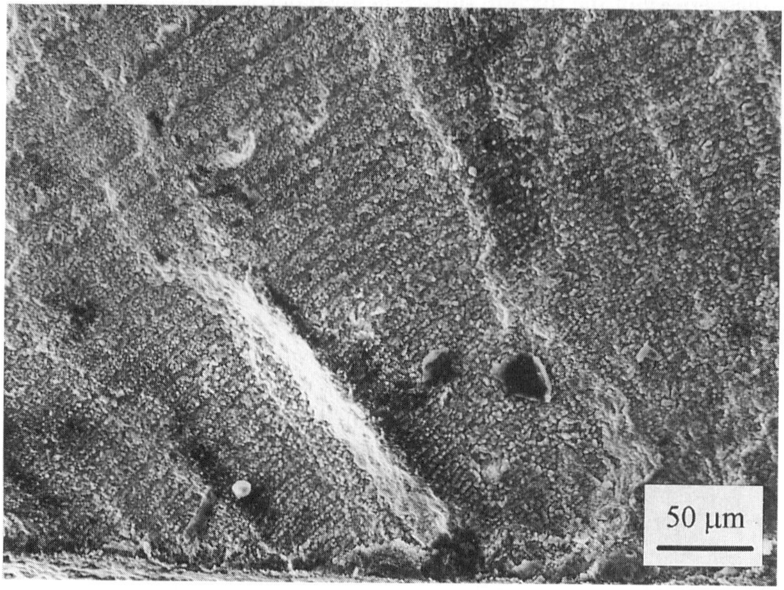

Figure 11. Aspect of the fracture surface: OP cycle, $\Delta\varepsilon = 0.9\%$, N= 157.

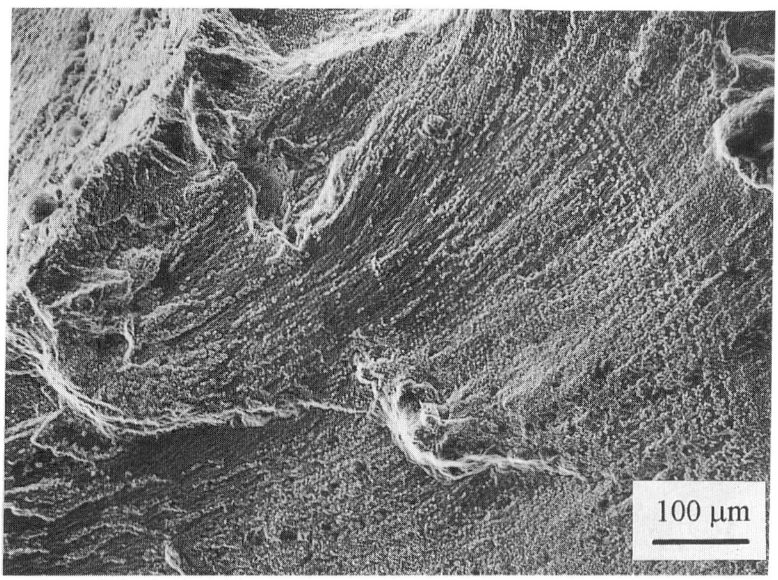

Figure 12. Fracture surface of a hollow specimen: D cycle , $\Delta\varepsilon = 0.9\%$, N= 200.

3.3. DISCUSSION

The experimental results obtained on MA 6000 alloy have shown a strong influence of the cycle shape on the fatigue life. The short lives observed during IP and OP cycling can be ascribed to their greater damaging effect. In fact, the maximum temperature is coincident with the maximum tensile strain and the maximum compressive strain respectively.

The characteristics of the cycle shape are reflected by the plot of stress amplitude versus imposed mechanical strain, evaluated at half life (Fig. 13). The stress response for IP and OP cycles is lower than that of a D cycle, confirming a lower stress carrying capability of the material due to the extreme temperature in correspondence to the maximum and minimum strain imposed. For the hollow specimen tested with a D cycle, the stress response is comparable to that of the solid specimen, and the same fatigue life would be expected. The sensible life reduction observed may be ascribed either to the increased - surface area (external and internal) at which cracks can incubate, or to a different stress distribution possibly due to a different temperature profile. These conditions can markedly reduce the time to crack nucleation.

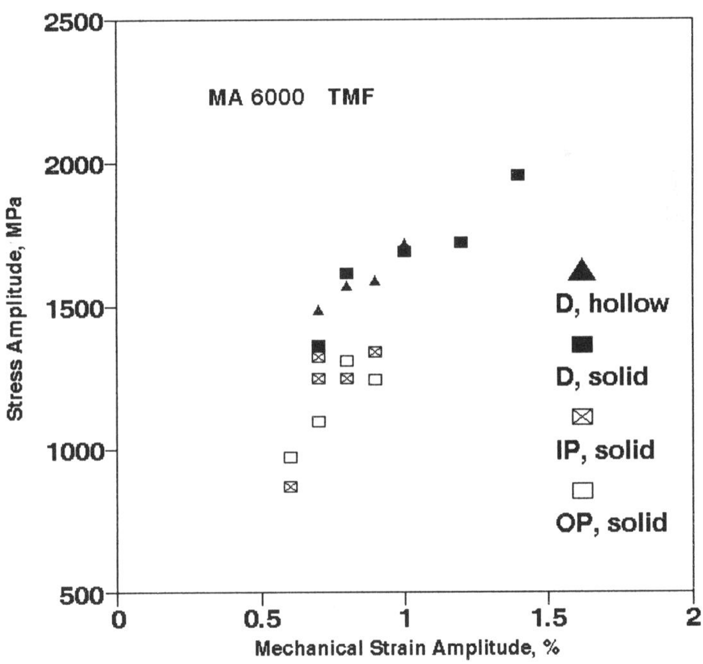

Figure 13. Stress amplitude versus strain range evaluated at half life.

4. Conclusions

The analysis of the TMF testing performed on MA 6000 alloy in the temperature range of 550°C - 1050°C with different cycles and specimen geometries leads to the following conclusions:

- The IP and OP cycling on solid specimens results in a reduction of fatigue life and in an increase of the scatterband with respect to the D cycle.
- The fatigue life of hollow specimens is sensibly lower than that of solid specimens tested with the same D cycle, probably due to a reduced time of crack incubation and to different temperature distribution.
- Cracks generally initiate at the external surface and propagate in the transgranular mode with the presence of fatigue striations; in the specimens tested with IP cycling secondary cracks and the absence of striations are observed; in hollow specimens cracks frequently start at the internal surface.

Acknowledgements - Authors gratefully acknowledge Mr. E. Signorelli for his helpful activity on SEM observations of fracture surfaces.

References

1. Rau, C.A., Jr, Gemma, A.E. and Leverant, G.R., *Fatigue at Elevated Temperature*, ASTM STP 520, American Society for Testing and Materials, 1973, 166-178.
2. Hopkins, S.W., *Low Cycle Thermal Mechanical Fatigue Testing*, ASTM STP 612, Spera D.A. and Mowbray D.F., (Eds.), American Society for Testing and Materials, 1976, 157-169.
3. Guèdou, J.Y. and Honnorat, Y., *Thermomechanical Fatigue Behaviour of Materials*, ASTM STP 1186, Sehitoglu H. (Ed.), American Society for Testing and Materials, 1993, 157-175.
4. Malpertu, J.L. and Rémy L., *Low Cycle Fatigue*, ASTM STP 942, Solomon, H.D. *et al.* (Eds.), American Society for Testing and Materials, 1988, 657-671.
5. Shi, H.J., Robin, C. and Plevinage, G., *Advances in Fatigue Lifetime Predictive Techniques*, second Volume, ASTM - STP 1211, Mitchell, M.R. and Landgraf R.W. (Eds.), American Society for Testing and Materials 1993, 105-116.
6. Bernstein, H.L. *et al.*, Prediction of thermal-mechanical fatigue life for gas turbines in electric power generation, ASTM - STP 1186, Sehitoglu, H. (Ed.), American Society for Testing and Materials, 1993, 212-238.
7. Hack, G.A.J., Inconel Alloy MA 6000 A new material for high temperature turbine blades. *Metal Powder Report,* **36**, (1981).
8. Janson, J. and Samuelsson, H., Anisotropic Elastic Analysis of uncooled ODS vane, Cost 501 Progress Report, September 1991.
9. Jaske, C.E., Thermal-mechanical, Low Cycle Fatigue of AISI 1010 Steel, ASTM STP 612, Spera D.A. and Mowbray, D.F. (Eds.), American Society for Testing and Materials, 1976, 170-198.
10. Marchionni, M., Bressers, J., Joos, R., Rémy, L., Timm, J., Vasseur, E., Low cycle and thermomechanical fatigue behaviour of an advanced ODS nickel-base superalloy, *Proc. of the Conference Materials for Advanced Power Engineering,* 1994, Liege, Vol. II, 1994, pp. 989-998.

HIGH-CYCLE FATIGUE OF HIGH-TEMPERATURE ALLOYS UNDER THERMAL CYCLING CONDITIONS

V.T. TROSHCHENKO
Institute for Problems of Strength
National Academy of Sciences of Ukraine
2 Timiryazevskaya str., Kiev 14
252014 Ukraine

1. Introduction

In the literature one can generally find the description of the investigations of strength and fracture behaviour of materials subjected to one type of loading, e.g. short- or long-term static loading, low-cycle loading, high-cycle loading, etc.

Yet in real structures, for instance in aircraft gas turbine engines, the elements are subject to a combined action of static, low-cycle and high-cycle loads under conditions of variable high temperatures. The laws of deformation and ultimate failure criteria of materials under these conditions have not been adequately studied.

At the Institute for Problems of Strength extensive investigations were performed into high-cycle fatigue in bending of nickel and titanium alloys, used in gas turbine industry, under conditions of fast heating up inducing high thermal stresses and simultaneously applied additional static tensile stresses [1-6]. The paper presents the results of these investigations.

2. Materials

The investigations were performed on high-temperature nickel and titanium alloys. Deformable nickel alloys EI598, EI698, EP693, EP962 are used to manufacture gas turbine blades which operate at 1073 to 1123 K. Cast nickel alloy ZhS6KP is used for service temperatures up to 1450 K.

High-temperature titanium alloys VT6 and VT9 are used to manufacture compressor discs and blades operating at temperatures up to 773 K. Chemical composition and mechanical properties of the materials studied are listed in Table 1. In this table T is the test temperature, $\sigma_{0.2}$ is 0.2% offset yield stress, σ_u is the ultimate strength, δ is the relative elongation, ψ is the relative reduction in the cross-section area, σ_{-1} is the fatigue limit under rotating bending [4].

J. Bressers and L. Rémy (eds.), Fatique under Thermal and Mechanical Loading, 179–188.

TABLE 1. Mechanical properties of the alloys studied (average values).

Material	T K	$\sigma_{0.2}$ MPa	σ_u MPa	σ_{-1} MPa	δ %	ψ %
	Nickel alloys					
EI598 (17.5% Cr, 1.35% Al,	293	685	1000	340	18.5	20.5
2.25% Ti, 5.0 % Mo, 0.9 % Nb)	1073	540	740	330	8.0	13.0
EI698 (14.5% Cr, 1.5% Al,	293	750	1150	340	27.5	27.5
2.5% Ti, 3.0% W, 1.55 % Nb)	923	630	1050	420	13.0	27.5
EP693	293	610	1020	440*	46.0	38.0
	1073	610	650	270	15.5	19.0
EP 962	293	950	1350	493	—	—
	1073	860	1010	455	3.3	6.9
ZhS6KP (10% Cr, 2.6% Ti,	293	825	1325	520	15.0	18.5
4.3% Al, 3.0% W, 5.0% Mo)	1173	475	775	360	11.0	17.5
	Titanium alloys					
VT6 (6.0% W, 4.0% V)	293	970	1000	510	10.0	30.0
	773	450	580	252	—	—
VT9 (6.4% Al, 2.4% Mo,	293	1030	1200	620	11.0	30.0
0.25 Fe)	773	660	850	540	14.0	60.0

* Test temperature is 673 K

2. Experimental Procedure and Processing of Results

2.1. EXPERIMENTAL EQUIPMENT

The scheme of the test set-up which enables high-cycle fatigue investigation under isothermal and nonisothermal conditions is shown in Fig. 1. Here the specimen (1) is gripped in the collets of the spindles which are set in rotation by the electric motor (3). The rotating specimen is subjected to bending by weights (2) at a frequency of about 50 Hz. The hydraulic system (4) allows static and low-cycle tensile loads up to 10 kN. Hour-glass specimens from nickel and titanium alloys were used with the diameter of the working section of 5 and 6 mm, respectively.

The specimen was cyclically heated with a high-frequency (440 kHz) generator (5) and an induction heater (6). In order to increase the temperature difference within a cycle in some tests the specimen was cooled with cooler (7). Prior to the realization of a specified thermal cycle for each material and loading regime the calibration of the specimen temperature was done with the use of thermocouples welded to the specimen and of a current collector which enables picking up signals from the rotating specimen during testing.

The data obtained were used for the computation of thermal stresses and for choosing the time of specimen heating and cooling in a cycle, which was controlled automatically during the test.

The temperature during the test was controlled with an optical pyrometer and a recording device. The pyrometer enables the threshold temperature to be specified, the excess of which makes the specimen heating system to be automatically de-energized, thus inhibiting the possibility of specimen overheating.

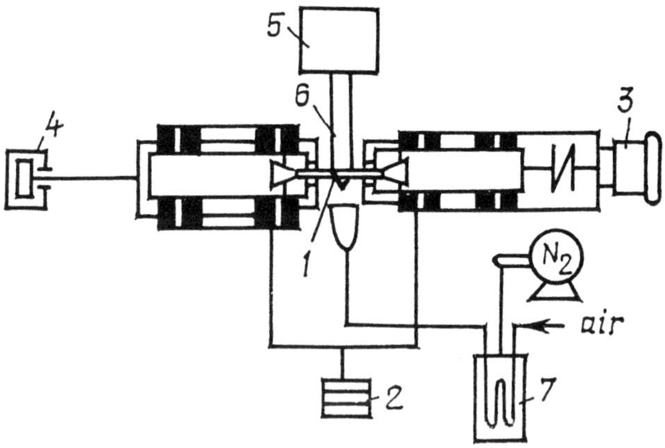

Figure 1. Schematic of the test set-up.

In isothermal tests a radiant heating furnace was used with a heater made of a high-temperature alloy, in combination with a system for automated maintaining of the specified temperature and its control.

Substantial attention was also paid to the calibration of the loading systems. Here the main issue is to establish the relation between the magnitude of the weight (2) in Fig. 1 and the bending moment in the specimen cross-section in the presence of an axial tensile force. With an increase of the tensile force, the bending moment decreases, with the magnitude of the weight being the same. The value of the additional torque, which was induced by the friction forces in the bearings, was also determined in the specimen cross-section. This is particularly important in the presence of an axial load.

2.2. THERMAL STRESS STATE OF SPECIMENS

The thermal stress state of the specimens was studied using the finite element method. First the non-stationary thermal conductivity problem was solved, where specific features of induction heating were taken into account, as well as the dependence of the material thermal-physical properties on the temperature. The computations were based upon the temperature values measured by thermocouples attached along the specimen perimeter. Next the quasistatic thermoelasticity problem was solved for specific moments in time considering the acting loads and nonuniform temperature field. This calculation procedure is described in [2, 4].

Fig. 2a gives a schematic representation of the axial σ_z, tangential σ_θ and radial σ_r thermal stress distribution along the specimen radius for a heating half-cycle. Fig. 2b shows the variation of axial thermal stresses on the specimen surface ($\sigma_{z,s}$) and in its centre ($\sigma_{z,o}$) depending on time t. With the thermal cycling regimes used, thermal tensile stresses occurring on the specimen surface in the cooling half-cycle were insignificant.

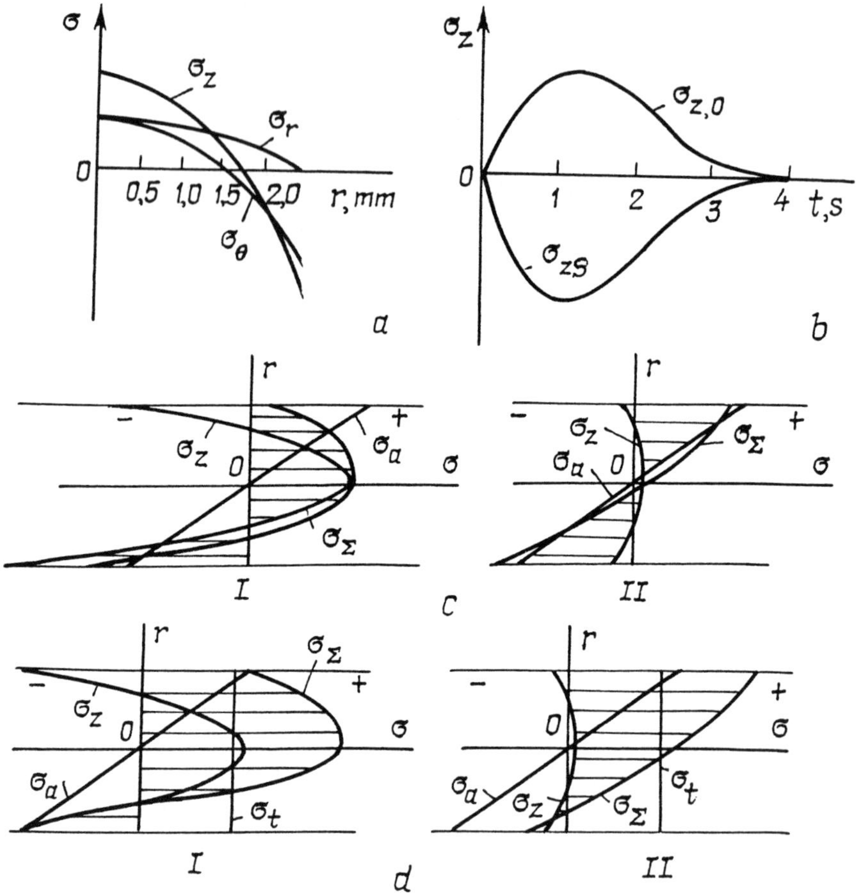

Figure 2. Scheme of thermal stress variation in a specimen (a, b) and distribution of stress components along the specimen diameter for different stages of temperature cycle and different values of an axial load (c, d) for the EI598 alloy.

Fig. 2c presents typical distributions of thermal stresses σ_z, bending stress amplitudes σ_a and total stresses σ_Σ along the specimen cross-section at superimposed tensile stress $\sigma_t=0$ for a heating half-cycle (I) and for a cooling half-cycle (II). Fig. 2d shows similar data for $\sigma_t>0$. From these figures it follows that under thermomechanical loading a very complex pattern of stress distribution along the specimen cross-section is generated which changes appreciably with time.

The analysis performed revealed that the maximum range of total stresses within a temperature cycle is observed on the specimen surface where, according to the analysis of the fracture surfaces, fatigue cracks are generally nucleated. For this reason, the results of the investigation of high-cycle fatigue were analyzed using the characteristics of the stress state in the surface layer of the specimen working section. The maximum σ_z value within a cycle was taken as the thermal stress characteristic, which is designated σ_T from here on. The analysis made in [2] revealed that for the stress state, which occurs on the specimen surface during its fast heating, the equivalent stresses calculated by the theories of maximum normal stresses, maximum shear stresses and the shape change energy are close to the $\sigma_{z,s}$ value. The cycle temperature (T_{min}, T_{max}), the heating time (t_h), the time of the temperature cycle (t_c) and the absolute maximum values of thermal stresses on the surface $[\sigma_T]_{max}$ are listed in Table 2, where σ_R and $(\sigma_R)_c$ are the experimental values of the fatigue limit for 10^7 cycles and the values calculated by means of the linear damage summation rule, respectively.

TABLE 2. Results of the investigation of high-cycle fatigue for alloys under thermal cycling.

Material	Regime of thermal cycling				σ_t	σ_R	$(\sigma_R)_c$	$\sigma_R/(\sigma_R)_c$
	T K	t_h s	t_c s	$[\sigma_T]_{max}$ MPa	MPa	MPa	MPa	
			Nickel alloys					
EI598	473—1073	2	60	206	0	170	330	0.52
	473—1073	3	60	138	0	180	—	—
	473—1073	395	1250	0	0	325	340	0.96
	673	—	60	0	40—250	240	—	—
	973	—	60	0	40—250	230	—	—
	1073	—	60	0	40—250	200	—	—
EI698	173—923	2	250	398	0	360	—	—
	173—923	2	250	398	300	100	—	—
	513—923	1.0	20	375	0	312	343	0.97
	513—923	1.0	20	375	300	145	190	0.76
	513—923	2.4	40	240	0	350	355	0.98
	513—923	2.4	40	240	300	185	200	0.93
EP693	473—1073	3	60	138	0	180	330	0.55
EP962	513—1073	1.2	60	340	0	312	343	0.97
	513—1073	1.2	60	340	300	232	315	0.74
	513—1073	3.0	60	170	0	383	445	0.86
	513—1073	3.0	60	170	300	260	320	0.81
ZhS6KP	473—1173	3.0	60	235	0	250	440	0.57
			Titanium alloys					
VT9	473—773	1	70	219	200	230	415	0.55
	473—773	3	70	72.5	200	340	400	0.85
	473—773	105	595	0	0	520	550	0.95
	293	—	120	0	0—200	380	—	—
	573	—	120	0	0—200	420	—	—
	673	—	120	0	0—200	360	—	—
	773	—	120	0	0—200	240	—	—
VT6	473—773	1.0	50	120	0	357	395	0.90
	473—773	1.0	50	120	100	294	355	0.83
	473—773	2.0	70	75	0	390	407	0.95
	473—773	2.0	70	75	100	346	370	0.94

2.3. PROCEDURE OF LIFE CALCULATION USING THE LINEAR DAMAGE SUMMATION HYPOTHESIS

For convenience of calculation, the curves of the temperature and thermal stress variation can be substituted by stepped lines. For this case the linear damage summation hypothesis can be written in the following form:

$$\sum_{i}^{s} \frac{t_i}{t_{fi}} = 1, \tag{1}$$

where t_i is the accumulated operating time at T_i, σ_{ai} and σ_{mi} ($\sigma_{mi} = \sigma_t + \sigma_{Ti}$); t_{fi} is the time to fracture under isothermal loading at T_i, σ_{ai} and σ_{mi}; s is the total number of steps over which the damages are summed. Eq. 1 can be transformed into the following form:

$$t_f = \frac{t_c}{\sum_{i}^{n} \frac{t_{ci}}{t_{fi}}}, \tag{2}$$

where t_f is the time to fracture under conditions of thermal cycling, t_c is the time of the temperature cycle, t_{ci} is the accumulated operating time at T_i within one temperature cycle, and n is the number of steps within one temperature cycle.

The number of mechanical loading cycles to fracture will be:

$$N_f = t_f \cdot f, \tag{3}$$

where f is the frequency of high-cycle loading.

The ultimate surfaces in the coordinates "σ_a—σ_m—t" for the specified values of T_i, which are needed for calculations by Eq. 2, were constructed from the diagrams in the coordinates "σ_a—σ_m" corresponding to two specified lifetimes.

The σ_a vs σ_m diagrams were described by the following equation:

$$\sigma_a = \alpha + \beta\sigma_m + \gamma\sigma_m^2 , \tag{4}$$

where coefficients α, β and γ were determined by three experimental points: $\sigma_a = \sigma_{-1}$ at $\sigma_m = 0$; $\sigma_a = \sigma_{ai}$ at $\sigma_m = \sigma_{mi}$; $\sigma_a = 0$ at $\sigma_m = \sigma_l$. Here σ_{-1} is the fatigue limit at a fully reversed loading cycle; σ_l is the long-term strength for a corresponding time to fracture; σ_{ai}, σ_{mi} are the intermediate values of σ_a and σ_m. To construct the ultimate surfaces, the author used both the results of his own and data from the literature.

The ultimate surfaces for tensile and compressive mean stresses were assumed to be identical. Using Eqs. 2 and 3 the lifetimes for several σ_a values were calculated, by which the magnitude of the fatigue limit for 10^7 cycles was extrapolated.

A detailed description of the procedure for the construction of ultimate surfaces and of the calculations involving the linear damage summation hypothesis is given elsewhere [4, 7].

3. Experimental Results and Discussion

The results of the investigation into the fatigue limits are listed in Table 2, where $[\sigma_T]_{max}$ are the maximum absolute values of thermal stresses within a cycle; σ_R, $(\sigma_R)_c$ are the experimental values of the fatigue limit for 10^7 cycles and those calculated by the linear damage summation hypothesis, respectively.

As it follows from the data listed in Tables 1 and 2, at stresses equal to the fatigue limit for 10^7 cycles the total stresses did not exceed the yield stress, which confirms the possibility of using the formulas of elasticity theory. The results presented in Table 2 allow to suggest the following. Thermal cycling according to the specified regimes leads to a reduction of the fatigue limits of the alloys studied as compared to the results obtained with the linear damage summation hypothesis. This reduction varies over a wide range for different materials and loading regimes.

The author analyzed the dependence of the results obtained on such factors as specific features of the induction heating, the magnitude of residual stresses in the surface layer and the kinetics of their variation during thermal cycling, changes in the structure of the surface layer during thermal cycling, temperature variation and load bifrequency which takes place in this case.

The influence of specific features of the induction heating was studied in [6], where it was shown that no noticeable difference is observed between the fatigue curves of high-temperature alloys under isothermal conditions obtained by induction and radiant methods of heating. Analogous results were obtained in [7].

Residual stresses in the surface layer were studied by layer-by-layer etching of the specimen surface layers and by measuring the specimen deflection during etching with a subsequent computation of residual stresses by the known formulas [5]. This method was used to study residual stresses in the specimens after their fabrication and the kinetics of their changes in the process of thermal cycling.

Fig. 3 shows the variation of axial residual stresses in the surface layer of the EI698 alloy specimens in the process of loading under thermal cycling conditions (Fig. 3a) and during isothermal loading (Fig. 3b). Thermal cycling was carried out according to the following regime: 473⇔923 K with the heating time t_h=1.2s and t_c=111s. In Fig. 3a, curve 1 is the initial distribution of residual stresses, curve 2 is after 3 thermal cycles, curve 3 is after 10 cycles, and curve 4 after 100 cycles. Loading under isothermal conditions was performed at the same mechanical load at test temperatures of 473 K (curve 2), 623 K (curve 3) and 923 K (curve 4) with the test time corresponding to the number of cycles of mechanical loading during 100 thermal cycles. Curve 1 in Fig. 3b corresponds to initial residual stresses [3, 5].

As it follows from Fig. 3, thermal cycling introduces no essential changes in the residual stress pattern. Much the same results were obtained for the EP692 and VT6 alloys.

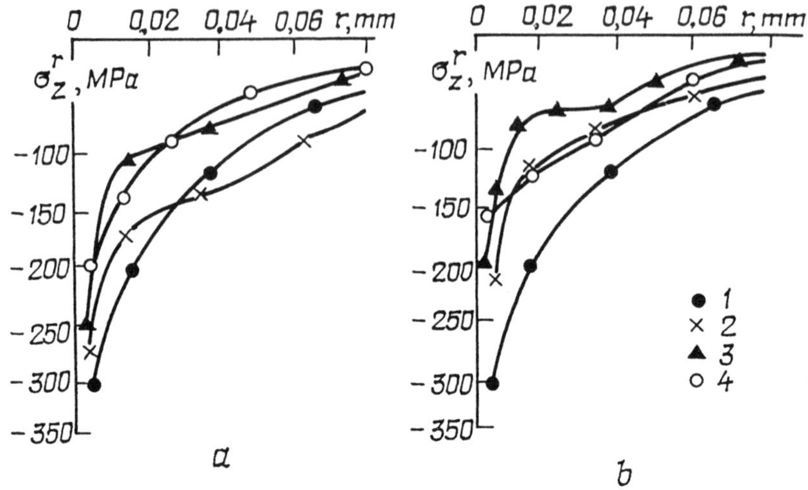

Figure 3. Variation of residual stresses in the surface layer of the EI698 alloy specimens in thermo-cyclic (a) and isothermal (b) tests.

The condition of the specimen surface layers after different testing regimes was studied by microhardness measurements and microstructure analysis using an optical microscope. It was found that with an increase in the isothermal test temperature the microhardness of nickel alloys EP962 and EP698 decreased, whereas that of titanium alloy VT6 increased. There is a fixed relation (Fig. 4) between the stabilized level of the axial residual stresses and the magnitude of microhardness in the specimen surface layer during mechanical loading under isothermal conditions for the VT6 (Fig. 4a), EI698 (Fig. 4b), and EP962 (Fig. 4c) alloys studied. The points which characterize the relationship between residual stresses and microhardness under thermal cycling conditions lie on the same curves in Fig. 4, which suggests that thermal cycling induces no specific features [5]. The analysis of the alloy microstructures revealed that thermal cycling causes no specific changes in the structure as compared to isothermal testing. A more detailed analysis of changes in the microstructure of the alloys studied is presented elsewhere [3, 5].

In the case when thermal cycling induces no thermal stresses, which corresponds to the regimes with long heating times for alloys EI598 and VT9 in Table 2, a fair agreement between the calculated and experimental results is observed.

The loading regime can be classified as a bifrequency one, whereby the low-frequency stresses correspond to thermal stresses or to the sum of thermal and axial tensile stresses ($\sigma_T + \sigma_t$), and high-frequency stresses correspond to bending stresses. The ratio of frequencies is about 1:1200 to 1:3500 and that of stress amplitudes corresponding to the fatigue limit is 1:2 to 1:5. As is shown by the data from the literature [8, 9] for such stress frequency and amplitude ratios a decline in life is observed under conditions of bifrequency loading.

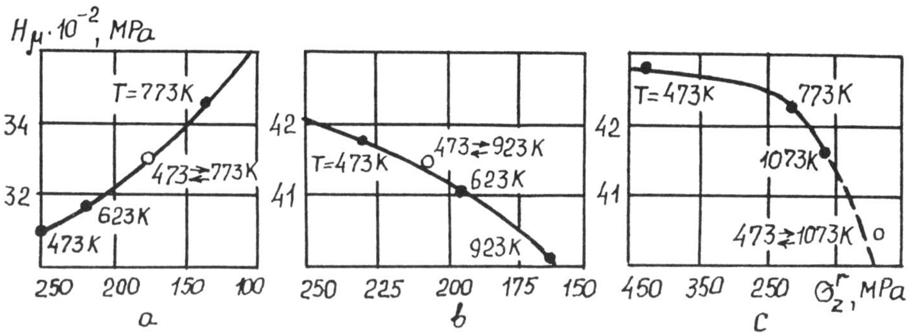

Figure 4. Relationship between residual stresses and microhardness.

The influence of bifrequency mode of loading upon high-cycle fatigue of the EI598 and VT9 alloys was studied on a test set-up analogous to that considered above (Fig. 1) under isothermal conditions, wherein bending stresses remained constant and the axial component was changing according to a trapezoidal law close to that which governs the changes of stresses in the surface layer in the presence of thermal stresses and the axial load. The results of these investigations are listed in Table 2. They suggest that bifrequency has an appreciable effect on the fatigue limit of the alloys studied.

4. Conclusions

A method and the equipment have been developed for the investigation of high-cycle fatigue of high-temperature alloys under conditions of fast heating. High-cycle fatigue of high-temperature nickel and titanium alloys has been studied under conditions of high heating rates which induce considerable thermal stresses on the specimen surfaces. This testing regime is shown to cause a reduction of the fatigue limits of the alloys studied as compared to those calculated by the linear damage summation hypothesis. The degree of the reduction is different for different alloys. The experimental data have been used to analyze the influence of the following factors upon the results obtained: induction heating, cyclic temperature variation which induces no thermal stresses, the kinetics of the variation of residual stresses, microhardness and the surface layer material structure, load bifrequency. A conclusion has been made that the reduction of the fatigue limits in high-cycle fatigue under complex thermo-mechanical loading compared to those obtained by the linear damage summation hypothesis is attributed primarily to the load bifrequency.

References

1. Troshchenko, V.T. and Zaslotskaya, L.A., Fatigue strength of superalloys subjected to combined mechanical and thermal loading, in K.Miller and R.Smith (eds.), *Mechanical Behaviour of Materials,* Pergamon Press, Oxford, 1979, pp.3-13.
2. Troshchenko, V.T., Zaslotskaya, L.A., Thermal stress state of cylindrical specimens under thermal cycling, *Strength of Materials* **9** (1977), 1188-1197.
3. Troshchenko, V.T., Gryaznov, B.A. and Yamshanov, Yu.B., Fatigue resistance of high-temperature alloys under cyclic temperature variation, *Strength of Materials* **26** (1994), N3, 175-181, N4, 253-259.
4. Gryaznov, B.A., Gorodetsky, S.S. and Nalimov, Yu.S. et al., *Fatigue of High-Temperature Alloys and GTE Blades,* Naukova Dumka, Kiev (in Russian), 1992.
5. Yamshanov, Yu.B., Development of methods for predicting life of high-temperature alloys under cyclic loading and nonisothermal conditions, Synopsis of the thesis, Institute for Problems of Strength of the National Ac.Sci. of Ukraine, Kiev (in Russian), 1992.
6. Troshchenko, V.T., Gryaznov, B.A. and Strizhalo, V.A. et al., *Methods for the Investigation of Metals Resistance to Deformation and Fracture under Cyclic Loading,* Naukova Dumka, Kiev (in Russian), 1974.
7. Harper, D.L., Feilbach, W.H. and Libsch, J.E., Application of induction heating to high-temperature fatigue, *Proc. ASTM* **63** (1963), 684-691.
8. Zaitsev, G.Z. and Aranson, A.Ya., *Fatigue Strength of Hydroturbine Elements,* Mashinostroyeniye, Moscow (in Russian), 1975.
9. Buglov, E.G., Filatov, M.Ya. and Kolikov, E.A., Fatigue resistance of materials under two-frequency loading (A Review), *Strength of Materials* **5** (1973), 527-535.

THE EFFECT OF ENVIRONMENT ON THE THERMAL FATIGUE BEHAVIOR OF GAMMA TITANIUM ALUMINIDE

W. WEI[1], W. DUNFEE[2], M. GAO[2] and R.P. WEI[2]
[1] *University Twente, Dept. of Mechanical Engineering, NL-7500 AE Enschede, The Netherlands*
[2] *Lehigh University, Dept. of Mechanical Engineering and Mechanics, Bethlehem, PA 18015 USA*

1. Introduction

Titanium aluminide intermetallics, in particular gamma (γ)-based alloys, continue to be of considerable interest for a number of medium to high temperature turbine engine applications in the aerospace and energy production industries [1-3]. This is based on a combination of properties including light weight, good oxidation resistance with respect to conventional titanium alloys, and good high temperature mechanical properties. These materials are also becoming critical to the development of future hydrogen driven energy systems including hypersonic transport [4].

Extensive research and development work has been conducted towards improving the low temperature ductility of these materials, and towards optimizing critical high temperature (isothermal) properties, in particular, creep and oxidation behavior. The behavior of these materials under operational loads and in aggressive environments is, however, of equal importance for component design and lifing. For hydrogen based systems, an understanding of the behavior of titanium aluminides under the combined influence of thermal cyclic loading and hydrogen environments is especially critical.

It is well known that titanium alloys are sensitive to hydrogen, where the primary embrittlement mechanism involves the formation of brittle hydrides, e.g. [5]. Hydride formation leading to a loss in ductility has also been observed in α_2-Ti_3Al intermetallics [3]. Hydrogen embrittlement of other intermetallics has been reported as well [6].

The effect of hydrogen on the behavior of γ-TiAl is somewhat unclear. A number of studies on single (γ) and dual phase alloys ($\gamma + \alpha_2$) report the formation of hydrides and a reduction in ductility, whereas other researchers have found little hydrogen uptake and no effect of hydrogen on mechanical behavior in conventional tensile testing and hydrogen testing [2-3,7-12]. However, under thermal cyclic loading, it has been clearly shown that hydrogen causes a drastic loss in thermal fatigue life and a major change in fracture behavior of a γ-TiAl alloy (Ti-48Al-2Cr in at.%) [13-16].

J. Bressers and L. Rémy (eds.), Fatique under Thermal and Mechanical Loading, 189–198.

This indicates that the combination of cyclic (thermal), rather than static, loading and hydrogen can be detrimental to the life of TiAl components. This paper summarizes the results of this work to date and indicates important areas required for further study for Ti-intermetallics before they can be used in future high temperature hydrogen energy systems.

2. Experimental Procedure

Thermal cyclic testing was conducted on a two-phase γ-TiAl alloy with the nominal composition of Ti-48Al-2Cr (at.%). The material was supplied in the form of a hot-isostatically-pressed (HIPed) ingot with a two-zone casting microstructure, namely, a thin layer of chill zone and a well-developed columnar zone (Fig. 1a). A section through the columnar zone shows a deformed duplex microstructure consisting of regions of (α_2+ γ) lamellae and equiaxed γ-grains (Fig. 1b). The grain size range was 50-300 μm.

Knife-edged specimens designed for thermal fatigue testing were electro-discharge machined from the ingot in and perpendicular to the columnar zone and then polished to remove surface defects. The specimen geometry and orientation in the ingot with respect to the microstructure is shown in Fig. 2. The specimen gage length thus had the microstructure shown in Fig. 1. This geometry provided a high surface area-to-volume ratio to maximize the environmental exposure area for reaction, allowed for rapid forced convective cooling, and ensured a uniform temperature distribution throughout the sample thickness.

Testing was conducted in a system developed to perform environmentally assisted thermal fatigue with a preloaded fixed-grip condition on the test specimens. The test system, shown schematically in Fig. 3, used direct electrical resistance heating to rapidly heat the specimen and a chilled gas jet for rapid cooling. Testing was conducted in high purity hydrogen (99.99%), air, or high purity helium (99.995%) as an inert reference. Thermal cycling was performed between room temperature and either 750° or 900°C. The specimens were tested with a prestress of 241 MPa (50% of room-temperature yield strength). The prestress was chosen to avoid compressive loading and buckling of the specimens at the peak temperature. Out-of-phase cyclic stresses were obtained, i.e. the maximum stress (241 MPa) occurred at the lowest temperature, and vice-versa (e.g. 34 MPa at 900°C). Specimens were tested to failure or to an arbitrarily determined maximum number of cycles. Further details of the test system and procedures are reported elsewhere [17].

Post-test examination of the specimens was conducted to determine the effect of environment on cracking behavior. Scanning electron microscopy (SEM) and metallo-graphy were performed in order to characterize fracture surface morphology of failed specimens, to determine the relationship between microstructure and cracking in failed and unfailed specimens, and to correlate this information with the cycles to failure data. An initial mechanistic model for the observed behaviour was developed.

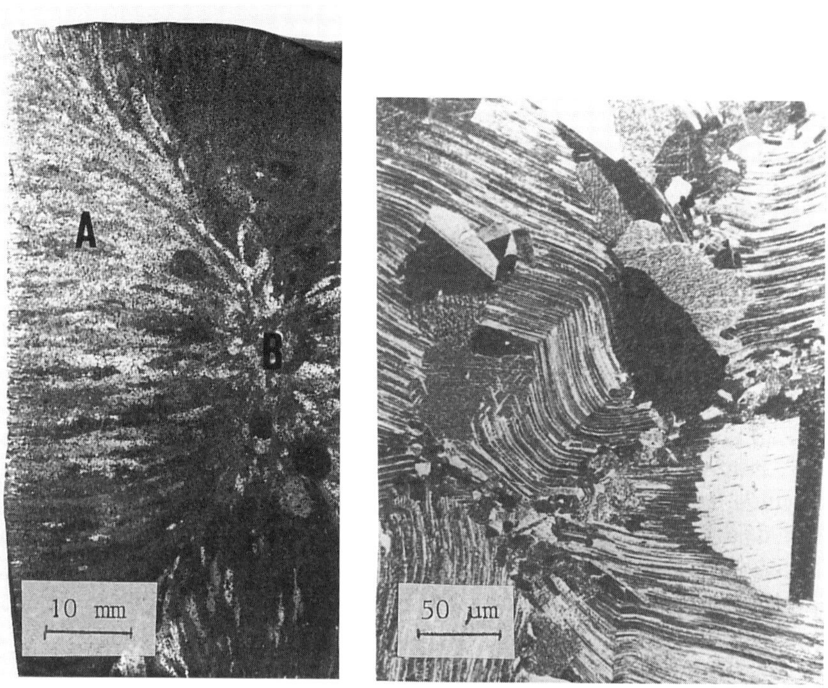

(a) (b)

Figure 1. Typical microstructure of Ti-48Al-2Cr ingot.
a) Casting microstructure (A-columnar zone, B-center of ingot)
b) Cross section through columnar zone [13,16] (also microstructure of the flat side of the specimen gage length).

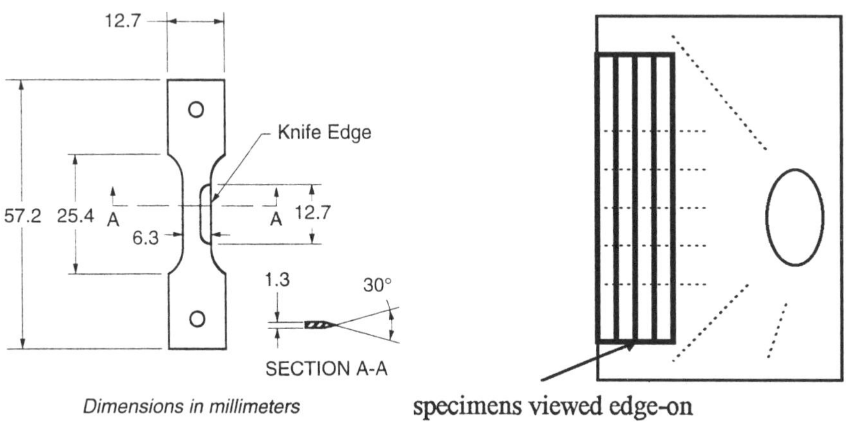

Figure 2. Specimen geometry and orientation within the ingot (schematic of Fig. 1a).

3. Results

The results of the thermal fatigue testing are summarized in Table 1 in terms of cycles to failure as a function of test atmosphere and temperature. It can be seen that hydrogen causes a drastic reduction in thermal fatigue life at 900°C with respect to air and helium. Thermal fatigue lives are also reduced in air with respect to helium, but there is considerable scatter in the data, where three of the six specimens reached more than 4000 cycles, two of those tests being terminated without failure.

Under thermal cycling at 750°C, hydrogen also appears to reduce thermal fatigue life, but there is considerable scatter here as well. None of the specimens tested in air or helium failed after up to over 6000 cycles at 750°C.

Typical micrographs from specimens tested in the three environments are shown in Figs. 4-6. For the specimens tested in hydrogen at 900°C and for the two short-lived specimens tested at 750°C, thermal fatigue cracking occurred in a brittle manner. Zigzag-shaped surface cracks with considerable branching were observed over the entire gage length, Fig. 4a. Cracking appears to have followed microstructural features of the material and three distinct paths were identified [13]. These include delamination along α_2/γ interfaces in the lamellar structure with some translamellar cracking; cleavage cracking through γ grains along what appear to be crystallographic planes; and cracking along grain boundaries, either between γ grains, between adjacent lamellar colonies, or between γ grains and lamellar colonies. The fracture surface morphology showed this microstructural influence on cracking, as well as the near absence of plastic deformation. No distinct crack initiation, propagation and rapid fracture zones were observed. This implies that fracture by thermal cycling in hydrogen at 900°C was a multiple crack initiation and growth process, rather than the result of the growth of a single crack.

For the three specimens tested in hydrogen at 750°C with lives over 1000 cycles, only crack traces and limited surface attack could be found. *In-situ* observations of crack growth using a travelling optical microscope attached to the test chamber showed that multiple zigzag-shaped surface cracks always formed and developed during the first 30 cycles of testing. However, unless the specimen fractured, crack growth nearly stopped by about the 40th or 50th cycle, and a thin surface oxide could be observed visually.

The fracture surface of the one specimen which did fracture (in 1828 cycles), Fig. 4b, revealed a different morphology than the short-lived specimens. Three regions are apparent. The crack initiated at the transition from the knife-edge to the main section of the specimen, then propagated in a semi-circle to form a relatively flat propagation zone, followed by final overload failure. The fracture surface was found to be slightly oxidized. The differences in fracture surface morphology between the short and long-lived specimens indicate that different fracture mechanisms may have been operating.

TABLE 1. Results of Thermal Cyclic Testing in Hydrogen, Air, and Helium.

Temperature Range	Number of Cycles to Failure		
	Hydrogen	Air	Helium
20-900°C	3	2106	> 4000
	10	2647	> 4145
	11	2782	> 4164
	24	4786	> 4626
	30	> 4100	> 6647
		> 6330	
20-750°C	36	> 4000	> 4108
	46	> 5626	> 4230
	1828	> 6033	
	> 1430	> 6262	
	> 3000		

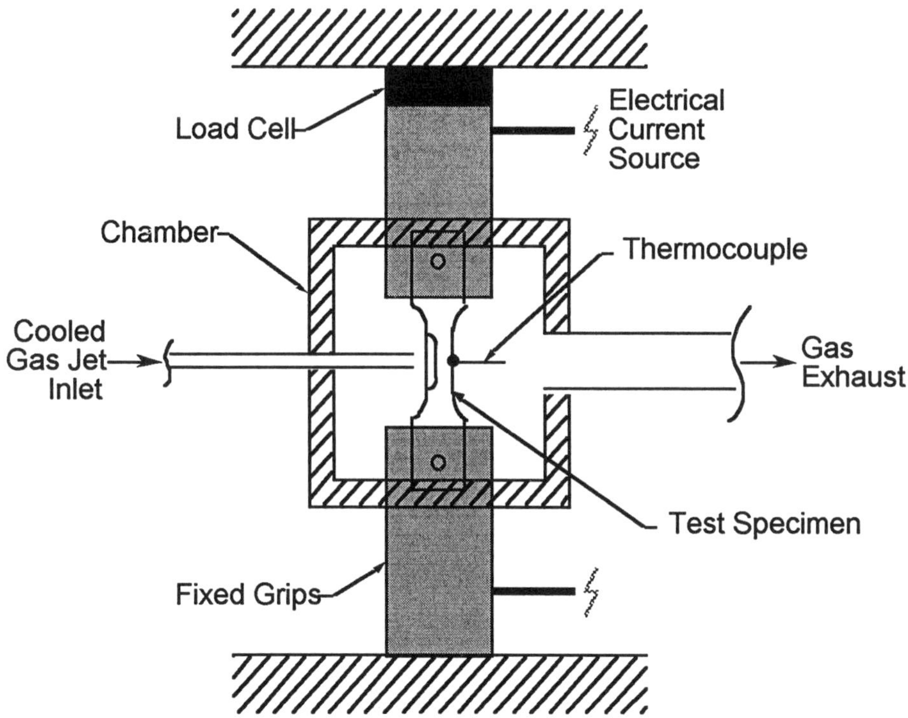

Figure 3. Schematic diagram of environmental chamber for thermal cyclic testing [13].

The three specimens which fractured in air at 900°C also showed a brittle nature. Unlike failures in hydrogen, fracture produced in air resulted from the growth of a single, main crack. Few secondary cracks were observed. The fracture also showed a three-zone feature as seen in Fig. 5a. The crack was seen to initiate at a severely oxidized region in one corner of the specimen near the knife edge, then propagated to form a relatively flat, semicircular propagation zone. A thick layer of yellowish TiO_2-type oxide, showing local buckling and flaking, was found on the specimen surfaces and on the fracture surfaces.

The once-straight knife-edges of the specimens were always found to be eroded into a slightly concave shape. This corrosion (oxidation)-erosion phenomenon was not seen in the other environments. The crack paths appeared to be transgranular in nature and insensitive to the microstructure, Fig. 5b, although some indications of cleavage-type fracture were found. Because of the heavy oxidation, little further information could, however, be obtained about the actual fracture surface morphology or crack path.

In helium, no specimens failed and only limited surface cracking was found, e.g. Fig. 6. These cracks propagated from grain to grain perpendicular to the loading axis without appreciable change in direction. The cracking in helium thus appears to be purely mechanical in nature and unrelated to the microstructure.

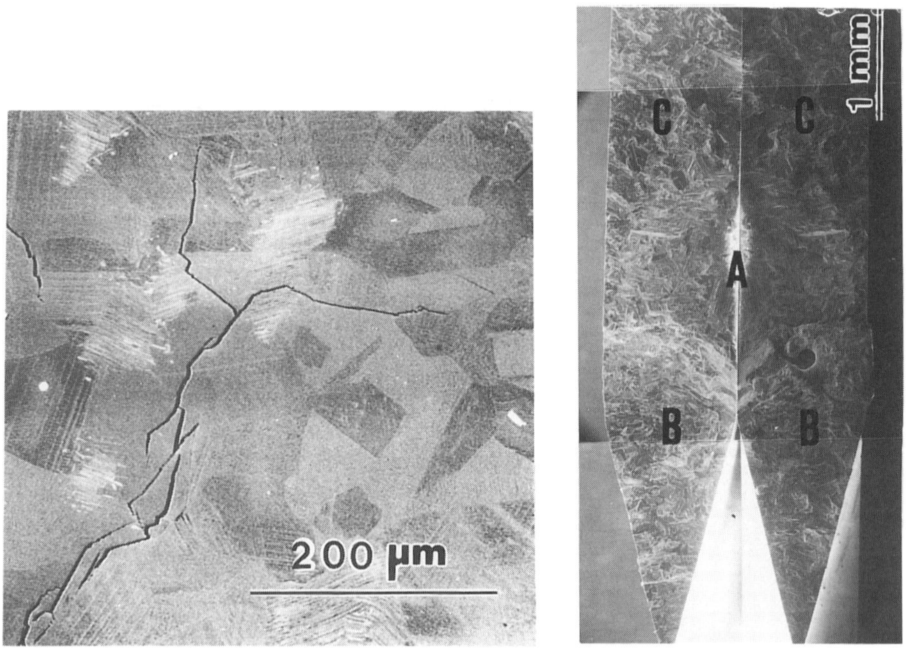

(a) (b)

Figure 4. Typical crack and fracture morphologies of specimens tested in hydrogen.
a) Surface cracking along microstructural features (900°C)
b) Fracture surface of specimens which failed at 1828 cycles (750°C)
A-crack initiation, B-crack propagation, C-final overload fracture.

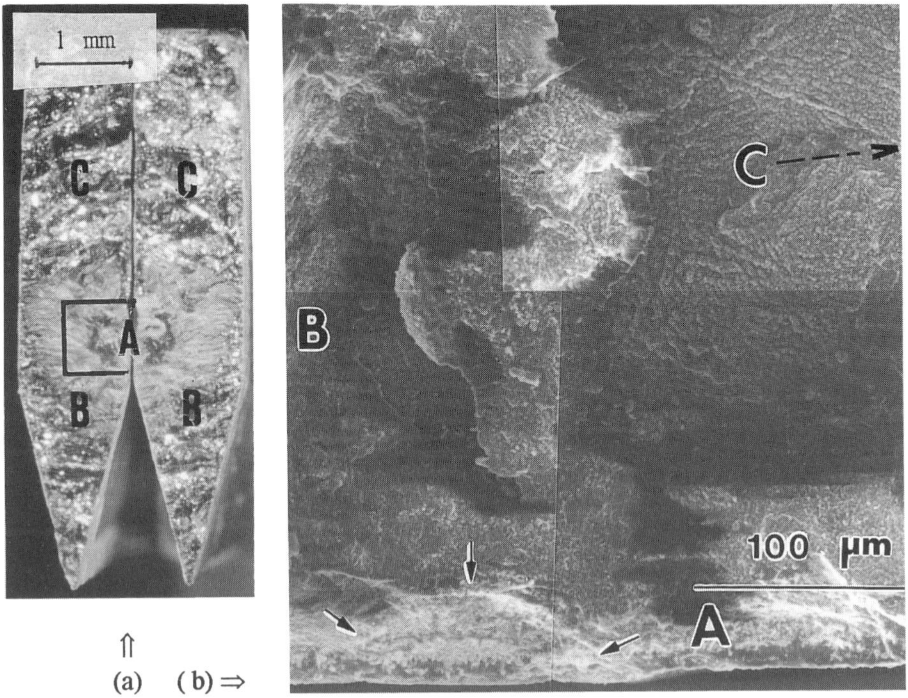

Figure 5. Typical fracture morphology of specimens tested in air (900°C) [14].
a) Mating fracture surfaces of specimen which failed at 2782 cycles
b) Typical fracture surface morphology near crack initiation site
A- initiation site (box in a, arrows in b), B- crack propagation,
C- final overload fracture.

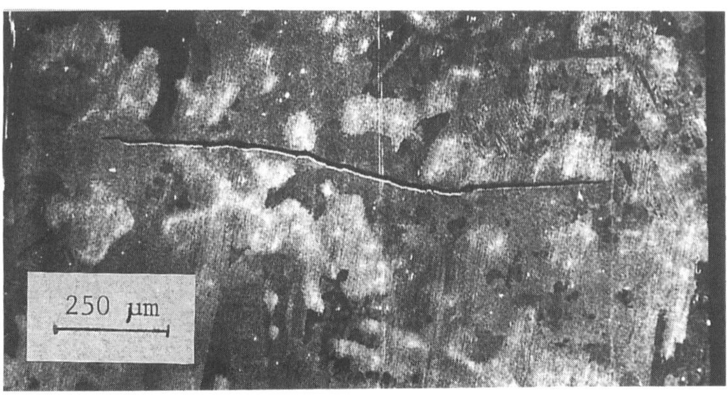

Figure 6. Typical surface crack observed for specimens tested in helium (900°C) [13].

4. Discussion

The results of this work show that hydrogen and air each have a significant effect on the thermal fatigue behavior of the γ-TiAl alloy investigated. At 900°C hydrogen clearly leads to embrittlement of the material. Fracture is strongly microstructure dependent and cracking follows several types of interfaces. This indicates that hydrogen may segregate preferentially to these interfaces which are already weaker than the bulk. By further decreasing bond strength at these interfaces, crack initiation is facilitated there. It should be noted, however, that although initial studies on hydrogen diffusion and solubility in γ-TiAl have not found conclusive evidence for the formation of brittle hydrides, this mechanism for embrittlement cannot be ruled out. In particular, it should be recalled that α_2-Ti_3Al does indeed form hydrides, and this could contribute to embrittlement in dual ($\gamma+\alpha_2$) phase TiAl alloys such as that tested in this work.

In air at 900°C the formation of a brittle surface oxide appears to lead to crack initiation and failure of the material. At this temperature, the oxide is a non-protective mixture of TiO_2 and Al_2O_3, e.g. [18], which under extreme thermal cyclic loading as well as in the flow of the cooling gas jet easily cracks and spalls. Oxidation is known to embrittle titanium alloys at high temperatures (above 500-600°C) and this would be the case for this material at correspondingly higher temperatures. Under thermal cyclic loading, oxide cracking can serve as fatigue crack initiation sites. As oxide cracking is a statistical process, it is not surprising that there is a large scatter in the cycles to failure data. At lower temperatures, e.g. 750°C, a stable, protective Al_2O_3 layer is formed. Crack initiation is thus more difficult and the specimens did not fail within the testing time limit. The mechanism for the effect of air on thermal cyclic cracking and life thus appears to depend on oxidation chemistry and kinetics, and thermal fatigue life is controlled by crack initiation.

In hydrogen at 750°C there is also considerable scatter in the data. It was found that the one specimen which failed in over 1000 cycles showed signs of oxidation, as did the two which did not fail. The two specimens which failed in less than 50 cycles did not show evidence for oxidation. It is known that dense, stable oxides can act as permeation barriers for hydrogen, e.g. for nickel base superalloys, e.g. [19]. It thus appears that the formation of a thin, stable Al_2O_3 layer may have retarded or prevented hydrogen uptake in some of the specimens, that is those with longer lives. It may be that contamination within the environmental chamber, e.g. residual water vapor or oxygen, was sufficient to suppress hydrogen effects leading to the conflicting results. This is also suspected to be the cause of the previously mentioned discrepancies in the literature concerning hydrogen diffusion and hydride formation in γ-TiAl alloys.

5. Conclusions

The effect of environment on the thermal fatigue behavior of a γ-based Ti-48Al-2Cr alloy was investigated by cycling between room temperature and 750° or 900°C. The results showed that hydrogen can severely attack the γ-alloy with lives as short as three cycles, while no failures were observed in helium for test durations of over 4000 cycles. The severity of hydrogen attack strongly depends on the upper limit of the temperature cycled and the cleanliness of the hydrogen. Specifically, the large scatter of life times at 750°C appears to have resulted from the competition between surface oxidation and hydrogen attack. High temperature oxidation in air can also reduce thermal cyclic lives depending on the oxidation chemistry and kinetics. At 900°C, the formation of unprotective mixed oxides can lead to early crack initiation, while at 750°C a protective oxide leads to thermal cyclic behavior similar to that in inert helium. The results suggest that further work is necessary in the area of thermal cyclic testing of γ-based titanium aluminides under simulated operating conditions before these materials can be applied in advanced high temperature and hydrogen-fuel energy systems.

References

1. Lipsitt, H.A., Titanium Aluminides - An Overview", in *Proceedings of MRS Symposium on High Temperature Ordered Intermetallic Alloys*, Vol. 39, Materials Research Soc., Boston, 1985, pp. 351-364.
2. Kim, Y.-W., Intermetallic Alloys Based on Gamma Titanium Aluminide, *Journal of Metals* **41**, (1989), 24-30.
3. Froes, F.H., et al.: Review of Synthesis, Properties, and Applications of Titanium Aluminides, *J. Mater. Sci.* **27** (1992), 5113-5134.
4. Jackson, R., et al., Hypersonic Structures and Materials, *Aerospace America* (1987), 24-30.
5. Birnbaum, H.K., Hydrogen Related Second Phase Embrittlement of Solids, in *Hydrogen Embrittlement and Stress Corrosion Cracking*, American Society for Metals, Metals Park, OH, 1984, pp. 153-177.
6. Stoloff, N.S., Shea, M. and Castagna, A., Hydrogen Embrittlement of Intermetallic Compounds and Their Composites, in *Environmental Effects on Advanced Materials*, The Metallurgical Society of AIME, Warrendale, PA, 1991, pp. 3-19.
7. Takasugi, T. and Hanada, S., Environmental Embrittlement of Gamma Titanium Aluminide, *J. Mater. Res.* **7**, (1992), 2739-2745.
8. Liu, C. T. and Kim, Y.-W., Room-Temperature Environmental Embrittlement of Gamma Titanium Aluminide, *Scripta Metall. Mater.* **27** (1992), 599-603.
9. Chan, K.S. and Kim, Y.-W., Rate and Environmental Effects on Fracture of a Two-Phase TiAl Alloy, *Metall. Trans. A* **24**, (1993), 113-125.
10. Nakamura, M., Hashimoto, K. and Tsujimoto, T., Environmental Effect on Mechanical Properties on TiAl Base Alloys, *J. Mater. Res.* **8**, (1993), 68-77.
11. Gao, M., Boodey, J.B., Wei, R.P. and Wei, W., Hydrogen Solubility and Microstructure of Gamma Based Titanium Aluminides, *Scripta Metall. Mater.* **27**, (1992), 1419-1424.
12. Boodey, J.B., Gao, M., Wei, W. and Wei, R.P., Hydrogen Occlusion and Hydride Formation in Tita-nium Aluminides, to be published in *Proc. Int. Symp.on Gamma Titanium Aluminides*, TMS Annual Meeting, Las Vegas, NV, 12-16 Feb. 1995, The Metallurgical Society of AIME, Warrendale, PA, 1995.
13. Gao, M., Dunfee, W., Wei, R.P. and Wei, W., Thermal Fatigue of Gamma Titanium Aluminide in Hydrogen, in *Fatigue and Fracture of Ordered Intermetallic Materials: I*, The Metallurgical Society of AIME, Warrendale, PA, 1994, pp.225-237.
14. Gao, M., Dunfee, W., Wei, R.P. and Wei, W., Thermal Mechanical Fatigue of Gamma Titanium Aluminide in Hydrogen and Air, to be published in *Fatigue and Fracture of Ordered Intermetallic Materials: II*, The Metallurgical Society of AIME, Warrendale, PA, 1995.

15. Gao, M., Dunfee, W., Wei, R.P. and Wei, W., Environmentally Enhanced Thermal-Fatigue Cracking of Gamma-Based Titanium Aluminide Alloy, to be published in *Proc. Int. Symposium on Gamma Titanium Aluminides*, 124th TMS Annual Meeting, Las Vegas, NV, 12-16 Feb. 1995, The Metallurgical Society of AIME, Warrendale, PA, 1995.
16. Dunfee, W., Gao, M., Wei, R.P. and Wei, W., Hydrogen Enhanced Thermal Fatigue of γ-Titanium Aluminide, submitted to *Scripta Metall. Mater.* (1995).
17. Gao, M., Dunfee, W., Miller, C., Wei, R.P. and Wei, W., Thermal Fatigue Testing System for the Study of Gamma Titanium Aluminides in Gaseous Environments, to be published in *Thermal-Mechanical Fatigue Behavior of Materials: 2nd Volume*, ASTM STP 1263, American Society for Testing and Materials, Philadelphia 1995.
18. Welsch, G. and Kahveci, A.I., Oxidation Behavior of Titanium Aluminide Alloys, in *Oxidation of High-Temperature Intermetallics*, The Metallurgical Society of AIME, Warrendale, PA, 1989, pp. 207-218.
19. Stehlow, R.A. and Savage, H.C., The Permeation of Hydrogen Isotopes Through Structural Material at Low Pressures and Through Metals with Oxide Film Barriers, *Nuclear Technology* **22** (1974), 127.

CHARACTERIZATION OF CREEP FATIGUE CRACKING IN TYPE 304 STAINLESS STEEL

R. OHTANI, T. KITAMURA, N. TADA and W. ZHOU
Department of Engineering Physics and Mechanics,
Kyoto University, Kyoto 606-01, Japan

1. Introduction

The behaviour of initiation and growth of small cracks in high-temperature push-pull low-cycle fatigue (LCF) was investigated on a Type 304 stainless steel. The characterization of cracking was done both on the basis of the precise observation on the surface of as well as inside the specimens which were subjected to a wide range of isothermal loading conditions, and on the basis of the numerical simulation for initiation and growth of small cracks. Similar fundamental properties must be found in thermal fatigue and in other kinds of polycrystalline steels and alloys.

2. Behaviour of Small Cracks [1,2]

2.1. CLASSIFICATION OF TYPE OF CRACKING IN HIGH TEMPERATURE FATIGUE

The observation of smooth bar specimens tested under several strain waveforms, strain rates and temperatures leads to the conclusion that at least three types of cracking can be recognized in the LCF regime: Type Fs is the pure fatigue type which shows the initiation by persistent slip band cracking and the growth by transgranular cracking. Type Cs is the creep-fatigue interaction type with grain boundary wedge type cracks at the surface of specimens. Type Ci is the monotonic creep type characterized by creep cavities and their coalescing cracks nucleating throughout inside the material. Fig. 1 shows photographs of these three types on the specimen surface (A), near the surface on the longitudinal section (B) and at the center (C).

2.2. LOADING CONDITIONS

As shown in Fig. 1, Type Fs is revealed when specimens are subjected to cyclic loading at a fairly low temperature or a very high strain rate cycle (fast-fast) at high temperatures. Type Cs is found under strain waveforms with slow strain rates in tension (slow-slow or slow-fast). Type Ci is generated while specimens are exposed at higher temperatures and/or lower tensile strain rates (very slow-fast). Fig. 2 shows

J. Bressers and L. Rémy (eds.), Fatique under Thermal and Mechanical Loading, 199–208.

the creep fatigue fracture mode map which indicates the domain of the three types of cracking in three coordinates of tensile strain rate, compressive strain rate and temperature.

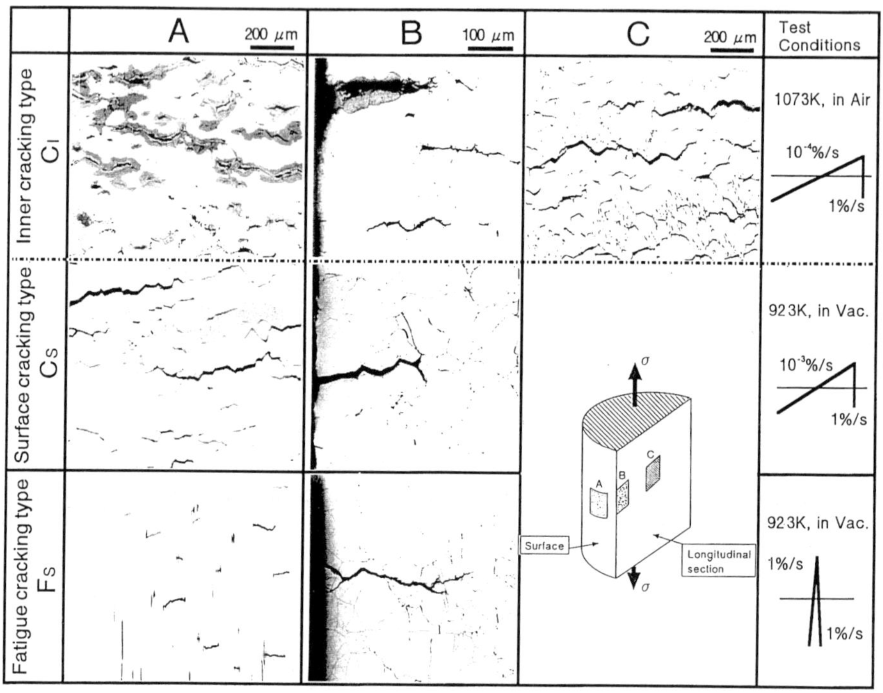

Figure 1. High temperature fatigue cracks classified into three types.

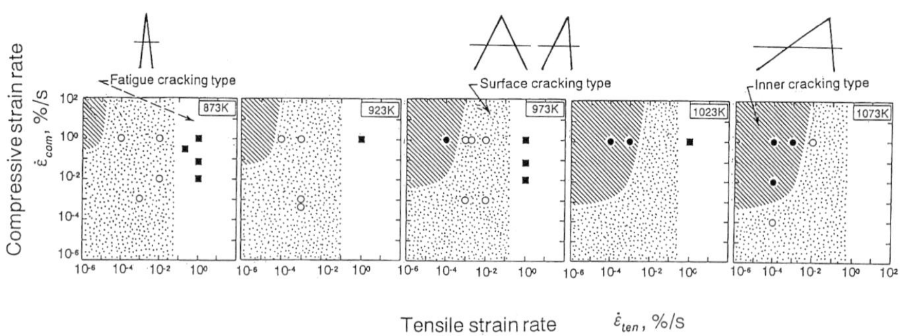

Figure 2. Creep fatigue fracture mode map.

2.3. DIFFERENCE BETWEEN SURFACE CRACKING AND INNER CRACKING IN CREEP FATIGUE

Type Fs fracture surface is typical of a fatigue striation pattern while Type Cs and Type Ci exhibit intergranular fracture. The morphologic difference between Type Cs and Type Ci in creep fatigue fracture is also evident; Type Cs shows smooth facets of grain boundaries generated by wedge type cracking and Type Ci shows dimple facets generated by creep cavitation. Therefore, it can be inferred from the loading conditions and the cracking morphology that grain boundary sliding being easier to occur at the free surface results in the surface cracking of Type Cs, and that grain boundary diffusion (Coble creep) which contributes to the cavity nucleation and growth, controls the inner cracking of Type Ci.

Distributions of cracking angles perpendicular to the stress axis in Domains A, B and C in Fig. 1 are shown in Fig. 3, where the angles on the surface (A) center around 0° while the inward angles at the surface (B) concentrate from 30° to 35° for both Types Cs and Ci. On the other hand, the inner cracks on the longitudinal section (C) for Type Ci also have angles around 0°.

The behavior of crack initiation and growth in Types Cs and Ci was examined by means of the data on the number of cracks and the mean crack length measured for the failed specimens after creep fatigue tests. First, the number of cracks per unit area (areal density) on the specimen surface (Domain A), n_s , and that on the longitudinal section (Domain C), n_p, are measured. Next, measured are the average crack lengths

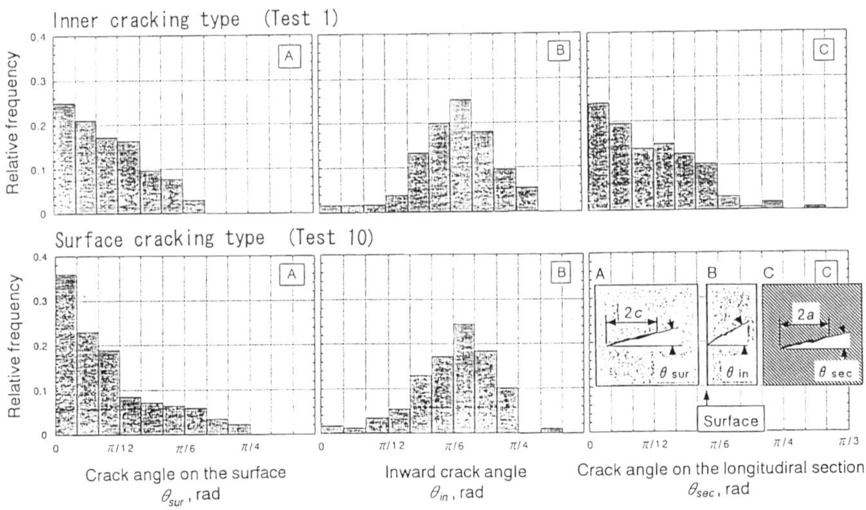

Figure 3. **Distribution of cracking angles in the domains A (on surface), B (near surface on longitudinal section) and C (at center on longitudinal section).**

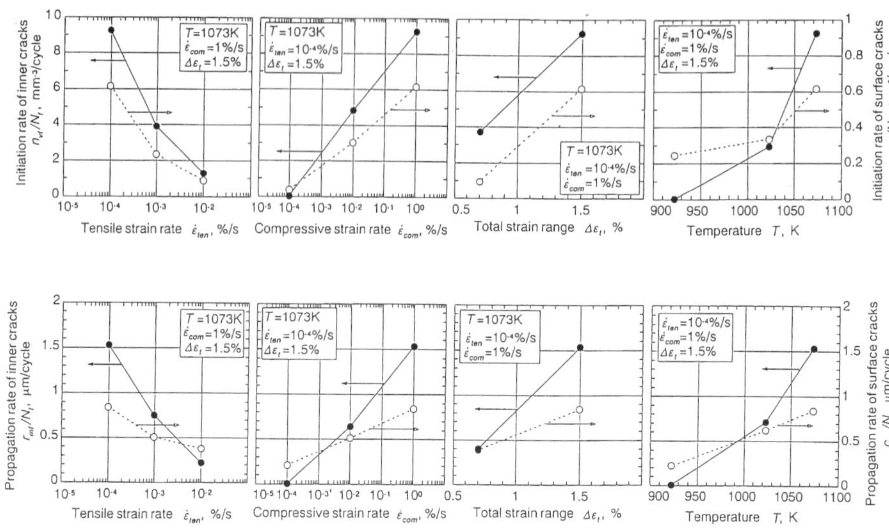

Figure 4. Effects of loading conditions on average increasing rate of crack density (upper) and that of crack length (lower).

on Domain A, c_m, and on Domain C, a_m. The number of inner cracks per unit volume (volumetric density), n_v, and the average radius of the inner cracks, r_m, are estimated by the following equations [3];

$$n_v = n_p \frac{\pi}{8a_m}, \quad r_m = a_m \frac{4}{\pi} \tag{1}$$

on the assumption that a number of penny shaped cracks of the same size are distributed in the specimens. By dividing these values by the number of cycles to failure, N_f, the following characteristic values can be obtained.

n_s/N_f: average increasing rate of surface crack density
n_v/N_f: average increasing rate of inner crack density
c_m/N_f: average increasing rate of surface crack length
r_m/N_f: average increasing rate of inner crack length

The values of n_s/N_f and n_v/N_f represent the easiness of initiation of distributed grain boundary cracks of Types Cs and Ci, respectively, and are termed "average crack initiation rates". The values of c_m/N_f and r_m/N_f represent the easiness of their growth, and are termed "average crack growth rates". These four characteristic values are shown in Fig. 4 as a function of the strain rate in tension, the strain rate in compression, the total strain range and the temperature. It is found that the temperature dependence of the above four values of Type Ci is stronger than that of Type Cs, which indicates an increasing preference for Type Ci cracking as the higher the temperature is. The dependence on the other factors also tends to be stronger in the case of Type Ci than in the case of Type Cs. The failure life of specimens damaged by Type Ci cracks is practically shorter than that of specimens with Type Cs cracks and it is the shortest of the high temperature LCF lives in the $\Delta\varepsilon$- N_f diagram.

3. Numerical Simulation of Random Cracking in Creep Fatigue [4-6]

3.1. MODELLING

Assumption 1: Crack initiation occurs discretely at each grain boundary (GB). Each GB has its own magnitude of fracture resistance, R. When the GB is subjected to the driving force for crack initiation, F, the fracture of the ith GB takes place according to the following rule.

$$F_i\,(t_c)_i = R_i \qquad \therefore (t_c)_i = R_i / F_i \tag{2}$$

where t_c is the crack initiation time. In general, the magnitude of R_i corresponds to the GB energy of the material, while the magnitude of F_i is given as a function of the fatigue loading conditions and the deformation characteristics.

Assumption 2: Crack growth also occurs discretely in a similar manner as crack initiation except that the driving force for crack growth, $K{\cdot}a$ (a is the crack length), is required in addition to F. After the crack has initiated at the ith GB according to Eq. (2), the adjacent (i+1)th GB is subjected to the driving force, $F_{i+1} + K_{i+1}\, l_i$ (where $a = l_i$: half the length of the ith GB), and the time to crack growth from the ith to the (i+1)th GB is given as

$$(t_c)_{i+1} - (t_c)_i = \frac{(R_{i+1}/F_{i+1}) - (R_i/F_i)}{1 + (K_{i+1}\cdot l_i / F_{i+1})} \tag{3}$$

and the smoothing crack growth rate is represented by

$$\left(\frac{da}{dt}\right)_{i+1} = \frac{l_{i+1}}{(t_c)_{i+1} - (t_c)_i} \tag{4}$$

where l_{i+1} is the length (or the diameter) of the (i+1)th GB.

In Eqs. (2) and (3), the ratio R/F pairs the variables R and F. The reciprocal, F/R, represents the fracture driving force per unit time or the rate of increase in the damage in each GB. The above model, therefore, follows a damage mechanics concept for each GB. The ratio K/F, which is the characteristic value for estimating the difficulty of crack growth, also pairs the variables, K and F; when K/F is small, many cracks initiate readily before some of them grow, and when it is large, the early cracking is easily succeeded by its growth. Once a crack propagates over distances such that the driving force for crack growth converges to $K{\cdot}l$, it corresponds to the stress intensity factor controlling the crack growth rate (for example, creep J-integral, J^*). Therefore, the above model also follows a fracture mechanics concept for relatively large cracks [7].

3.2. METHOD OF NUMERICAL SIMULATION

The values of R, F and K in Eqs. (2) and (3) are thought to be stochastic variables acquiring different magnitude in each GB. It is, however, difficult to estimate these values, especially R and F, from any simple experiment. Then, it is assumed that F and K are both deterministic while R is a statistic quantity.

In the present simulation, uniform random numbers from 0 to 1 are attributed to R. The reason is that the aforementioned "crack initiation rate", n_s/N (n_s is the areal density of surface cracks and N is the number of cycles) is actually observed to be constant in most interrupted tests and so the crack initiation time, t_c, in Eq. (2) is found to distribute uniformly from the early stage to the latter half of the fatigue tests. On the other hand, the magnitude of F and K is determined by fitting calculated crack initiation curves (n_s-N or n_p-N) and crack growth curves (c_m-N or a_m-N) with measured ones, respectively.

Fig. 5 shows how to proceed with the numerical calculation of surface cracking of Type Cs. The GB structure in Fig. 5 is constructed numerically in the computer on the basis of an isotropic grain growth model [8]. In the model, nuclei distributed at random locations are assumed to grow at a constant velocity in all directions, and each GB is formed where two adjacent grains meet. Adopting proper number of nuclei and grain growth velocity, the identical GB structure with that of the Type 304 stainless steel tested can be obtained.

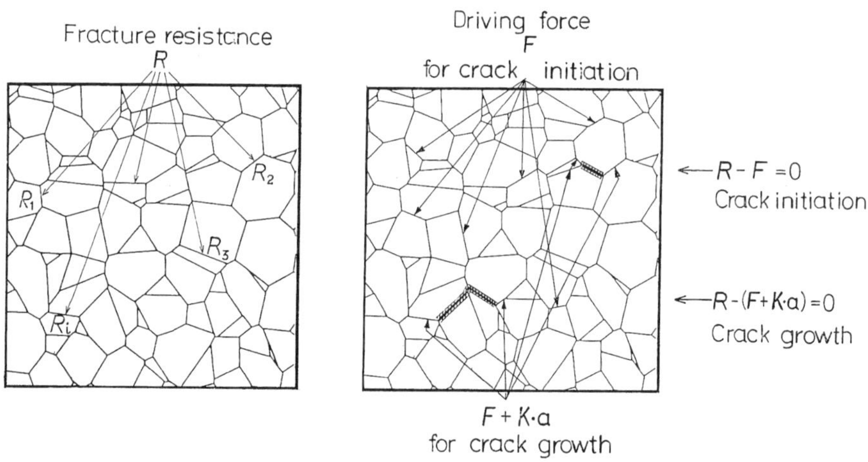

Figure 5. Method to assign fracture resistance, R, driving forces for crack initiation, F, and for crack growth, K, to every grain boundary for simulating surface cracks.

Fig. 6 shows the 2D model for inner cracking simulation, where GB facets in the material are represented by the projection on a plane perpendicular to the stress axis and their stratification.

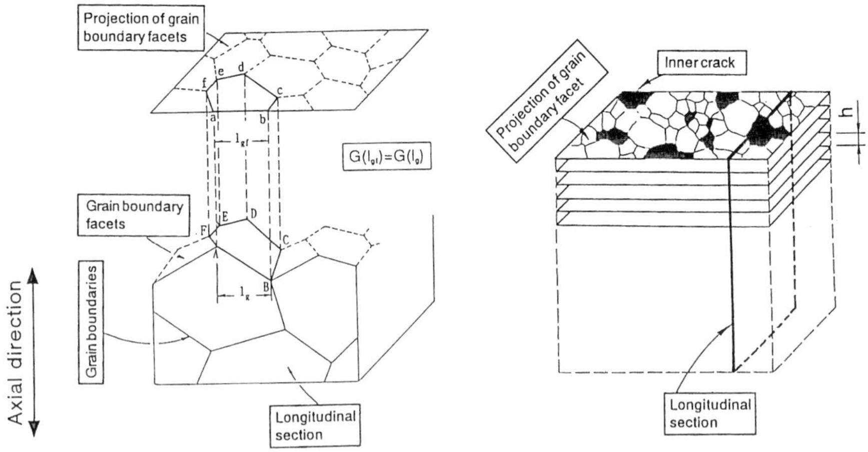

Figure 6. **2D-model for inner cracking simulation where grain boundaries are represented by facets projected on a plane perpendicular to the stress axis (left) and their stratification (right).**

3.3. RESULTS

Fig. 7 shows the comparison of the numerical simulation with the experimental observation of (a) the areal crack density, n_s, on the surface A for Type Cs surface cracking, (b) the areal crack density, n_p , on the longitudinal section C for Type Ci inner cracking and (c) the volumetric crack density, n_v, inside the material D for Type Ci. The n_s -curve of Type Cs declines from the middle stage of fatigue life mainly due to crack coalescence. The simulation curve based on the 3D model agrees well with the experimental data. Good correspondence is also found for the n_p -curve of Type Ci when F=7x10^{-4} and K=3x10^{-2} are used for the calculation, although it is not at the latter half of the fatigue life for K= 0. Using these values of F and K, the inner crack density, n_v, can be estimated. The result is shown in Fig. (c). It has been clarified by a geometrical analysis and also from Eq. (1) that the areal crack density, n_p, is dependent both on the volumetric crack density, n_v, and on the mean crack radius, r_m. n_p, increases with r_m even when n_v is constant. Therefore, the dependence of n_p on the value of K is usually different from that of n_v.

Fig. 8 indicates the change in (a) the mean crack length, a_m, on the longitudinal section C and (b) the mean crack radius, r_m, inside the material D for Type Ci inner cracking. The values of a_m and r_m are constant until the late stage of the fatigue life. However, they increase drastically at the end. This also suggests that crack coalescence takes place frequently near the end of the fatigue life.

Fig. 9 is an example of the visual expression of a part of simulated inner cracks. It is clear that facet-by-facet cracking dominates the spatial and temporal distributions of crack initiation and growth.

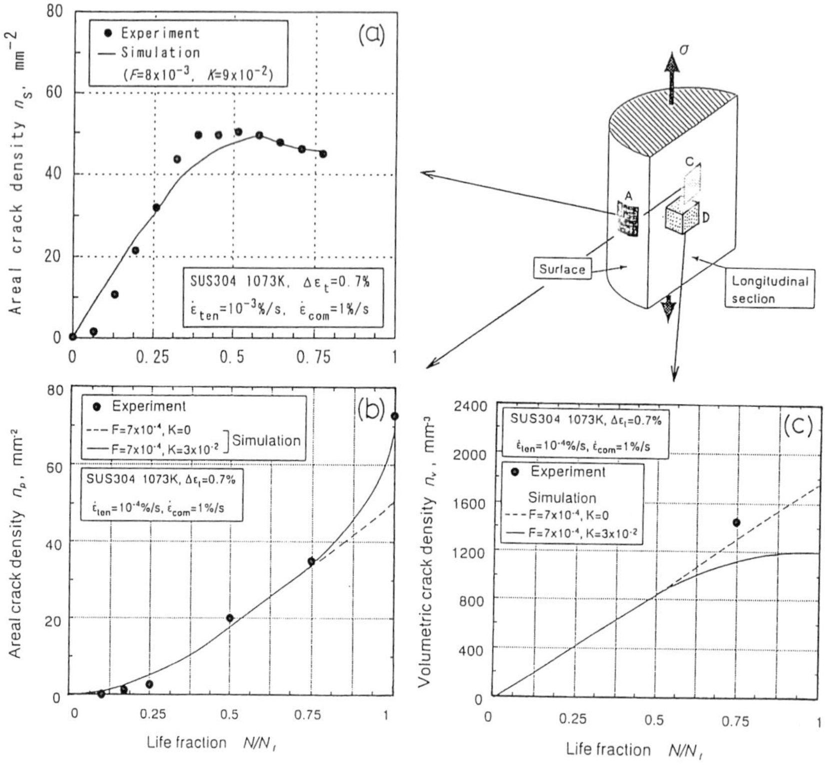

Figure 7. **Comparison of simulation and observation on the crack density. (a) Areal crack density on the surface A for Type Cs surface cracking, (b) areal crack density on the longitudinal section C for Type Ci inner cracking and (c) volumetric crack density inside D for Type Ci.**

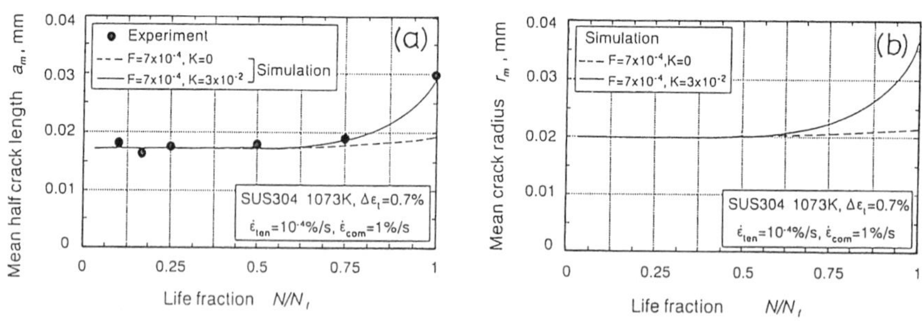

Figure 8. **Comparison of simulation and observation on the crack length for Type Ci inner cracking. (a) On the longitudinal section C and (b) inside the material D.**

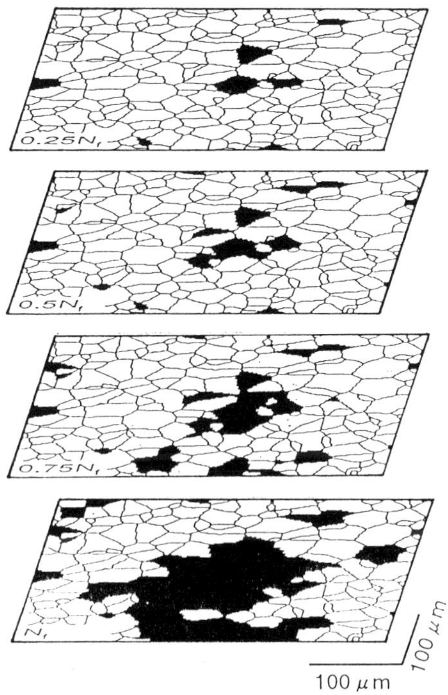

Figure 9. Visualization of simulated inner cracks showing facet-by-facet initiation and growth.

4. Concluding Remarks

In the experimental results in this paper, emphasis is put on the effect of cyclic loading at high temperatures. Three types of crackings, pure fatigue type (transgranular surface cracking, Fs), creep-fatigue interaction type (intergranular surface cracking, Cs) and monotonic creep type (intergranular inner cracking, Ci), are found to exist in the high-temperature LCF regime, and their appearance is dependent on tensile strain rate, compressive strain rate, their combination (strain waveform) and temperature. Among them, inner cracking type fracture tends to be most dangerous because the rate of its initiation is larger than that of other types of cracking and the rapid growth is brought about by the frequent crack coalescence. Attention should be paid to this type of cracking encountered under creep dominated thermal fatigue conditions in long term service.

The numerical analysis carried out in this study is an inverse problem analysis. The inverse analysis, by means of which input values for numerical calculation (the fracture resistance, R, the driving force for crack initiation, F, and for crack growth, K) can be evaluated by fitting the calculated results with the actual cracks revealed on the surface, is found to be very effective in particular for the inner cracking type which is difficult to measure experimentally. The simulation is, therefore, of great advantage to the estimation of randomly distributed cracking in components of high temperature service for the purpose of their life diagnosis [9].

References

1. Ohtani, R., Kitamura, T. and Tada, N., Cracking Behaviour of Heat-resisting Steels, Alloys and a Carbon-fibre-reinforced Polymer at Elevated Temperatures, *Materials Science and Engineering*, **A143** (1991), 213-222.
2. Ohtani, R., Kitamura, T., Tada, N. and Zhou, W., Experimental Mechanics on Initiation and Growth of Distributed Small Creep-Fatigue Cracks, *Recent Advances in Experimental Mechanics*, J.F.Silva Gomes et. al. (eds.), Balkema, Rotterdam, (1994), pp.1173-1179.
3. Tada, N., Ohtani, R., Kitamura,.T. and Yamada, M., Inverse Analysis of Distribution of Internal Small Defects, *JSME International Journal*, Ser.A, **37** (1994), 450-455.
4. Ohtani, R., Kitamura, T. and Tada, N., Numerical Simulation of Initiation and Early Propagation of Creep-Fatigue Small Cracks Based on a Model of Random Fracture Resistance of Grain Boundaries, *Structural Design for Elevated Temperature Environments-Creep, Ratchet, Fatigue, and Fracture*, C.Becht, R. Ohtani, L.K. Severud and S.Y. Zamrik (eds.), PVP-**163**, ASME, (1989), pp.123-127.
5. Kitamura, T., Tada, N. and Ohtani, R., Stochastic Simulation of Initiation and Early Growth of Small Cracks in Creep-Fatigue, *Advances in Structural Reliability Methods* (1993 IUTAM Symposium), P.D. Spanos and Y.-T. Wu (eds.), Springer, Berlin, (1994), pp.301-318.
6. Tada, N., Zhou, W., Kitamura, T. and Ohtani, R., Creep-Fatigue Intergranular Fracture of Inner Cracking Type in Type 304 Stainless Steel-Numerical Simulation on Initiation and Growth of Small Cracks, *J.of Society for Materials Science*, Japan (in Japanese) **44** (1995), 84-89.
7. Ohtani, R. and Kitamura, T., Creep-Fatigue Interaction Under High-Temperature Conditions, *Handbook of Fatigue Crack Propagation in Metallic Structures*, A.Carpinteri(ed.), Elsevier, Amsterdam, **2**, (1994), pp.1347-1383.
8. Mahin, K.W., Hanson, K. and Morris, J.W., Jr., Comparative Analysis of the Cellular and Johnson-Mehl Microstructures through Computer Simulation, *Acta Metallurgica*, **28** (1980), 443-453.
9. Ohtani, R., Kitamura, T. and Tada, N., Methodology of Remaining Life Assessment in High Temperature Applications Based on a Monte Carlo Simulation of Grain Boundary Cracking, *Mechanical Behavior of Materials* (Proc.ICM-6), M.Jono and T.Inoue (eds.), Pergamon, London, **2**, (1991),pp.205-212.

CREEP-FATIGUE BEHAVIOUR UNDER SERVICE-TYPE STRAIN CYCLING

J. GRANACHER and A. SCHOLZ
Institut für Werkstoffkunde, TH Darmstadt
Grafenstraße 2, D-64283 Darmstadt, Germany

1. Introduction

High temperature components are normally operating under variable loading conditions [1,2,3]. Load changes lead to temperature transients and thus at the heated surface of the components strain cycling can be critical. Additionally to fatigue, creep occurs due to relaxation in hold phases. To simulate this type of loading a service-type strain cycle has been developed earlier [1,4] (Fig. 1). It contains a compressive strain hold phase 1 simulating start-up condition, a zero strain hold phase 2 characterizing constant load and a tensile strain hold phase 3 referring to shut-down. Anisothermal strain cycling is of special interest because it is nearer to practice than isothermal testing. With regard to the design life of power plants up to 200 000 h long-term testing is important. As demonstrated earlier [2,3,5] creep-fatigue life under such conditions can be analysed with the aid of the generalized damage accumulation rule if internal stress and a damage interaction concept are considered. An extension of this concept to longer test durations and from single-stage to three-stage strain cycling is of interest.

2. Single-Stage Service-Type Strain Cycling

On two typical heat resistant steels single-stage service-type strain cycling was carried out with constant total strain range $\Delta\varepsilon$. Cylindric test pieces of 10 mm diameter and 35 mm gauge length were tested. The number of cycles to failure N_f was defined by a 1.5 %-drop of the stress range from its linear course, corresponding to a crack depth of 0.5 to 1 mm. The strain rate at the ramps between the hold times was $\dot{\varepsilon}_r = 6$ % / min. In isothermal package-type (pa) tests a longest time to failure $t_f = N_f\, t_p = 45\,000$ h was reached on a bainitic 1Cr-1Mo-0.7Ni-0.3V-steel at 525°C (Fig. 2a) and $t_f = 32\,000$ h was reached on a martensitic 12Cr-1Mo-V-steel at 550°C (Fig. 2b). Anisothermal (an) tests with compensated thermal strain [4] were carried out up to 8 000 h (Fig. 2a). They give only insignificantly higher numbers of cycles to failure than comparable isothermal (iso) tests [6,7] (Fig. 3a). The package-type tests can economically replace isothermal tests in the long-term region [4,7] (Fig. 3b). They comprise short strain cycling packages periodically inserted into

J. Bressers and L. Rémy (eds.), Fatique under Thermal and Mechanical Loading, 209–214.

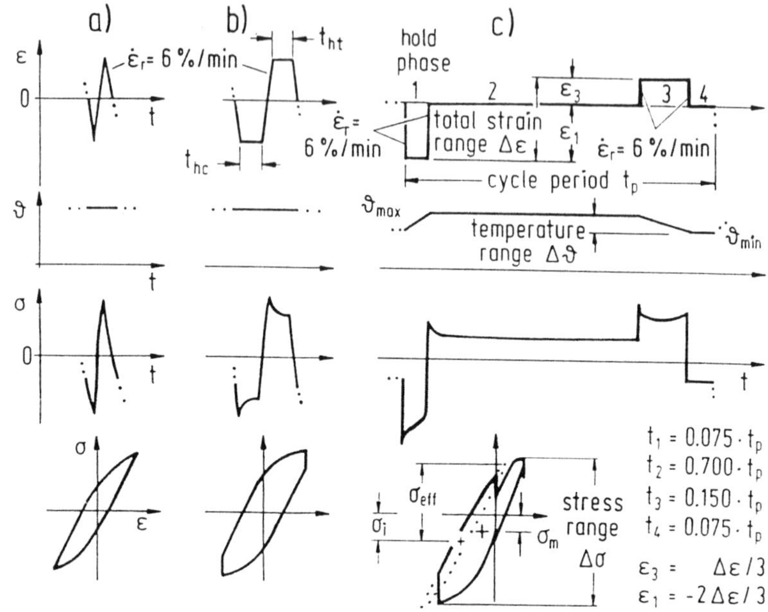

Figure 1. **Different strain cycles: standard cycle without (a) and with (b) hold times as well as service-type cycle (c) of a heated surface [1,4].**

creep packages (Fig. 4). In the long-term region of these tests the total strain ranges are below 0.4% and the cyclic softening of the two steels is almost completely disappearing [5,7].

The cyclic deformation can be explained on the basis of an effective stress concept (Fig. 1c) governing plasticity and creep. The effective stress σ_{eff} [5] is the difference between a measured external stress and an internal stress σ_i. A special value there of is the mean stress σ_m. It is situated in the centre of a fictitious hysteresis loop without hold times which envelopes the measured hysteresis loop.

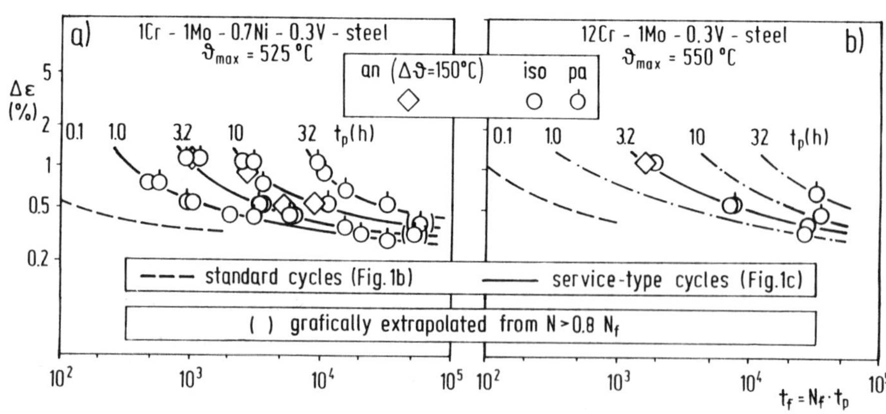

Figure 2. **Isothermal creep-fatigue failure curves on 1Cr-1Mo-0.7Ni-0.3V-steel (a) and 12Cr-1Mo-V-steel (b).**

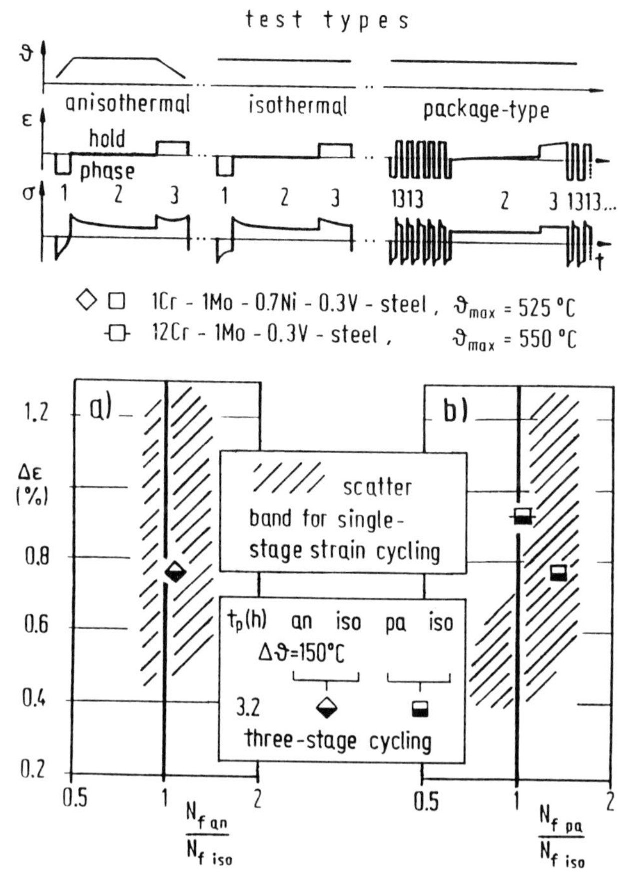

Figure 3. Ratio of the number of cycles to failure for comparable service-type test conditions, scatter bands for single-stage tests with (a) anisothermal (an) cycles and isothermal (iso) cycles and (b) with package-type (pa) testing and isothermal cycles as well as first results of the three-stage tests.

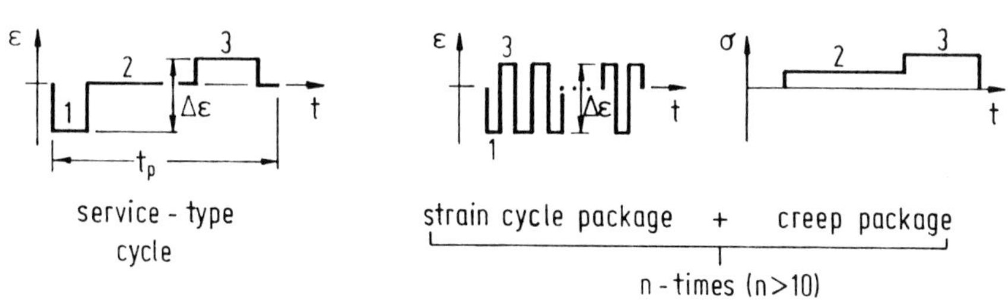

Figure 4. Simulation of long-term isothermal service-type testing by isothermal package-type tests [4].

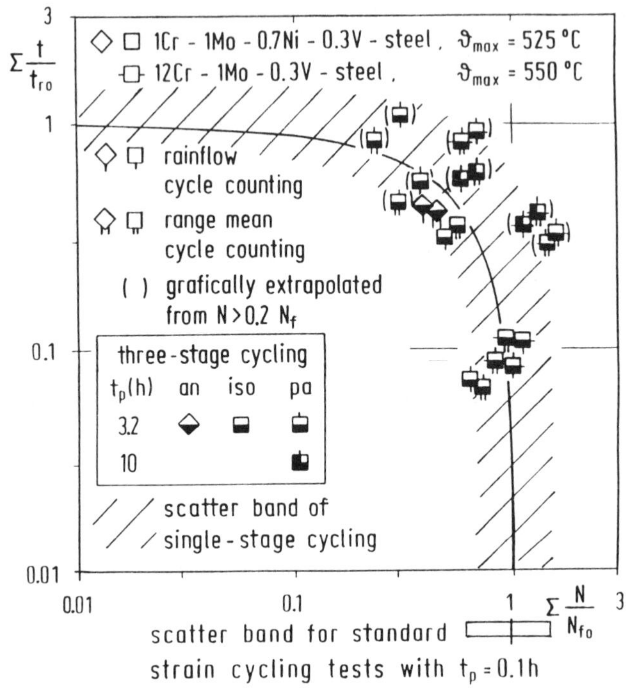

Figure 5. **Result of the creep-fatigue life analysis of service-type strain cycling, scatter band for single-stage testing and first results of three-stage testing.**

For the analysis of the creep-fatigue life (Fig. 5) of the service-type strain cycling tests a special method was developed [5,7]. It is based on the generalized damage accumulation rule

$$\Sigma N/N_{fo} + \Sigma t/t_{ro} = L \tag{1}$$

which combines the Miner rule for fatigue damage and the life fraction rule for creep damage. With respect to the accumulation of fatigue damage, the number of cycles to failure N_{fo} was taken for standard strain cycling with short hold times (Fig. 1b). The shifting of N_{fo} due to the creep-fatigue interaction was considered in function of prior service-type strain cycling. Further, the influence of mean stress σ_m on N_{fo} was taken into consideration by applying the Smith, Watson and Topper parameter [5,8]. For the evaluation of the creep life, the rupture time t_{ro} was taken for the effective stress σ_{eff} and cyclic softening was considered as well as shifting of t_{ro} due to prior service-type strain cycling. The result of the analysis of the single-stage service-type strain cycling tests shows a relative small scatter band with a mean relative creep-fatigue life of L ≈ 1. Microstructural analyses confirmed (Fig. 6) that long-term strain cycling leads to similar defects as long-term creep. Thus the application of Eq. 1 in the long-term region is justified.

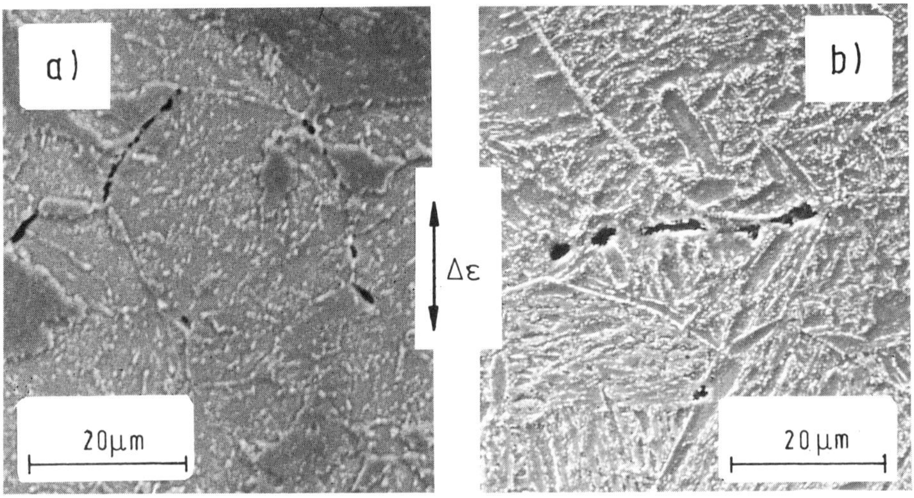

Figure 6. Pores and microcracks near the main crack tip after isothermal single-stage long-term service-type testing of 1Cr-1Mo-0.7Ni-0.3V-steel at $\vartheta = 525°C$, $\Delta\varepsilon = 0.54$ %, $t_f = 31\,300$ h (a) and 12Cr-1Mo-V-steel at $\vartheta = 550°C$, $\Delta\varepsilon = 0.35$ %, $t_f = 26\,100$ h (b).

3. Three-Stage Service-Type Strain Cycling

A loading collective with each three different values of temperature range and total strain range was arranged to simulate a typical combination of cold starts, warm starts and hot starts of power plants with the frequencies indicated in Fig. 7. First anisothermal, isothermal and package-type 3 000 h-tests lead to similar results (Fig. 3) as observed for single-stage testing. As a consequence package-type tests were started which have achieved up to 9 000 h test duration. A first creep-fatigue

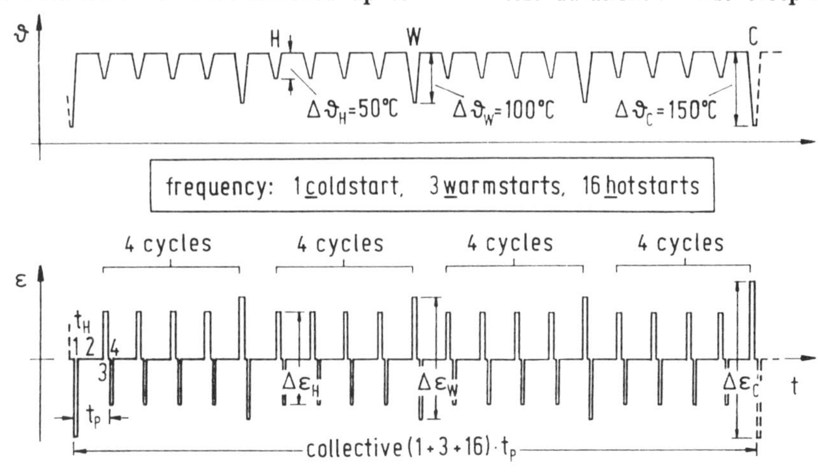

Figure 7. Three-stage service-type strain cycling.

life analysis (Fig. 5) shows similar life values as for single-stage strain cycling. The application of the rainflow cycle counting or the range mean cycle counting gives only small differences in the creep-fatigue life.

4. Conclusions

The strain cycling at the heated surface of components can be simulated by service-type creep-fatigue tests. Anisothermal tests on a bainitic and a martensitic steel lead only to insignificantly higher numbers of cycles to failure as comparable isothermal tests. In the long-term region isothermal tests can be replaced by more economical package-type tests. A creep-fatigue life analysis demonstrates the applicability of the generalized damage accumulation rule, if internal stress, stress relaxation, cyclic softening as well as pre-loading influences due to fatigue and creep are considered. Besides single-stage cycling three-stage cycling was investigated. The latter can also be tested with isothermal package-type tests and the first results there of are comparable to those of single-stage strain cycling.

Acknowledgements - Thanks are due to the "Bundesminister für Wirtschaft" (AIF-No 8 608), to the "Verein Deutscher Eisenhüttenleute" and to the "Arbeitsgemeinschaft Warmfeste Stähle" for their support of the work.

References

1. Kloos, K.H., Granacher, J., Barth, H. and Rieth, P., Dehnwechselverhalten eines warmfesten niedriglegierten chmiedestahles bei betriebsähnlichem Beanspruchungsablauf, in K.-T. Rie / E. Haibach (eds.), *Int. Symposium on Low-cycle fatigue strength and Elasto-plastic behaviour of materials*, Stuttgart, 1979, pp. 63-72.
2. Batte, A.D., Creep-Fatigue Life Predicitions, in R.P. Skelton (ed.), *Fatigue at High Temperature*, London, 1983, pp. 365-401.
3. Tremmel, D. and Mayer, K.H., Determination and appraisal of useful life experiency of HP and IP rotor shafts, EPRI-Seminar "Life Assessment and Improvements of Turbo-Generator Rotors for Fossil-Plants", Raleigh, 1984.
4. Kloos, K.H., Granacher, J. and Rieth, P., Long-term behaviour of heat resistant steels under variable creep-rupture conditions and under strain controlled service-type fatigue, *Int. Conf. of Engineering Aspects of Creep*, Sheffield, 1980, **1**, pp. 91-96.
5. Granacher, J. and Scholz A., Creep-fatigue behaviour of heat resistant steels under service-type long-term conditions, in K.-T. Rie (ed.), *Third Int. Conf. on Low-Cycle Fatigue and Elasto-Plastic Behaviour of Materials*, Elsevier Appl. Science, London, New York, 1992, pp. 235-241.
6. Forrest, G.P., The use of Strain Cycling Tests for Assessing Thermal Fatigue Resistance, *Applied Materials Research* (1965), 239-246.
7. Kloos, K.H., Granacher, J. and Scholz, A., Langzeitverhalten einiger warmfester Stähle unter betriebsähnlicher Kriechermüdungsbeanspruchung, *Mat.-wiss. u. Werkstofftech.* **24** (1993), 409-417.
8. Smith, K.N., Watson, P. and Topper, T.H., A Stress-Strain Function for the Fatigue of Metals, *Journal of Materials* **4** (1970), 767-778.

SENSITIVITY OF FATIGUE CRACK GROWTH IN A REACTOR STEEL TO THERMOMECHANICAL AGEING IN HYDROGEN ENVIRONMENT

H.M. NYKYFORCHYN, O.Z. STUDENT and B.P. LONIUK
Department of Corrosion and Metal Protection
Karpenko Physico-Mechanical Institute of the NAS of Ukraine
5 Naukova Str., 290601 Lviv, Ukraine

1. Introduction

The simultaneous effect of sustained mechancal load, elevated temperatures (450°C) and hydrogen environment on the oil hydrocracking reactor shells may cause the degradation of the steel microstucture. On the one hand, hydrogen accelerates the diffusion processes in the material [1]. On the other hand, it may cause a specific hydrogen-induced corrosion effect [2] (hydrogen-carbon interaction) resulting in decarburization and formation of high pressure methane. These factors induce the microstructural changes, the process of damage accumulation and concurrent decrease of the original metal properties. In such cases, the extent of steel degradation is controlling in terms of the prediction of the component life. However, conventional tensile properties usually are not sensitive enough to this type of materials degradation [3,4]. In contrast, fracture mechanics parameters (describing the local fracture characteristics) are expected to be strongly sensitive to the microstructural changes and damage induced by hydrogen attack.

The subject of this paper is to study the sensitivity of fatigue crack growth parameters to the degradation of oil hydrocracking reactor shell steel caused by hydrogen-induced ageing.

2. Experimental Procedure

The experiments were carried out on specimens that were cut from a plate of a Cr-Mo-V steel. The wt.% composition is as follows: 0.15% C; 2.80% Cr; 0.40% Ni; 0.70% Mo; 0.27% V; 0.50% Mn; 0.305 Si; 0.015% S; 0.012% P. The tensile properties of this steel are shown in Table 1.

According to the thermomechanical ageing technique, 10x25 mm and 5x25 mm planar specimens were subjected to the tensile load in a hydrogen test chamber with a pressure of 0.5 MPa and a temperature of 450°C, similar to that in a reactor shell. The temperature was achieved by heating the whole external surface of the chamber. In order to estimate the hydrogen effect a similar ageing procedure was carried out in air. The maximum load level (240 MPa) during the ageing sumulation test is

J. Bressers and L. Rémy (eds.), Fatigue under Thermal and Mechanical Loading, 215–220.

approximately equal to the operating stress. At this stress level creep deformation does not occur. The ageing time intervals were 780 hours and 2200 hours.

TABLE 1. The tensile properties of the Cr-Mo-V steel at various temperatures.

Temperature, °C	σ_y, MPa	σ_{UTS},MPa	Elongation, %	Reduction area, %
20	470	610	25	79
450	390	470	18	76

According to the procedure described in [5], single-edge notches of 2 mm wide, 5 mm deep and of radius 0.1 mm were made on the aged specimens. After that a fatigue crack at the notch bottom was initiated by cantilever bending. The stress intensity factor range ΔK was gradually decreased to the threshold level ΔK_{th}. The crack size was ≈3 mm, eliminating the notch effect on the stress field at the crack tip. Such preparation of aged specimens for fatigue testing lasted 7-10 days. Then the specimens were tested in fatigue crack growth by cantilever bending at room temperature, frequency f = 10 Hz, and load ratio R = 0; and at 450°C, f = 0.05 Hz, and R = 0.1 (the specimen temperature levels were achieved by means of electric heating). The crack size increment (≈ 0.2 mm) was periodically measured on both lateral sides of the specimens with a microscope with a accuracy of ± 0.01 mm. The test time was within 14-17 days. Based on these experiments, the fatigue crack growth rate da/dN - ΔK graphs were plotted. The calculation of the stress intensity factor value for such specimen geometry and such loading type was made using the relationships [6].

The fatigue crack closure was evaluated only at room temperature using the compliance technique [7]: a strain-gauge transducer was fixed at the crack tip and its signal was periodically measured during crack propagation. The effective stress intensity factor range ΔK_{eff} was evaluated, and da/dN - ΔK_{eff} graphs were plotted.

3. Experimental Results

The investigated steel has a needle-type ferrite-pearlitic structure (Fig. 1a) in the virgin state. At half the value of the maximum stress (120 MPa), this structure did almost not change after the high-temperature ageing in hydrogen environment irrespective of the ageing period. However, under the conditions of load 240 MPa and ageing time 2200 hours, the needle-type structure disappeared and a more homogeneous conglomerate of ferrite and carbides formed (Fig. 1b).

Fig. 2 illustrates the result of fatigue crack growth experiments carried out at room temperature on the specimens that have been aged in H_2 at 120 MPa; the graphs show that the ageing retards the fatigue crack propagation, in particular in its near-threshold region, and this influence increases with increasing duration of the hydrogen treatment. Taking crack closure into account does not change the relative positions of the da/dN - ΔK_{eff} graphs.

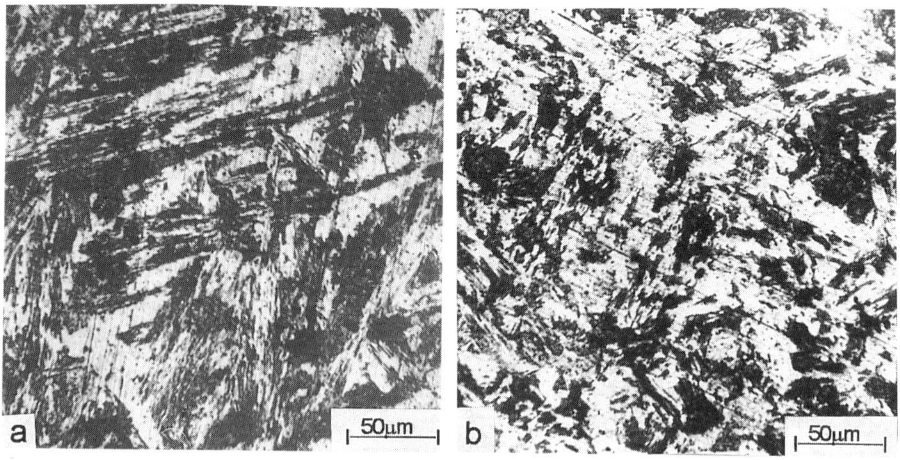

Figure 1. Microstructure of the steel in (a) the virgin state; (b) aged in H_2 at σ=240 MPa during 2200 hours.

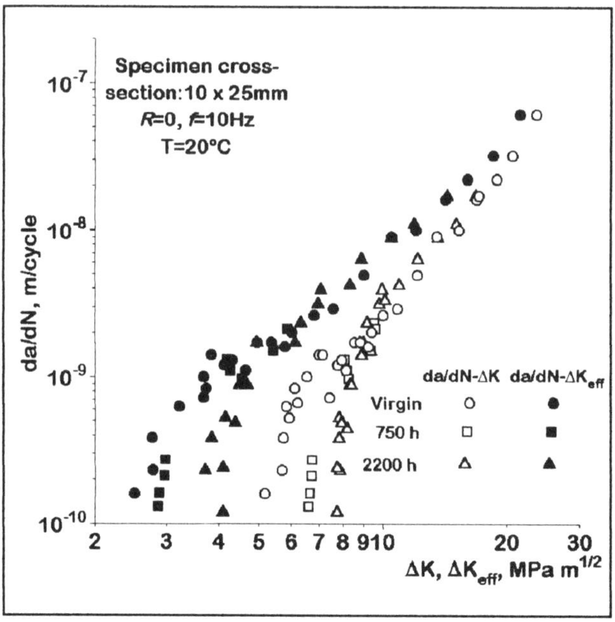

Figure 2. Fatigue crack growth rate graphs for the virgin steel and the hydrogen-aged one at σ=120 MPa.

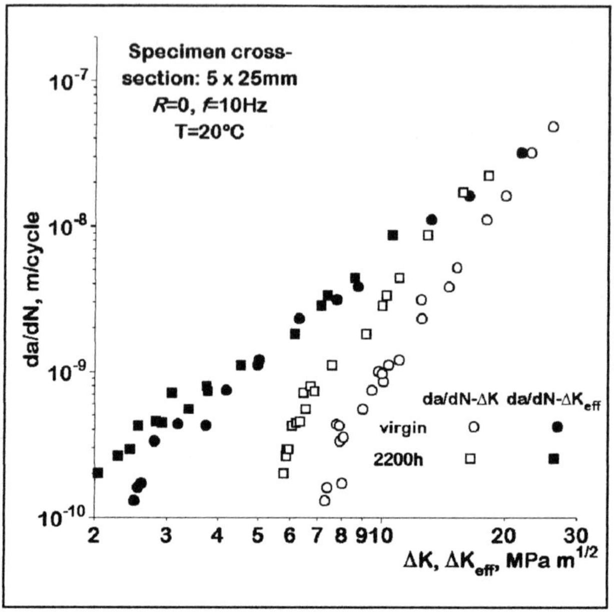

Figure 3. Fatigue crack growth rate graphs for the virgin material and the hydrogen-aged one at $\sigma = 240$ MPa.

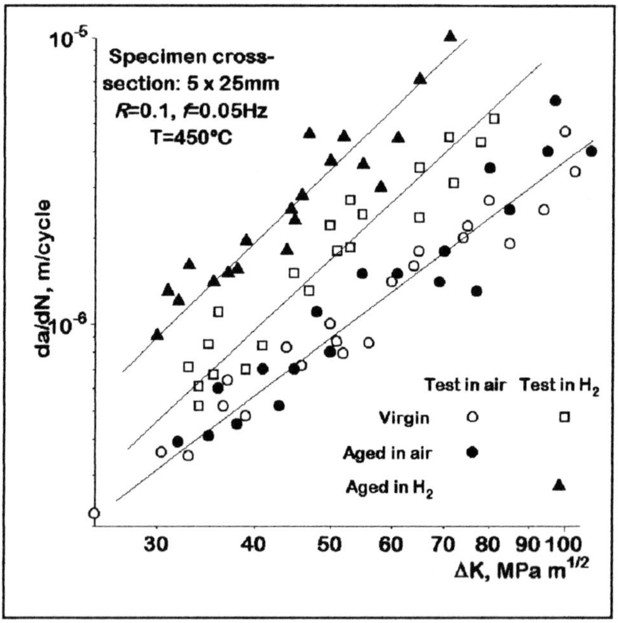

Figure 4. Fatigue crack growth rate graphs for the virgin material and after ageing in air and in hydrogen at $\sigma = 240$ MPa.

In the case of ageing in H_2 at the load 240 MPa, the opposite effect was found (Fig. 3): the resistance of aged steel to the fatigue crack growth is lower than in the virgin material. The same effect is observed in the *da/dN* - ΔK_{eff} graphs.

The experiments carried out at 450°C concern only high crack growth velocities (Fig. 4). The steel ageing in air at a stress of 240 MPa has no significant effect on the crack growth in air. At the same time, the tests of the virgin steel in hydrogen environment show an increase of the fatigue crack growth rate in comparison with that in air. The thermomechanical ageing in hydrogen at a load of 240 MPa resulted in a more significant reduction of the fatigue crack growth resistance relative to the virgin material.

4. Discussion

The thermomechanical ageing in hydrogen at load levels typical for service conditions substantially changes the steel microstructure, though the ageing period (2200 hours) is comparatively small. The fatigue crack growth velocities both at room and operating temperatures change, too. Taking into account the negligible influence of such short-term ageing of the material in air on its fatigue behaviour (Fig. 4), we can argue that the observed ageing effects are caused by hydrogen.

The analysis of a wide range of crack growth rates at room temperature shows that the fatigue threshold ΔK_{th} is the most sensitive to ageing. The variations of fatigue crack growth rate and ΔK_{th} result from:

- microstructural changes and damage;
- material hydrogenation;
- change of fatigue crack closure.

The *da/dN*-ΔK_{eff} dependencies and the effective threshold $\Delta K_{th\ eff}$ reflect the effect of the first two factors. Moreover, for vessels under pressure, the estimation of fatigue crack growth is more important at high load asymmetry when there is no crack closure, and thus it does not affect substantially the faigue crack growth. As it can be seen from the near-threshold areas of the *da/dN* graphs for the aged material at low stress (Fig. 2), these ageing conditions cause the increase of $\Delta K_{th\ eff}$. The reason is that physically the fatigue threshold is determined by cyclic microyielding resistance, while hydrogen increases the microshear resistance at room temperature [8]. In this case, the strengthening influence of hydrogen dominates over the possible strength reduction caused by microstructural changes and material damage. However, the more severe ageing conditions (at higher stress level, Fig. 3) already cause the opposite effect of high-temperature hydrogen treatment on the near-threshold crack growth: $\Delta K_{th\ eff}$ value is lower than that in the virgin steel. Thus the material degradation resulting from the ageing under these conditions is so intensive that it eliminates the strengthening action of hydrogen on the near-threshold crack growth.

The disparity between the *da/dN* - ΔK relationship for the specimens with thickness 5 mm and 10 mm (compare Figs. 2 and 3) in the virgin steel is, perhaps, caused by the crack closure phenomenon that is responsible for the effect of specimen thickness even under plane strain conditions [9] if the crack closure results from the oxidation processes.

Another hydrogen effect mechanism at high temperatures consists in promoting the plastic deformation and accelerating the fatigue crack growth rate [8] (Fig. 4). The crack closure in the ΔK upper range is negligible. The material ageing under conditions similar to the operating conditions causes the additional crack growth velocity increase. It should by mentioned that application of the linear fracture mechanics approaches to the specimens of chosen size at elevated temperature and high ΔK level is not quite correct due to the presence of a big plastic zone at the crack tip. Nevertheless, a qualitative estimation of the hydrogen and ageing effect on the fatigue crack growth resistance can be made.

5. Conclusions

Among the fatigue crack growth parameters evaluated at room temperature, the fatigue threshold is the most sensitive to the thermomechanical hydrogen-induced ageing. In this case, two opposite factors affect the change of ΔK_{th}: hydrogen-induced strengthening and microstructural degradation.

Under the conditions similar to that in oil hydrocracking plant, hydrogen accelerates the degradation of Cr-Mo-V steel structure and decreases the fatigue crack growth resistance.

References

1. Sidorenko, B.M., Fedorov, B.B. and Pokhmurskyi, V.I., Acceleration of self-diffusion processes in metals due to dissolved hydrogen effect, *Physicochemical Mechanics of Materials* **13**, No. 6 (1977), 27-30. (In Russian).
2. Nelson, H.G., The hydrogen embrittlement, in C.L. Briant and S.K. Banerji (eds.), *Treatise on Materials Science and Technology, vol. 25: Embrittlement of Engineering Alloys,* Academic Press, New York, etc., 1983, pp. 256-333.
3. Masaoka, I., Kinoshita, K., Chiba, R. *et al.*, Hydrogen attack limit of 2 1/4 Cr-1Mo steel, *Welding Research Council Bulletin* **305** (1985), 1-8.
4. Molinié, E., Piques, R. and Pineau, A., Behaviour of a 1Cr-Mo-0.25V steel after long-term exposure. 1. Charpy impact toughness and creep properties, *Fatigue & Fracture of Engineering Materials and Structures* **14** (1991), 531-545.
5. Yarema, S.Ya., *Test Method for Determination of Crack Growth Rate and Crack Extension Resistance under Cyclic Loading,* Karpenko Physico-Mechanical Institute, Lviv, 1994.
6. Nakasa, K., Takei, H., Itoh, H. and Kobyashi, M., Crack propagation assisted with hydrogen under cyclic stress in Ni-Cr-Mo steel, *Transctions of the Japan Institute of Metals* **27** (1986), 260-269.
7. Nykyforchyn, H.M., Andrusiv, B.M., Voldemarov, A.B. and Kutsyn, M.A., The evaluation of fatigue crack closure effect. *Physicochemical Mechanics of Materials* **18**, No. 5 (1982), 100-103 (in Russian).
8. Nykyforchyn, H.M., Schaper, M., Student, O.Z. and Skrypnyk, I.D., Fatigue ceack growth kinetics and mechanism in steel at elevated temperatures in gasoaus hydrogens, in JK-H. Schwalbe and C.Berger (eds.), *Structural Integrity: Experiments - Models - Applications, Proc. 10th Biennial European Conference on Fracture,* EMAS, Warley, England, 1994,pp. 1265-1270.
9. Nykyforchyn, H.M., Some ambiguities of experimentally evaluated parameters of fracture mechanics, in A. Bakker (ed.), *7th International Conference on Mechanical Behaviour of Materials, book of abstracts,* Delft University Press, Delft, 1995, pp. 357-358.

CRACK GROWTH RESISTANCE OF ROLLING ALLOYS UNDER THE INFLUENCE OF THERMOMECHANICAL LOADING AND COOLING WATER ENVIRONMENT

O.M.ROMANIV and B.D.VASYLIV
Karpenko Physico-Mechanical Institute of the National Academy of Sciences of Ukraine
5 Naukova St.
Lviv 290601, Ukraine

1. Introduction

In service conditions materials of rollers for hot rolling are subjected to a complex influence of mechanical (primarily contact) stresses and stresses which arise from the inhomogeneous and local heating. The latter have very complex thermal origins and they are repeated cyclically with every turn during operation of the rollers in the phase of contact of the roller surface with the hot rolled products [1]. To prevent overheating of the rollers and also in the case of thermomechanical treatment of the products, water cooling of the rollers is used. Such working medium lowers the temperature of the material but stimulates its corrosion.

Such phenomena of thermomechanical nature stimulate the early stages of surface deterioration which causes crack initiation followed by propagation under the influence of mechanical, thermal and corrosive factors. This paper presents an attempt to evaluate the process of crack propagation using the approaches and methods of fracture mechanics [2].

2. Materials and Methods of Investigation

Three rolling materials were investigated (Table 1): grey nodular cast iron *SSh 1Cr-1Ni*, preeutectic steel *0.5C-1Cr-1Ni* and overeutectic steel *1.4C-1Cr-1Mn-2Si-1Ni-1Mo*.

TABLE 1. Chemical content of investigated materials.

Grade of materials	Elements content, %							
	C	Cr	Mn	Si	Ni	Mo	S	P
SSh 1Cr-1Ni	3.90	0.86	0.80	1.50	1.10	-	0.160	0.300
0.5C-1Cr-1Ni	0.51	1.02	0.43	0.25	1.13	-	0.011	0.012
1.4C-1Cr-1Mn-2Si-1Ni-1Mo	1.43	0.67	0.41	1.37	0.76	0.34	0.035	0.018

J. Bressers and L. Rémy (eds.), Fatique under Thermal and Mechanical Loading, 221–226.

Prismatic precracked specimens as shown in Fig. 1a were tested on the specially designed testing machine, the principle scheme of which is shown in Fig. 2.

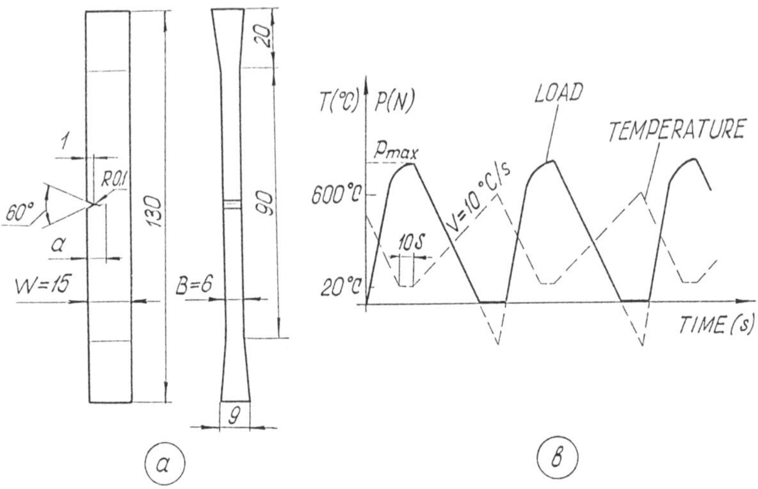

Figure 1. Specimen configuration (*a*) and thermomechanical loading regime of specimens (*b*).

Two thick plates 1 and columns 2 form the rigid frame which resists to mechanical loading. Such loading arises during heating of the specimen 3 which is fitted in wedge grips 4. The upper grip is fitted in series with a rigid cylindrical high sensitivity tensodynamometer 5. To avoid misalignment of the dynamometer and specimen the guides 6 are provided. The device 7 automatically registers the level of mechanical loading. By displacement of the nut 8 a static mechanical tensile stress in the specimen is created. So during periodical heating and cooling a cyclic change of the stresses is created, and the loading history has the form as in Fig. 1b. The specimen is heated by an electric current of 1300 A, the temperature being controlled by the chromel-alumel (XA) thermocouple which is connected with the control unit. The unit automatically performs the assigned heating regimes of the specimen, which is cooled by water using sprayer 9. The general view of the device is shown in Fig. 3.

The cyclic heating and cooling of the specimen was achieved in the temperature range 20...600°C with a frequency about 0.01 Hz and a stress ratio R=0 (pulsating cycle). Observations of crack growth were made with an optical device for distance crack length measurement KM-8. The amplitude of the stress intensity factor was determined using the Equation

$$\Delta K = \frac{\Delta P}{BW}\left[1.99 - 0.41(a/W) + 18.7(a/W)^2 - 38.48(a/W)^3 + 53.85(a/W)^4\right]\sqrt{a}$$

where a - crack length, B - thickness of the specimen, W - specimen width, and ΔP - loading amplitude.

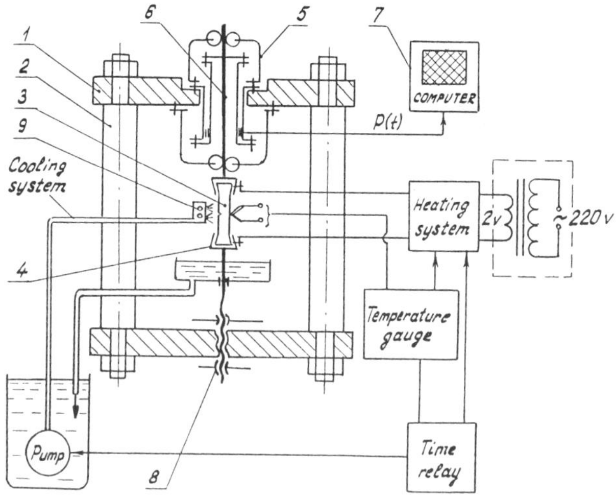

Figure 2. Schematic of testing device.

The testing results were presented analytically as the dependence $V=da/dN=C(\Delta K)^n$, and the coefficients C and n were evaluated statistically by the least squares method.

For comparison the fatigue crack growth was studied also at ambient temperature under pure mechanical loading.

3. Results

The fatigue crack growth behaviour was investigated in the crack growth range of 10^{-6} to 10^{-3} m/cycle which corresponds to the Paris region of the crack growth curve, and also to the region of the accelerated crack growth. These results are shown in Fig. 4.

Control testing of the investigated materials was made at room temperature in air. This showed that *0.5C-1Cr-1Ni* structural steel has a higher fatigue crack growth resistance than typical rolling materials such as *SSh 1Cr-1Ni* nodular cast iron and *1.4C-1Cr-1Mn-2Si-1Ni-1Mo* overeutectic steel.

Compared to nodular cast iron the high carbon overeutectic steel is also characterized by higher contact strength as shown by the designers of this kind of new rolling material [3].

Testing of the selected materials in thermocycling conditions using electric contact heating and water cooling leads to a drastic drop of the crack growth resistance of the rolling alloys, manifested by a sharp shift of the crack growth curves to the left (Fig. 4).

In the case of testing in air, the traditional rolling material (*SSh 1Cr-1Ni* nodular cast iron) showed the lowest level of crack growth resistance. High-strength

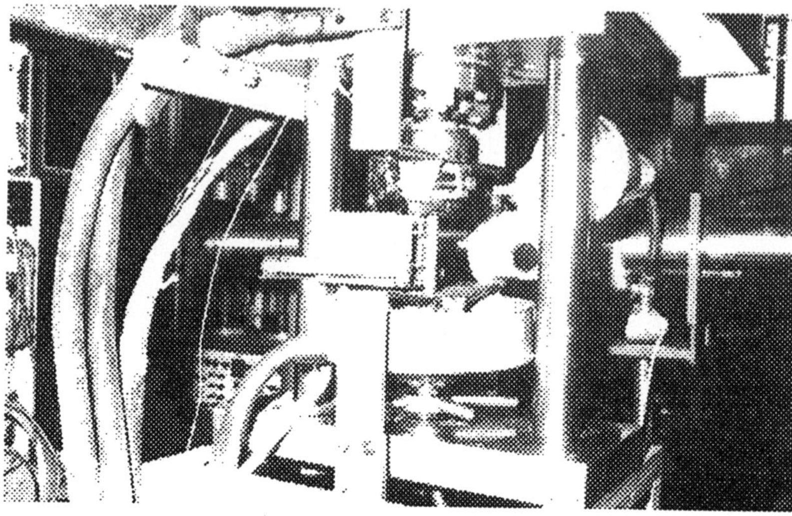

Figure 3. General view of the testing device.

0.5C-1Cr-1Ni steel showed substantially higher fracture resistance. But without any doubt the highest resistance to crack growth in conditions of thermomechanical cycling and corrosive medium influence is shown by overeutectic rolling *1.4C-1Cr-1Mn-2Si-1Ni-1Mo* steel.

4. Discussion

A drop of the resistance to fatigue crack growth under the influence of thermomechanical cycling and corrosive medium may be regarded first of all as a consequence of a fracture toughness decrease at elevated temperature. It was shown [4] that the fracture toughness of usual carbon mildly alloyed steel drops by 40 percent when the testing temperature increases from 20 to 600°C. Because of the time-varying temperature within every cycle the mechanical loading and induced stresses have a specific inhomogeneous character. Consequently the maximum crack growth rate does not correspond to the maximum level of tensile stress but to the time intervals when relatively high tensile stress combines with relatively high temperature in the vicinity of the crack tip (Fig. 1). Similar effects of influence of the temperature and mechanical stresses were observed during out of phase TMF of various heat resistant steels [5]. For *0.5C-1Cr-1Ni* steel at ambient temperature the corrosive medium accelerates crack growth in the Paris region [6]. There are also reasons to consider that this influence is much greater in a case of increased temperature, because of the creation of the conditions for dissociation of water and chemisorption of hydrogen in the vicinity of the crack tip [7]. Proceeding from such considerations, the factor of hydrogen metals degradation in conditions of thermo-mechanical cycling is probably one of the most important in the process of crack growth acceleration.

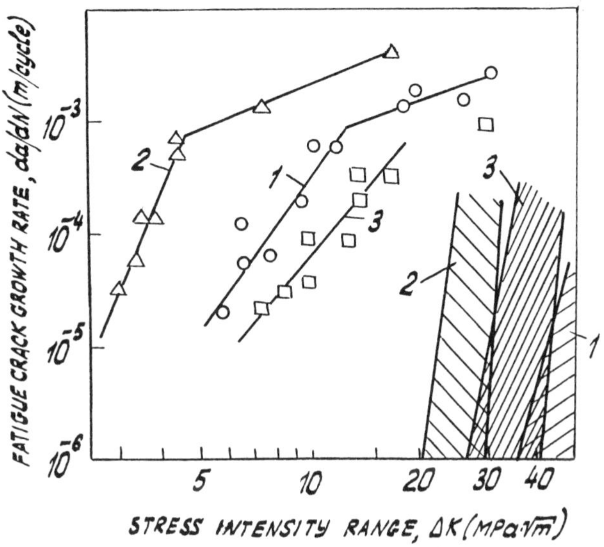

Figure 4. Fatigue-crack propagation rates in preeutectic *0.5C-1Cr-1Ni* steel (1), *SSh 1Cr-1Ni* nodular cast iron (2) and *1.4C-1Cr-1Mn-2Si-1Ni-1Mo* overeutectic steel (3) under thermomechanical loading as a function of the nominal stress-intensity range, ΔK, at R=0. Crack growth-rate data for these materials at R=0 and ambient temperature are also shown for comparison (right part of the figure).

The thermocycling under mechanical stresses contributes to the structural changes in the steels and in the cast iron which should be also taken into account. The investigations of the microstructure of materials near the crack tip showed that such changes actually occurred. They are connected with the decomposition of cementite in steels and coagulation of graphite in cast iron. The smallest changes were observed in high carbon *1.4C-1Cr-1Mn-2Si-1Ni-1Mo* steel which can be explained by the high temperature stability of carbides of chromium and molybdenum. It can be assumed that the target oriented alloying of overeutectic steels, which ensures an increase of dispersive special carbides, is an important factor in imparting higher stability to rolling alloys under the common influence of thermocycling and corrosive environment.

We carried out fractographic investigations to explain the tendency to fatigue crack growth retardation in the area of high stresses (formation of the unusual bends in the curves 1,2 - Fig. 4). Such a tendency is caused by thermoplastic blunting of the crack tip, which is related to the change of the micromechanism of crack growth with the formation on the fracture surface of typical ductile dimple relief [8]. In the case of tests at ambient temperature such plastic blunting and increasing of crack displacement are not observed.

The presented results elucidate tendencies of the combined influence of factors of temperature, mechanical stresses and corrosive environment on the mechanical stability of rolling alloys, and create the basis for searching new thermomechanically stable alloys.

References

1. Tylkin, M.A. *Increase of durability of the metallurgical equipment*, Metallurgia Publishers, Moscow 1975.
2. Zalessky, B.I., Korneyev, D.M., Crack formation under cyclic heating and cooling of steels, in *Production and Treatment of Steel*, Metallurgizdat Publisher, Moscow, (1984), pp.332-340.
3. Scoblo, T.S., Rudyuk, S.I, Voronina, V.A., Overeutectic steel for rolling-mills, in *Sci.-Tech. Inform. Paper* **6**, Ferrous Metallurgia Publishers, Moscow, (1974), pp. 48-49.
4. Romaniv, O.N., *Fracture Toughness of Structural Steels*, Metallurgia Publishers, Moscow 1979.
5. Granacher, J., Scholz, A., Creep-fatigue behaviour of heat resistant steels under service-type long-term conditions, in K.T. Rie (ed.), *Proc. on Third Int. Conf. on Low-Cycle Fatigue and Elasto-Plastic Behaviour of Materials*, Elsevier Appl. Science, London, New York, (1992), pp. 235-241.
6. Romaniv, O.N, Tkach, A.N., Siminkowich, V.N., Structure-metallurgical approach to design of steels with high crack propagation resistance, in *Proc. 6th Intern. Conf. on Fract.*, New Delhi, (1984), pp. 3591-3598.
7. Agladze, T.R., Kolotyrkin, Ya.M., Romaniv, O.N., Nikiforchyn, G.N., The effect of adsorption - chemical interaction for metal-environment system in stress corrosion cracking, in *Characterization and Control of Metal Surfaces and Surface Films in Corrosion*, Proc. 4th Jap.-USSR Sem. on Cor., Tokyo, (1985) pp. 256-270.
8. Romaniv, O.M., Structural aspects of corrosion fatigue of metals, in *Proc. 8th Intern. Conf. on Fract.*, Kyiv, (1983), pp. 292-293.

THERMO-MECHANICAL FATIGUE OF A COMPOSITE STEEL

L. LINDÉ and P.J. HENDERSON
Swedish Institute for Metals Research
Drottning Kristinas väg 48
S-114 28 Stockholm, Sweden

1. Introduction

In black liquor recovery boilers (BLRB) used in the paper industry, the materials have to withstand a combination of high temperatures and very corrosive environments. The material studied in this paper is used in pipes for applications in a part of the BLRB functioning as a large heat exchanger, where chemicals are recovered from the pulping process and at the same time organic waste is converted to energy. In the boiler black liquor, containing Na_2SO_4 with a pH of 12 to 14, is cleaned from the waste originating from the process of converting wood to fibre for paper production. The cleaning process includes reducing the Na_2SO_4 to Na_2S and burning off the waste, resulting in a high temperature at the pipes in which water is circulated to recover the heat. As a result the pipes are subjected to a high temperature and an aggressive alkaline environment due to the black liquor.

Traditional carbon steels only last for a couple of years in such an application due to corrosion. On the other hand, austenitic stainless steels having a much better corrosion resistance are sensitive to stress corrosion cracking. By using a combination of the two steel types in a composite material, one combines the advantages of the two materials while the disadvantages are reduced considerably. This has been utilised in pipes with a load bearing carbon steel on the inside and a stainless steel on the outside for protection against corrosion. This type of composite material has been used successfully in BLRB since the 1970's [1].

In a recovery boiler the material is subjected to varying temperatures resulting in thermal stresses and strains. However, little is known about the fatigue properties of a composite such as the one described here, especially during non-isothermal conditions. As a means to study this, thermo-mechanical fatigue (TMF) testing was performed on a material used in BLRB at the prevailing temperature range.

J. Bressers and L. Rémy (eds.), Fatique under Thermal and Mechanical Loading. 227–236.

2. Experimental

2.1. MATERIAL

The material in this study was supplied by AB Sandvik Steel as a hot extruded composite steel bar with a plain carbon steel core of SS 14 35 (DIN St 45.8) and an outer layer of Alloy 825, an austenitic corrosion resistant Ni-Fe-Cr alloy. The chemical compositions of the two materials are given in Table 1 below. The diameter of the inner ferritic component was 12.8 mm and the outer diameter of the bar was 19 mm. Testing was performed not only on the composite material but also on the carbon steel and on semi-coated material where the gauge length was only partially coated with Alloy 825. After specimen manufacture the austenite layer in the all-coated composite material was 0.3 mm thick and in the semi-coated the austenite thickness varied from 0.1 mm to zero. All the test specimens were hollow with a straight gauge length. All-ferritic and semi-coated specimens had an outer diameter of 13.0 mm while the composite specimens were 13.5 mm in diameter. All specimens had an 6.0 mm inner diameter. The inner surface was not coated.

TABLE 1. Chemical composition (nominal) of the carbon steel and Alloy 825 in weight% according to AB Sandvik Steel (* maximum content).

	C	Si	Mn	Cr	Ni	Mo	Al	P	S	Cu	Ti
DIN St 45.8	0.19	0.25-0.35	0.6-0.8				0.003*	0.020*	0.020*		
Alloy 825	0.030*	0.5*	0.8	21.5	40.0	3.0				1.7	0.8

2.2. TESTING PROCEDURE AND EVALUATION

Testing was performed in a servohydraulic test machine equipped with a high frequency generator and an induction coil for heating of the specimen. The temperature was monitored by type S thermocouples welded to the gauge length. All tests were performed in total (mechanical+thermal) strain control by the means of a side entry extensometer with a gauge length of 12.5 mm. Collection and evaluation of test data were carried out by a computerised data acquisition system developed at the Swedish Institute for Metals Research [2]. It performs a real-time recording and processing of test data, making the information available instantly.

The different material combinations were tested in TMF between 300 and 600°C with a strain rate of $1.0 \cdot 10^{-4}$ 1/s. During testing the strain and temperature were cycled in-phase (IP) with each other. Out-of-phase (OP) testing with a 180° shift between strain and temperature was tried but no reliable results were obtained due to a specimen shape instability resulting in a "barrelling" or "double-necking" effect and failure outside the gauge length.

Crack initiation was determined as the first deviation from the extrapolated maximum stress behaviour, A in Fig. 1, giving the number of cycles to crack initiation, N_i. As a measurement of the fatigue life the number of cycles to a 10%

drop from the stable linear behaviour in tensile peak stress, N_{90}, was used, B in Fig. 1. The number of cycles to failure, N_f, which results in a total separation of the specimen, the traditional measurement of the fatigue life, was not considered suitable since it depends on the spacial distribution of initiated cracks which can lead to a large scatter in N_f values.

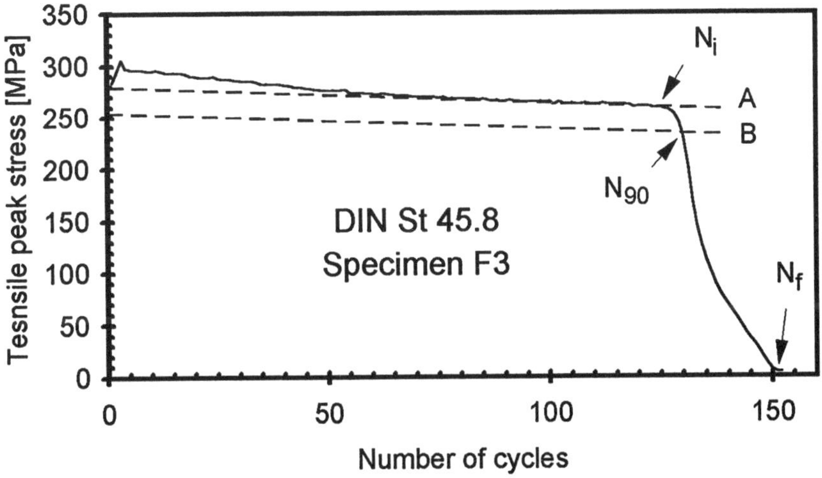

Figure 1. Example of evaluation of N_i, N_{90} and N_f for specimen F3 from the tensile peak stress versus number of cycles plot.

Testing conditions were evaluated from the hysteresis loop closest to half the number of cycles to crack initiation, $0.5{\cdot}N_i$. For the calculations of elastic strain during testing the following expression was used, based on standardised values in [3]:

$$E = 212 - 8.1{\cdot}10^{-2} \cdot T - 2.8{\cdot}10^{-5} \cdot T^2 \qquad (1)$$

where E is Young's modulus in GPa and T the temperature in °C.

Three specimens, one of each type tested at approximately the same strain range, were examined metallographically by light optical microscopy (LOM) and in a scanning electron microscope (SEM). Investigated specimens were sectioned longitudinally, polished to 1 μm and etched at room temperature in a solution consisting of 20 g picric acid ($C_6H_3N_3O_7$) in 100 ml HCl.

3. Results

3.1. METALLOGRAPHY AFTER TESTING

None of the examinations in either LOM or SEM revealed any visible cracking in the interface between the ferritic steel and the Ni-Fe-Cr alloy. A photograph of the interface is shown in Fig. 2. In the all-coated specimen no fatigue cracks initiated in the outer austenitic layer and failure was caused by intergranular cracks growing through the ferrite from the inner bore. Some cracks that were observed in the ferrite part in all specimen types were perpendicular to the stress axis and had no connection with the surface. One example of such a crack can be seen in Fig. 3. Due to the fact that the cracks did not exhibit any oxidation it was assumed that the cracks initiated internally. This is a type of crack often observed after creep testing of ferritic steels [4, 5]. In the semi-coated specimen some cracks in the coating continued into the ferritic steel where the crack surfaces oxidised, as shown in Fig. 4. In the all-ferritic specimen there were numerous cracks on both the outer surface as well as at the inner bore which all exhibited oxidation. There were approximately 30% more cracks at the inner bore than the outer surface. The ferrite surfaces were oxidised and had become roughened, but the austenite surfaces remained smooth.

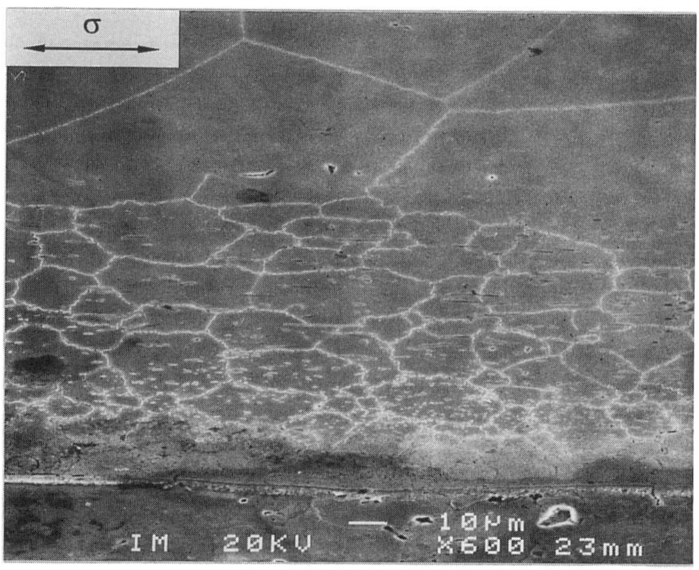

Figure 2. Interface between Alloy 825 (upper) and DIN St 45.8 (lower) in composite specimen AF2 tested in TMF-IP with $\Delta\varepsilon_t$=1.47% between 300 and 600°C.

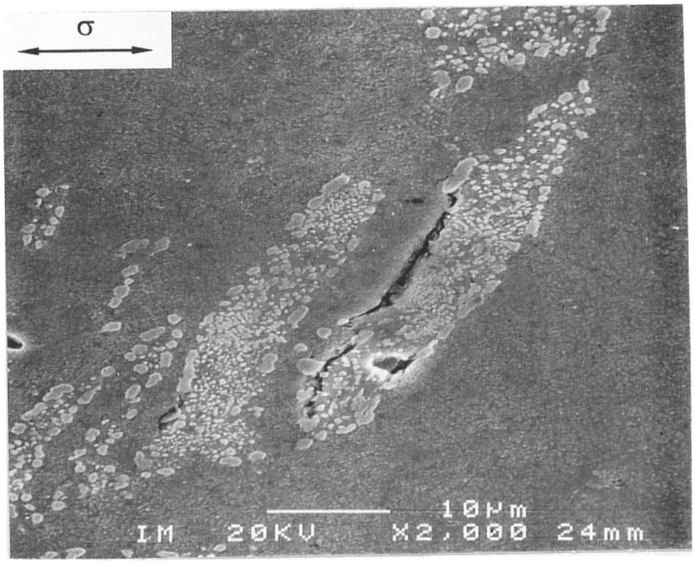

Figure 3. Internal crack in the ferritic DIN St 45.8 steel, specimen F3, after TMF-IP testing between 300 and 600°C with $\Delta\varepsilon_t$=1.49%.

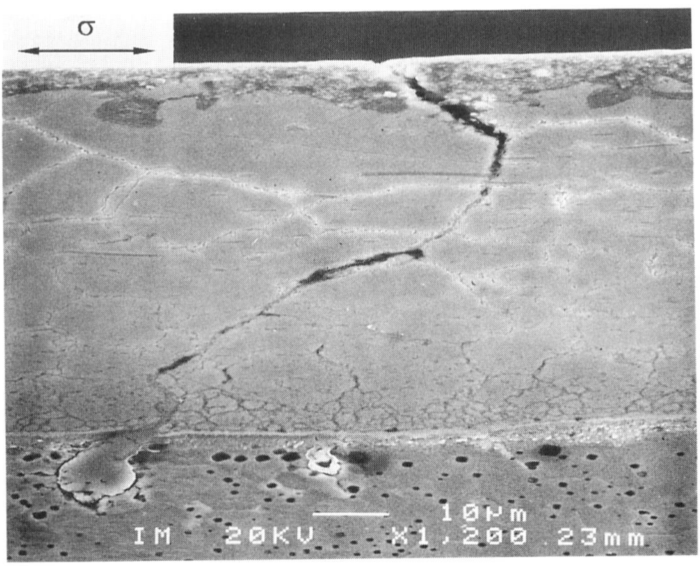

Figure 4. Intergranular crack in the austenitic layer propagating into the ferritic core in semi-coated specimen C2 tested in TMF-IP with $\Delta\varepsilon_t$=1.44% between 300 and 600°C.

In the composite specimen the composition of the substrate and coating materials was investigated after the test which lasted for 115 hours. It was found that there was a small increase in the chromium content of the ferritic steel close to and at the interface compared to the bulk material. However, the nickel content was not increased in the ferrite close to the interface.

3.2. FATIGUE TESTING

The results from the fatigue testing are shown in Table 2 for all TMF-IP tests. As can be seen in Fig. 5 the fatigue life for composite material was longer than for uncoated and semi-coated material when considering total strain range, $\Delta\varepsilon_t$, versus N_{90}. It can also be seen that the lifetimes for the ferritic and semi-coated materials were very similar.

TABLE 2. Results from TMF-IP testing.

Material type	Test nr.	$\Delta\varepsilon_t$ [%]	$\Delta\varepsilon_p$ [%]	$\Delta\varepsilon_e$ [%]	$\Delta\sigma$ [MPa]	N_i	N_{90}	N_f
Ferritic	F1	0.98	0.68	0.30	565	240	265	376
	F2	0.72	0.43	0.29	541	525	536	593
	F3	1.49	1.12	0.36	675	123	129	151
Semi-coated	C1	0.99	0.67	0.32	589	210	217	900
	C2	1.44	1.09	0.35	647	130	133	153
Composite	AF1	0.97	0.67	0.30	554	412#		
	AF2	1.47	1.13	0.34	620	211	226	354
	AF3	0.70	0.41	0.29	538	826	829	908

#) Test interrupted before crack initiation.

The lines in Fig. 5 correspond to the summation of the Coffin-Manson, Eq. 2, and Basquin, Eq. 3, relationships:

Coffin-Manson $$\Delta\varepsilon_p = A \cdot N_{90}^{\alpha} \tag{2}$$

Basquin $$\Delta\varepsilon_e = B \cdot N_{90}^{\beta} \tag{3}$$

where $\Delta\varepsilon_p$ and $\Delta\varepsilon_e$ are the plastic and elastic strain ranges, respectively, and A, α, B and β are material constants. The corresponding values for the constants are shown in Table 3.

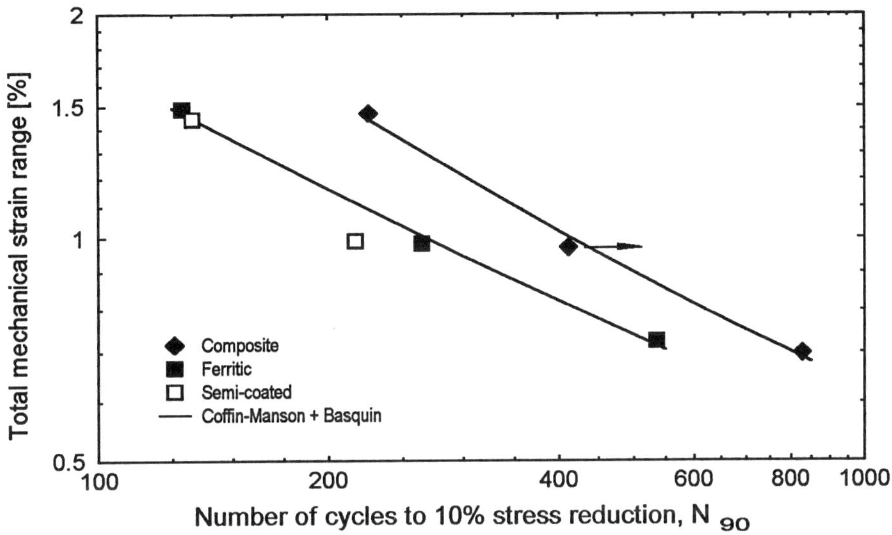

Figure 5. Total mechanical strain range versus number of cycles to 10% stress reduction for tested materials. The lines correspond to a summation of the Coffin-Manson and Basquin equations, Eqs. 2 and 3 respectively.

One TMF-OP test was performed on an all-ferritic specimen which resulted in an increase in the specimen outer diameter (barrelling) along the gauge length and a decrease in the diameter (necking) directly outside the gauge length. Failure occurred at one of the necked regions. The maximum diameter of the barrelled region was 13.8 mm and the minimum necked diameter 11.5 mm compared to the original 13 mm diameter.

TABLE 3. Values of constants in Eqs. 2 and 3.

Material type	A	α	B	β
Ferritic	29.1	-0.67	0.75	-0.15
Composite	80.0	-0.79	0.58	-0.10

4. Discussion

4.1. COMPOSITE FATIGUE LIFE

An outer layer of the corrosion resistant Alloy 825 increased the fatigue life of the specimens with at least 50% under the test conditions, see Fig. 5. In the composite steel, failure occurred by cracks which had initiated at the inner bore in the ferritic. In the all-ferritic specimens cracks propagated from both the inner and outer surfaces. Crack initiation and propagation was intergranular and it is likely that oxidation played an important role in the nucleation of cracks. Fatigue cracks seemed to initiate easier in thin austenite layers than in a thicker one. In the composite specimens, with a coating thickness of 0.3 mm, no cracks were seen on the surface while in the semi-coated material, where the coating thickness was 0.1 mm or less, cracks were observed on the surface when tested at the same strain range. One possible reason for this can be observed in Fig. 2 as the grain size in the Ni-Fe-Cr alloy becomes considerably smaller close to the interface, approximately 10 μm, than at the outer surface where the grain size was approximately 75 μm. The specimens were manufactured by progressively removing the outer austenite layer so that in the thin coating with only small grains remaining, the number of grain boundaries in the surface was considerably higher than in a thick coating. The crack shown in Fig. 4 initiated at a grain boundary at the surface and propagated in an intergranular manner. Therefore, the semi-coated specimens showed no improvement in life over the all-ferritic specimens. In an investigation of Alloy 800H [6], which is similar to Alloy 825 but with a lower nickel content, the fracture mode was intergranular crack initiation and propagation during TMF-IP testing between 200 and 800°C.

In the ferritic base material unoxidised internal cracks perpendicular to the stress axis were observed that could be associated with creep damage. It is reasonable to assume that those internal cracks participate in the final fracture process.

4.2. COMPARISON WITH OTHER RESULTS

It has been reported for other materials that the fatigue life during TMF-OP often differs from that during IP [7-10]. In some ferritic steels and nickel-base superalloys studied, the IP life was found to be longer than the OP life [7, 8] whereas in an austenitic steel and an ODS alloy, the reverse was found [9, 10]. Since no OP tests could be performed on the material in this study, due to shape instabilities, a comparison with an another ferritic steel of 1Cr0.5Mo type from Ref. 7 has been made. As can be seen in Fig. 6 the IP tests had a marginally longer fatigue life than the OP tests between 200 and 525°C, although the difference was not very large. Sometimes TMF conditions are simulated by isothermal low cycle fatigue (LCF) tests at the highest TMF temperature. For the material in Ref. 7 the difference in fatigue life was large, about half an order of magnitude but there was a 100-fold difference in strain rate. When the differences in strain rate were taken into account the differences in cyclic life diminished and use of the Ostergren damage approach, which includes a frequency factor, reduced the OP, IP and LCF data to a single curve with a scatter of 1.5 in lifetime. All the fatigue tests reported in Ref. 7 had failed by transgranular initiation and propagation and no internal intergranular cracking was seen, indicating little or no creep damage. In cases where IP testing produces intergranular cracking it often occurs that IP lives are considerably shorter than OP lives, [9, 10]. The effect of barrelling was reported before on soft or ductile materials and it is thought to be due to a non-uniform specimen geometry [11] or, which is more likely in this case, an uneven distribution in temperature [11, 12].

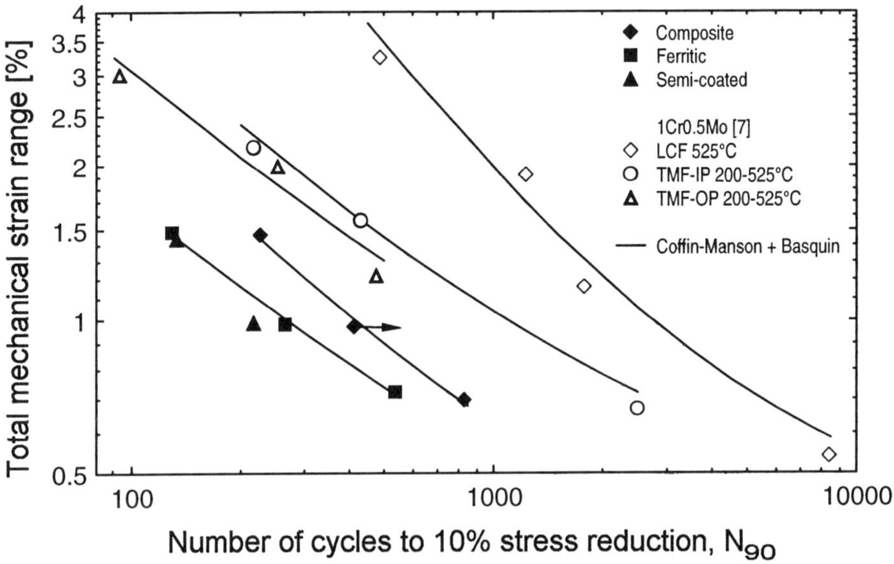

Figure 6. Comparison between TMF-IP results between 300-600°C for DIN St 45.8 with and without an outer layer of Alloy 825 from this study and results for a 1Cr0.5Mo steel [7]. The strain rate for the TMF tests was $1 \cdot 10^{-4}$ 1/s and for the LCF tests $1 \cdot 10^{-2}$ 1/s.

5. Conclusions

In this study a composite steel with an inner core of the ferritic steel DIN St 45.8 and an outer layer of the austenitic Alloy 825 has been studied during TMF-IP testing between 300 and 600°C. TMF-OP could not be performed because of severe barrelling effects. For comparison specimens of uncoated and semi-coated ferritic steel were tested under similar conditions. It was found that an austenitic coating increased the fatigue life by at least 50%. This was due to the reduction of surface oxidation effects. Further it was also found that the fatigue resistance of the coating decreased as the grain size decreased due to intergranular crack initiation and propagation. In the ferritic steel core internal cracks emerged that were related to creep damage. The adhesion at the austenite/ferrite interface was very good, with no evidence of cracking seen.

Acknowledgements - The authors thank Jacek Komenda for metallographic assistance. The work presented in this paper was financed by the Advanced Materials Group of Sweden's High Temperature Materials Committee (Varmhållfasthetskommittén). Participating companies in this group are ABB STAL AB, Avesta Sheffield AB, Celsius Materialteknik AB, Kanthal AB, AB Sandvik Steel, Sydkraft AB, Vattenfall Energisystem AB and Volvo Aero Corporation. The material was supplied by AB Sandvik Steel.

References

1. Odelstam, T., Performance of composite furnace tubes in recovery boilers, proc. conf. on *4th International Symposium on Corrosion in the Pulp and Paper Industry*, Swedish Corrosion Institute, Stockholm, 1983, pp. 64-67.
2. Lindblom, J. and Lindé, L., Data collection and evaluation during thermo-mechanical fatigue testing, Swedish Institute for Metals Research, report IM-3102, 1994.
3. Swedish Standard, *Stainless Steel - SS steel 14 14 35*, SIS-MNC, 1985.
4. Wu, R., Storesund, J., Sandström, R. and Von Walden, E., Creep properties of 1Cr0.5Mo steel welded joints with controlled microstructures, *Welding in the World* **30** (1992), 329-336.
5. Storesund, J., Seitisleam, F. and Borggreen, K., Iso-stress creep testing of nickel plated miniature specimens of an 0.5Cr0.5Mo0.25V steel, Swedish Institute for Metals Research, report IM-3170, 1994.
6. Ivarsson, B.G. and Henderson, P.J., Thermo-mechanical fatigue properties and their relation to creep for Alloy 800H, Swedish Institute for Metals Research, report IM-2556, 1990.
7. Lindé, L, Ivarsson, B., Henderson, P. and Lindblom, J., Thermo-mechanical, low cycle and thermal fatigue behaviour of a 1Cr0.5Mo steel, Swedish Institute for Metals Research, report IM-2840, 1992.
8. Gayda, J., Gabb, T.P., Miner R.V. and Halford, G.R., Bithermal low-cycle fatigue behaviour of a NiCoCrAlY-coated single crystal superalloy, *1987 TMS-AIME annual meeting*, The Metallurgical Society Inc., Denver, 1987, pp. 179-198.
9. Henderson, P.J., Sandström, R. and Ivarsson, B.G., Thermo-mechanical fatigue of the austenitic stainless steel 253 MA, in H. Nordberg and J. Björklund (eds.), proc. conf. on *Applications of stainless steels '92*, Jernkontoret, Stockholm, 1992, pp. 94-103.
10. Lindé, L. and Henderson, P., Thermo-mechanical and low cycle fatigue of the oxide dispersion strengthened alloy MA 754, Swedish Institute for Metals Research, report IM-3114, 1994.
11. Coffin, L.F., Instability effects in thermal fatigue, in Spera D.A., Mowbray D.F. (eds.) *ASTM STP 612*, American Society for Testing and Materials, 1976, pp. 227-238.
12. Castelli, M.G., Ellis, J.R. and Bartolotta, P.A., Improved thermomechanical testing techniques in the presence of cyclic instabilities, *5th TMF Workshop*, NASA Lewis Research Centre, 1989.

TMF LIFE AND DAMAGE MECHANISM MAPS FOR TITANIUM MATRIX COMPOSITES

R.W. NEU
George W. Woodruff School of Mechanical Engineering
Georgia Institute of Technology
Atlanta, Georgia 30332-0405, USA

1. Introduction

Thermomechanical fatigue (TMF) life prediction is quite complex, because life depends on the active damage progression mechanism. The active mechanism is dependent on many factors including the type of cycle, temperature, stress, strain, environment, and cyclic frequency. Consequently, the active damage mechanism is not always easily identifiable for a given stress (or strain)-temperature-time history.

One way to visualize the effect of stress and temperature on the life and damage mechanisms is through the use of maps that allow visualization of two independent variables at once. Maps have been found to be useful for illustrating the effects of stress and temperature on the deformation mechanisms in many monolithic materials [1] as well as the effects of stress and notch length on the bridging and crack growth mechanisms in titanium matrix composites [2].

This paper describes the procedure for constructing TMF life and damage mechanism maps for a titanium matrix composite. This 4-ply composite is composed of unidirectional silicon carbide fibers (SCS-6) in a beta-titanium matrix (Timetal 21S). The behavior of this material under stress-controlled TMF is described in Refs. [3, 4].

2. Construction of the Maps

The procedure for constructing the maps involves identifying all the possible damage progression mechanisms through experiments and analyses. The complexity of the damage progression modeling is reduced by focusing on the mechanisms that govern the steady-state damage rate. The variables that influence the damage rate are identified for each possible steady-state mechanism. Either an empirical or semi-empirical expression that represents the average damage rate as a function of the relevant variables is adopted for each mechanism. The damage mechanisms are generally dependent on the fiber and/or matrix response in composites. Therefore, micromechanics analyses are needed to determine the response of the constituents.

J. Bressers and L. Rémy (eds.), Fatique under Thermal and Mechanical Loading, 237–247.

Once damage expressions have been adopted and the constants have been determined from key experiments, the life and damage mechanism maps can be constructed. The variables for the axes are the maximum stress, S_{max}, and maximum temperature, T_{max}, of the cycle. Secondary variables such as the temperature range, stress ratio, and frequency remain constant, but these effects can be examined by constructing a series of maps. For each possible stress-temperature combination, a life analysis is conducted to obtain the cycles to failure and the dominant damage mechanism. The life analysis involves determining the constituent response and then computing the damage rate for all possible damage mechanisms. The life is given by the damage expression that gives the highest damage rate. This information is used to generate lines of constant life on the maps and to partition the map into regions showing the dominant damage mechanisms.

2.1. DAMAGE MECHANISMS

Damage mechanisms are identified through a number of experimental techniques and measurements. For SCS-6/Timetal 21S, these include strain accumulation [3, 4], modulus degradation [3, 4], acoustic emission [5], residual strength [4, 6], fractography and microscopy [3, 6], numerical analyses [7, 8], and simple life analyses [3, 9]. Three primary damage mechanisms were identified in a SCS-6/titanium aluminide composite system [10]. Similar mechanisms were identified in the SCS-6/Timetal 21S system [11]. The mechanisms are (1) time-independent fatigue of the matrix, (2) surface-initiated fatigue-environment damage, and (3) fiber-dominated damage.

2.1.1. Fatigue of the Matrix

This mechanism is the time-independent fatigue of the matrix. The matrix response is controlled by both the applied mechanical load and the mechanical strain induced by the mismatch in the coefficient of thermal expansion (CTE) between the fiber and matrix. The fatigue mechanism is analogous to low cycle fatigue of the monolithic matrix material.

2.1.2. Surface-Initiated Fatigue-Environment Damage

This mechanism describes the synergistic influences of the environmental attack and matrix fatigue. The matrix undergoes strain cycling similar to the fatigue of the matrix mechanism. The surface that is exposed to the environment becomes embrittled. The maximum strain that the embrittled surface layer is capable of supporting becomes smaller with time of exposure. Once this embrittled surface layer cracks, crack growth continues with cycling since the crack tip can easily become embrittled through the exposure of the crack tip to the environment. Compared to the previous matrix fatigue mechanism, multiple initiation sites are generally found and consequently many more matrix cracks perpendicular to the loading axis are observed when the material undergoes this mechanism.

2.1.3. Fiber-Dominated Damage

This mechanism describes a progression of fiber fractures. During the very early cycling (<0.1 N_f), matrix stress ratchets down, with fiber stress increasing due to the viscoplastic behavior of the matrix [7]. Fiber failures occur periodically on subsequent cycles, promoted by the increasing fiber stress, as well as by the reduction in fiber strength. The mechanism causing fiber strength degradation is not completely clear, but may be either frictional rubbing at the fiber/matrix reaction layer or carbon-rich coating [12] or an environmental attack to the components of the reaction and carbon layers [13] or possibly a combination of these mechanisms [14]. As a function of percent life, the fiber fracture events occur at a near constant rate [5]. Failure occurs after a sufficient number of fiber breaks have occurred and the matrix can not withstand the additional load.

2.2. MICROMECHANICS

The constituent behavior is determined through a rule-of-mixtures model that incorporates thermoelastic and viscoplastic material models [8]. For this system, the fiber is thermoelastic and the matrix is viscoplastic and modeled using the Bodner-Partom unified constitutive theory [15, 16]. First, the processing cool-down is simulated to obtain the initial residual stress state. Then the analysis is run for 10 fatigue cycles to capture the effect of ratchetting on the redistribution of the fiber and matrix stress. The response at cycle 10 is used as the approximate stabilized behavior of the composite.

2.3. PREDICTION MODEL

Of the possible damage mechanisms, the one that results in the highest rate of damage accumulation controls life and is termed the dominant damage mechanism. Since the mechanism is generally not known *a priori*, the average damage per cycle, D, for each possible mechanism is computed, and the one producing the most damage controls life:

$$D = \max(D^{Mfat}, D^{Menv}, D^{F}) \tag{1}$$

where D^{Mfat} is the damage per cycle for the fatigue of the matrix mechanism, D^{Menv} is the damage per cycle for the surface-initiated fatigue-environment mechanism, and D^{F} is the damage per cycle for the fiber-dominated mechanism. In the case of damage growing linearly with the number of cycles, N, the number of cycles to failure, N_f, is given by

$$N_f = 1/D \tag{2}$$

Basing TMF prediction modeling on different possible damage mechanisms was found to be critical for general TMF life prediction [9]. Synergism between any two of these mechanisms is not included in Eq. (1). However, synergism between fatigue, creep, environment, phasing, and time is incorporated in the damage expressions for each mechanism.

Semi-empirical or purely empirical relationships are adopted to describe the damage rate of each identified mechanism. Generally, damage is a nonlinear function of N. However, determining the nonlinear relationship for each mechanism is challenging. Therefore, the average steady-state damage rate is chosen to be independent of the amount of damage. This simplification allows the focus of the modeling to be placed on the factors that control the damage progression during the majority of the life. Events occurring during the initial and final cycles are indirectly incorporated in the empirical constants.

2.3.1. Fatigue of the Matrix

The fatigue of the matrix mechanism is dependent on the mechanical strain range of the matrix, $\Delta\varepsilon_m^{(m)}$, and mean stress in the matrix, $\sigma_m^{(m)}$. It is assumed that temperature does not significantly affect this mechanism. Since D^{Mfat} is controlled by crack initiation during most of the life, one possible relationship is the generalized Coffin-Manson relationship with mean stress correction [17],

$$\frac{\Delta\varepsilon_m^{(m)}}{2} = \frac{\sigma_f' - \sigma_m^{(m)}}{E^{(m)}}\left(\frac{2}{D^{Mfat}}\right)^b + \varepsilon_f'\left(\frac{2}{D^{Mfat}}\right)^c \tag{3}$$

The empirical constants, $\sigma_f', b, \varepsilon_f'$, and c, are determined from low cycle fatigue tests on the matrix material conducted at room temperature, and $E^{(m)}$ is the modulus of the matrix at room temperature. The constants for SCS-6/Timetal 21S are given in Ref. [11].

2.3.2. Surface-Initiated Fatigue-Environment Damage

This term is dependent on the matrix mechanical strain range, $\Delta\varepsilon_m^{(m)}$, describing the fatigue process, as well as on time, temperature, and the kinetics of oxidation describing the severity of the environmental attack. The time-dependent processes are described through the cycle time, t_c, and an effective oxidation constant, D_{eff}. Synergistic effects due to the stress-temperature phasing are incorporated through Φ^{Menv},which is a function dependent on the ratio of the thermal and mechanical strain rates. A mechanistic-based expression was given in Ref. [10]. For SCS-6/Timetal 21S, the expression becomes

$$D^{Menv} = A^{Menv}\left(\Delta\varepsilon_m{}^{(m)}\right)^{4.44}\left(\Phi^{Menv}D_{eff}\right)^{1.85}\left(t_c\right)^{0.259} \tag{4}$$

where A^{Menv} is a scaling parameter. The exponents and coefficients are determined from key out-of-phase (OP) TMF and isothermal fatigue tests [9].

2.3.3. Fiber-Dominated Damage

The fiber-dominated damage mechanism is primarily dependent on the fiber stress, $\sigma^{(f)}$. The fiber stress is influenced by the time-dependent redistribution of stress between matrix and fiber under the cyclic mechanical and thermal loading. In addition, degradation of the fiber strength occurs with time at temperature. This may involve diffusion of oxygen through the matrix or matrix cracks to the fibers or by the piping of oxygen through exposed fiber ends. In the case of the SCS-6 fiber, oxidation of the carbon-rich coatings occurs. An Arrhenius factor, $\exp[-Q^F / RT]$, treats this effect. The phasing of stress and temperature can also influence the rate of damage accumulation through Φ^F. The expression that was adopted to describe this mechanism is [10]

$$D^F = A^F \int_0^{t_c} \phi^F \exp\left[-\frac{Q^F}{RT}\right]\left(\frac{\sigma^{(f)}}{\sigma_T}\right)^{16.6} dt \qquad (5)$$

where A^F is a constant and σ_T is the average fiber strength. The constants and exponent are primarily determined from key in-phase (IP) TMF tests [9].

3. Results

3.1. OUT-OF-PHASE TMF

OP TMF maps are shown in Fig. 1. Each map shows the effects of S_{max} and T_{max} of the cycle on the life and dominant damage mechanism. The effect of temperature range (ΔT) is examined through a series of maps. Figs. 1(a)-1(c) are maps for ΔT = 500°C, 250°C, and 50°C, respectively. The stress ratio (R) is 0.1, the frequency (f) is 0.00556 Hz (3 min. cycle period), and the atmosphere is air. The labels on the life contours (n) represent $N_f = 10^n$ cycles. The shaded region near the top of the maps is unattainable, since fracture would have occurred during the first loading. The line delineating this region is the approximate ultimate tensile strength (UTS) of the material. It is dependent on the temperature at S_{max} of the cycle.

The maps are most accurate in the region where T_{max} is between 550°C and 815°C and S_{max} is greater than 400 MPa. In this region the experimental data is generally within a factor of 3 of the predicted life for OP TMF and within about a factor of 10 for IP TMF [9]. Outside this region the damage expressions are essentially extrapolated. The room temperature response was based on the fatigue behavior of monolithic Timetal 21S. The actual room temperature composite life may be lower because local stress risers in the composite are not established at this time.

Failure is generally defined as the cycle when the specimen separates into two pieces. When the maximum applied stress is very low, separation may not occur. However, the average damage rate may be high enough to attain a critical level of damage. Recall that the damage rate is based on the mechanism that occurs during majority of the life. The rate controlling process for the *Menv* mechanism is the growth of matrix cracks. At high applied stress, fiber degradation and individual fiber failures occur as a last step before complete specimen separation. When the maximum applied stress is low, the final transient resulting in the individual fiber failures may not occur, yet the distribution of matrix cracks is indicative of failure.

The maps help classify the complex TMF behavior, especially observed in advanced composites. Much information can be obtained from these maps. For example, the change in slope in the life contours near 400°C in Fig. 1(a) is indicative of a change in the damage mechanism. The mechanism changes from fatigue of the matrix (labeled *Mfat* in the figure) to surface-initiated fatigue-environment damage (labeled *Menv*) as T_{max} increases. It is clearly apparent from the map that this change in mechanism is controlled more by T_{max} than S_{max}.

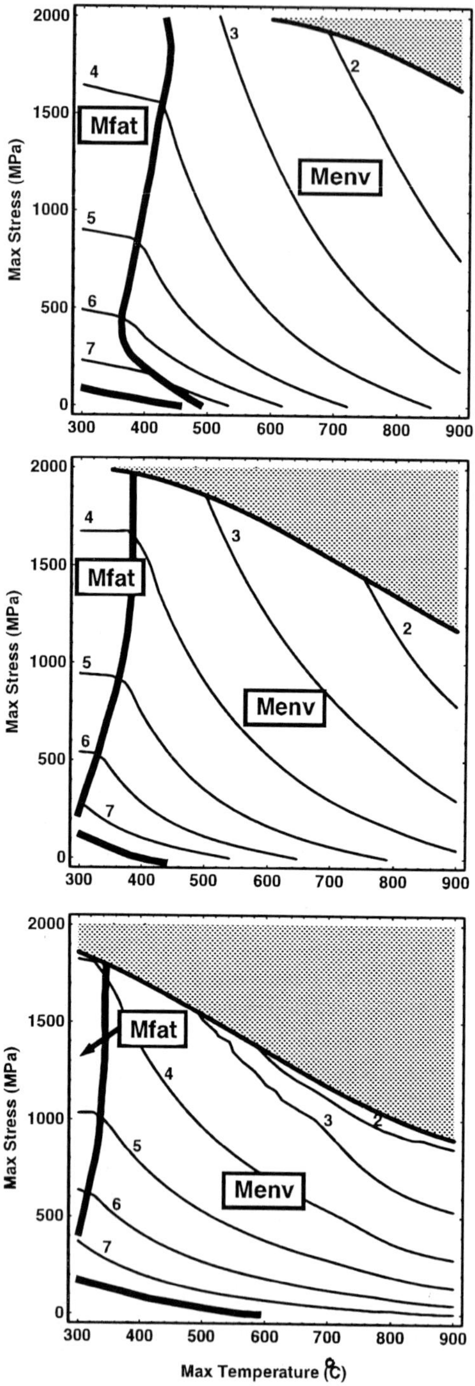

(a) ΔT = 500°C.

(b) ΔT = 250°C.

(c) ΔT = 50°C.

Figure 1. Life and damage mechanism maps for out-of-phase TMF with R = 0.1 and f = 0.0056 Hz in air atmosphere.

As ΔT increases, the shift from the *Mfat* to *Menv* mechanism occurs at a higher T_{max}. This is explained by the reduction of the cycle time that experiences significant environmental degradation.

Also as ΔT increases, N_f for a given S_{max} and T_{max} can either decrease, stay about the same, or even increase. At very low S_{max}, N_f decreases with ΔT increasing. The mechanical stress/strain in the matrix is controlled by mismatch in the CTE between the fiber and matrix during the thermal cycling. The strain from thermal cycling, ΔT, is more dominant than the mechanical strain imposed from ΔS. At high S_{max} and intermediate temperatures, the reverse trend is observed: N_f increases with increasing ΔT. Under OP TMF, S_{max} occurs at T_{min}, which is decreasing with increasing ΔT. The likelihood of failure is reduced, because the strength of the matrix increases with decreasing temperature. In addition, the fraction of the cycle undergoing significant environmental degradation is reduced.

3.2. IN-PHASE TMF

IP TMF maps for ΔT = 500°C, 250°C, and 50°C are shown in Figs. 2(a)-2(c), respectively. Compared to the OP TMF maps, the IP TMF maps contain a larger region where life is greater than 10^7 cycles. However, the UTS curve is lower on the IP TMF maps, because S_{max} under IP TMF occurs at the higher temperature of the cycle when the strength of the matrix is lower. Although the differences in the IP and OP TMF maps are significant, the maps must coincide for pure thermal cycling ($S_{max} = 0$).

All three dominant damage mechanisms appear on the IP TMF maps. The fiber-dominated mechanism (labeled *Fiber*) occurs when S_{max} is relatively high. Most IP TMF tests that have been conducted in the laboratory fall in this region. The life lines are close together indicating that the life is especially sensitive to S_{max}, and therefore have fairly flat stress-life curves. Life is less sensitive to S_{max} when the *Mfat* mechanism operates at lower T_{max}. On the higher ΔT maps (Figs. 2(a) and 2(b)), the boundary between the *Fiber* and *Mfat* mechanisms is a diagonal and therefore dependent on both S_{max} and T_{max}.

Also on the higher ΔT maps, the *Menv* mechanism operates at high T_{max} and low S_{max}. Within this region, N_f decreases with decreasing S_{max}. This seemly contradictory region is easily explained. Recall that the matrix strain is controlled by thermal cycling, ΔT, at low S_{max}. Since the matrix has a higher CTE than the fibers, the residual stress in the matrix after the processing cool-down is positive. Consequently, during the heating portion of the cycle, the matrix strain decreases. Therefore, the matrix experiences an OP cycle, and the *Menv* mechanism has a propensity to operate. As ΔT becomes smaller, the matrix OP cycle is lessened, and the size of the *Menv* region in the lower right corner of the map is reduced.

At very high T_{max}, the life is longest when $S_{max} \approx 350$ MPa for ΔT = 500° C and $S_{max} \approx 250$ MPa for ΔT = 250°C. The matrix strain induced by the applied mechanical cycling is in-phase with temperature and the matrix strain induced by thermal cycling is out-of-phase with temperature. At this location on the map, these

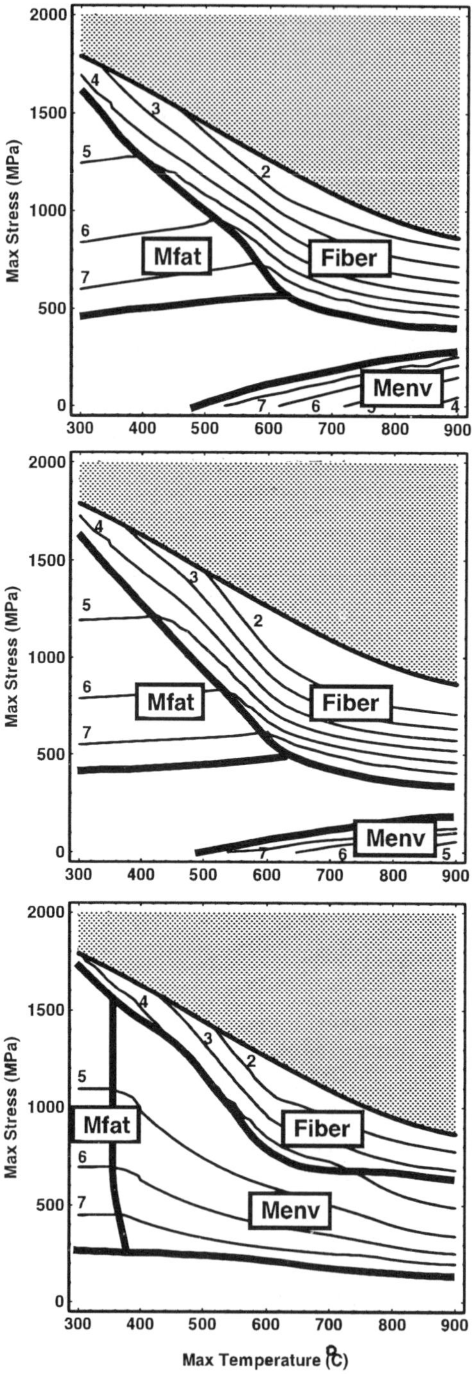

(a) ΔT = 500°C.

(b) ΔT = 250°C.

(c) ΔT = 50°C.

Figure 2. Life and damage mechanism maps for in-phase TMF with R = 0.1 and f = 0.0056 Hz in air atmosphere.

two matrix strains cancel resulting in no net cyclic strain and therefore no driving force for matrix fatigue. Since the fiber stress is also small, the life is long.

At $\Delta T = 50°C$, the *Menv* mechanism is more prevalent at higher S_{max}. The matrix experiences an IP cycle in this region, but the strength of this thermal cycle is small because of the small ΔT. As will be apparent in the next section, as ΔT becomes small, the behavior begins to look like isothermal fatigue.

3.3. ISOTHERMAL FATIGUE

A map for isothermal fatigue conducted at the same frequency as the TMF tests is shown in Fig. 3(a). Isothermal fatigue is essentially a special case of TMF. It is the case when $\Delta T = 0$. This map is similar to the IP TMF map with the smallest ΔT (Fig. 2(c)). Note that the temperature scale has changed. Isothermal fatigue appears to be less influenced by the *Fiber* mechanism but slightly more influenced by the *Menv* mechanism compared to IP TMF at $\Delta T = 50°C$.

The map for isothermal fatigue under a frequency 100 times lower is shown in Fig. 3(b). The two time-dependent mechanisms reduce predicted lives at higher T_{max}. The degradation due to the *Fiber* mechanism shows a greater effect on frequency, which is consistent with experiments [9]. The UTS curve is expected to decrease as frequency decreases; however, the curve was not adjusted for frequency, because the loading rate effect on UTS was not available.

4. Summary and Conclusions

To construct life and damage mechanism maps, a clear understanding of the damage micromechanisms is required, so that an expression for the damage rate for each possible dominant damage mechanism may be adopted.

The relationships among IP TMF, OP TMF, and isothermal fatigue are clarified through the maps by showing the whole picture. They illustrate that one needs to be careful when associating a damage mechanism with a particular cycle type. For example, IP TMF, which is typically considered fiber-dominated, can be controlled by one of three damage mechanisms depending on the stress-temperature-time history. The life is dependent not only on the cycle type, but also on the other controlling parameters such as stress, temperature, and environment.

The cycle type that is more life limiting depends on the stress-temperature-environment-time history. Considering a particular S_{max} and T_{max} of a TMF cycle, either OP or IP TMF may be more life limiting, or they may be nearly equivalent. At higher T_{max} and higher S_{max}, IP TMF is the more severe cycle. At higher T_{max} and lower S_{max}, OP TMF is a more severe cycle. At lower T_{max} for all S_{max}, OP TMF is a more severe cycle. Generally a larger ΔT exaggerates the damage mechanism associated with that cycle type.

Maps help reduce the complexity of TMF behavior. They can be used as a first assessment for comparing the severity of different stress-temperature combinations.

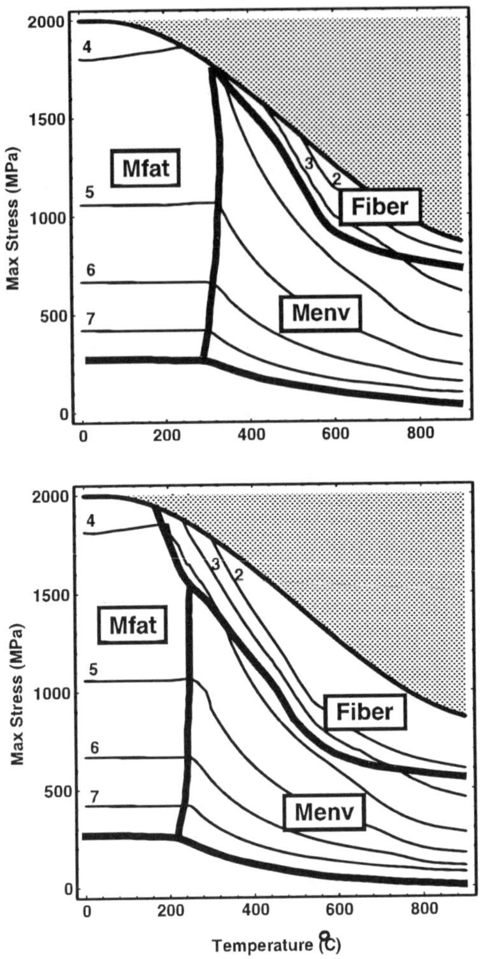

(a) f = 5.6 x 10^{-3} Hz.

(b) f = 5.6 x 10^{-5} Hz.

Figure 3. Life and damage mechanism maps for isothermal fatigue with R = 0.1 in air atmosphere.

In addition, maps of different material systems can be compared as a potential tool for material selection in applications where thermal and mechanical cycling is critical.

Acknowledgements - This work was initiated when the author was an NRC Associate at Wright Laboratory Materials Directorate, Wright-Patterson Air Force Base, Ohio. The support extended by the National Research Council, Washington, D.C., through their Associateship Program is gratefully acknowledged. Dr. T. Nicholas as well as numerous others in the Materials Behavior Branch are thanked for their fruitful discussions, data, and support.

References

1. Frost, H.J. and Ashby, M.F., *Deformation-Mechanism Maps*, Pergamon Press, Oxford, 1982.
2. Zok, F.W., Connell, S.J., and Du, Z.Z., Fatigue maps for titanium matrix composites, in W.S. Johnson, J.M. Larsen, and B.N. Cox (eds.), *Life Prediction Methodology for Titanium Matrix Composites, ASTM STP 1253*, American Society for Testing and Materials, Philadelphia, 1995.
3. Neu, R.W. and Nicholas, T., Effect of laminate orientation on the thermomechanical fatigue behavior of a titanium matrix composite, *Journal of Composites Technology & Research* **16** (1994), 214-224.
4. Castelli, M.G., Characterization of damage progression in SCS-6/Timetal 21S $[0]_4$ under thermomechanical fatigue loading, in W.S. Johnson, J.M. Larsen, and B.N. Cox (eds.), *Life Prediction Methodology for Titanium Matrix Composites, ASTM STP 1253*, American Society for Testing and Materials, Philadelphia, 1995.
5. Neu, R.W. and Roman, I., Acoustic emission monitoring of damage in metal matrix composites subjected to thermomechanical fatigue, *Composites Science and Technology* **52** (1994), 1-8.
6. Revelos, W.C., Jones, J.W., and Dolley, E.J., Thermal fatigue of a SiC/Ti-15Mo-2.7Nb-3Al-0.2Si composite, *Metallurgical and Materials Transactions* (in press).
7. Kroupa, J.L. and Neu, R.W., The nonisothermal viscoplastic behavior of a titanium-matrix composite, *Composites Engineering* **4** (1994), 965-977.
8. Neu, R.W., Coker, D., and Nicholas, T., Cyclic behavior of unidirectional and cross-ply titanium matrix composites, *International Journal of Plasticity* (submitted).
9. Neu, R.W. and Nicholas, T., Methodologies for predicting the thermomechanical fatigue life of unidirectional metal matrix composites, in M.R. Mitchell and R.W. Landgraf (eds.), *Advances in Fatigue Lifetime Predictive Techniques (3rd Symposium), ASTM STP 1292*, American Society for Testing and Materials, Philadelphia, 1996.
10. Neu, R.W., A mechanistic-based thermomechanical fatigue life prediction model for metal matrix composites, *Fatigue and Fracture of Engineering Materials and Structures* **16** (1993), 811-828.
11. Neu, R.W., Thermomechanical fatigue damage mechanism maps for metal matrix composites, in M.J. Verrilli and M.G. Castelli (eds.), *Thermo-Mechanical Fatigue Behavior of Materials: 2nd Volume, ASTM STP 1263*, American Society for Testing and Materials, Philadelphia, 1995.
12. Walls, D.P. and Zok, F.W., Interfacial fatigue in a fiber reinforced metal matrix composite, *Acta Metall. Mater.* **42** (1994), 2675-2681.
13. Gambone, M.L. and Wawner, F.E., The effect of elevated temperature exposure of composites on the strength distribution of the reinforcing fibers, in J.A. Graves, R.R. Bowman, and J.J. Lewandowski (eds.), *Intermetallic Matrix Composites III*, Volume 350, MRS, Pittsburgh, PA, 1994, pp. 111-118.
14. Blatt, D., Karpur, P., Stubbs, D.A., and Matikas, T.E., Observations of interfacial damage in the fiber bridged zone of a titanium matrix composite, *Scripta Metallurgica et Materialia* **29** (1993), 851-856.
15. Chan, K.S., Bodner, S.R., and Lindholm, U.S., Phenomenological modeling of hardening and thermal recovery in metals, *Journal of Engineering Materials and Technology* **110** (1988), 1-8.
16. Neu, R.W., Nonisothermal material parameters for the Bodner-Partom model, in L.A. Bertram, S.B. Brown, and A.D. Freed (eds.), *Material Parameter Estimation for Modern Constitutive Equations*, MD-**43**/AMD-**168**, ASME, 1993, pp. 211-226.
17. Morrow, J., Cyclic plastic strain energy and fatigue of metals, in *Internal Friction, Damping, and Cyclic Plasticity, ASTM STP 378*, 1965, pp. 45-87.

EFFECTS OF GRAIN BOUNDARY PHASE VISCOSITY ON HIGH TEMPERATURE CYCLIC FATIGUE BEHAVIOUR OF A Si_3N_4

G. ROEBBEN[1,2], M. STEEN[1], J. BRESSERS [1] and
O. VAN DER BIEST[2]
[1] *Institute for Advanced Materials,*
Joint Research Centre of the European Commission
P.O. Box 2, 1755 ZG Petten, The Netherlands
[2] *Department of Metallurgy and Materials Engineering,*
Leuven University
De Croylaan 2, 3001 Heverlee, Belgium

1. Introduction

1.1. FATIGUE IN CERAMICS

The term fatigue is used incorrectly by ceramicists to describe the degradation of the ceramic's mechanical properties with time, irrespective of whether the applied load is cyclic or not. The most common fatigue mechanism for monolithic ceramics in this sense is stress corrosion cracking. This process consists of the stress promoted absorption and reaction of gaseous molecules (especially H_2O, [1]) at the tip of surface cracks, and it results in slow crack growth, often roughly according to the empirical Eq. 1:

$$\frac{da}{dt} = A\,K_I^n \tag{1}$$

In Eq. 1 a stands for crack length, K_I for the stress intensity factor, and A and n are material and environment dependent constants obtained from experiments. Slow-crack-growth degrades the strength of a ceramic, according to the fracture mechanics relation of Eq. 2, which relates the fracture strength σ_f to the critical stress intensity factor K_{Ic} , the length a_c of the most critical crack and an appropriate shape parameter Y.

$$\sigma_f = \frac{K_{Ic}}{Y\sqrt{a_c}} \tag{2}$$

Since the application of technical ceramics as structural materials is spreading, the need to improve their resistance to cyclic loads is increasing. This is reflected in the rapid development of a cyclic fatigue (as opposed to static fatigue) branch in the research field of mechanical properties of ceramics. To comply with the more broadly

J. Bressers and L. Rémy (eds.), Fatique under Thermal and Mechanical Loading, 249–258.

accepted ISO-definition of fatigue ("the changes in properties which can occur in a material due to the repeated application of stresses or strains"), in the remainder of this text the term fatigue will be used instead of cyclic fatigue.

The search for fatigue mechanisms has been triggered by the discovery of differences between lifetimes and crack growth rates under static and cyclic loads at room temperature. As a basis for comparing the measured lifetimes and crack growth rates, Evans [2] proposes using the empirical crack growth rate equation Eq. 1. Once the parameters A and n have been determined for a specific ceramic under a static load, one can calculate lifetime by integrating the crack growth rate between an estimated initial crack length a_0 and a critical crack length a_c. This integration can be performed for static loads (analytically) as well as for cyclic loads (usually numerically), and thus permits the prediction of the lifetime under cyclic fatigue conditions if the failure mechanism under static and cyclic loads is the same. If the experimentally determined fatigue lifetimes do not correspond with the predictions by the integration method, one can conclude the existence of a specific positive (= lifetime lengthening) or negative (= lifetime shortening) fatigue mechanism.

1.2. LOW TEMPERATURE FATIGUE IN MONOLITHIC CERAMICS

For monolithic ceramics (such as SiC, Si_3N_4, ZrO_2, Al_2O_3, ...) the fatigue phenomena occurring at low temperature (below approximately 1000°C) have now been sufficiently documented to draw some first conclusions. Most of the room temperature fatigue tests on monolithic ceramics reveal a decrease in lifetime or an increase in crack growth rate when compared to tests under static loads. This negative fatigue effect is more easily distinguished in ceramics which show R-curve behaviour, i.e. those ceramics for which the fracture resistance increases with crack length. The origin of R-curve behaviour lies in the shielding of the crack tip from the applied stress intensity by either crack wake bridging features, or by a damage zone around the crack tip. Both of these make the ceramic more prone to fatigue. Crack wake bridges can fail or degrade through friction when the crack wake closes during the unloading part of a fatigue cycle. Elaborate crack-wake bridge degradation models have been proposed [3,4]. Crack tip damage zones also can promote crack growth during the unloading half of the cycle, due to their mechanical incompatibility with the surrounding undamaged material. This phenomenon occurs under pure compression fatigue as well, as described and modelled by Ewart and Suresh [5].

1.3. EFFECTS OF HIGH TEMPERATURE ON SILICON NITRIDE CERAMICS

1.3.1. The Presence of Amorphous Grain Boundary Phases

The application of ceramics as structural materials is only cost-effective at high temperatures and/or under severe environmental loading conditions, where, due to the high bond strength of the ionic-covalent crystal structure, their mechanical properties remain largely unaffected, whereas the competing metallic materials melt, deform or corrode. However, monolithic ceramics have a maximum application temperature, as they are susceptible to creep damage; some of them even at temperatures not far above

1000°C. In the case of Si_3N_4, successful sintering requires a small amount of oxide additives (e.g. Al_2O_3, Y_2O_3, MgO, ...). Together with the SiO_2 layer often present on the silicon nitride powder before sintering, these additives form a secondary glassy silicate phase, which promotes the diffusion necessary to sinter to a fully dense ceramic. Unfortunately, the remaining glassy layer between the grains of the sintered ceramic affects the deformation and oxidation resistance of the bulk at higher temperatures. To limit the effect of the grain boundary phases on the high temperature properties attempts to decrease the amount of sintering additives needed by developing new sintering (hot isostatically sintering [6]) and powder processing (silicon nitride powder coating with sintering additives [7]) techniques have been explored. Crystallizing the glassy intergranular phases by an appropriate heat treatment has also been examined [8]. It has been shown that a very fine amorphous layer nevertheless always remains [9].

1.3.2. Viscous Phases and the Resistance of Si_3N_4 to Static and Cyclic Loads

Under static loads, the flow of silicate-glasses along the grain boundaries of two-phase monolithic ceramics leads to an increased creep rate, until the viscous fluid is squeezed out of the grain boundaries along which the silicon nitride grains glide. This creep mechanism is more important under tensile than under compressive loads, and leads to the well known asymmetric creep behaviour [10]. Phenomenological viscoplastic models, incorporating this asymmetry, have been developed to calculate the creep deformation of simple ceramic structures (e.g. bending specimens, [11]). These models are based on time-*in*dependent properties of the viscous grain boundary phase and do not take into account that the creep resistance as well as the resistance to subcritical crack growth considerably increase when the glassy intergranular phase devitrifies [12,13].

The specific effect of a glassy intergranular phase on the fatigue resistance of monolithic ceramics is not as well documented. The small amount of high temperature fatigue results show that the effect of fatigue load on lifetime or crack growth rate can be positive, when compared to the effect of static load. Fig. 1 [14] summarizes the results of an earlier JRC-IAM-research project on high temperature fatigue of a silicon nitride. The figure combines lifetime versus stress-curves (for tests under static loads), and cycles to failure versus stress amplitude-curves (for tests under cyclic loads with R = -1) obtained on a silicon nitride containing a partially amorphous grain boundary phase, tested at 1200°C in air. The time to failure under static load is much shorter than that under cyclic load, thus manifesting a positive fatigue effect. A constant stress of 320 MPa leads to failure after 100 seconds, whereas a cyclic load of this amplitude is sustained over 430 hours without leading to fracture. In fact, 320 MPa seems to be the fatigue limit for this silicon nitride tested under alternating load.

A number of specific high temperature fatigue mechanisms have been proposed to explain the observed positive fatigue effects. All of them relate to the presence of an amorphous grain boundary phase, which is supposed to cause a frequency dependent energy dissipation and to have an increased load bearing capacity at higher loading rates [15,16].

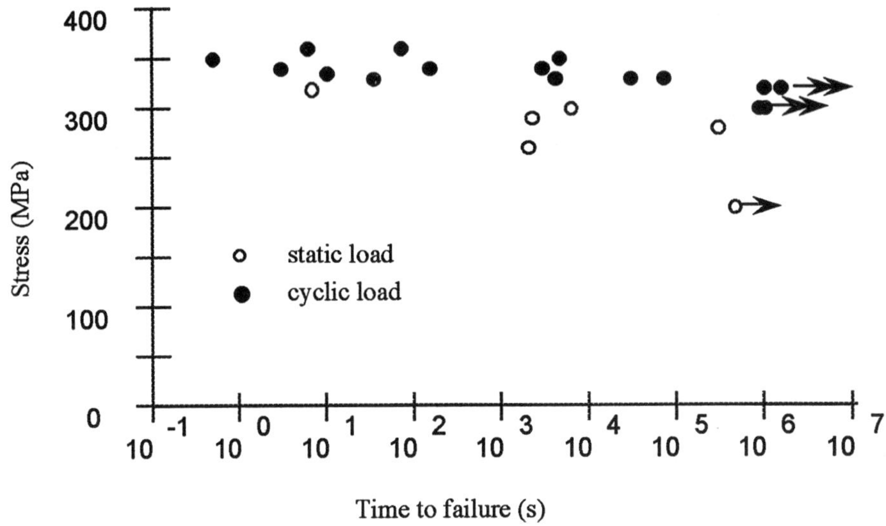

Figure 1. Comparison of lifetime under cyclic (R= -1) and static load for a hot-pressed silicon nitride at 1200°C.

2. Experimental Procedure

In order to clarify the high temperature fatigue issue, the authors have undertaken uniaxial fatigue tests on smooth silicon nitride specimens. Because the amount and viscosity of intergranular phases in silicon nitride is of primary concern with respect to the high temperature mechanical properties, a reliable test method was developed first in order to enable to assess the viscous contribution of the glassy phase to the stress-strain behaviour, before, after or even during a long term mechanical test at high temperature. Then the possible frequency dependence of creep strain accumulation and of the fatigue resistance was studied. Very precise cyclic strain measurements are required to investigate the underlying deformation and damage mechanisms.

2.1. MATERIALS

The material studied is a silicon nitride, pressureless sintered with the addition of 5 wt% Al- and 6 wt% Y-containing powders. During sintering, β-silicon nitride is formed. Since some of the aluminium atoms substitute for silicon atoms in the Si_3N_4 grains, and an equal amount of oxygen atoms substitute for nitrogen atoms, the resulting material is called "SiAlON". The substitution level z (used in the formula $Si_{6-z}Al_zO_zN_{8-z}$) for the studied material is close to 0.4.

Two batches (G for glassy and C for crystalline) have been studied. The as sintered G-batch contains both a glassy and a crystalline intergranular phase. The crystalline phase is determined by TEM and XRD as Yttrium Aluminium Mellilite

($Y_4Al_2O_9$ or YAM) which forms upon cooling from the sintering temperature. The amount of this metastable phase varies from specimen to specimen and even within a specimen. The C-batch has undergone a post-sintering heat treatment for 5 h at 1420°C, and corresponds to the commercially available Syalon 201 ceramic (Syalons Ltd.). The heat treatment transforms the YAM phase and the glassy intergranular residues into Yttrium Aluminium Garnet ($Y_3Al_5O_{12}$ or YAG), with the exception of a few residual glass pockets at triple points between YAG and sialon grains, and of a very thin amorphous intergranular film.

2.2. TEST PROCEDURE

Uniaxial creep and fatigue stress controlled (R = 0, R = -1) tests were run on an electromechanical test machine, according to the procedures described elsewhere [17]. Water-cooled hydraulic collet-type grips were used to clamp the flat specimen (outer dimensions 160 mm x 20 mm x 6 mm). The reduced uniform gage section measures 30 x 8 x 6 mm^3. During the clamping procedure, parasitic bending stresses, caused by test machine misalignment and by specimen and grip misalignment, are monitored on-line via two sets of four strain gauges. The bending strains were reduced to less than 5% of the uniaxial strain at 100 MPa. A 90 mm high induction furnace heats the reduced section of the specimen through a SiSiC susceptor. A contacting rod extensometer with Si_3N_4 rods was used to measure the deformation.

3. Results and Discussion

3.1. FREQUENCY DEPENDENT PHENOMENA

3.1.1. Stress-Strain Hysteresis

Some of the specimens underwent a series of sinusoidal load cycles (σ_{max}= 150 MPa, σ_{min}= -150 MPa) at different temperatures (from 750°C to 1350°C) and frequencies (from 1 to 0.001 Hz). Fig. 2 demonstrates the temperature and frequency dependence of the stress-strain hysteresis for a G specimen, while Fig. 3 shows the same for a C specimen. Fig. 4 compares the hysteretic loops for two different G specimens.

The results have to be interpreted in view of the microstructural characteristics of the two batches. The as sintered sialon (Fig. 2) does not show inelastic deformation (reflected by the opening of the stress-strain loops) below 1000°C. For the heat treated specimens on the other hand, the onset of inelastic deformation begins at 1200°C (Fig. 3). For both batches, increasing temperature and decreasing frequency result in an opening of the stress-strain curves. The lower volume fraction and/or higher viscosity of the intergranular phases in the heat treated sialon increase the temperature at which the hysteretic phenomenon occurs, as compared to the as sintered sialon. The hysteretic area at a given temperature and frequency of different

G-specimens varies considerably (Fig. 4). This confirms the findings of the microstructural observations which showed that the volume fraction of crystalline versus amorphous grain boundary of the as-sintered material varies from specimen to specimen.

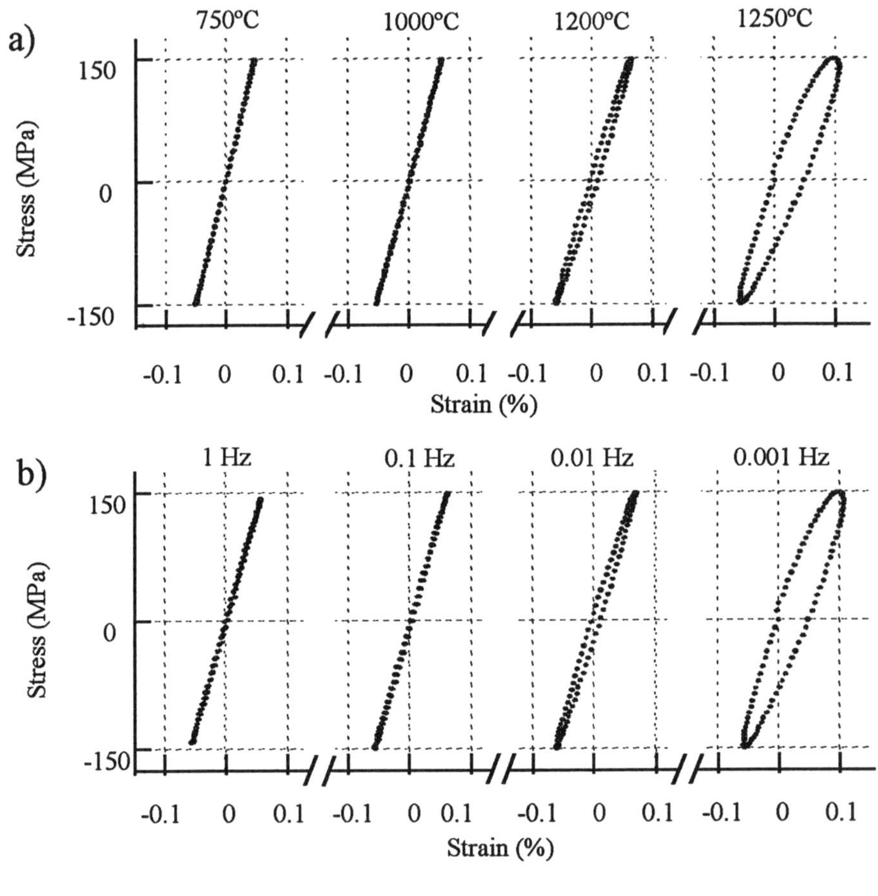

Figure 2. a) **Temperature dependence (frequency = 0.001 Hz) and b) frequency dependence (temperature = 1250°C) of stress-strain hysteresis in G-specimen.**

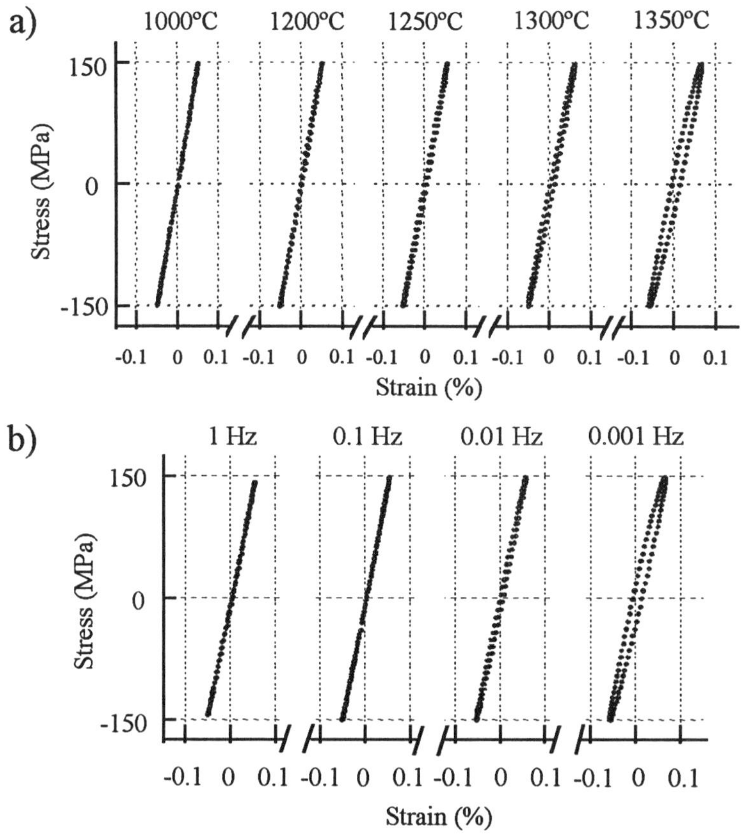

Figure 3. a) Temperature dependence (frequency = 0.001 Hz) and b) frequency dependence (temperature = 1350°C) of stress-strain hysteresis in C-specimen.

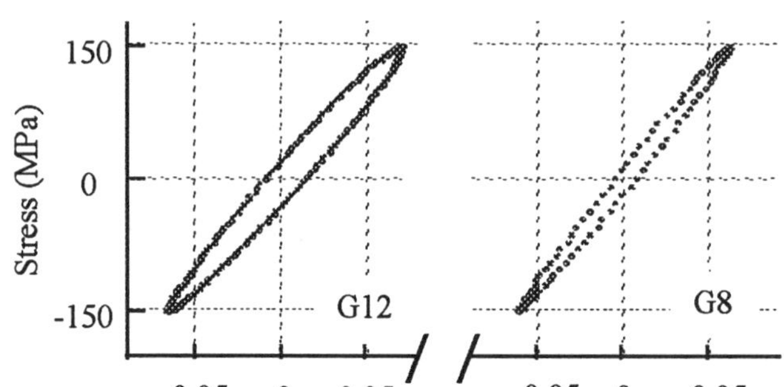

Figure 4. Comparison between stress-strain hysteresis of two different G-specimens (temperature = 1200°C, frequency = 0.001 Hz).

3.1.2. Strain Accumulation under Static and under Cyclic Loads

In Fig. 5 the evolution of the accumulated strain in a sequence of fatigue and creep loading on a single G specimen is shown. In the case of fatigue loads the strain corresponds to the maximum strain in each cycle. Before the application of the 100 MPa static load (after 75 hours at 1250°C), the specimen had undergone several creep and fatigue loading cycles at lower temperatures and lower stresses, during which a residual strain of 0.01% was generated.

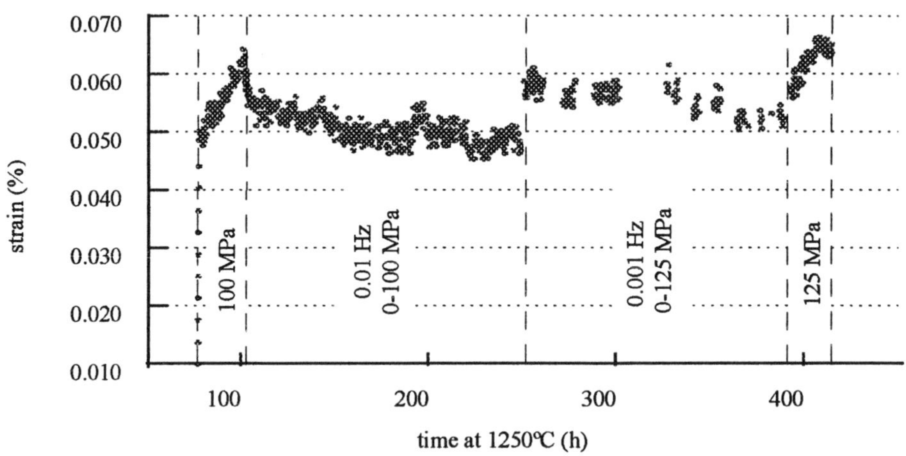

Figure 5. The strain accumulation during a sequence of fatigue and creep tests on a G specimen.

It can be observed that the inelastic strain accumulated during the 100 MPa creep stage is almost totally recovered, when a fatigue load cycling between 0 and 100 MPa is applied. Moreover, when the peak-stress of the stress-cycle is increased to 125 MPa, additional non-elastic strain is still not measured. On the other hand, when the load is kept at 125 MPa, strain accumulates again. These observations show that the rate of inelastic strain accumulation under cyclic loading at a load ratio R=0 cannot compensate for the decrease induced by creep strain recovery. This indicates that the strain accumulation rate under cyclic loading is well below that under static load, as also evidenced when pure creep and pure fatigue loading at the same maximum stress are compared. At present it is not yet possible to establish accurately the specific relation between test frequency and load ratio and strain accumulation rate under cyclic loading, since the measured strain rates are near the resolution of the equipment ($\sim 10^{-10}$ s^{-1}).

3.1.3. Implications for Fatigue Resistance

In the absence of other hysteresis causing phenomena such as the opening and closing of bridged cracks, the hysteretic area can be assumed to represent the amount of energy dissipated in the viscous grain boundary phases under cyclic loading conditions. Therefore, this dissipated energy is not available as a driving force for deformation accumulation. This can explain the observed decrease in strain accumulation rates under cyclic loads at high temperatures, when compared to strain rates obtained in static load experiments where this hysteresis is absent. Similarly, the dissipated energy is also not available for creating extra crack wake surface. The hysteretic stress-strain behaviour therefore seems to offer sufficient ground to understand delayed failure under fatigue loads as compared to tests under static load.

3.2. TIME-DEPENDENT PHENOMENA

3.2.1. Evolution of Stress-Strain Hysteresis during Long Term High Temperature Tests

In long-term fatigue tests on as-sintered sialon, the hysteretic area at a given temperature and frequency decreases with time (Fig. 6). This is caused by progressive crystallization of the intergranular phases during the test. As the volume fraction of crystalline grain boundary increases, the Young's modulus also increases as evidenced by a progressive decrease of the strain amplitude at constant stress amplitude of the individual fatigue cycles.

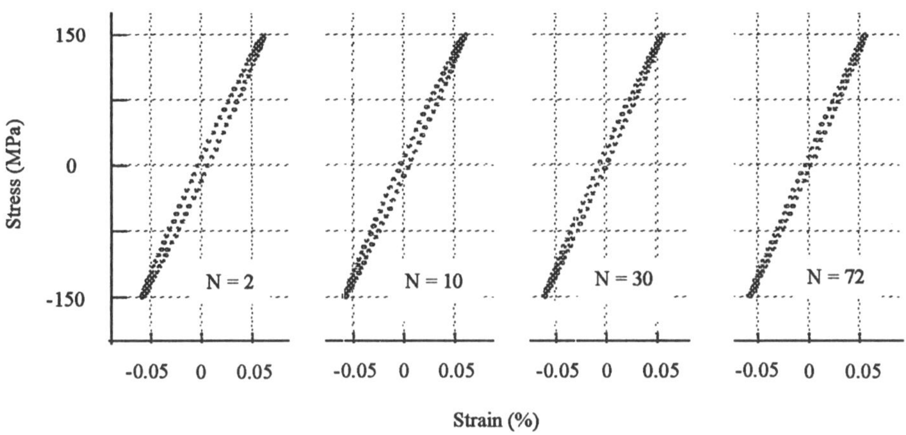

Figure 6. Evolution with time of stress-strain hysteresis loops in a 0.001 Hz-fatigue test at 1200°C on an as sintered specimen.

3.2.2. Consequences of Grain Boundary Devitrification for Fatigue Resistance

The crack growth resistance of ceramics depends on the amount of amorphous intergranular phases. Therefore, the fatigue resistance of a ceramic in which progressive devitrification occurs is changing concurrently. As the expected lifetimes under cyclic loading ar3e larger than under static loading (*3.1.3.*), the time-dependent influence of grain boundary phase crystallization becomes more important. Progressive devitrification may therefore lead to a further increase in lifetime in the "metastable" silicon nitride studied, in addition to the aforementioned frequency dependent contribution.

4. Conclusions

A newly proposed method to systematically characterize the influence of amorphous intergranular phases on the high temperature deformation properties in monolithic ceramics was successfully applied. Also, the evolution with time of the crystallinity of intergranular phases during fatigue and creep tests is monitored during long-term tests. Both the presence of the amorphous grain boundary phase, causing frequency dependent energy dissipation, and the evolution of its volume fraction with time at temperature are invoked to explain the observed positive (lifetime lengthening) effect of fatigue loading, compared to static loading.

References

1. Lawn, B.R., Physics of fracture, *J. Am. Ceram. Soc.*, **66** (1983), 83-91.
2. Evans, A.G., Fuller, E.R., Crack propagation in ceramic materials under cyclic loading conditions, *Metall. Trans.* **5A** (1974), 27-33.
3. Gilbert, C.J., Petrany, R.N., Ritchie, R.O., Dauskardt, R.H., Steinbrech, R.W., Cyclic fatigue in monolithic alumina: mechanisms for crack advance promoted by frictional wear of grain bridges, *J. Mat. Sci.* **30** (1995), 643-654.
4. Li, M., Guiu, F., Subcritical fatigue crack growth in alumina, *Acta metall. mater.* **43** (1995), 1859- 1884.
5. Ewart, L., Suresh, S., Crack propagation ceramics under cyclic loads, *J. Mat. Sci.* **22** (1987), 1173- 1192.
6. Tanaka, I., Pezzotti, G., Okamoto, T., Miyamoto, Y., Hot isostatic press sintering and properties of silicon nitride without additives, *J. Am. Ceram. Soc.* **72** (1989), 1656-1660.
7. Djuricic, B., Davies, I.J., Pickering, S., McGarry, D., Bullock, E., Verwerft, M., Bronsveld, P.M., De Hosson, J.Th., Coating of particles for the design of intergranular phases in engineering ceramics, Proc. 3rd Intl. Conf. Ceramic - Ceramic Composites, October 1994, Mons, Belgium.
8. Lewis, M.H., Barnard, P., Oxidation mechanisms in Si-Al-O-N ceramics, *J. Mater. Sci.* **15** (1980), 443-448.
9. Clarke, D.R., On the equilibrium thickness of intergranular glass phases in ceramic materials, *J. Am. Ceram. Soc.* **70** (1987), 15-22.
10. Dryden, J.R., Kucerovsky, D., Wilkinson, D.S., Watt, D.F., Creep deformation due to a viscous grain boundary phase, *Acta Metall.* **37** (1989), 2007-2015.
11. Croizet, D., Etude expérimentale et numérique du comportement à haute température d'un nitrure de silicium, Dissertation Ecole des Mines de Paris, (1992).
12. Saraiva Martins, C., Steen, M., Guerra Rosa, L., Evaluation of strength toughness and subcritical crack extension of a commercial HP Si_3N_4 exposed to a sulfidizing environment, *J. Mater. Sci.* **29** (1994), 5158-5164.
13. Quinn, G. D., Braue, W. R., Fracture Mechanism maps for advanced structural ceramics: part 2, sintered silicon nitride, *J. Mater. Sci.* **25** (1990), 4337-4392.
14. Bolsch, D., Untersuchung zum Wechselfestigkeitsverhalten von Siliziumnitrid bei hohen Temperaturen, Dissertation University of Karlsruhe (1991).
15. Lin, C. K. J., Socie, D.F., Static and cyclic fatigue of alumina at high temperatures, *J. Am. Ceram. Soc.* **74** (1991), 1511-18.
16. Fett T., Himsolt G., Munz D., Cyclic fatigue of hot-pressed Si_3N_4 at high temperatures, *Adv. Ceram. Mater.* **1** (1986) 179-84.
17. Roebben G., Steen M., Van Der Biest O., Visco-elastic energy dissipation in a sialon ceramic : quantification and implications for fatigue resistance, *Scripta Metall. Mater.* **33** (1995) 1417-21.

THERMAL FATIGUE OF THERMOSTRUCTURAL COMPOSITES

S. KHARRAT, B. COUTAND, J.P. LARTIGAU, E. MARTIN
and J.M. QUENISSET
Laboratoire de Génie Mécanique
Université Bordeaux I - 33405 Talence Cedex, France

1. Introduction

During the last decade, technological advances have been expected from the development of refractory composites based on either carbon, ceramic, glass or intermetallic matrices. However, a lack of reliability could be impeding with respect to various potential applications. Since their use temperatures range between room temperature and temperatures higher than 600°C, an understanding of the physicochemical and thermomechanical incompatibilities between components of these heterogeneous materials is particularly important.

The behavior of refractory composites under thermal cycling is one of the main concerns regarding these materials. Indeed, thermal fatigue simultaneously combines the effects of chemical modifications particularly in the vicinity of fiber/matrix (F/M) interfaces, and the effects of mechanical fatigue related to the cycling of thermally induced residual stresses. Moreover, these coupled phenomena can interact with the environment which can be inert, oxidizing or vacuum. Although these materials usually undergo simultaneous thermal and mechanical loadings, the mechanisms involved in thermal fatigue have been considered to be sufficiently complex and difficult to clarify, thus justifying the exclusion of any mechanical loading for an investigation in this field.

The present study, devoted to a laminated 2D glass-ceramic matrix composite, focuses on pointing out the predominant effects of thermal cycling, whose mechanisms contribute to composite damage and performance reduction.

After a brief presentation of the 2D-SiC/MAS-L composite and a concise description of the thermal fatigue and damage characterization conditions, particular attention will be devoted to the assessment of residual stress levels in the matrix and fiber. Their importance will be taken into account to propose mechanisms which attempt to explain the experimental results reported.

J. Bressers and L. Rémy (eds.), Fatigue under Thermal and Mechanical Loading, 259–268.

2. Experimental

2.1. MATERIAL

The 2D-SiC/MAS-L composite was fabricated by hot pressing at approximately 1300°C a stacking of 0°/90° unidirectional layers of SiC fibers (NICALON NLM 202) in the presence of an oxide powder derived from a gel at 500°C [1]. After processing, the glass-ceramic matrix of the laminate does not exhibit any microcracks and reveals the existence of five distinct phases: β–eucryptite with an estimated negative coefficient of thermal expansion (CTE) $\alpha \approx -6.4 \cdot 10^{-6}\,K^{-1}$ [2], mullite ($\alpha \approx 4.5 \cdot 10^{-6}\,K^{-1}$) [3], β–spodumene (($\alpha \approx 0.9 \cdot 10^{-6}\,K^{-1}$) [2], α–cordierite ($\alpha \approx 2.6 \cdot 10^{-6}\,K^{-1}$) [4] and an amorphous phase identified as $LiAlSi_3O_8$ [5]. Furthermore, high resolution transmission electron microscopy and microanalyses have shown the nanostructure and microstructure of the composites and particularly the presence of free carbon in the F/M interfacial zone [5-6].

After processing at such a high temperature (1300°C), the composite is particularly suited to the high temperatures of use anticipated for this type of material. Its main characteristics are reported in Table 1, where ε^d and ε^r stand for the first deviation of linearity and rupture strain, respectively, and σ^r is the rupture stress.

At room temperature, this type of composite loaded under tension exhibits non brittle behavior related to dissipative mechanisms such as microcracking, and F/M sliding with friction. This behavior is illustrated after fracture by a particularly extended fiber pull-out phenomenon.

TABLE 1. Mechanical characteristics of $(0/90)_{6s}$ SiC/MAS-L composite before thermal cycling.

E (GPa)	ε^d (%)	ε^r (%)	σ^r (MPa)
82	0.18	0.41	161

2.2. THERMAL FATIGUE TESTING

Rectangular specimens (120 x 10 x 3 mm^3) of $(0/90)_{6s}$ SiC/MAS-L composites were thermally cycled in a chamber either under vacuum, in air or in argon. The apparatus, based on an RF heating device, has been described elsewhere [7]. Since the composite is an electric insulator, a conductive susceptor had to be interposed between the specimen and the heating coil. A graphite susceptor could be used under vacuum or in argon whereas a molybdenum disilicide ($MoSi_2$) susceptor was required in air. In both cases, the specimens were heated and cooled by radiation with controlled speeds of $10Ks^{-1}$ and $-5Ks^{-1}$ respectively as illustrated in Fig. 1.

The maximum temperature gradients in the specimen central area were found to be lower than $2Kmm^{-1}$ during thermal cycling between 500°C and 1150°C, which limits stresses induced by non uniform distributions of temperature.

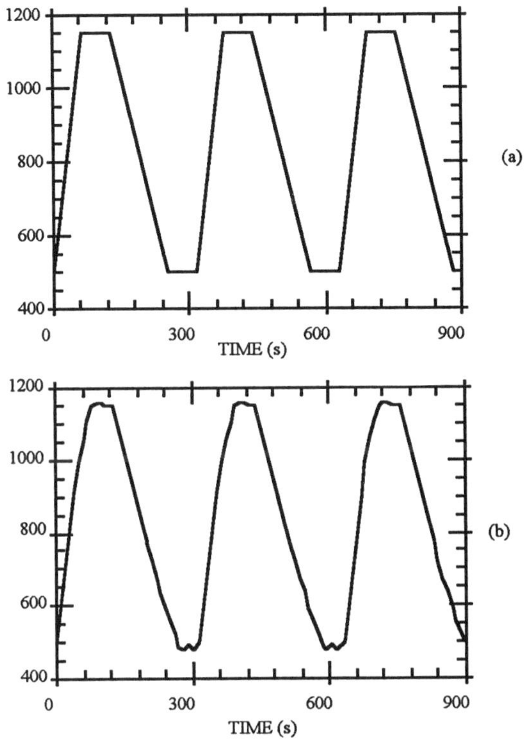

Figure 1. Thermal cycles: a- RF driving signal; b- Thermocouple measurement.

2.3. DAMAGE CHARACTERIZATION

The damage and the resulting decrease in performance of the composites after thermal fatigue were assessed through deviations in stiffness and density as a function of the number of thermal cycles. Also, the analysis of tensile stress-strain curves, with the help of simultaneous record of acoustic emission signals and the micrographic examination of specimen faces and fracture surfaces have contributed to the interpretation of the deviations in behavior after thermal cycling. It is noteworthy that the tensile tests were performed with the two adjacent central plies under transverse tension (90°plies).

3. Thermal Stresses

At a macroscopic scale, the anisotropy of the unidirectionally reinforced plies gives rise to residual stresses in the 2D $(0/90)_{6s}$ composite during the cooling step after hot pressing. Although the processing temperature reaches 1300°C, the state of residual stresses in the 2D $(0/90)_{6s}$ composite is assumed to appear below 1000°C. Thus, residual stresses were computed for a temperature deviation ΔT = -1000°C with the thermomechanical characterization of 1D plies, as reported in Table 2.

TABLE 2. Mechanical and physical characteristics of 1D-SiC/MAS-L ply.

V_f (%)	E_T (GPa)	E_L (GPa)	ν_{LT}	ν_{TL}	α_L (10^{-6} K^{-1})	α_T (10^{-6} K^{-1})
33	95	120	0.28	0.22	3.4	1.7

After processing, the plies are under an average transverse stress of -72 MPa and a longitudinal tension which leads to tensile stresses of 120 MPa in the fibers and 45 MPa in the matrix.

In addition to the state of residual stress induced by ply anisotropy, a complementary state of residual stress is generated at a microscopic scale by CTE mismatch between the fibers and matrix. It corresponds to residual stresses which develop during a 1D ply cooling. The use of the fiber and matrix physical characteristics as reported in Table 3, and a classical computation method described elsewhere [8] leads to tensile stresses in the reinforcement of 150 MPa in the longitudinal direction and 43 MPa for the hoop and radial stresses. In contrast, the longitudinal and hoop stresses in the matrix correspond to compression of -34 MPa and -74 MPa respectively .

TABLE 3. Mechanical and physical characteristics of SiC fibers and MAS-L matrix [10].

	d (µm)	E (GPa)	ν	σ^r (MPa)	α (10^{-6} K^{-1})
SiC fiber	15	200	0.25	2800	3.1
MAS-L	-	75	0.25	-	2.1

A first approximation of the previous stress state superposition can be represented by tensions in the axial direction of 270 MPa in the reinforcement and 11 MPa in the matrix. During thermal cycling between 500°C and 1150°C these stresses fluctuate between 135 MPa and -41 MPa in the fibers and 5 MPa and -2 MPa in the matrix. Since these levels of stress are definitely lower than the loadings acceptable by the reinforcement and matrix (they are estimated after processing to be 950 MPa and 90 MPa respectively), the residual stresses related to the stratification of anisotropic plies and to the F/M coupling should not play a significant role in the damage mechanisms of thermal fatigue. However, the conditions of load transfer at the F/M interface which can be represented by low interfacial shear stress (ranging between 3 and 7 MPa [4, 9]) can still be lowered by thermally induced mechanical fatigue of the F/M interface.

Significant mismatch between the CTE of the different matrix phases is expected to generate high thermal stresses (section 2.1). Nevertheless, a lack of descriptive informations about the volume fraction and morphology of each phase in the MAS-L matrix makes it difficult to assess these thermally induced stresses. Surely, after stress relaxation at high temperature by viscous sliding, the stress at low temperature could initiate microcracks at grain boundaries between phases exhibiting very different CTE and sharp morphology. Consequently, examination of the matrix after thermal cycling is an important concern.

4. Results

Comparison between the tensile stress-strain curves obtained before and after thermal cycling (Figs. 2 and 3) reveals a significant degradation of the composite in terms of stiffness, strength and work of fracture; these results were independent of the atmosphere and the corresponding mass deviations (Fig. 4).

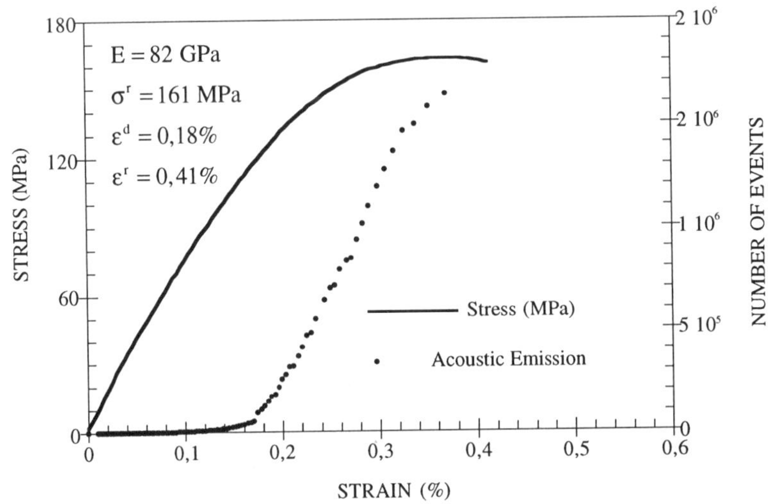

Figure 2. Tensile curve of non cycled 2D-SiC/MAS-L composite.

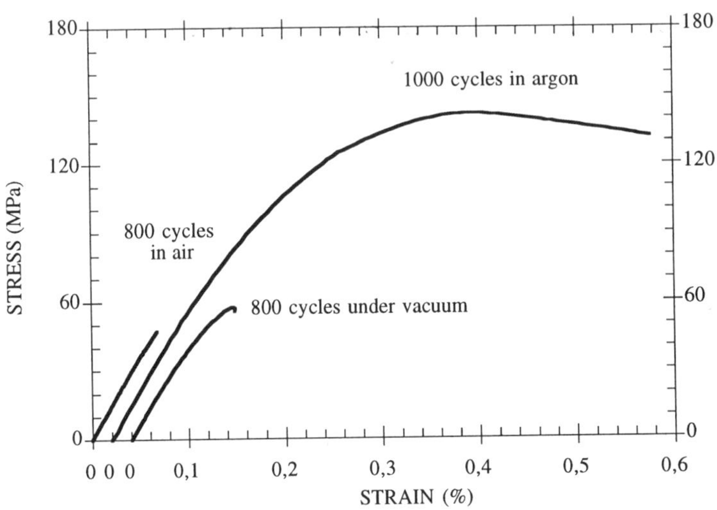

Figure 3. Tensile curve of 2D-SiC/MAS-L composite after thermal cycling.

4.1. DEVIATION IN ELASTIC PROPERTIES

Under vacuum and in argon, thermal cycling leads to a decrease in modulus of about 13% which stabilizes after about 300 cycles. This limited phenomenon of damage has to be connected to the micrographic observations made on specimens before and after thermal fatigue. After processing the composites do not show any microcracks, thermal cycles give rise to a microcrack network as illustrated in Fig. 5. It is worthwhile to note that the microcracks correspond partly to phase boundaries.

The decrease in stiffness related to thermal cycling in air is not more significant during the first hundreds of cycles than in an inert atmosphere. However the stiffness continues to decrease after 300 cycles even though this lowering is not as significant as during initial cycling. Thus, in air, thermal fatigue damage appears to combine at least two phenomena: (1) a short-range microcracking phenomenon similar to that illustrated in Fig. 5 and (2) a long-range phenomenon which could be attributed to chemical modifications of the MAS-L matrix in interaction with the oxidizing atmosphere.

4.2. DEVIATION IN THE MASS

Only small decreases in mass (less than 0.2%) occurred after thermal cycling in argon or in air, however a mass loss of about 1% could be measured after 100 hours (1200 cycles) of thermal fatigue under vacuum (Fig. 4).

The quasi negligible mass loss during the first hundred cycles seems to be accentuated by the development of the microcrack network in the matrix.

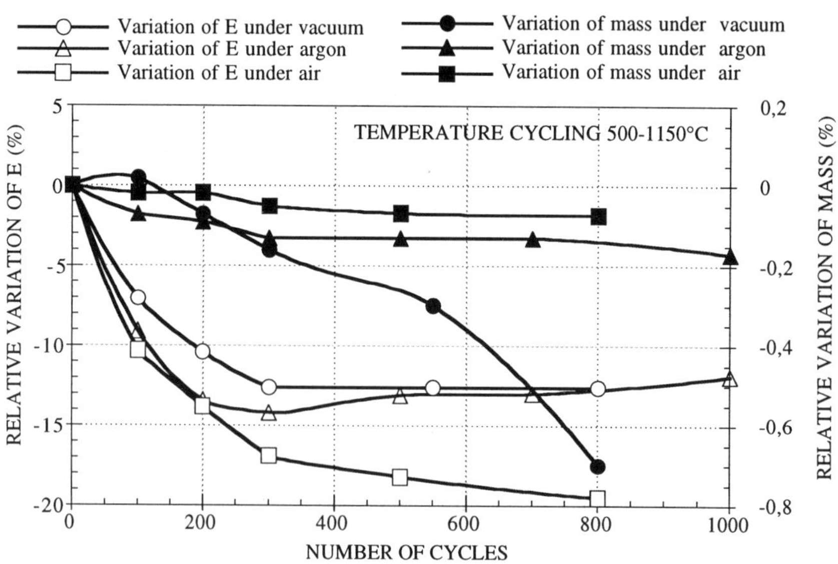

Figure 4. **Relative Young's modulus and mass variations of 2D SiC/MAS-L composites during thermal fatigue testing.**

4.3. STRENGTH DROP

As illustrated in Fig. 3 and in Table 4 significant mass losses and modulus decreases are associated with drops in strength and work of fracture:

(1) the larger decrease in modulus of specimens submitted to thermal fatigue in air can be correlated to a 70% drop of ultimate tensile strength,

(2) the unambiguous mass loss related to thermal fatigue under vacuum can be correlated to a 73% drop in strength,

(3) an 18% decrease in performance illustrates the composite stability under an inert atmosphere (argon) despite the occurrence of early microcracking of the matrix.

These observations indicate that the initial microcraking phenomenon is not the only important factor of degradation for this type of composite subjected to thermal fatigue.

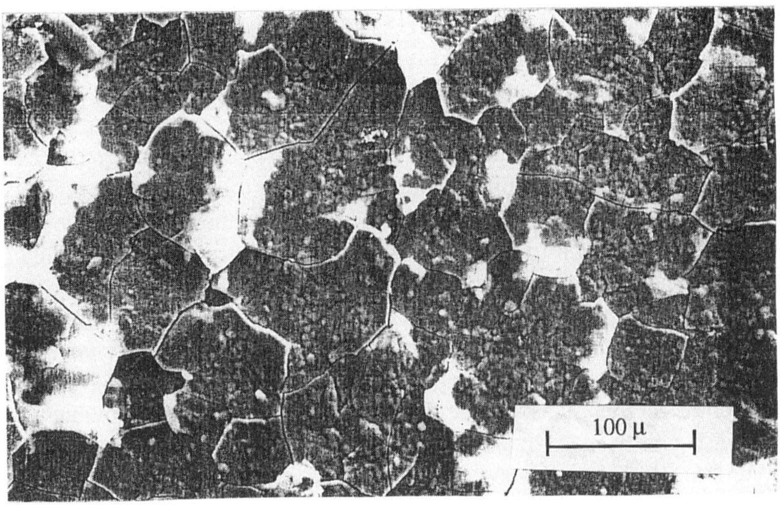

Figure 5. Matrix microcracking in a 2D SiC/MAS-L composite submitted to thermal cycling between 500 and 1150°C under argon atmosphere.

TABLE 4. Mechanical characteristics of $(0/90)_{6s}$ SiC/MAS-L composite after thermal cycling.

Numbre of cycles	Atmosphere	E (GPa)	ε^d (%)	ε^r (%)	σ^r (MPa)
800	vacuum	68	0.04	0.11	58
800	vacuum	70	-	0.05	36
1000	argon	71	0.13	0.56	132
800	air	70	0.04	0.07	52

ε^d: yield strain, ε^r: rupture strain and σ^r: rupture strength.

5. Degradation Mechanisms

Although the mechanism of matrix microcracking is not the main cause of performance drop during thermal cycling, it occurs whatever the gaseous environment during the tests and could facilitate the interaction between the inside of the composites and the environment. Thus the origin of this microcracking mechanism can be attributed to the following effects :

(1) the thermal gradient in the direction orthogonal to the most extended radiative surfaces of the specimen can induce tensile stresses in the external composite plies. However, modelling and computation have estimated the temperature gradient in the range of 1 - 2Kmm^{-1} [10]. Thus, despite the insulating properties of the MAS-L matrix and the rather high cooling speed of the specimens, the corresponding residual stresses are so small that microcracking is not likely to be initiated by the heterogeneity of the temperature distribution in the specimens,

(2) in contrast, it has been shown in Section 3 that the residual stresses induced by ply anisotropy and F/M CTE mismatch are not able to initiate matrix microcracking while the CTE mismatch between the various phases of the matrix is so significant that microcracks are able to develop in the matrix. This is confirmed by the location and directions of microcracks along phases boundaries (Fig. 5).

This mechanism, which stops as soon as matrix residual stresses are released, can be considered as limited damage which occurs during initial cycling. It affects moderately the composite stiffness and yield strain (ε^d) but should not modify the rupture performance (ε^r, σ^r), (Table 4). Nevertheless, tensile tests have shown that more or less significant drops in performance are related to thermal fatigue, depending on the environment.

Concerning thermal fatigue in argon, the moderate decrease in composite strength can be related to the performance loss of the fibrous reinforcement during its exposure at high temperature as reported elsewhere [11 - 14]. Above 1000°C, the microstructure of the SiC Nicalon fiber is affected by an unfavorable coarsening effect. Consequently, the combination of the initial microcracking mechanism and fiber evolution leads to moderate reductions of E, ε^d and σ^r but the work of fracture remains almost unchanged because of an increase in rupture strain (0.56% compared to 0.41% before thermal fatigue). This extension of the damage domain can be correlated to the mechanical fatigue of the F/M interface resulting in an increase in fiber pull out.

Under vacuum, similar behavior to the results obtained for specimens treated in argon would be expected, but in fact the drastic drop of all the mechanical characteristics suggests that other mechanisms are involved. In addition to the fiber degradation caused by a coarsening effect, a slight damage to the SiC Nicalon surface is the expected cause. An oxidation reaction of SiC gives rise to SiO and CO. The release of these gases may be accelerated by the chamber dynamic vacuum and the alternating microcracks opening and closure. That leads to significant mass loss. Thus, the kinetics of fiber degradation is so much increased under vacuum that the reinforcement fractures at low strains ranging between 0.19% and 0.25%. Thus, taking into account the tensile residual stresses induced in the fibers by thermomechanical coupling, the reinforcement degradation leads to more or less brittle composites with failure strain ranging between 0.05% and 0.11%.

In air, the mechanisms involved are still more complex and can be schematically described as follows. The initial microcracking phenomenon enables the penetration of the oxidizing atmosphere up to F/M interfacial zones. Consequently, the carbon interphase is rapidly eliminated by CO release whose evacuation by microcrack opening and closure facilitates further fiber oxidation and the formation of silica. Consequently, the mass loss related to the interphase oxidation is approximately compensated by the silica formation whose presence renders the composite more brittle. However, CO release is not so active in air as under vacuum, which limits slightly the fiber degradation. The embrittlement effect of the silica leads to a significant drop in failure strain and composite strength.

6. Conclusion

The study of a 2D-SiC/MAS-L composite submitted to thermal fatigue under various atmospheres has shown:

(1) An early microcracking phenomenon which reaches saturation after approximately 300 thermal cycles. This microcracking is attributed to CTE mismatch between the various phases constituting the matrix rather than the thermomechanical coupling between fiber and matrix or between anisotropic plies. This mechanical damage does not strongly decrease the composite strength but enhances the chemical degradations of interfacial zones and fibers.

(2) An evolution of the fiber microstructure by a coarsening effect which leads to a fiber strength decrease.

(3) A release of CO and SiO from fiber oxidation in air and particularly in a vacuum atmosphere. This mechanism is enhanced by the thermally induced mechanical fatigue.

(4) An elimination in air of the carbon interphase replaced by a silica layer giving rise to an embrittlement effect.

Finally, 2D-SiC/MAS-L laminates loose 18% of their strength under thermal fatigue in argon, the drop in strength increases to approximately 70% under vacuum and in air.

Acknowledgements - The authors want to thank Aerospatiale for supplying the materials and the European Economic Community for its financial support.

References

1. Larnac, G. , Péres, P., Donzac, J.M., Elaboration et caractérisation du composite à matrice vitrocéramique SiC/MAS-L. *Revue des Composites et des Matériaux Avancés*, **3**-n° hors série (1993), 27-41.
2. Strnad, Z., Glass ceramic materials, *Glass Sci. and Tech.* Ed. Elsevier, **8**, (1986), 174-179.
3. Bleay, S.M., Scott, V.D., The relationship between interface microstructure and bulk properties in glass-ceramic matrix composites, in T.W. Clyne (ed.) *Proc. of the 2nd Conf. on Advanced Materials and Processes*, University of Cambridge, (1991), pp. 300-306.
4. Labrugère, C., Influence de l'évolution physico-chimique des fibres et de la matrice et de la zone interfaciale fibre/matrice sur le comportement mécanique des composites SiC/C/SiC et SiC/MAS-L après vieillissement thermique sous atmosphère contrôlée". Thèse de l'Université Bordeaux I, (1994), N° d'ordre 998.
5. Kervadec, D., Chermant, J.L., Les composites à matrice vitrocéramique. *Revue des Composites et des Nouveaux Matériaux.* **1**, n°1 (1991), pp. 7-49.
6. Monthioux, M., Nano et microstructures de composites SiC/LAS-M. *Revue des Composites et des Matériaux Avancés*, **3**-n° hors série (1993), pp. 69-90.
7. Kharrat, S., Coutand, B., Martin, E., Nardo, M., Testing procedure for characterizing the thermal fatigue resistance of thermo-structural composites". *European Conference on Composites Testing and Standardization*, ECCM-CTS - September 1992, Amsterdam, pp. 329-333.
8. Warwick, C.M., Clyne, T. W., Development of composite coaxial cylinder stress analysis model and its application to SiC monofilament systems., *J. Mat. Sci.* , **11** (1991), pp. 3817-27.
9. Benoit, M., Brenet, P., Rouby, D., Comportement des interfaces dans des composites céramique-céramique. *Revue des composites et des matériaux avancés*, **3**-n° hors série (1993), pp. 235-251.
10. Kharrat, S., Comportement en fatigue thermique des matériaux composites réfractaires. *Thèse de l'Université Bordeaux I*, Février 1995, N° d'ordre 1277.
11. Yajima, S., Okamura, K., Hayashi, J., Omori, M., Synthesis of continuous SiC fibers with high tensile strength. *J. Amer. Cer. Soc.*,. **59** N°7-8, (1976) pp.324-27.
12. Mah, T.I., Hecht, N.L., McCullum, D.E., Hoenigman, J.R., Kim, H.M., Katz, A.P., Lipsitt, H.A., Thermal stability of SiC fibres (Nicalon), *J. Mat. Sci.* , **19** (1984), pp. 1191-201.
13. Le Coustumer, P., Monthioux, M., Oberlin, A., Mécanisme de dégradation d'une fibre NICALON série 200. Proc. AMAC CODEMAC, Bordeaux 29/30 Mars 1990, pp. 43-54.
14. Johnson, S.M., Brittain, R.D., Lamoreaux, R.H., Rowcliffe, D.J., Degradation mechanisms of silicon carbide fibers. *J. Amer. Cer. Soc.*, 1988, **71**, n°3, pp. c132-35.

FATIGUE BEHAVIOUR OF SIC FIBRE TOUGHENED BARIUM MAGNESIUM ALUMINOSILICATE GLASS-CERAMIC LAMINATES

E. VANSWIJGENHOVEN, M. WEVERS and O. VAN DER BIEST
Department of Metallurgy and Materials Engineering
Katholieke Universiteit Leuven
de Croylaan 2 B-3001 Leuven, Belgium

1. Introduction

Ceramic fibre glass-ceramic matrix composites combine a potential as high temperature material with a relatively easy processing route [1, 2]. The carbon rich interface layer in as-produced SiC fibre aluminosilicate matrices ensures both controlled damage development and failure through interface debonding, fibre fracture and fibre pull-out [1, 2, 3].

The fatigue behaviour of these composites in air has been extensively studied but is not fully documented yet [4 - 9]. In particular, the relationship between stress, temperature, and laminate lay-up on the one hand and fatigue damage initiation, damage propagation, and failure on the other hand has not been investigated systematically. The purpose of this paper is to report on the fatigue behaviour of SiC/BMAS laminates at high temperatures in air. First, it provides evidence of damage tolerance during fatigue below a material embrittlement temperature. Second, the danger of brittle fatigue failure at high temperatures is demonstrated.

2. Material

The continuous Tyranno silicon carbide (SiC) fibre toughened barium magnesium aluminosilicate (BMAS) matrix was produced by AEA Technology Harwell using the slurry infiltration - hot pressing technique [10]. Twelve or sixteen layers of pre-preg were stacked together in lay-up sequences as given in Table 1. The material was consolidated at temperatures below which crystallisation was very slow and then crystallised in a treatment between 1200 and 1300°C. The crystalline phases identified in the matrix were barium-osumilite, celsian and cordierite but some residual glass remained present. During production, a carbon rich reaction layer developed at the fibre/matrix interface [10, 11].

J. Bressers and L. Rémy (eds.), Fatigue under Thermal and Mechanical Loading, 269–278.

TABLE 1. Matrix composition and composite lay-up.

<table>
<tr><th colspan="5">matrix composition</th></tr>
<tr><td colspan="5">50 wt% SiO_2, 28 wt% Al_2O_3, 7 wt% MgO, 15 wt% BaO</td></tr>
<tr><td>uni-directional
(UD)</td><td colspan="2">cross-plied
(CP)</td><td>angle-plied
(AP)</td><td>quasi-isotropic
(QI)</td></tr>
<tr><td>$(0)_{16}$</td><td>$(0_2,90_2)_{2s}$</td><td>$(0,90)_{3s}$</td><td>$(+45_2,-45_2)_{2s}$</td><td>$(0_2,+45_2,-45_2,90_2)_s$</td></tr>
</table>

3. Experimental Set-up

All three point bending, tensile, and tension-tension fatigue experiments were performed on a MTS 810 servohydraulic system. The load was measured with a 10 kN load-cell and the displacement with a 10 mm LVDT (Linear Variable Differential Transducer). Three point bending was done at a displacement rate of 1 mm/min on rectangular bar specimens with a length of 25 mm and a width and thickness of about 3 mm using a standard jig with a span of 20 mm. This resulted in a span to depth ratio of about 7. Tensile and tension-tension fatigue experiments were performed with flat dogbone specimens ($150*12*3$ mm^3). The specimens were shaped from the pressed plates using a CNC machine equipped with diamond coated fingermills. The specimens were clamped in rigid water-cooled hydraulic clamps and misalignment was minimised using strain-gauged dummy specimens and the MTS alignment kit. The absolute residual misalignment stresses were lower than 10 MPa and remained approximately constant during tensile testing. The longitudinal strain was measured using strain gauges or a MTS low contact force extensometer resistant to high temperatures. For high temperature experiments, a MTS short furnace was used. The temperature was measured with three thermocouples. The three zone control reduced the temperature difference along the gauge length of specimens to about 10°C. Air humidity was not controlled but varied between 50 and 60% relative humidity. The fatigue tests were sinusoidally run at a frequency of 3 Hz and with a stress ratio of 0.1. Due to the restricted amount of material available and the extent of the test program, the traditional fatigue testing methodology was not used. Specimens were not submitted to only one thermomechanical fatigue test condition up to failure or up to a given number of fatigue cycles. Instead, specimens were tested for a set of thermomechanical fatigue conditions with increasing severity (higher temperature and/or higher stress). Fatigue survival of the test with the more harsh conditions was believed to imply fatigue survival of the less severe conditions and the influence of prior loading was believed to be negligible. Fatigue run out was arbitrarily defined as 10^5 cycles and the fatigue limit was defined as the fatigue stress below which no fatigue failure occurred. If no damage developed in the first 10^5 cycles, the fatigue stress was assumed to be lower than the fatigue damage initiation stress. Furthermore, the initiation of damage was believed to be reflected in the mechanical behaviour of specimens through a decrease in stiffness.

4. Results

The SiC/BMAS material was first characterised using three point bending and tensile testing. The first two sections summarise these test results. The third section gives a description of the observed fatigue behaviour.

4.1. THREE POINT BENDING BEHAVIOUR

The three point bending behaviour of the as produced uni-directional, cross-plied, angle-plied and quasi-isotropic laminates was similar. The initial part of the curve showed proportionality between load and deflection and the load maximum was followed by a load decrease characteristic of controlled damage development and failure. The failure mode was a combination of shear and tensile failure due to the small value of the span to depth ratio. The three point bending strength decreased from about 1000 MPa for UD to 500 MPa for CP, 400 MPa for QI and 200 MPa for AP material. Three point bending of specimens previously exposed to air at high temperatures indicated that exposure to air above 500°C embrittled the composite. The carbon rich interface zone responsible for the tough behaviour disappeared through interface reaction and/or oxidation [11]. However, an up-quenching pre-treatment (placing the material in a furnace at 1100°C and keeping it at temperature for 1 hour) protected the material from embrittlement by oxygen (Fig. 1). The pre-treatment sealed the specimen surface due to glass flow and/or fibre end oxidation. This pre-treatment however only proved to be effective with pore-free and crack-free material. Interconnected pores and cracks sometimes provided easy infiltration paths for oxygen, resulting in a brittle material after up-quenching.

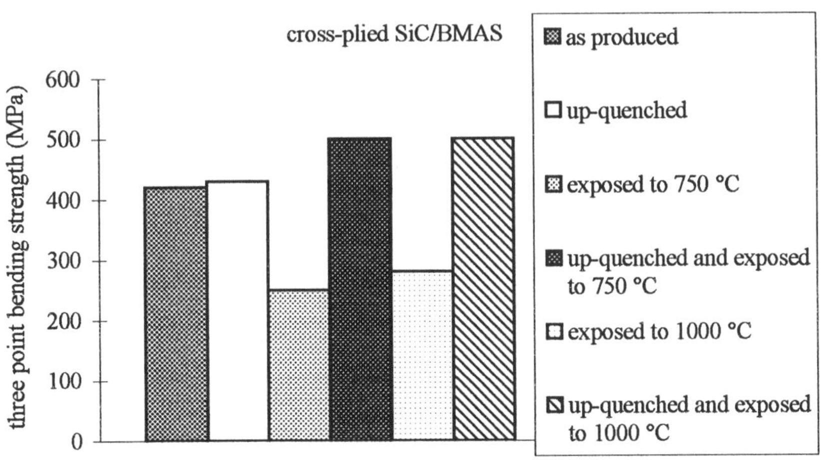

Figure 1. Three point bending strength of cross-plied SiC/BMAS showing degradation by exposure to hot air and protection against embrittlement by up-quenching.

4.2. TENSILE BEHAVIOUR

The stress-strain response during room temperature tensile testing was similar for UD, CP, AP and QI laminates. Initially, the relationship between stress and strain was linear. As damage developed, deviation of linearity occurred (Fig. 2). The damage development sequence was investigated using a travelling optical microscope. For uni-directional material, it consisted of matrix cracking accompanied by interface debonding followed by fibre fracture and pull-out. The damage development sequence in materials with fibres non-parallel to the loading direction was more complex. The major damage mechanism in 90°-plies was transverse matrix cracking while for 45°-plies interface debonding was observed. Tensile failure was governed by the failure of fibres in the longitudinal plies. The tensile strength was proportional to the fraction of fibres in the direction of load application. Table 2 summarises the values of the initial Young's modulus and tensile strength at room temperature. The initial Young's modulus and the tensile strength decreased with temperature. At 1000°C, the modulus of UD material was about 100 GPa and the tensile strength about 300 MPa (Fig. 2).

TABLE 2. Initial Young's modulus and tensile strength at room temperature.

	UD	CP	AP	QI
Young's modulus (GPa)	130	120	115	120
tensile strength (MPa)	650	300	100	150

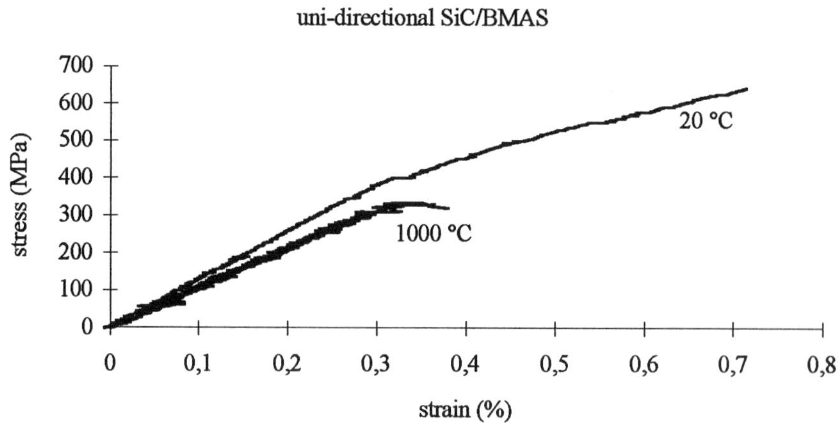

Figure 2. Stress-strain behaviour during tensile testing of UD composite material.

4.3. TENSION-TENSION FATIGUE BEHAVIOUR

The fatigue behaviour of different composite lay-ups was investigated at temperatures between 20 and 1100°C. Two types of material behaviour were observed. Tough and embrittled SiC/BMAS behaved distinctly different. When fatigue damage initiated in tough materials, stress redistribution took place. The stress-strain response became characterised by non-linearity and hysteresis and failure was non-catastrophic. Before and during failure an appreciable amount of energy was dissipated through matrix cracking, interface debonding, fibre fracture and fibre pull-out. These energy dissipating mechanisms were not observed in the embrittled material. The high stress concentrations associated with initial fatigue damage most often resulted in rapid fatigue failure. The stress-strain response of the embrittled material did not change greatly before failure. Failure was catastrophic and fracture surfaces showed only few signs of fibre pull-out (Fig. 3).

a) b)

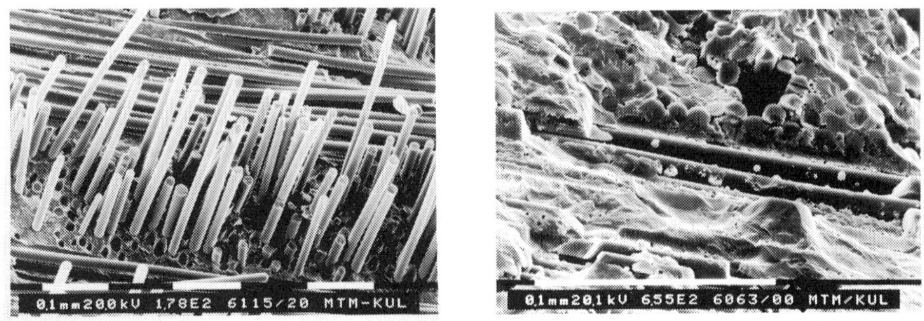

Figure 3. Secondary electron SEM images of fracture surfaces of tough and embrittled cross-plied SiC/BMAS a) after fatigue at 20°C, b) after fatigue at 1000°C.

Fig. 4 illustrates the fatigue behaviour of tough SiC/BMAS. The relationship between load and elongation during fatigue of uni-directional SiC/BMAS at 20°C is shown. Before damage development, the relationship was linear with no hysteresis. After damage development, the behaviour became non-linear and characterised by hysteresis. Damage development in the material resulted in a loss of stiffness, an increase in hysteresis and finally in failure. Fig. 5 illustrates the behaviour of embrittled SiC/BMAS. It shows the relationship between load and elongation during fatigue of uni-directional SiC/BMAS at 1000°C. The load-elongation response of the material was similar at the start of the test and just before failure. The initial stiffness was retained but the load drop associated with failure was sudden and complete.

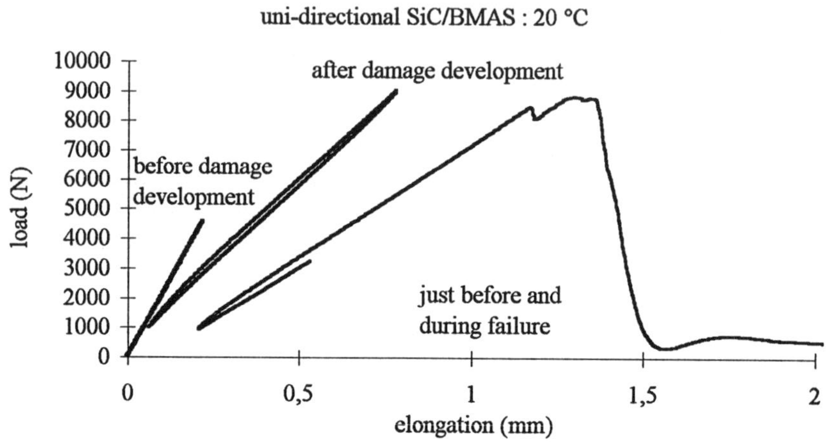

Figure 4. Relationship between load and elongation during fatigue of UD SiC/BMAS at 20°C.

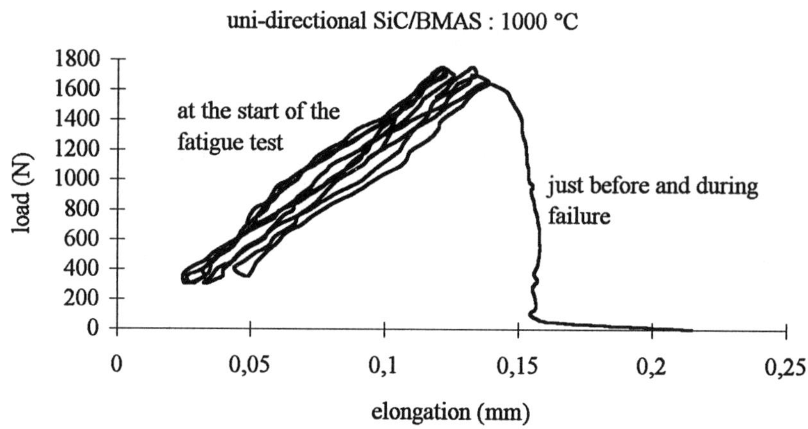

Figure 5. Relationship between load and elongation during fatigue of UD SiC/BMAS at 1000°C.

After evaluation of the fatigue behaviour of SiC/BMAS, the embrittlement temperature during fatigue of SiC/BMAS was estimated to be 500°C. This was the lowest temperature for which brittle fatigue failure was observed. At temperatures below the embrittlement temperature, three fatigue regimes were observed for all of the laminate lay-ups investigated : a 'no damage' regime in which no damage initiated within a number of cycles (10^5), a 'damage' regime in which damage developed but no failure occurred before 10^5 cycles and a 'failure' regime in which fatigue failures occurred. The fatigue damage initiation stress and the fatigue limit separated the different regimes. The limited amount of tests that was performed only allowed a rough estimation of both stresses (Table 3). The damage development during room temperature fatigue was investigated using an optical travelling microscope. Small microcracks were observed in uni-directional material above 250 MPa. These cracks however did not influence the stress-strain response. At about 400 MPa, extensive matrix cracking resulted in stiffness loss. In cross-plied material, transverse cracks developed at about 60 MPa. At higher stress levels, matrix cracks in the longitudinal plies were observed. Interface cracks were the major type of damage in angle-plied material. Most of the fatigue damage developing in the 'damage' regime took place during the first cycles. The stiffness loss was then most pronounced. As the number of cycles increased, fatigue damage development slowed down and a characteristic fatigue damage state was often reached. Fig. 6 shows the evolution of the Young's modulus during fatigue of angle-plied SiC/BMAS at room temperature. During cycling up to 20 MPa, no damage developed and the Young's modulus remained constant. During the first cycles up to 40 and 60 MPa, the Young's modulus dropped to reach a constant value. Some high cycle fatigue tests resulted in stiffness recovery. The latter was observed because an interaction between fatigue cracks and fatigue debris resulted in an increase of the minimum strain during fatigue. Fatigue testing to high enough stresses resulted in failure. The stress-strain behaviour just before failure indicated rapid fatigue damage development and after failure, a pull-out 'tail' was often observed. The fracture surfaces of the longitudinal plies showed signs of fibre pull-out, while the fracture surfaces of plies with fibres not parallel to the load application direction followed the fibre/matrix interfaces (Fig. 3).

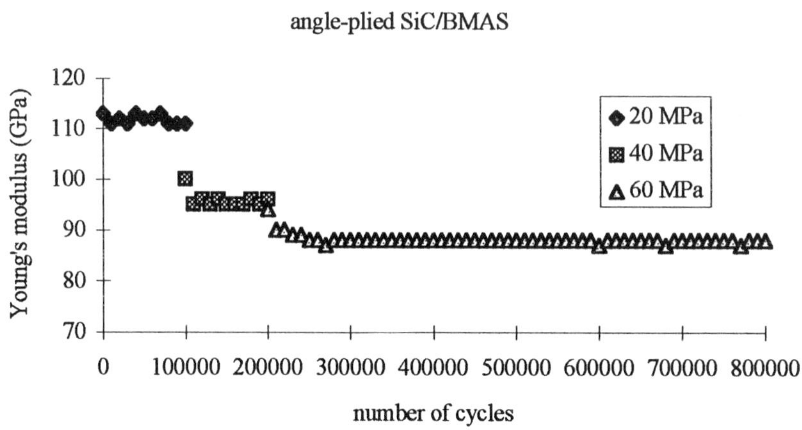

Figure 6. Evolution of the Young's modulus during fatigue of AP SiC/BMAS at 20°C.

TABLE 3. Fatigue damage initiation stress and fatigue limit for the different SiC/BMAS laminates.

	UD	CP	AP	QI
fatigue damage initiation stress (MPa)	250	60	30	40
fatigue limit (MPa)	400	200	60	120

As mentioned above, the composite material lost its damage tolerance above the embrittlement temperature of 500°C. To illustrate the relationship between fatigue stress and temperature on one hand and fatigue survival (10^5 cycles) and fatigue failure on the other hand, the tables 4 and 5 are used. Trying to synthesise the fatigue behaviour, no information concerning the number of tests performed, the exact number of cycles run without or up to failure or the ratio between fatigue failure and fatigue survival is given. Only the results of tests with the uni-directional and cross-plied material are summarised by the tables but angle-plied and quasi-isotropic material behaved similarly. The tables show that at room temperature only the application of high fatigue stresses resulted in failure. Fatigue failure occurred at lower stresses at higher temperatures. The fatigue limit decreased with increasing temperature. Not shown in the tables is the fact that the number of cycles run up to failure decreased with an increase in fatigue stress and that at a 1000°C, fatigue failure followed fatigue damage initiation rapidly. Some specimens failed at rather low stresses during room temperature fatigue due to embrittlement caused by pre-treatment. Some cross-plied specimens with damage in the 90°-plies survived fatigue cycling well above the embrittlement temperature. The 0°-plies of these specimens were most probably not embrittled since no oxygen could infiltrate the intact 0°-plies. The tables give no direct information concerning the evolution of the fatigue damage initiation stress as a function of stress. For uni-directional material, the fatigue damage initiation stress decreased with an increase in temperature to reach a value of about 100 MPa at 1000°C. This was mainly caused by the decrease in residual compressive stress in the matrix. For laminates with fibres not parallel to the applied stress, the fatigue damage initiation stress was only slightly influenced by the temperature.

TABLE 4. Fatigue failure and run-out of UD material as a function of stress and temperature.

MPa\°C	20	300	400	500	600	700	800	900	1000	1100
50	●	●		●		●		●		●
100	●	●	●	●	●	●	●	●	⊗	●
150	●					●				
200	●	●	■	●	⊗	⊗				
250	■				⊗					
300	■			⊗						
350	●									
400	●		⊗							
450	■									

● = no fatigue failures
■ = at least one fatigue failure
⊗ = all fatigue failures

TABLE 5. Fatigue failure and run-out of CP material as a function of stress and temperature.

MPa\°C	20	400	450	500	550	600	700	800	850	1000	1100
25	●	●				●		●		●	●
50	●			●	●	●	●	■	⊗	■	⊗
75	●	●		●			⊗				
100	■	●	●	■	●	■					
125	●	⊗		⊗							
150			⊗								
175	●										
200	■			⊗							

● = no fatigue failures
■ = at least one fatigue failure
⊗ = all fatigue failures

5. Conclusions

Fig. 7 summarises the fatigue behaviour of SiC/BMAS laminates.

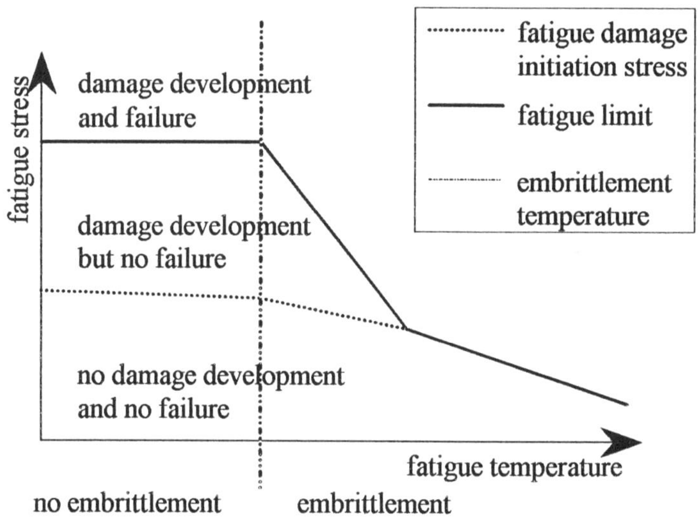

Figure 7. Schematic description of the fatigue behaviour of SiC/BMAS.

At room temperature and at temperatures below the embrittlement temperature (500°C), three fatigue regimes were observed. The first corresponded to a combination of thermomechanical conditions and material properties not resulting in fatigue damage development within a number of fatigue cycles. In the second regime, thermomechanical conditions and material properties were such that damage developed but no failure occurred. In the third regime, the application of a high fatigue stress resulted in fatigue failure. An increase in the number of cycles under consideration resulted in a decrease of the importance of the regimes with no failure. Above the embrittlement temperature, the regime in which damage developed without resulting in fatigue failure lost its importance. The material embrittled and the fatigue failure resistance decreased. The higher the temperature, the more rapid fatigue failure followed fatigue damage development.

The observed behaviour indicated that the use of Tyranno SiC/BMAS as a structural material should be limited to temperatures below the embrittlement temperature of 500°C. Above this temperature, the material lost its damage tolerance. This might result in brittle failure after damage initiation.

Acknowledgements - This work is supported by the Belgian Ministry of Science Policy under IUAP Contract N° 41 and is carried out in the framework of the European Commission's Brite-Euram Program BE 4610-09.

References

1. Davidge, R.W., Fibre-reinforced ceramics, *Composites* **18** (1987), 92-98.
2. Prewo, K.M., Fiber-reinforced ceramics: new opportunities for composite materials, *Ceramic Bulletin* **68** (1989), 395-400.
3. Evans, A.G., The mechanical performance of fiber-reinforced ceramic matrix composites, *Materials Science and Engineering* **107** (1989), 227-239.
4. Prewo, K.M., Fatigue and stress rupture of silicon carbide fibre-reinforced glass-ceramics, *Journal of Materials Science* **22** (1987), 2695-2701.
5. Holmes, J.W., Influence of stress ratio on the elevated-temperature fatigue of a silicon carbide fiber-reinforced silicon nitride composite, *J. Am. Ceram. Soc.* **74** (1991), 1639-45.
6. Zawada, L.P., Butkus, L.M., Hartman, G.A., Tensile and fatigue behaviour of silicon carbide fiber-reinforced aluminosilicate glass, *J. Am. Ceram. Soc.* **74** (1991), 2851-58.
7. Thomas, W.A., Kopp, M.W., Sanchez, J.M., Stress-strain behaviour of nicalon-fiber-reinforced calcium aluminosilicate composites under tensile fatigue conditions, *J. Am. Ceram. Soc.* **76** (1993), 2175-79.
8. Rouby, D., Reynaud, P., Fatigue behaviour related to interface modification during load cycling in ceramic-matrix fibre composites, *Composites Science and Technology* **48** (1993), 109-118.
9. Holmes, J.W., Wu, X., Sørensen, B.F., Frequency dependence of fatigue life and internal heating of a fiber-reinforced/ceramic-matrix composite, *J. Am. Ceram. Soc.* **77** (1994), 3284-86.
10. Churchman-Davies, J.J.R., AEA Harwell UK, personal communication.
11. Plucknett, K.P. et al., Environmental ageing effects in a silicon carbide fibre-reinforced glass-ceramic matrix composite, *Journal of Microscopy* **177** (1995), 251-263.

EFFECT OF TEMPERATURE CHANGE ON DELAMINATION CRACK GROWTH OF UNIDIRECTIONAL CFRP UNDER CYCLIC LOADING

Y. NAKAI
Department of Mechanical Engineering,
Kobe University
1-1, Rokkodai, Nada, Kobe 657, Japan

1. Introduction

Since the dominant fracture mechanism of carbon fiber reinforced plastics (CFRP) laminates in most cases is delamination along the interface of prepregs, a fracture mechanics approach to the delamination crack growth is required to assure the integrity of structures and machine components made of CFRP. Since structures and machine components are usually subjected to cyclic or variable loading conditions, information about the delamination behaviour under cyclic loading is particularly important. Most of the fatigue crack growth tests on CFRP were conducted under constant load range tests or load shedding tests, which have been widely employed for fracture mechanics approaches to fatigue crack growth experiments of metallic materials [1-6]. The delamination fatigue crack growth of CFRP, however, is not always controlled by the stress intensity range, ΔK, because of the existence of fiber bridging. Therefore, delamination crack growth rate is not always a unique function of the stress intensity range, ΔK [7].

In the present paper, constant ΔK tests were conducted to study the effects of test temperature and loading frequency and to clarify the mechanisms of delamination fatigue crack growth of CFRP.

2. Experimental Procedure

The unidirectional CF/epoxy laminates used in the present experiments were made from prepregs of Toray P3060E-15, which consisted of carbon fibers (Torayca T300B) and epoxy resin (Toray #3601). The elastic moduli of the laminate are E_1 = 143 GPa, E_2 = 8.43 GPa, and ν_{12} = 0.30, respectively, where the coordinates 1 and 2 are the fiber direction and the thickness direction, respectively [5,8]. Double cantilever beam (DCB) specimens, shown in Fig. 1, were employed for the delamination crack growth tests. Two holding-blocks with holes, which were made of an aluminum alloy, were bonded to the specimens. To introduce an initial notch,

J. Bressers and L. Rémy (eds.), Fatique under Thermal and Mechanical Loading, 279–284.

Teflon film of 50 μm thick was inserted between prepregs in the midsection before processing.

To conduct ΔK controlled tests, fatigue crack growth was monitored by using a compliance method, where the distance between two loading grips was measured by a linear variable differential transformer (LVDT). Although the front of the actual crack was curved in a thumbnail shape, the experimentally obtained relation between the average crack length and the compliance agreed with the relation obtained by the finite element method [5,6].

A computer controlled electrodynamic loading system was employed for the fatigue crack growth experiments. The crack growth behaviour was examined under pseudo-constant ΔK condition by having the control computer automatically reducing the load range after each 50 μm increment of crack growth. Tests were conducted from 24°C to 80°C in air. Fatigue crack growth tests were carried out over a range of frequencies from 0.05 Hz to 20 Hz at a load ratio, R, of 0.5 (where R is the ratio of minimum to maximum load during one fatigue loading cycle).

3. Experimental Results and Discussion

3.1. EFFECT OF PRE-CRACKING

The crack growth behaviour under constant ΔK condition (0.44 MPam$^{1/2}$) at 24°C is shown in Fig. 2, where Δa is the crack extension from the tip of the initial notch or from the pre-crack. Circular marks in the figure indicate the growth behaviour of a fatigue crack that initiated from an initial notch. Rectangular marks show the growth behaviour of a fatigue crack after pre-cracking. In this specimen, pre-cracking was accomplished by monotonic loading. The length of the pre-crack was 15 mm.

In metallic materials, it is well known that the retardation of fatigue crack growth is observed after overloading. In the present material, however, this kind of behaviour was not observed after the overloading (monotonic loading). The crack growth rate was almost constant until it grew by 32 mm. For the crack initiated from an initial notch (Teflon film), the growth rate was not constant until the crack grew about 3 mm. After this initial stage, the growth rate was almost constant. It has been considered that the initial transitional behaviour is observed in the epoxy resin enriched region, which is formed at the root of the initial notch (Teflon film). In this

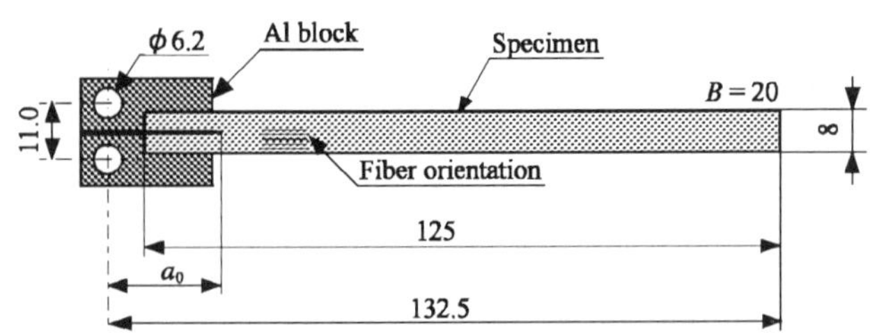

Figure 1. Test specimen (Dimensions in mm).

specimen, however, the size of the region must be much smaller than 3 mm because the thickness of the film was 50 μm. As already described in the previous section, the crack front was curved as a thumbnail shape, and the crack proceeded at the midsection. For the crack extension at the midsection less than 3 mm, the crack growth behaviour shows the transitional behaviour and the growth rate is smaller than the steady state value because te stress intensity factor for this semi-elliptical

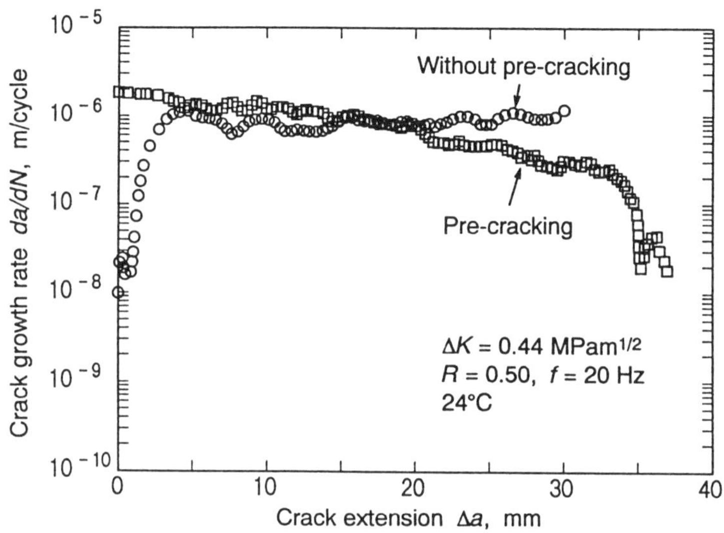

Figure 2. Interlaminar fatigue crack growth under constant ΔK condition.

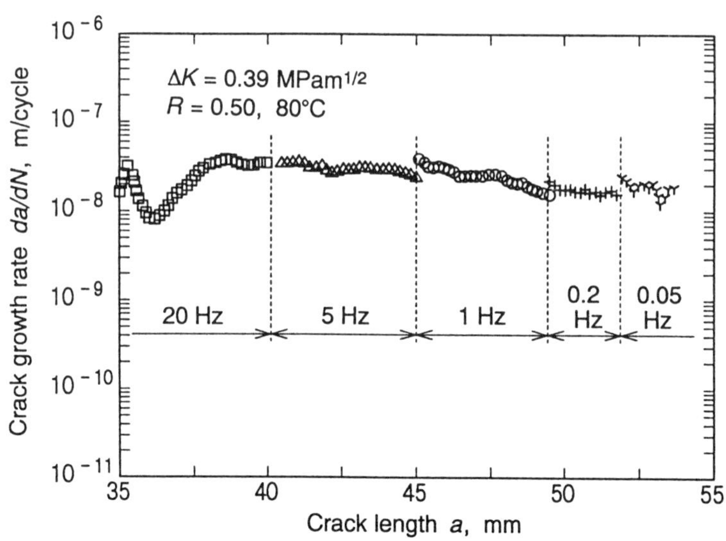

Figure 3. Effect of loading frequency on interlaminar fatigue crack growth.

crack emanating from the notch root is smaller than that for an equivalent through thickness crack.

After this initial transition and op to a crack extension, $\Delta\alpha$, of 32 mm, the crack growth rate is almost constant, and it is controlled by the stress intensity factor. Fiber bridging may be responsible for the deceleration of the crack growth fo $\Delta\alpha > 32$mm.

3.2. EFFECT OF LOADING FREQUENCY

To examine the influence of frequency, the cyclic load frequency was changed after each 5 mm increment in crack growth. The experiment was conducted in the steady state crack growth regime ($\Delta\alpha < 32$ mm). The results obtained at 80°C under a ΔK of 0.39 MPam$^{1/2}$, are shown in Fig. 3, where the crack growth rate is plotted against total crack length measured from the loading line. The crack growth rate is almost constant and independent of the loading frequency for the range of frequencies employed in this experiment, and the interlaminar crack growth behaviour is cycle dependent. Uematsu *et al.* [9] reported for CF/PEEK composite laminate at 200°C that the delamination crack growth was cycle dependent for the frequencies higher than 0.05 Hz, while it was time dependent for lower frequencies. The interlaminar crack growth in the present material may be time dependent for lower frequencies, but tests al low frequencies could not be achieved because of time limitations.

3.3. EFFECT OF TEST TEMPERATURE

To examine the effect of test temperature, the temperature was incrementally changed after each 5 mm crack growth. This experiment was also conducted in the steady state crack growth regime ($\Delta\alpha < 32$ mm). The results obtained at 20 Hz under a ΔK of 0.41 MPam$^{1/2}$, are shown in Fig. 4, where the crack growth rate is plotted against total crack length measured from the loading line. The crack growth behaviour did

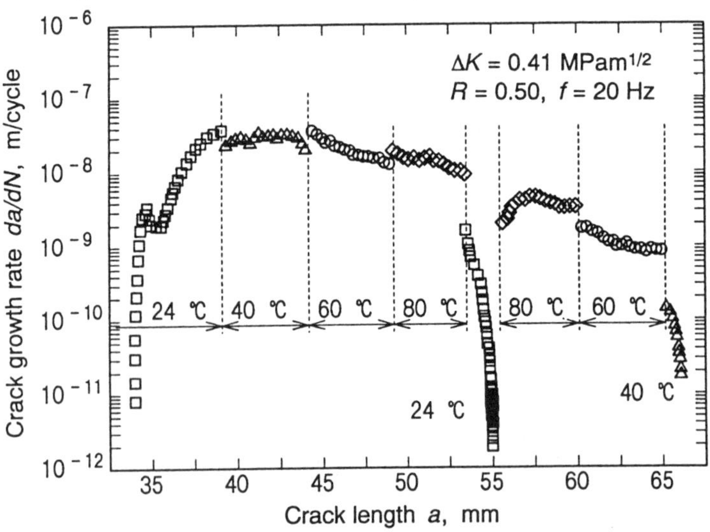

Figure 4. Effect of test temperature on interlaminar fatigue crack growth. (Highest temperature 80°C).

not change when the test temperature increased from 24°C to 80°C. On the other hand, the crack growth rate drastically decreased after changing the temperature from 80°C to 24°C. When the temperature was increased agains to 80°C, the crack growth rate was almost unchanged. When the temperature decresed from 80°C to 60°C, the growth behaviour was not affected. When the temperature decreased further to 40°C, the growth rate decreased again. As seen in Fig. 5, the same behaviour was observed for the highest temperature of' 60°C. When the test temperature van changed from 60°C to 24°C or 40°C, the growth rate decreased. When the test temperature was further raised from 24°C or 40°C to 60°C, the growth rate reached the same value of the isothermal experiment.

To summarize the above results, the crack growth drastically decreased only when the test temperature decreased from higher than 50°C to lower than 50°C. This phenomenon can be explained by supposing that there is large difference in the strength of the epoxy resin or of the interfaces around 50°C. When the temperature is lower than 50°C, the strength of the resin or of the interface is enough high, and therefore, the bridging hardly takes place. At high temperature, fibers can pull out from the resin easily, but the resin cannot support high load because its strength or that of the interface is low, and the bridging force by fibers is small. When the temperature decreases from higher than 50°C to below 50°C, fibers that pull out from the resin at high temperature bridge the fracture surface at low temperature, and can support enough load to reduce the true stress intensity factor at the crack tip because the strength of the resin or the interface is high enough at low temperature.

It is not clear why the transition in the strength of the epoxy resin or the interface takes place at 50°C, which is much lower than the glass transition temperature of this epoxy resin, which is 240°C. The temperature at the crack tip may be much higher than the ambient temperature.

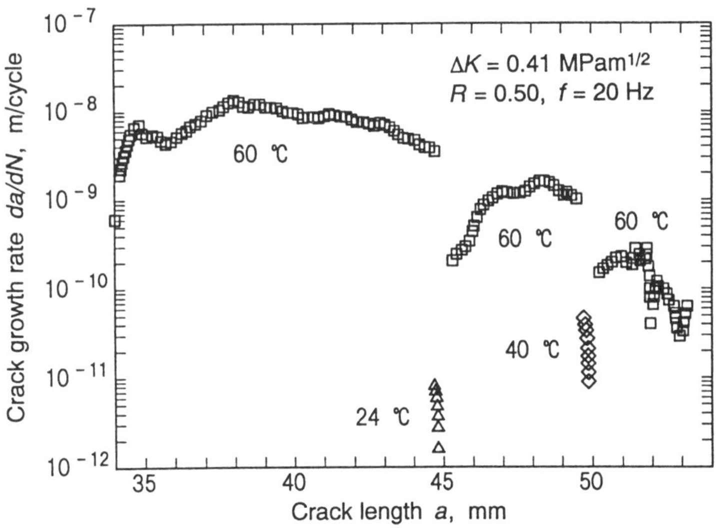

Figure 5. Effect of test temperature on interlaminar fatigue crack growth. (Highest temperature: 60°C).

4. Conclusions

The effects of loading frequency and test temperature on the delamination crack growth were examined with double cantilever beam (DCB) specimens for a CF/epoxy laminate under constant ΔK condition. The following result were obtained.

(1) The interlaminar crack growth rate is constant under constant ΔK conditions. Since an effect of loading frequency was not observed, an interlaminar crack propagates by cycle dependent mechanism(s).

(2) The interlaminar crack growth rate is unaltered when the test temperature is changed from lower to higher temperature. On the other hand, the interlaminar crack growth rate drastically decreases after decreasing the temperature from higher than 50°C to lower than 50°C.

Acknowledgements - Support of this work by Grant-in-Aid for Scientific Research (C) (Project No. 06650109) by the Ministry of Education, Science and Culture is gratefully acknowledged. The author also expresses his appreciation to Toray Industries, Inc. for providing the material used in this study.

References

1. Tanaka, K., Tanaka, H. and Yamagishi, K., Deformation and propagation of mode I fatigue cracks with crack-bridging in unidirectional CFRP, *Preprint of the 70th JSME Fall Annual Meeting*, No. 920-78, Japanese Soc. Mech. Eng. (1992), pp.139-141.
2. Hojo, M., Ito, H., Aoki, T. and Takahashi, J., Simple test method for delamination fatigue threshold in CF/PEEK laminates, *Preprint of the 70th JSME Fall Annual Meeting*, No. 920-78, Japanese Soc. Mech. Eng. (1992), pp. 133-135.
3. Koterazawa, R., Nose, M., Kawai, Hojo, M. and Aramaki, K., Fatigue crack growth and fatigue damage development in fiber reinforced composite under variable amplitude stresses, *J. Mat. Sci., Japan,* **42** (1994) 46-51.
4. Sembokuya, H., Hojo, M., Nagasawa, C., Kemmochi, K. and Maki, H., Comparison of fracture mechanisms between tension fatigue behaviour and mode I delamination fatigue crack propagation behaviour in unidirectional CF/epoxy laminates, *J. Mat Sci., Japan,* **42** (1993), 52-58.
5. Nakai, Y., Yamamori, H., Nakamura, M. and Ohji, K., Effects of frequency and temperature on delamination crack growth of unidirectional CFRP under cyclic loading, *J. Mat. Sci., Japan,* **42** (1993), 384-390.
6. Nakai, Y. and Ohji, K., Effects of frequency and temperature on delamination fatigue crack growth in unidirectional CFRP, in Bailon, J.-P. and Dickson, J.I. (eds.), *Fatigue* 93, Engineering Materials Advisory Service, U.K. (1993), pp.1379-1384.
7. Nakai, Y. and Yamashita, M., Effects of frequency and temperature on delamination fatigue crack growth of unidirectional CFRP under constant ΔK condition, *J. Mat. Sci., Japan,* **43** (1994), 1258-1263.
8. Tsai, S.W. and Hahn, H.T., *Introduction to Composite Materials,* Technomic Publishing, Westport (1980), Chapter 9.
9. Uematsu, Y., Kitamura, T., Ohtani, R. and Ohtsuka, H. Delamination crack growth of uniderectional carbon reinforced thermoplastic polymer under high temperature fatigue, *J. Mat. Sci., Japan,* **42** (1994), 59-64.

DAMAGE DEVELOPMENT DUE TO CYCLIC THERMAL LOADING IN CROSS-PLY CARBON/EPOXY LAMINATES

C. HENAFF-GARDIN, J.L. DESMEUZES and D. GAILLOT
Laboratoire de Mécanique et de Physique des Matériaux,
URA 863, ENSMA, Site du Futuroscope, B.P. 109,
86960 Futuroscope Cédex, France

1. Introduction

In the space environment, advanced composite materials may experience thermal cyclic loading [1]. Large thermal strains and stresses may develop in composite structures due to the mismatch in the coefficients of thermal expansion of adjacent plies at different orientations [2].

Thermal fatigue of long fibre composite laminates implies biaxial in-plane strains in each layer of the specimen (whatever their orientation). So, thermal exposure is likely to enhance damage similar to that observed under mechanical loading [3]. Under uniaxial mechanical loading (static or fatigue), a characteristic pattern of matrix cracks develops very early in the specimen life, these cracks running parallel to the fibre direction in the most disoriented plies with respect to the loading axis. Subsequently, other damage modes usually occur, the damage history depending on the material constituents, on the loading history and on the laminate stacking sequence [4, 5].

Under thermal loading, crack patterns in cross-ply laminates consist of cracks in both 0° and 90° plies [6, 7]. Delaminations may also occur at the crossing of these cracks. During cooling down of different symmetric laminates, Fang *et al.* [8] and Herakovich *et al.* [9] showed that both rates and amplitudes of the thermal excursions affect the extent and form of damage. In the case of cross-ply asymmetric [0°/90°] laminates, residual thermal stresses may be sufficient to create a regular array of cracks in 0° and 90° layers. Roussy [10] has described the evolution of damage during cooling down of different cross-ply composite materials.

Few thermal fatigue experiments have been conducted on composite laminates. Tests were generally undertaken for one temperature amplitude [6, 11] and authors were focusing their attention on material influence [3] or on the influence of damage on thermal expansion coefficients [12]. Moreover, these last authors have proposed a shear lag model for transverse crack propagation under thermal fatigue.

The present study consists in following the damage evolution in a cross-ply composite laminate submitted to thermal cyclic loading : we focus our attention on the thermal fatigue history, with different temperature amplitudes. A fracture mechanics analysis will then be applied to the present test conditions, in order to explain the influence of temperature amplitude on damage evolution.

J. Bressers and L. Rémy (eds.), Fatique under Thermal and Mechanical Loading, 285–293.

2. Material and Experimental Procedures

A $[0_4/90_4]_S$ cross-ply carbon/epoxy T300/914 laminate, with a nominal ply thickness of 0.12 mm has been studied. In this stacking sequence, there are the same number of 0° and 90° plies, with two external layers of four 0° plies, and one internal layer of eight 90° plies. Before machining, no damage was present in the material.

Because of the experimental procedure, the test coupons had to be very small : 20 mm square (with a 2 mm thickness). These coupons were cut from the laminates using a water-lubricated diamond saw; nevertheless, some cracks were present before the beginning of thermal cycling, but only in the 0° layers (the outer).

The specimens were alternatively cooled in a liquid nitrogen cold chamber and heated in a circulating air hot chamber. This entails manually immersing the coupon in liquid nitrogen for 15 minutes and then placing it in an air-oven for 15 minutes.

A thermocouple, placed on the specimen surface, allowed to record the temperature evolution during cycling (maximum temperature rate approximately 8°C/min) (see example of temperature variation with time on Fig. 1). Note that no information on the temperature gradient in the sample could be obtained.

The effect of the maximum temperature is investigated (between +200°C and +130°C), with a minimum temperature remaining constant (equal to -200°C). Four different tests at different amplitudes have been conducted :

- -200°C to +20°C (room temperature)
- -200°C to +50°C
- -200°C to +90°C
- -200°C to +130°C.

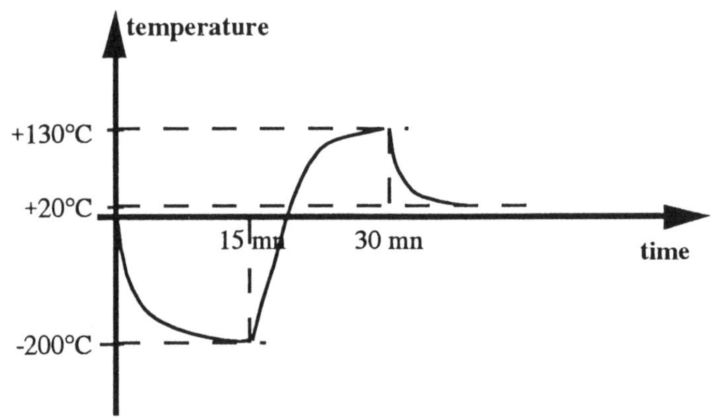

Figure 1. Schematic representation of temperature evolution during one thermal cycle.

One specimen has been tested for each temperature amplitude. The -200°C/+50°C and -200°C/+90°C coupons have undergone 30 cycles, the -200°C/+20°C and -200°C/+130°C tests have lasted 40 cycles. The damage development, particularly the evolution of 0° - and 90° cracks, was investigated throughout the tests by means of penetrant enhanced X-ray radiography.

3. Experimental Results

3.1. DAMAGE EVOLUTION : X-RAY RADIOGRAPHS

Fig. 2 shows the damage evolution with cycle number for the four different temperature amplitudes. Note that cracks are present before the first cycle : these cracks are due to machining.

When observing the radiographs throughout cyclic tests, it appears that cracks appear almost simultaneously in 0° and 90° layers along the fibres. These cracks initiate, span the entire layer thickness, grow in number more or less rapidly and may reach a saturation stage. The crack multiplication is faster in the 90° plies than in the 0° ones. Cracks are generally regularly spaced. Most of the cracks are spanning the entire specimen width or length (i.e. cracks are 20 mm long).

A second damage type is clearly observed in the -200°C/+90°C and the -200°C/+130°C tests : it consists of delaminations between 0° and 90° layers, along pre-existing cracks. Delaminations appear very late in the -200°C/+50°C coupon, and don't initiate at all in the -200°C/+20°C one. Moreover, it appears that this second damage mode initiates when cracks begin to saturate in number. Such an observation has been made under mechanical fatigue loading for various $[0_n/90_m]_s$ carbon/epoxy laminates [5]; moreover, the type of the second damage depended on the stacking sequence.

The delaminations appear in two steps : the first initiate along 0° cracks, followed by others along 90° cracks.

3.2. CRACK DENSITY EVOLUTIONS

On each of the radiographs, we have determined the crack densities in 0° and 90° plies. The density in one ply orientation is defined as the crack number per millimetre in <u>one layer</u> of the given orientation.

Figs. 3 and 4 present the crack density evolution with cycle number for the different tests. Thermal cycling between -200°C and +20°C didn't present any crack or other damage initiation before 40 cycles. The stresses induced by thermal cycling in each layer may be too small to initiate any damage, even after a large cycle number. For the three other tests, longitudinal and transverse cracks appeared from the first cycles, and they rapidly increased in number before reaching a saturation stage. The crack density evolutions are quite analogous for the -200°C/+90°C and -200°C/+130°C cyclic tests. The densities reach rapidly a saturation value (in a few thermal cycles). The saturated crack densities differ with the layer orientation :

$$d_{s90} = 0.45 \text{ crack/mm} \qquad d_{s0} = 0.30 \text{ crack/mm.}$$

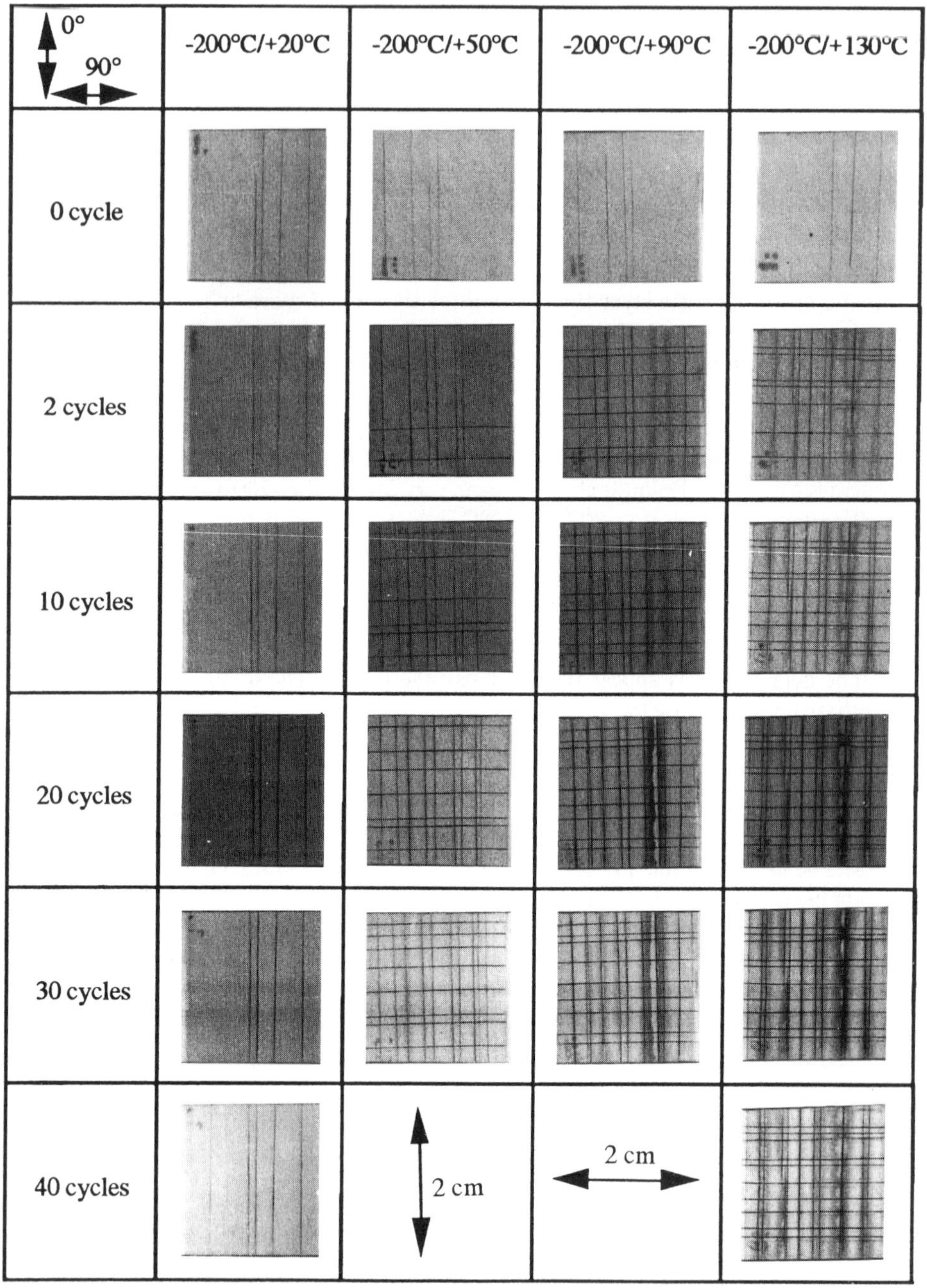

Figure 2. Damage evolution for the four studied temperature amplitudes (0° direction is vertical, 90° is horizontal).

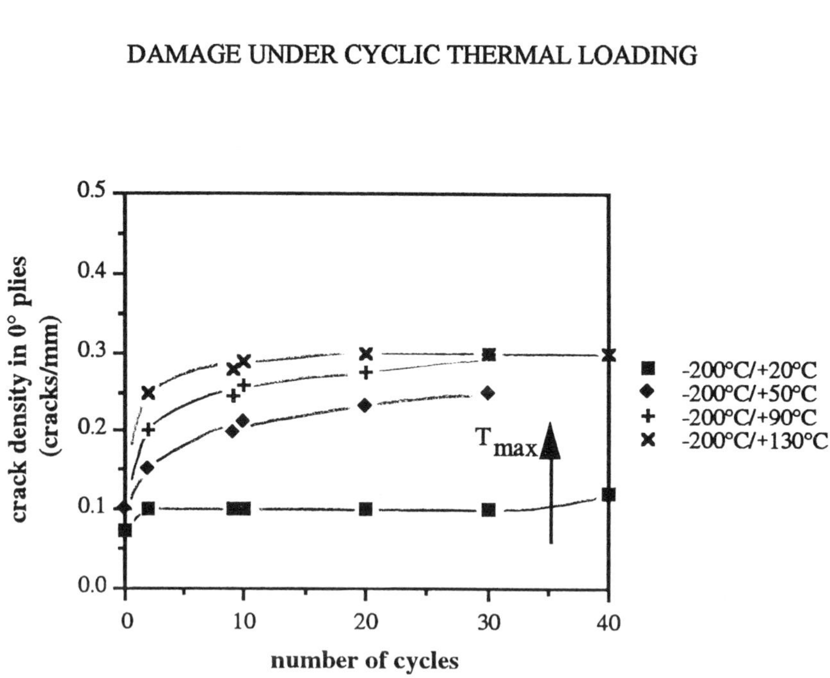

Figure 3. Crack density evolution in 0° plies for the four studied temperature amplitudes.

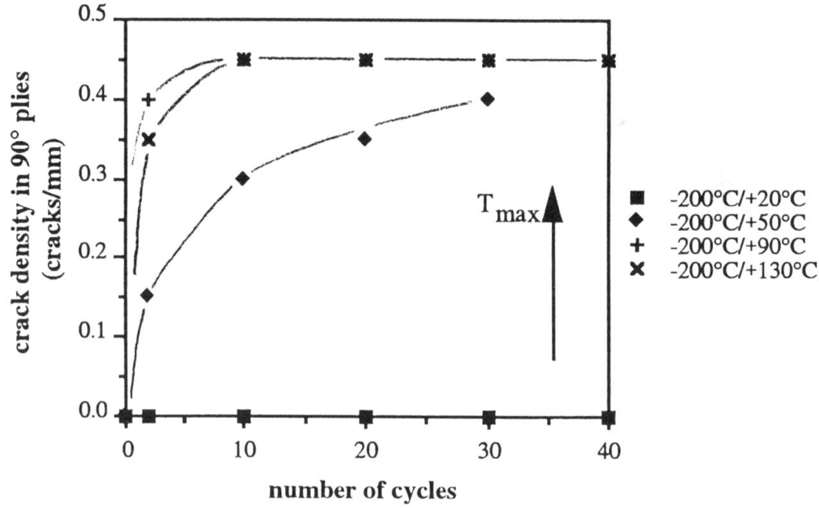

Figure 4. Crack density evolution in 90° plies for the four studied temperature amplitudes.

It should be noted that previous results obtained under uniaxial tensile loading [4] in cross-ply $[0_n/90_m]_s$ laminates (with different values of coefficients n and m) have shown that the thicker the inner 90° layer is, the less numerous the transverse cracks are. But here, when comparing the ultimate densities in inner and outer plies, it appears that in the thinner layers (i.e. 0° ones), the crack density is lower than in the thicker 90° layer. So, it appears here that outer plies don't behave in the same manner as the inner layers, probably because of free-edge effects : crack initiation seems to be more difficult in outer 0° layers. Experiments are already in progress to study the influence on damage of the layer location in the stacking sequence.

For the -200°C/+50°C test, the density evolution is rather slow : after 30 thermal cycles, the longitudinal and transverse densities are still below the ultimate values ; a few additional cycles are needed to reach the saturation regime.

In the three tests where a crack evolution is observed, crack multiplication is faster in the 90° layer than in the 0° ones. This tendency is in accordance with previous results on cross-ply carbon/epoxy laminates submitted to uniaxial tensile fatigue loading : the thicker the cracked ply is, the faster the crack evolution is [13].

The higher the temperature is, the faster the crack evolution is. It seems that there exists a threshold temperature (between +20°C and +50°C) below which no damage may be initiated. Moreover, above a critical temperature (between +50°C and +90°C), the crack densities reach their saturation value after the first cycles.

4. Modelling and Discussion

A fracture mechanics analysis based on the strain energy release rate has been developed previously [4, 14]. This model predicts the transverse crack evolution in a cross-ply $[0_n/90_m]_s$ laminate submitted to uniaxial mechanical tension (along the 0° layers). The laminate is supposed to have a cracked inner 90° layer (with a varying d_{90} density), whereas the outer 0° layers remain uncracked. Below a critical value G_c of the strain energy release rate, many fatigue cycles are necessary to create new cracks, and above this value, cracks are initiated after one fatigue cycle. In the case of T300/914 carbon/epoxy composite, this critical value G_c is equal to 140 J/m^2 [15].

In a first approach, this model will be applied to the present experiments, but it will not take into account crack initiation and multiplication in the outer 0° layers. Work is currently in progress to model, in an analogous way, a cross-ply laminate submitted to biaxial loading with cracks in both 0° and 90° layers.

The strain energy release rate G is a function of the stacking sequence geometry, of the material elastic characteristics, of the applied loading and of the transverse crack density, as follows :

$$G = K_1 \frac{\exp(\lambda/d)^2 - \dfrac{2\lambda}{d}\exp(\lambda/d) - 1}{(\exp(\lambda/d)+1)^2} \frac{1}{\left(1 + K_2 \dfrac{\tanh(\lambda/2d)}{(\lambda/2d)}\right)^2} \varepsilon^2$$

where d is the transverse crack density, ε is the applied loading. λ, K_1, K_2 are functions of the stacking sequence geometry and of the material elastic characteristics but they don't depend on damage. The evolution of the strain energy release rate is then associated only with the variations of crack density and with applied loading. For more details on the employed methodology and on the different parameters of this analysis, please refer to [4, 14].

In order to apply this model to the present test conditions, and to analyse only the 90° ply cracking, it has been necessary to make an analogy between mechanical and thermal loading. In a unidirectional ply, strains induced by temperature variations are far less important along the fibers than perpendicualar to the fibers. This mismatch may result in tension or compression strains when stacking up layers of different orientations (Fig. 5). The theory proposed by Datoo [16] allows to calculate the strains in each ply of an uncracked laminate submitted to temperature variations. In our $[0_4/90_4]_s$ laminate, thermal strains are equal to -1.89 10^{-5} ΔT along the fibres, and 2.06 10^{-5} ΔT perpendicular to the fibres. In these conditions, we choose an equivalent mechanical loading : $\varepsilon = 2.06\ 10^{-5}\ \Delta T$ along 0° direction.

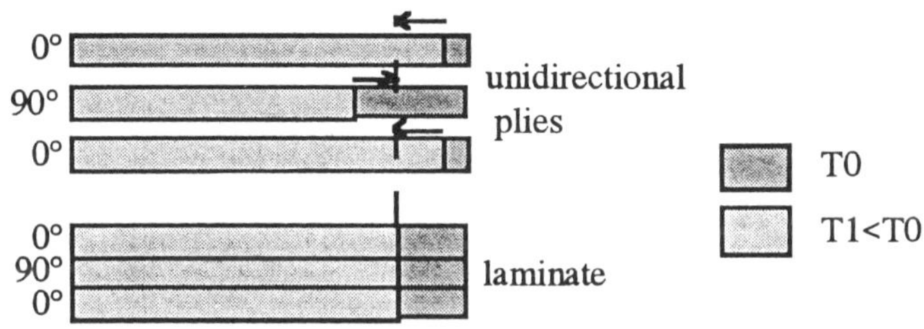

Figure 5. Representation of the thermal strains in unidirectional 0° and 90° plies, and in a cross-ply laminate.

The evolution of the strain energy release rate G versus the crack density in the 90° layer has then been calculated for the four different temperature amplitudes (Fig. 6).

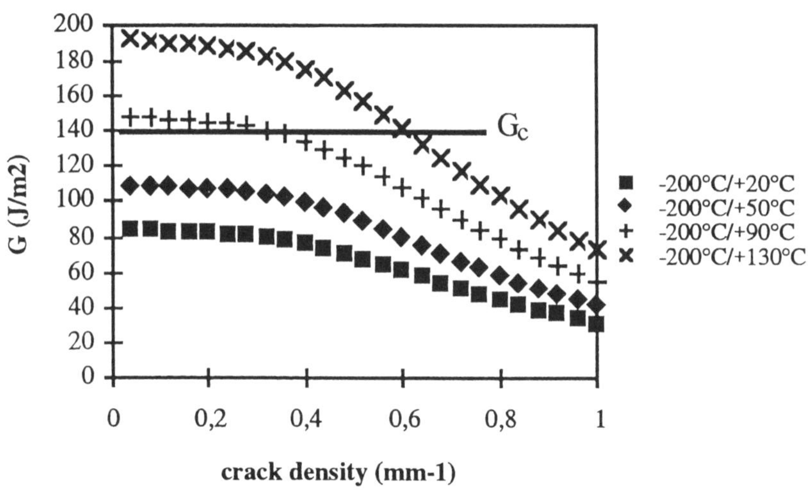

Figure 6. Strain energy release rate evolution versus 90° crack density for the four studied temperature amplitudes.

Throughout the tests, G decreases slowly with increasing crack density. The initiation of new 90° cracks becomes more and more difficult, leading to a saturated number of cracks. Such a decrease in crack density rate has been experimentally obtained (refer to Fig. 4).

For the two -200°C/+20°C and -200°C/+50°C tests, the values of the strain energy release rate are always below the G_c value : a slow damage evolution should be observed. During the -200°C/+50°C test, after 30 thermal cycles, the ultimate 90° crack density has not been really reached yet. Moreover, because of too small G values, no damage evolution at all has been observed in -200°C/+20°C.

For the two tests with larger temperature amplitudes, the initial G values are higher than the G_c value : cracks may appear from the first thermal cycle. Very few cycles are necessary to obtain the saturation in crack density. The higher the strain energy release rate is, the earlier the saturation stage is reached.

5. Conclusions

This study is concerned with the damage evolution under thermal cyclic loading of a $[0_4/90_4]_s$ T300/914 laminate.

Under thermal biaxial loading, cracks initiate in both 0° and 90° plies, followed by delaminations along the cracks, at the 0°/90° interface.

The influence of the temperature amplitude has been clearly put in light. Below a threshold amplitude, no damage evolution at all is observed (after 40 cycles). Above a critical value, cracking appears from the first thermal cycle, and doesn't evolve in the following cycles. Between these two values, a slow cracking evolution is obtained. The ultimate crack densities are independent of the temperature amplitude.

A fracture mechanics analysis which has been presented previously has been applied to the present test conditions. The evolution of the strain energy release rate with 90° ply density is in good accordance with the experimental observations.

Acknowledgements - The authors wish to thank Professor M.C. Lafarie-Frenot for fruitful discussions about this work.

References

1. Paillous, A., Pailler, C., Degradation of multiply polymer-matrix composites induced by space environment, *Composites*, **25**(4), (1994), 287-295.
2. Aswendt, P., Hofling, R., Speckle interferometry for analysing anisotropic thermal expansion - application to specimens and components, *Composites*, **24**(8), (1993), 611-617.
3. Forsyth, D.S., Kasap, S.O., Wacker, I., Yannacopoulos, S., Thermal fatigue of composites - ultrasonic and SEM evaluations. *Journal of Engineering Materials and Technology - Transactions of the ASME*, **116**(1), (1994), 113-120.
4. Henaff-Gardin, C., Lafarie-Frenot, M.C., Brillaud, J., El Mahi, A., Influence of the stacking sequence on fatigue transverse cracking in cross-ply laminates, in *Damage detection in composite materials*, J.E. Masters, (Ed). 1992, American Society for Testing and Materials Philadelphia, pp. 236-255.

5. Lafarie-Frenot, M.C., Henaff-Gardin, C., Urwald, E., Specificity of fatigue damage behaviour in carbon/epoxy cross-ply laminates, *Fatigue 93*, ISF., Montréal, Québec, (1993), 1355-1360.
6. Jennings, M.T., Elmes, D. and Hull, D., Thermal fatigue of carbon fibre / bismaleimide matrix composites, in *Third international conference on composite materials*, Bordeaux, France, Elsevier Applied Science, (1989), pp. 563-569.
7. Boniface, L., Ogin, S., Smith, P., The effect of temperature on matrix crack development in crossply polymer composite laminates. *Fifth European conference on composite materials*, Bordeaux, EACM, (1992), pp. 139-160.
8. Fang, G.P., Schapery, R.A., Weitsman, Y., Thermally-induced fracture in composites, *Engineering Fracture Mechanics*, **33**(4), (1989), 619-632.
9. Herakovich, C.T., Davis, J.G., Mills, J.S., Thermal microcracking in Celion 600/PMR-15 graphite/polyimide, *Thermal stresses in severe environments*, Plenum Publishing Corporation, (1980), pp. 649-663.
10. Roussy, L., Etude du comportement des matériaux composites aux températures cryogéniques, 1994, PhD thesis Université de Franche-Comté.
11. Adams, D.S., Bowles, D.E., Herakovich, C.T. Thermally induced transverse cracking in graphite-epoxy cross-ply laminates, *Journal of reinforced Plastics and composites*, **5**, (1986), 152-169.
12. Boniface, L., Ogin, S.L., Smith, P.A., Fracture mechanics approaches to transverse ply cracking in composite laminates, in K. O'Brien, (Ed.), *Composite materials : fatigue and fracture, Third volume*, 1991, *American Society for Testing of Materials*, pp. 9-29.
13. Lafarie-Frenot, M.C., Henaff-Gardin, C., Tahiri, V.L., Evolution of fatigue damage in various carbon/epoxy laminates, *7th Int. Conf. on mechanical behaviour of materials*, 1995, May 28 - June 2, 1995, The Hague, The Netherlands.
14. Henaff-Gardin, C., Urwald, E., Lafarie-Frenot, M.C., Modelling of the 90° ply fatigue crack growth along the width of cross-ply carbon/epoxy coupons, *Physicochemical Mechanics of materials*, **30**(1), (1994), 109-112.
15. Bookholt, H., *et al.* The investigation and understanding of improved CFRP composite materials, Technical publication 060. 1990, GARTEUR ed. ONERA.
16. Datoo, M.H., Residual stresses, in *Mechanics of fibrous composites*, Elsevier Applied Science, (1991), pp. 367-420.

SOLIDIFICATION STRUCTURES AND POST-FATIGUE PRECIPITATE MORPHOLOGY IN A SINGLE CRYSTAL SUPERALLOY

D.J. ARRELL, K.M. OSTOLAZA, J.L. VALLÉS and J. BRESSERS
Institute for Advanced Materials
Joint Research Centre
1755 ZG Petten, The Netherlands

1. Introduction

Single crystal nickel-base superalloys, which are widely used in high-temperature structural applications have been developed with the aim of optimising creep resistance at elevated temperatures. Due to this, creep failure in aero-engine turbine blades is nowadays a rare event, and they are most likely to fail *via* a fatigue-based mechanism. In service conditions these components are exposed to cyclic changes in stress and temperature, which are best simulated in the laboratory by the so-called thermomechanical fatigue (TMF) test. In recent years, much effort has been devoted to the study of precipitate morphology under creep conditions [1,2], as it has an influence on the mechanical performance. A similar analysis of the changes in precipitate structure which develop under TMF is also highly relevant.

Since there is a large solidification temperature range in these single crystal alloys, the casting process results in the formation of a microstructure exhibiting a set of dendrites parallel to the growth axis. This study is focused on the influence that the solidification structures have on the development of local variations in the precipitate morphology.

2. Experimental Methods

A series of strain-controlled TMF tests with the mechanical strain axis parallel to the [001] casting direction were carried out on single crystals of the commercial nickel-based superalloy SRR99, which had been processed via a standard solution treatment and ageing route followed by a coating heat treatment of 1 hour at 1373 K and 16 hours at 1143 K. The composition of the alloy is shown in Table 1.

TABLE 1. Composition of superalloy SRR99 as reported by Hopgood and Martin [3].

	Al	Ti	Cr	Co	Ta	W	C	Ni
Atomic %	12.0	2.7	9.6	5.0	0.9	3.0	0.07	balance
Weight %	5.5	2.2	8.5	5.0	2.8	9.5	0.02	balance

J. Bressers and L. Rémy (eds.), Fatigue under Thermal and Mechanical Loading, 295–304.

The specimens selected for the present study, which correspond to tests carried out within the framework of a wider research programme on the mechanical performance of superalloys, are listed in Table 2.

TABLE 2. Testing conditions.

Specimen Number	R	$\Delta\varepsilon_m$ (%)	Cycles (x10^3)
1	0	0.8	5.7
2	0	0.8	2.8
3	0	0.7	7.2
4	0	0.5	37.0
5	-	-	3.0
6	-∞	0.7	10.6
7	-∞	1	0.9

The temperature cycle in the tests, extending from 573 K to 1323 K, lags by -135° with respect to the strain cycle. The samples were linearly heated and cooled at 25 Ks^{-1} and 12.5 Ks^{-1} respectively, resulting in a total cycle time of 90 seconds. For the strain ratio

$$R = \varepsilon_{\min} / \varepsilon_{\max} \tag{1}$$

values of both R = 0 and R = -∞ were considered, as shown in Fig. 1, with mechanical strain ranges $\Delta\varepsilon_m$ between 0.5 and 1%.

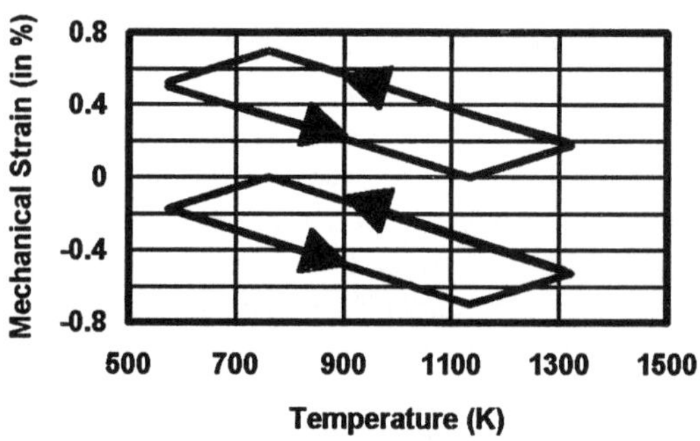

Figure 1. TMF cycles with $\Delta\varepsilon_m$ = 0.7% and R=0 (top) or R=-∞ (bottom).

In order to identify the different effects of the thermal and mechanical cycling, a thermally-fatigued specimen (sample 5) which had been cycled for a time equivalent to that of failure for a 0.8% strain, R=0 TMF test (sample 2) was also examined. Specimens sectioned parallel to the (001) and (010) crystal planes, *i.e.* perpendicular and parallel to the fatigue axis, were examined under SEM and electron probe micro-analysis (EPMA), in order to establish the precipitate coalescence behaviour in different parts of the sample and the corresponding chemical composition profiles.

3. Microstructural Observations

Optical microscopy of chemically etched samples clearly identified the boundaries of the casting dendrites, which suggested that a chemical inhomogeneity existed. More detailed examination using SEM allowed us to further differentiate the microstructure, and to categorize three distinct microstructural regions. These are the dendritic material which solidified along the casting axis, that which solidified in the cube directions perpendicular to the crystal growth axis, and finally the remainder of the casting. These regions will be referred to as primary dendrite cores, secondary dendrite arms and interdendritic material respectively.

As would be expected, in addition to the usual dispersion of cuboidal γ' particles within the γ matrix, the interdendritic region contains a small fraction (<1%) of other solidification products such as tungsten and tantalum carbides, as well as larger blocks of monolithic γ' and shrinkage pores. There is also a relatively large amount of sub-grain boundary material. The distribution of these phases is unaffected by the mechanical treatment.

Within the accuracy levels provided by EPMA, the chemical compositions of the dendrite cores and dendrite arms are indistinguishable. However, these two regions have a significantly lower γ' volume fraction than the interdendritic regions. In a typical example (sample 1), the γ' volume fraction in the dendrite arms is approximately 0.59 as compared with 0.66 interdendritically. EPMA shows that the major chemical difference between the dendritic and interdendritic regions is in the amount of tungsten, a γ volume fraction enhancing element. The average W content is approximately 2.5 atomic % in the interdendritic material and 3.4 atomic % inside the dendrites, as is shown in Fig. 2.

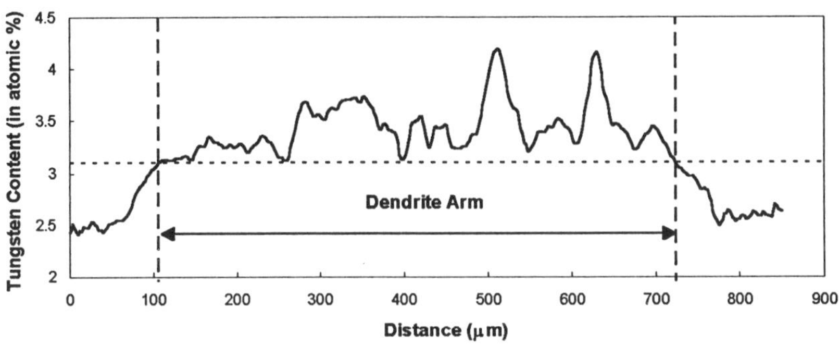

Figure 2. EPMA plot of tungsten distribution. The scan passes from an interdendritic region (left) across a secondary dendrite arm (centre) and into the next interdendritic region (right).

An examination of the thermally cycled specimen showed, as stated in Table 3, that some isotropic (*i.e.* triaxial) γ' coalescence occurred in the interdendritic regions, though after 75 hours there was only a relatively small amount. Biaxially coalesced γ' structures, usually known as plates, were found within the primary dendrite cores, normal to the growth axis. These structures are functionally identical to the plates widely reported as forming under creep in SRR99 and other similar alloys. Even more surprisingly, biaxially coalesced structures are also seen to form in the dendrite arms, but perpendicular to the secondary growth axes. These morphologies are shown schematically in Fig. 3.

TABLE 3. Coalescence behaviour in the different microstructural regions.

Specimen Number	Interdendritic	Dendrite Cores	Dendrite Arms
1	Slightly plate-like	Strongly plate-like	Needles normal to casting axis and dendrite arm axis
2	Slightly plate-like	Strongly plate-like	Needles normal to casting axis and dendrite arm axis
3	Slightly plate-like	Strongly plate-like	Plates normal to casting axis with weaker coalescence along arm axis
4	Triaxial	Plate-like	Plates normal to casting axis with weaker coalescence along arm axis
5	Triaxial	Slightly plate-like	Plates normal to dendrite arm axis
6	Slightly needle-like	Slightly plate-like	Strongly needle-like parallel to casting axis
7	Needle-like	Slightly plate-like	Needle-like parallel to casting axis

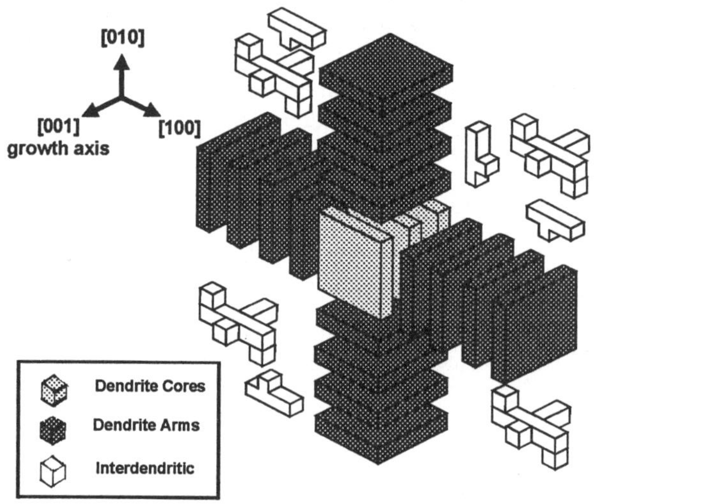

Figure 3. Schematic of coalescence in thermally cycled material.

The superposition of the mechanical strains produced by the TMF cycle markedly modifies the coalescence behaviour in the three microstructural regions. Specifically, R=0 cycling oblates the microstructure in the directions perpendicular to the applied stress (and crystal growth) axis, resulting in a stronger biaxial coalescence behaviour in the dendrite cores, a tendency towards plate formation in the dendrite arms at higher strain ranges and in biaxial structures in the interdendritic material. The effect of a low strain range cycle (sample 4) is shown in Fig. 4, and that of a higher strain range (sample 1) in Fig. 5. A schematic of this high strain range coalescence can be seen in Fig. 6. A secondary effect of the R=0 cycle is that at the highest strain range examined ($\Delta\varepsilon_m$ = 0.8%), in the interdendritic zones there is also a coalescence of γ' along <110>-type directions, which has been associated with the cutting of γ'

particles by creep-type dislocations [4]. This feature is not found within the dendrite arms.

The imposition of an $R=-\infty$ cycle has the opposite effect to that of the R=0 cycle. It results in the coalescence structures becoming more needle-like (or less plate-like) by means of an increase in the degree of coalescence along the casting axis and a reduction along the cube directions normal to the casting axis. At the higher strain range (sample 7), the interdendritic material coalesces to a greater degree than at the lower strain range (sample 6), despite the short duration of the test. The coalescence behaviour for $\Delta\varepsilon_m$ = 1.0% is shown schematically in Fig. 7.

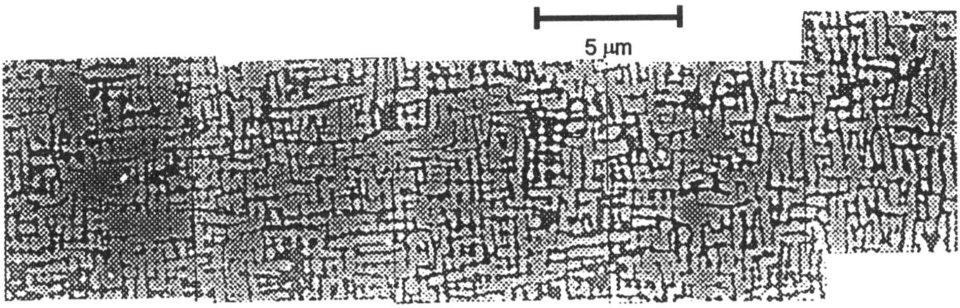

Figure 4. Micrograph showing typical coalescence behaviour under R=0 conditions for $\Delta\varepsilon_m$ = 0.5% (sample 4). The image shows the transition zone between a dendrite core (left) and a dendrite arm (right) viewed in a plane perpendicular to the applied stress axis.

Figure 5. Micrograph showing typical coalescence behaviour under R=0 conditions for $\Delta\varepsilon_m$ = 0.8% (sample 1). The image shows the boundary zone between a dendrite arm (left) and interdendritic material (right) viewed in a plane perpendicular to the applied stress axis.

4. Analysis Using a Coalescence Criterion

In the current study we have an inhomogeneity in chemical composition which is most clearly shown by the sharp difference between the tungsten content of the dendrites and interdendritic regions. This is significant for two particular reasons. Firstly, alloying with tungsten increases the lattice parameter of both the γ' and γ phases by around 1‰ per 1 weight % addition [5], and secondly, tungsten has been shown to enhance the tensile and creep strength of alloys with a composition very similar to SRR99 [6].

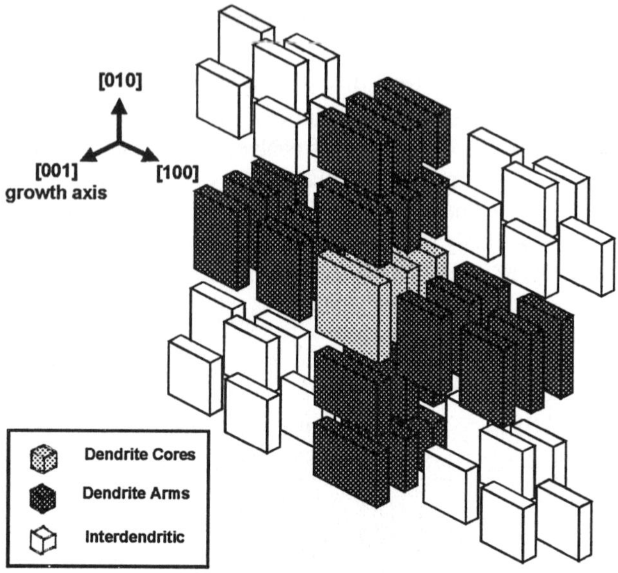

Figure 6. Schematic of coalescence under $\Delta\varepsilon_m$=0.7% and R=0 conditions.

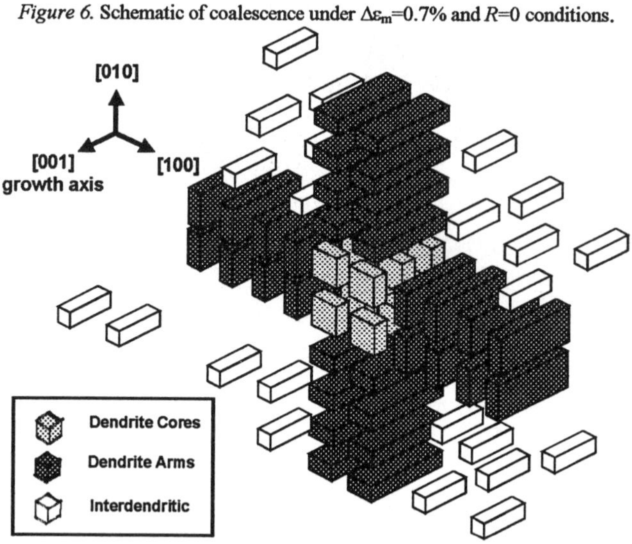

Figure 7. Schematic of coalescence under $\Delta\varepsilon_m$=1.0% and R=-∞ conditions.

It is well known that applied stresses produce non-isotropic coalescence of γ' particles during creep in superalloys. In alloys with a negative lattice mismatch

$$\delta = \frac{2\left(a_{\gamma'} - a_{\gamma}\right)}{a_{\gamma'} + a_{\gamma}} < 0 \tag{2}$$

such as SRR99 [7], one finds either bi-directional coalescence producing plates normal to the load (and casting) axis under tensile stresses or uni-directional coalescence forming needle-like structures parallel to the stress axis under compressive loads.

Let us first analyse the effect that the increase in the lattice parameters within the dendrite has on the single-crystal specimen. Since the material is nominally coherent throughout its structure, a residual stress distribution will arise between the dendrites and the interdendritic material, as shown schematically in Fig. 8. If we resolve these residual stresses into the three dendrite growth axes, we find compressive stresses inside the dendrite in all three axes. The stress component will be equal in the two secondary directions ([100] and [010]), since the stress in that plane is found to have tangential direction, and moreover one can assume axial symmetry around the dendrite axis. The stress component in the growth axis is expected to differ from the other two, though it is not obvious whether it will be larger or smaller.

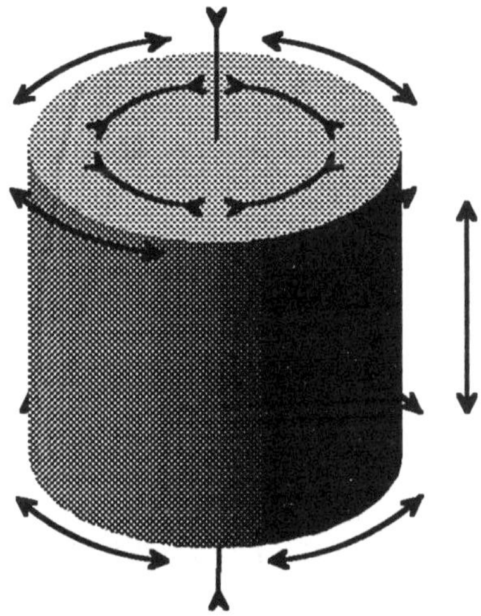

Figure 8. Schematic of residual stresses within a dendrite core.

We shall now consider the stress fields at the γ-γ′ interfaces in the three types of microstructural region in the light of the coalescence criterion developed by two of the authors [8]. This criterion, based on the anisotropy in the distribution of interfacial misfit dislocations, has recently been extended to triaxial conditions [9] and used to explain the effect of a coating on rafting behaviour [10]. The formulation of the extended criterion is reproduced in Eq. 3 and 4.

$$R_{xy} = \frac{U^{xz}}{U^{yz}} = \frac{\left(a_\gamma^2\right)_x N_x + \left(a_\gamma^2\right)_z N_z}{\left(a_\gamma^2\right)_y N_y + \left(a_\gamma^2\right)_z N_z} = \frac{f_x + f_y}{f_y + f_z} \tag{3}$$

$$R_{yz} = \frac{U^{xy}}{U^{xz}} = \frac{\left(a_\gamma^2\right)_x N_x + \left(a_\gamma^2\right)_y N_y}{\left(a_\gamma^2\right)_x N_x + \left(a_\gamma^2\right)_z N_z} = \frac{f_x + f_y}{f_x + f_z} \tag{4}$$

where

$$f_x = \left(1+\left(\sigma^x - \nu\left(\sigma^y + \sigma^z\right)\right)/E_\gamma\right)\cdot\left| a^o_{\gamma'} - a^o_\gamma \frac{1+\left(\sigma^x-\nu\left(\sigma^y+\sigma^z\right)\right)/E_\gamma}{1+\left(\sigma^x-\nu\left(\sigma^y+\sigma^z\right)\right)/E_{\gamma'}}\right| \quad (5)$$

$$f_y = \left(1+\left(\sigma^y - \nu\left(\sigma^x + \sigma^z\right)\right)/E_\gamma\right)\cdot\left| a^o_{\gamma'} - a^o_\gamma \frac{1+\left(\sigma^y-\nu\left(\sigma^x+\sigma^z\right)\right)/E_\gamma}{1+\left(\sigma^y-\nu\left(\sigma^x+\sigma^z\right)\right)/E_{\gamma'}}\right| \quad (6)$$

$$f_x = \left(1+\left(\sigma^z - \nu\left(\sigma^x + \sigma^y\right)\right)/E_\gamma\right)\cdot\left| a^o_{\gamma'} - a^o_\gamma \frac{1+\left(\sigma^z-\nu\left(\sigma^x+\sigma^y\right)\right)/E_\gamma}{1+\left(\sigma^z-\nu\left(\sigma^x+\sigma^y\right)\right)/E_{\gamma'}}\right| \quad (7)$$

The ratios R_{xy} and R_{yz} in these equations provide the surface energy ratios corresponding to precipitate faces perpendicular to the three cube directions. Only two equations are required since $R_{xz} = R_{xy}*R_{yz}$. These ratios can then be used to predict the tendency to anisotropic coalescence, as they indicate whether the surface areas in each of two directions are likely to grow or shrink with respect to each other, in order to reduce the total interfacial energy around a precipitate. Which coalescence direction is energetically favourable will depend on whether these ratios have values smaller than, equal to or larger than one. Specifically, when a ratio is larger than one this indicates a tendency to greater coalescence along the direction given by the second subindex.

The morphology predictions for SRR99 at elevated temperatures, using the triaxial extension of the coalescence criterion, can be summarised as follows. Firstly, coalescence is most favourable along the axes of the most compressive or least tensile stresses, and secondly, biaxial stress distributions which are symmetrical around a perpendicular axis have a similar effect as a uniaxial one of opposite sign along the perpendicular axis. The stress dependence of the coalescence behaviour is reversed for alloys with a positive lattice mismatch.

Thus, for an SRR99 specimen not subjected to external loading, γ' plates would be predicted to form perpendicular to a dendrite growth axis if the residual stresses are least compressive along that axis. Since this is the behaviour found inside the dendrites in the thermally cycled material (sample 5) one can deduce that in this case the compressive stresses along a dendrite growth axis are smaller than the stress components perpendicular to that direction.

We can now superimpose the mechanical strains produced by TMF onto the residual strains resulting from the chemical heterogeneity introduced by the casting process. If R=0 cycling is applied, in the dendrite arms the strain state ceases to be axially symmetric, with the component parallel to the casting direction becoming either less compressive or tensile and, as a consequence of the Poisson effect, with the other two axes becoming more compressive. This results in preferential coalescence along the axis perpendicular to both the dendrite arm and casting directions, as that is the direction of the most compressive stresses. This is in agreement with the experimental results. The application of $R= -\infty$ cycling combines the tendency to produce a needle-type microstructure (coalescence parallel to the casting axis) with that to produce the coalescence found in the absence of external loads. For the highest strain range ($\Delta\varepsilon_m$=1%) cycle the coalescence is greatly reduced, as the specimen failed after a short period of time (approximately 24 hours), which prevented the coalescence from attaining the degree expected.

Further, the experimental results allow us to estimate the size of the residual stresses, because the coalescence structures within the dendrite arms exhibit a progression from rectangular plates through to needles as the applied strain range increases. Whilst for R=0 at the highest strain range examined ($\Delta\varepsilon$ = 0.8%) the coalescence is almost entirely uniaxial, there is a significant amount of coalescence along the load axis at strain ranges of 0.5 and 0.7%. From this we conclude that, whilst being smaller than the applied ones, the residual strains are of the same order of magnitude, probably between 0.1 and 0.3%. This value can be compared, for example, to the 0.85% lattice parameter difference between two alloys based on MAR-M247 and containing 2.8 and 3.7 atomic % of tungsten [5], which can be considered as a theoretical upper bound.

In both of the R=0 tests with a 0.8% strain range, the γ' coalescence in the interdendritic material markedly differed from that found in the other tests. Specifically, coalescence also occurred in <110>-type directions, thus resulting in highly irregular precipitate structures. This was not the case inside the dendrites, and it indicates that plastic deformation may have taken place in the interdendritic regions, which is in line with results obtained in studies of crack growth in SRR99 [11]. The unexpected behaviour can be attributed to the difference in the tungsten content of the two microstructural regions, since Nathal and Ebert [6] found that at 1273 K a reduction by around 1 atomic % in the tungsten content can lower the 0.2% yield stress by as much as 120 MPa. As a consequence of the disparate behaviour in the different microstructural regions, the single crystal alloy could better be treated not as a purely monolithic system, but more like a composite.

5. Conclusions

The precipitate coalescence behaviour inside specimens of superalloy SRR99 which had undergone thermal fatigue or TMF tests was analysed both by microstructural characterisation techniques and in the light of a recently developed extension to a coalescence prediction criterion. Three distinct microstructural regions could be identified, namely dendrite cores, dendrite arms and interdendritic regions, each of which exhibited a clearly different coalescence behaviour.

The triaxial formulation of the prediction criterion based on interfacial energies was successfully shown to explain the disparate coalescence behaviour in the three morphologically-different regions.

The criterion has also been used in reverse, *i.e.* in order to determine the stress distribution required to produce a particular experimental coalescence configuration. This allowed us to subsequently estimate that the strains caused by the difference between the average lattice parameter of the dendrites and that of the interdendritic regions are around 0.2% .

In the case of high strain range tests with R=0, plastic deformation is found to occur, but it appears to be limited to the interdendritic regions. This favours the view that single crystal superalloys behave in a way that is more akin to a composite than to purely monolithic systems.

Acknowledgements - This work has been performed within the Specific Research and Development Programme of the European Commission.

References

1. Nathal, M.V., MacKay, R.A., The Stability of Lamellar γ-γ' Structures, *Materials Science and Engineering* **85** (1987) 127-138.
2. Nathal, M.V., MacKay, R.A, Miner, R.V., Influence of Precipitate Morphology on Intermediate Temperature Creep Properties of a Nickel-Base Superalloy Single Crystal, *Metallurgical Transactions A* **20A** (1989) 133-141.
3. Hopgood, A.A., Martin, J.W., Coarsening of γ'-precipitates in single-crystal superalloy SRR99, *Materials Science and Technology* **2** (1986) 543-546.
4. Feller-Kniepmeier, M., Link, T., Correlation of microstructure and creep stages in the <100> oriented superalloy SRR99 at 1253 K, *Metallurgical Transactions A* **20A** (1989) 1233-1238.
5. Nathal, M.V., Ebert, L.J., The influence of Cobalt, Tantalum, and Tungsten on the Microstructure of Single Crystal Nickel-Base Superalloys, *Metallurgical Transactions A* **16A** (1985) 1849-1862.
6. Nathal, M.V., Ebert, L.J., The influence of Cobalt, Tantalum, and Tungsten on the Elevated Temperature Mechanical Properties of Single Crystal Nickel-Base Superalloys, *Metallurgical Transactions A* **16A** (1985) 1863-1870.
7. Kuhn, H.A., Biermann, H., Ungár, T., Mughrabi, H., An X-ray study of creep-deformation induced changes of the lattice mismatch in the γ' -hardened monocrystalline nickel-base superalloy SRR99, *Acta metallurgica et materialia* **39**, (1991) 2783-2794.
8. Arrell, D.J., Vallés, J.L., Interfacial dislocation based criterion for the prediction of rafting behavior in superalloys, *Scripta Metall. Mater.* **30**, (1994)149-153.
9. Arrell, D.J., Vallés, J.L., (1995) Unpublished work
10. Bressers, J., Arrell, D.J., Ostolaza, K., Vallés, J.L. (1995)Effect of Coatings on Precipitate Rafting in Superalloys, presented at *7th International Conference on Mechanical Behaviour of Materials,* May 28 - June 2 in The Hague, The Netherlands
11. Johnson, P., Bressers, J., (1995) Characteristics of a population of naturally initiated cracks that evolve during the thermomechanical fatigue of bare and aluminide coated SRR99, presented at the *International Symposium on Fatigue under Thermal and Mechanical Loading,* May 22-24 in Petten (N.H.), The Netherlands

EFFECT OF SUBSTRATE CRYSTAL ORIENTATION ON THE MICROSTRUCTURE AND FRACTURE PROPERTIES OF NICKEL ALUMINIDE COATINGS ON SUPERALLOYS SUBJECTED TO TMF

K.M. OSTOLAZA[1,2], D.J. ARRELL[1] and J.L. VALLÉS[1]
[1] *Institute for Advanced Materials*
P. O. Box 2, 1755 ZG Petten, The Netherlands
[2] *INTA, Materials Department*
Carretera de Ajalvir Km 4; 28850 Torrejón de Ardoz-Madrid, Spain

1. Introduction

Single crystal nickel-based alloys for blades of aero gas turbines are frequently provided with a nickel aluminide diffusion coating, which produces an increase of the corrosion resistance. Thermomechanical fatigue (TMF) tests are used in the assessment of the performance of coated superalloys, as they simulate, under a controlled mechanical environment, the complex service conditions which turbine blades are subjected to.

Nickel aluminide diffusion coatings are more prone to fracture during out-of-phase than in in-phase cycling. However, a comprehensive understanding of the causes leading to their failure has not yet been reached, though both microstructural degradation and mechanical factors are known to be involved in the development of coating damage. Neither is there a consensus as to the influence of the coating on the life of the whole component [1, 2].

A wide range of microstructures has been reported in literature [3] in the case of nominally similar nickel aluminide diffusion coatings on the alloy studied here. Since their formation is governed by diffusion, two factors affecting the diffusion process are anticipated to significantly contribute to an explanation of the microstructural diversity of these coatings. The first one is the variables involved in the coating production, such as temperature, duration and activity of the packing agent, and the second is the orientation of the monocrystalline substrate.

The analysis of texture in NiAl diffusion coatings has received little attention to date. Most studies report coatings of polycrystalline character [4, 5], though there are instances of the presence of a highly textured microstructure [2, 6]. This apparent contradiction could also be a result of the factors mentioned above. Furthermore, the

J. Bressers and L. Rémy (eds.), Fatique under Thermal and Mechanical Loading, 305–313.

type and degree of crystallographic texture is generally believed to influence both crack initiation and propagation.

As coating process and substrate orientation can affect coating texture, this study concentrates on the influence of these variables on the microstructure. A study of this has the potential for optimising coating performance, particularly in improving crack tolerance.

2. Experimental Procedure

Eight specimens made of single crystal superalloy SRR99 have been selected from among those tested in a wider programme on thermomechanical fatigue of coated superalloys [7]. Each of them was machined from a separate 25 mm diameter casting and with the main specimen axis within 10° of the [001] crystallographic orientation. The specimens consisted of an 8mm gauge length region with a 12 mm*3 mm rectangular section and two threaded ends with a 16 mm diameter circular section. A pack aluminide coating process at 870°C and the standard heat treatment were applied. Two different values of θ in the gauge length region, where θ is defined as the angle between the direction normal to the coating and the [100] direction of the substrate (Fig. 1), were considered for this study, namely $\theta = 0°$ and $\theta = 35°$. The accuracy in θ for each specimen is estimated to be of $\Delta\theta = \pm 5°$. The specimens with $\theta = 35°$ in the region of rectangular section are denoted as batch A and those with $\theta = 0°$ as batch B. All specimens in the same batch were treated together during the application of the coating.

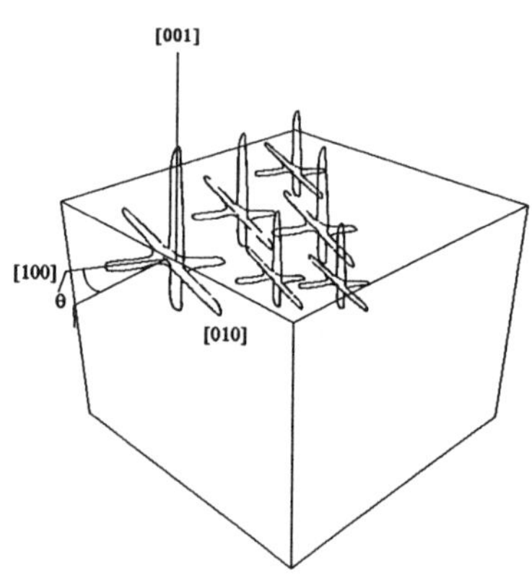

Figure 1. Definition of the angle θ relating coating and substrate orientation.

A total of six specimens (taken from both batches), the so-called mechanical samples, were fatigued under strain-controlled TMF. All the tests were continued until failure, with the same temperature cycle, extending from 573 K to 1323 K, and an R=0 mechanical cycle with a strain range $\Delta\varepsilon_m$ between 0.5 and 1%. The strain cycling lagged by 135° (out-of-phase) with respect to the temperature cycling. The test conditions for these specimens are listed in Table 1. Two other specimens (one of each batch) were used to characterise the microstructure of the coating in its as received (AR) condition.

TABLE 1. Test conditions and crack initiation.

Specimen	θ	$\Delta\varepsilon_m$ (%)	Crack initiation	
			σ_{max} (MPa)	Type
1	35°	1	≈ 860	line
2	35°	0.8	≈ 830	line
3	35°	0.7	≈ 850	line
4	35°	0.5	$\lesssim$ 500	not brittle
5	0°	1	900	line
6	0°	0.8	≈ 720	point

Coating thickness measurements using SEM` were carried out on the AR specimens, both in the gauge length region and in the end regions, in order to detect any possible variations associated to substrate orientation or the coating process. In order to correlate the differences in coating thickness with the variations in chemical composition, line scans using electron probe microanalysis (EPMA) were also performed. Additionally, X-ray difractometry using Cu K_α radiation provided texture pole figures for the family of $\{110\}_{\beta\text{-NiAl}}$ planes. Given that the penetration depth of such a radiation is typically around 25 μm, the microstructural information obtained is expected to correspond to a material layer up to that depth, which is still inside the coating. All pole figures were obtained in such a way that the main specimen axis (within 10° of $[001]_{L1_2}$) was used as reference (RD). When considering the analysis of the microstructural orientation, we describe a coating as having {hkl}-texture when the set of planes which is found to be preferentially oriented parallel to the coating surface is of the {hkl} family.

3. Microstructure of the As-Received Coating

A typical microstructure of a nickel aluminide diffusion coating (Fig. 2) consists of the following regions. The outermost layer or main coating (MC) is essentially β-NiAl. Below this layer, we find the subcoating diffusion zone (SC) where γ′, β-NiAl and carbide precipitates are detected. Deeper inside, one finds the coating-affected bulk zone (CB) which has basically the same microstructural constituents as the bulk (γ and γ′) but has experienced some degree of chemical interaction with the coating.

In this work, we shall mainly be concerned with the microstructural features found in the MC region of the AR specimens.

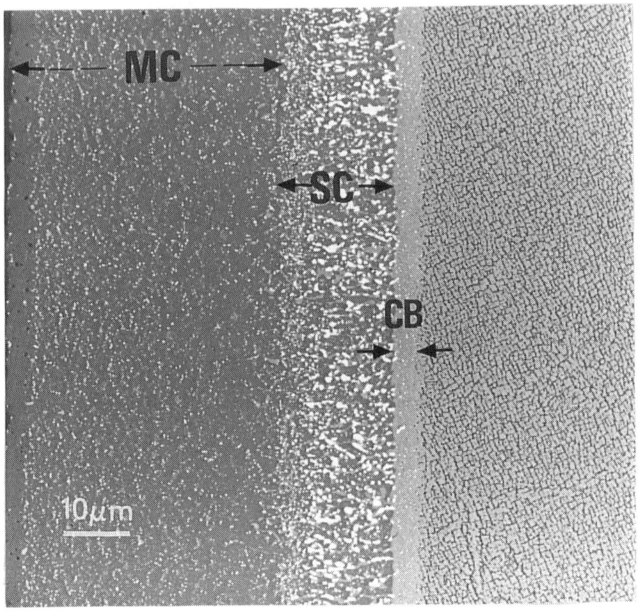

Figure 2. Typical microstructure of coating from batch A in as received condition.

The measurement of MC thickness at the specimen ends (round section) in the as-received condition showed a considerable difference between batches A and B, though no dependence on the angle θ was detected around the circular contour. Specifically, for batch A, the thickness of the MC region was 35±5 μm and in the case of batch B 52.5±5 μm. The variation in thickness between the two batches can be associated to differences in the aluminising process.

Some other microstructural differences between the two batches were also observed. Specimens from batch A had a thin γ' layer in the CB region, which was absent in those in batch B. The volume fraction and size of carbide precipitates in the SC region was also different in the two batches. Specifically, a lower degree of precipitation was detected in the coatings with a thinner MC zone.

The EPMA analysis of the chemical distribution in AR coatings (Fig. 3) showed that the coatings in batch A were Ni-rich (with atomic % ratio of Ni to Al ≈1.5), whilst the chemical composition of the coatings in batch B was closer to that of stoichiometric NiAl (atomic % ratio ≈1).

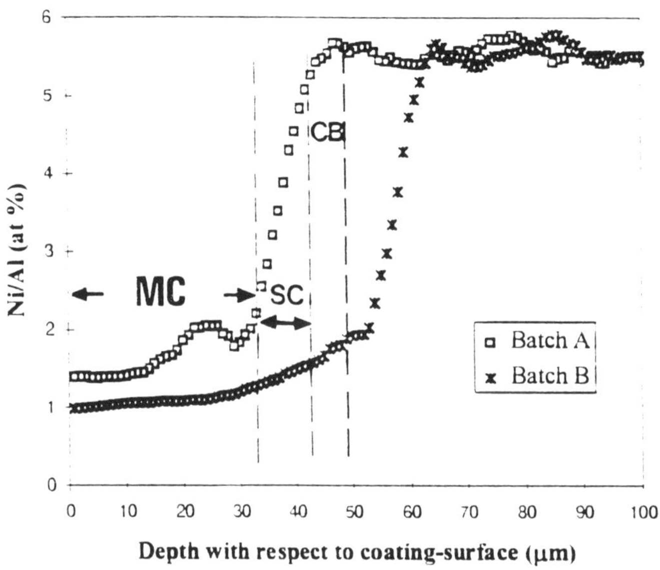

Figure 3. Typical EPMA composition ratio of Ni Al (in atomic %) versus distance to coating surface for batches A (marked) and B.

The texture analysis of the coating in the AR specimens from batches A and B (Figs. 4 and 5) showed a {111}-texture for batch A and a {011}-texture for batch B. This difference can be explained by considering the establishment of a Bain-type orientation relationship between the single crystal substrate and the coating.

At its simplest, a Bain-type of orientation relationship between a bcc phase, such as β-NiAl, and an fcc phase, such as the γ and γ′ phases in the substrate, can be understood to be equivalent to a π/4 rotation of the bcc cell (from a position initially parallel to the fcc cell) around any of the three <100> axes. For example, a +π/4 rotation around the [001] axis would result in the following orientation relationship

$$(001)_{B2} // (001)_{L1_2}; (1\bar{1}0)_{B2} // (100)_{L1_2}; (110)_{B2} // (010)_{L1_2}.$$

Since the flat surface of the coating is parallel to the $(100)_{L1_2}$ plane and a $\{110\}_{B2}$-texture is measured, in the case of batch B this rotation axis is found to be the main specimen axis. In the case of batch A (θ=35°) the surface of the coating is no longer parallel to the $(100)_{L1_2}$ plane. Thus, the observation of a $\{111\}_{B2}$ texture in the coating can only be understood by considering the fact that with a different value of θ one has a different rotation axis associated to the Bain type of orientation relationship. Specifically, the rotation axis changes from one parallel to the main specimen axis (effectively $[001]_{L1_2}$) to a perpendicular axis ($[010]_{L1_2}$).

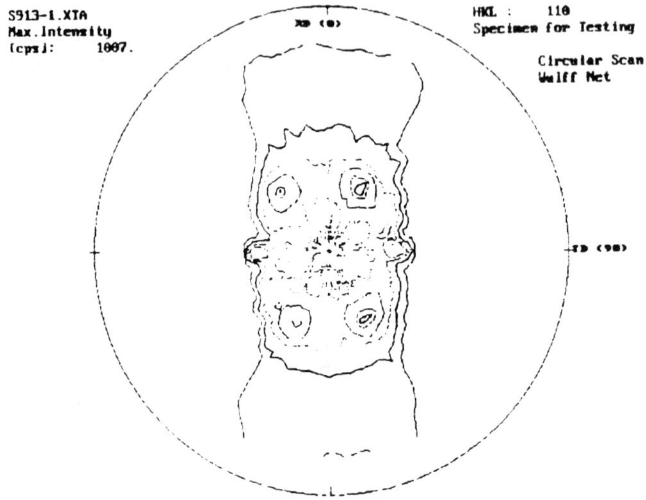

Figure 4. {110} Pole figure for AR coating of batch A (θ=35°) using the main growth axis as reference (RD).

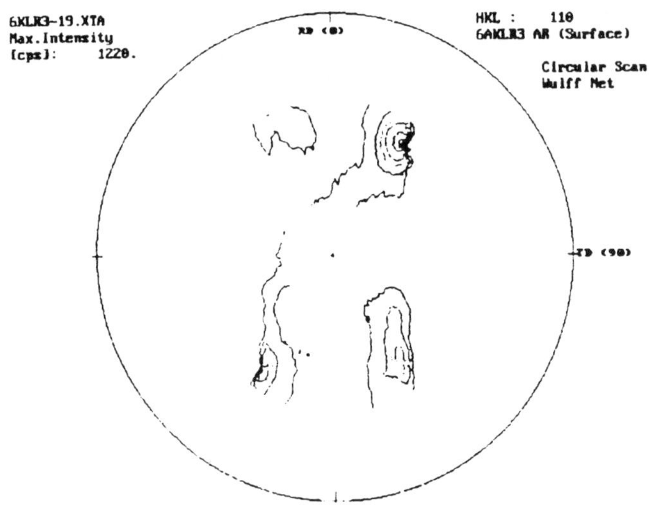

Figure 5. {110} Pole figure for AR coating of batch B (θ=0°) using the main growth axis as reference (RD).

In addition, in the $\{110\}_{B2}$ pole figure for batch A (Fig. 4) a number of extra poles is observed (a total of 6 instead of the 3 expected for a single texture). This can be explained by the presence of {111}<11$\bar{2}$> internal twinning which, in the case of a B2 structure, is related to the activation of a diffusionless (or martensitic) phase transformation [8]. Although the presence of a $L1_0$ ($c/a \approx 0.86$) phase in this coating has not yet been clearly established, these twin features are similar to those reported in hypostoichiometric NiAl [9, 10]. A formation of such an $L1_0$ phase in nickel aluminide coatings on $R= -\infty$ TMF tested superalloys, where it has been shown to have an influence on coating cracking, has recently been reported by some of the authors [11].

4. Coating Cracking under TMF

We shall now attempt to establish a relationship between the microstructure and the mechanical behaviour of the coating, particularly its fracture. To this aim, we concentrate on the initiation of brittle surface cracks under TMF. In specimens tested under R= 0 conditions, we observe minor differences in terms of the macroscopic stress for coating crack initiation between the coatings of batches A and B.

Under the assumption that the coating remains brittle for both batches and for either crack initiation type, coating fracture would occur when a critical stress is reached. Since the coating typically accounts for less than 5% of the specimen load-bearing section, it is unlikely to significantly affect the macroscopic stress to cracking. When the macroscopic stress for cracking of the coating is measured experimentally for batches A and B, one finds 850 and 900 MPa respectively (Table 1). This relatively small deviation may mask some relevant differences in the mechanism of the macroscopic response. In fact, the threshold value of the applied strain range, $\Delta\varepsilon_{th}$, above which line initiation is found can be seen in Table 1 to differ between the two batches. Specifically, in batch A the initiation type changes from line to not brittle at a value of the mechanical strain range $\Delta\varepsilon_{th} < 0.7\%$, which is lower than the $\Delta\varepsilon_{th} > 0.8\%$ found for batch B in which the transition is from line to point. Thus, the structure of the coatings in batch A is more prone to brittle fracture than in batch B. This is of particular significance, as brittle coating cracking has been shown to be highly detrimental to TMF life [1].

Two different types of crack propagation are observed in NiAl, intergranular fracture and {110}-cleavage [12-13]. The first case has been related to an inherent lack of ductility of NiAl, due to its high ductile-brittle transition temperature (DBTT). Small changes in the chemistry and microstructure of the intermetallic can modify its DBTT. In fact, an increase in ductility (as a consequence of a decrease in the DBTT) has been observed [13] when the composition deviates from that of stoichiometric NiAl or when the microstructure is refined. Whilst the coatings in both batches have a similar grain size, the more strongly non-stoichiometric coatings in batch A would be expected to be more ductile than those in batch B. Since this is not the case, the coating ductility behaviour cannot be explained by means of the DBTT effect [2].

In the case of a cleavage mechanism, the effect of the observed differences in coating texture can be considered using a simple monocrystalline model. When cracking taking place *via* $\{110\}_{B2}$-cleavage [12], a critical value of the stress σ_n normal to this crystallographic plane is required to be reached for nucleation, this normal stress is related to the applied stress, σ, by

$$\sigma_n = \sigma . M \tag{1}$$

where $M=\cos^2\lambda$ and λ is the angle between the load axis and the axis normal to any given $\{110\}$-type crystallographic plane. Alternatively, the cleavage threshold can be formulated as a function of strain. The critical macroscopic strain ε^* for triggering cleavage can be calculated using Eq. (1) as

$$\varepsilon^* = \frac{\sigma^*}{E} = \frac{\sigma_n^*}{E\,M} \tag{2}$$

where E is the elastic modulus along the load axis.

As a consequence of the differences in texture orientation discussed above, the angle λ between the applied stress and the normal to the $\{110\}_{B2}$ cleavage planes differs from batch A to batch B. Thus, the maximum value M_{max} of the angular factor M is different between the two batches. Specifically, for the coating textures found for batch A, where the load axis is effectively parallel to the $<\bar{1}01>_{B2}$ orientation (and to the $<11\bar{2}>_{B2}$ for the internally twinned structure), a maximum value of $M_{max} = 1$ can be achieved. This value is significantly higher than the $M_{max} = 0.5$ obtained for batch B, where σ // $[001]_{B2}$.

In order to produce a realistic estimate of the threshold strain for cleavage, values of the elastic modulus of NiAl are needed. Due to the lack of modulus data at elevated temperatures, we can only base our calculation on room temperature values [2]. For batch B with σ // $<001>_{B2}$ a low elastic modulus of 100 GPa is reported, and E M_{max} =50 GPa. In the case of the batch A (σ // $<\bar{1}01>_{B2}$) a much higher modulus is found, 190 GPa, and thus, E M_{max} = 190 GPa. Introducing these values into Eq.(2) shows that the microstructure of batch A has a much lower strain threshold for cleavage than that of batch B. In principle, this could also be related to the above mentioned differences in $\Delta\varepsilon_{th}$ between both coating batches. In the case of polycrystalline coatings, using the maximum value M_{max} = 1 (or even the average value M = 0.36) and E = 170 GPa, one obtains a strain threshold for cleavage which is higher than that for batch B. The presence in batch A of a significant amount of γ' has been detected, which could have an effect on the value of the elastic modulus, though this contribution was not included here.

The effect of secondary substrate orientation on coating failure, as reported in this work is of clear interest for the optimisation of turbine blade lives. Consequently, the differences in coating cracking behaviour shown here should be accounted for in the development of life prediction models.

5. Conclusions

The relationship between the microstructure of a nickel aluminide diffusion coating on superalloy SRR99 and the formation of brittle surface cracks during thermomechanical fatigue has been studied. Two factors have been shown to affect coating microstructure, namely the secondary orientation of the monocrystalline substrate and the coating process. They produce significant differences in the orientation and degree of the B2-NiAl texture. Texture has been found to have a strong influence on both the peak stress and the strain range threshold values for brittle coating cracking under out-of phase, strain-controlled R=0 TMF cycling.

Acknowledgements - We would like to thank J. Bressers for valuable discussions. This work has been performed within the Specific Research and Development Programme of the European Commission.

References

1. Bressers, J., Martínez-Esnaola, J. M., Martín-Meizoso, A., Timm, J., Arana-Antelo, M., Micro-cracking in a single crystal nickel based alloy during thermo-mechanical fatigue, in M. J.Verrilli and M. G. Castelli (eds.), *Thermo-Mechanical Fatigue Behavior of Materials*, **2**, ASTM STP 1263 (1995), American Society for Testing and Materials, Philadelphia.
2. Wood, M.I., Mechanical Interaction between coatings and superalloys under conditions of fatigue, *Surface and Coatings Technology* **39/40** (1989), 29-42.
3. Gale, W. F., King, J. E., Microstructural development in aluminide diffusion coatings on nickel-base superalloy single crystals, *Surface and Coatings Technology* **54/55** (1992), 8-12.
4. Gale, W. F., King, J. E., Microstructural degradation of aluminide coatings on single crystal nickel based superalloys during high temperature exposure, *Materials Science and Technology* **9** (1993), 793.
5. Gale, W. F., King, J. E., Decomposition of the B2-Type matrix of Aluminide Diffusion coatings on single Crystal Nickel Base superalloy substrates, *Metallurgical Transactions* **23 A** (1992), 2657.
6. Wood, M. I., Report **RAE TR 88-021**, Royal Aircraft Establishment, England, 1988.
7. Bressers J., J. Timm, S. Williams, A. Bennett, E. E. Affeldt, Effect of cycle type and coating on the TMF lives of a single crystal nickel-base gas turbine blade alloy, in M. J.Verrilli and M. G. Castelli (eds.), Proceedings of the International Conference *Thermo-Mechanical Fatigue Behavior of Materials*, **2**, ASTM STP 1263 (1995),, American Society for Testing and Materials, Philadelphia.
8. Hosford, W. F., *The Mechanics of Crystals and Textured Polycrystals*, The Oxford Engineering Science Series **32** (1993), Oxford University Press.
9. Enami, K., Nenno, S., Shimizu K., Crystal Structure and Internal Twins of the Ni-36.8 at % Al Martensite, *Trans. Journal of the Institute of Metals*, **14** (1973), 161-165.
10. Gale, W.F., Totemeier, T.C., King J.E., Modification of the B2-type matrix of aluminide diffusion coatings on nickel-base superalloys - Bulk aluminide analogues, *Metallurgical Materials Transactions A*, **26A** (1995), 949-955.
11. Ostolaza, K.M., Vallés, J.L., Bressers, J., Sánchez Pascual, A., Evolution of a nickel-aluminide diffusion coating on a single crystal superalloy under TMF, to appear in the Proceedings of Surface Treatment ' 95, Computational Mechanics Publications, Southampton, UK.
12. Chang, K. -M, Darolia, R., Lipsitt, H. A., Cleavage fracture in B2 aluminides, *Acta metallurgica et materialia*, **40** (1992), 2727-2737.
13. Ball, A., Smallman, R. E. , The deformation properties and electron microscopy studies of the intermetallic compound NiAl, *Acta metallurgica et materialia*, **14** (1966), 1349-1355.

CHARACTERISTICS OF A POPULATION OF NATURALLY INITIATED CRACKS THAT EVOLVES DURING THE THERMOMECHANICAL FATIGUE TESTING OF SRR99

P.K. JOHNSON and J. BRESSERS
Institute for Advanced Materials -JRC Petten
P.O.Box 2, 1755 ZG Petten
The Netherlands

1. Introduction

A determination of the relationship between the microstructure of single-crystal cast turbine blades and the behaviour of small fatigue cracks developed during strain-controlled thermo-mechanical fatigue (TMF) testing can allow an understanding of how some microstructural variables contribute to the scatter in TMF test lifetime. In order to model the statistical aspects of crack nucleation, growth and coalescence processes it is essential to identify which microstructural constituents, if any, are relevant to different stages of the TMF crack population's development. For example, regarding the fracture behaviour of directionally solidified IN738LC, McLean suggests that a correlation could exist between carbide spacings and crack mean-free-path [1]. With respect to polycrystalline materials it has been found that a random process of selecting crack initiation sites located individually either within grains [2] or at etch pits [3] appears to be satisfactory in simulations of the development of a crack population. In this work, the relationship between the microstructure of a single crystal superalloy and the crack behaviour induced during its TMF testing is explored.

2. Experimental

2.1. MATERIAL AND TESTS

Single crystals of superalloy SRR99 were supplied by Rolls Royce plc. in the form of cylindrical bars solidified along their main axis, which is within 10° of the <001> direction. The test pieces were subjected to a solution heat-treatment and ageing sequence. Light metallography was carried out upon both as-polished and etched samples of as-received material. Along the gauge length of the specimen the cross-section is rectangular (12 mm x 3 mm) and the sample is oriented during machining such that the gauge surfaces produced each contain one of the remaining <100> directions. After machining the sample surfaces are ground to a specified finish. One

J. Bressers and L. Rémy (eds.), Fatique under Thermal and Mechanical Loading, 315–326.

of the larger flat surfaces is monitored whilst testing by scanning it with a light-microscope based image aquisition system as described by Bressers *et al.* [4].

Strain-controlled TMF tests with mechanical strain ranges as indicated in Table 1 were performed with a 573K-1323K temperature range, and with heating and cooling rates of 25K/s and 12.5K/s respectively. The strain cycle follows the temperature-cycle with a phase shift of 135° (see Arana *et al.* [5]). Reflected light microscope images, showing surface crack initiation events, small crack growth and crack coalescence events, are obtained during the strain-controlled TMF tests. Images collected following pre-determined periods of cycling were stored in order to carry out post-test general examination and crack length measurements. Both load and strain are recorded throughout the tests as functions of temperature and time. Failure is reached when a 33% drop in the 'saturation' value of the maximum stress has occurred.

TABLE 1. Description of the test conditions

Sample	Mechanical strain range (in %)	R(strain)	Cycles to failure
1	0.7	$-\infty$	21820
2	1.0	0	1981
3	1.0	$-\infty$	1430

2.2. SURFACE SPACINGS BETWEEN MICROSTRUCTURAL CONSTITUENTS

A light metallography examination of the surface of the piece removed in the machining of the test-piece gauge reveals two main regions: the dendrite cores, nominally defect-free, and a region of dendrite arms and interdendritic material (Fig. 1). The regions can be differentiated both by mapping defect densities and by simply polishing and etching. The distribution of these areas is sensitive to small mis-orientations (i.e. <10°) from the surface corresponding to an ideal {100} plane (other than the {100} plane with a normal parallel to the cast direction). Hence, in the initial condition, the surface regions of the casting, in which fatigue cracks originate during the current tests, can reasonably be divided into sound and defective material.

An estimated mean primary dendrite spacing λ_1, found by counting dendrites in a known area of section perpendicular to the solidification direction and by treating them as having square profiles, gives a centre-centre spacing of 440±10 μm. The secondary dendrite arm spacing, λ_2, was estimated as ~130 μm. Both values were incorporated into a grid to aid in the interpretation of features in the images recorded of a surface of a sample that had been TMF tested.

Measurements of the spacing between carbide colonies were also made. In Fig. 2 typical carbide particle colonies are shown in the castings, and a few voids are also visible. The procedure to measure the spacings, using a light microscope, is sketched in Fig. 3a. The total length of the window used for the carbide spacings was 45 mm. The boundaries of a carbide colony were estimated by eye, and the distance between the centres was recorded. The resulting distribution of spacings is shown in Fig. 4.

Dendrite core and interdendritic material were differentiated via etching. Dendrite core diameters were estimated as shown in Fig. 3b. The results are shown in Fig. 5. The existence of connected groups of dendrites leads to a skewed set of spacings.

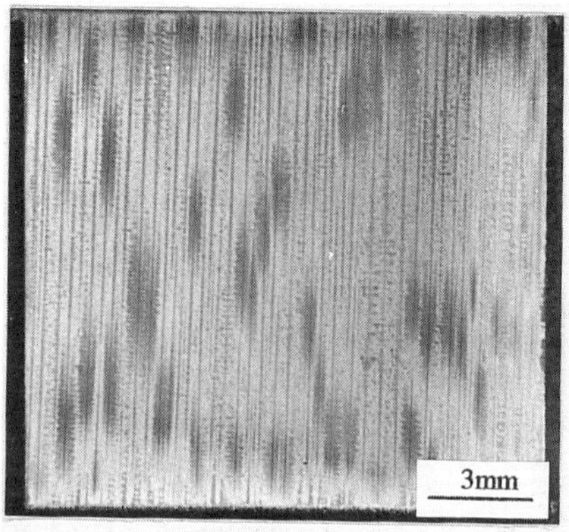

Figure 1. Etched surface of an un-tested off-cut taken from sample 1, previously mating the gauge surface, showing the dendrite cores (dark contrast) and interdendritic regions (light contrast).

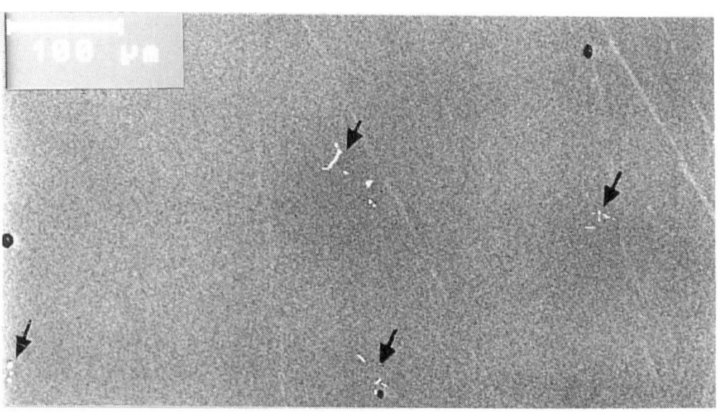

Figure 2. Backscattered SEM (scanning electron microscope) image of a section perpendicular to the load axis.Clusters of bright particles, indicated by the arrows, are carbide particle colonies developed during the casting process.

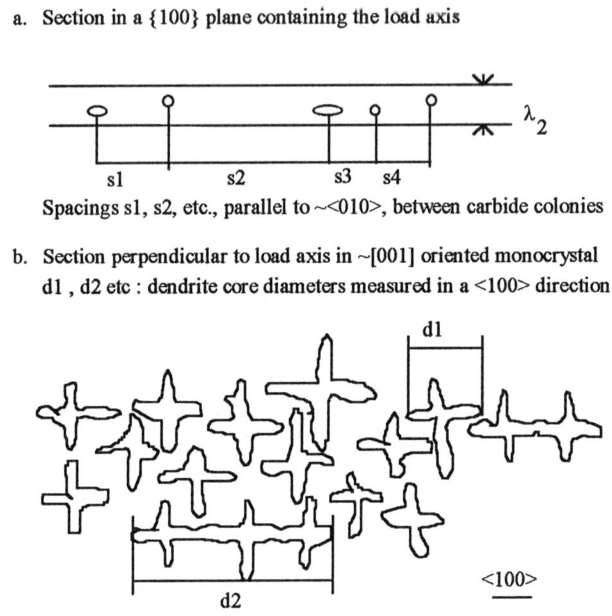

Figure 3. Measurement schemes: a) for defect spacings using the secondary dendrite arm spacing to define a limiting element of the surface and b) for dendrite core diameters..

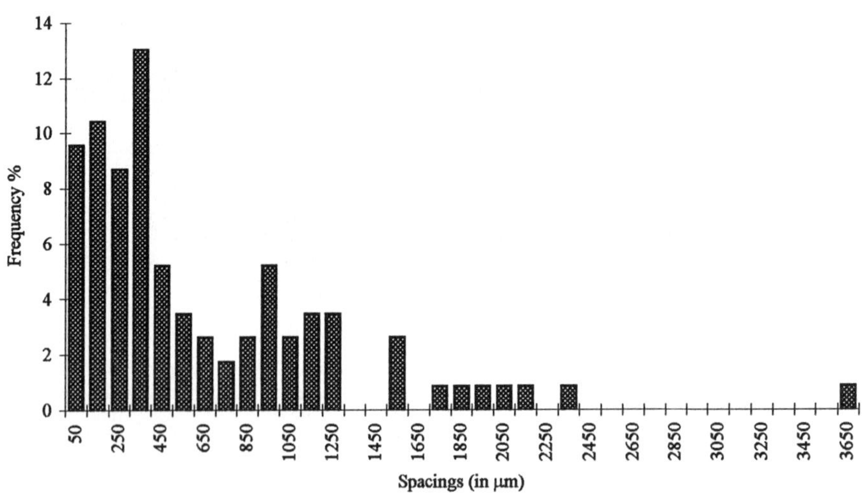

Figure 4. Distribution of spacings between carbide colonies.

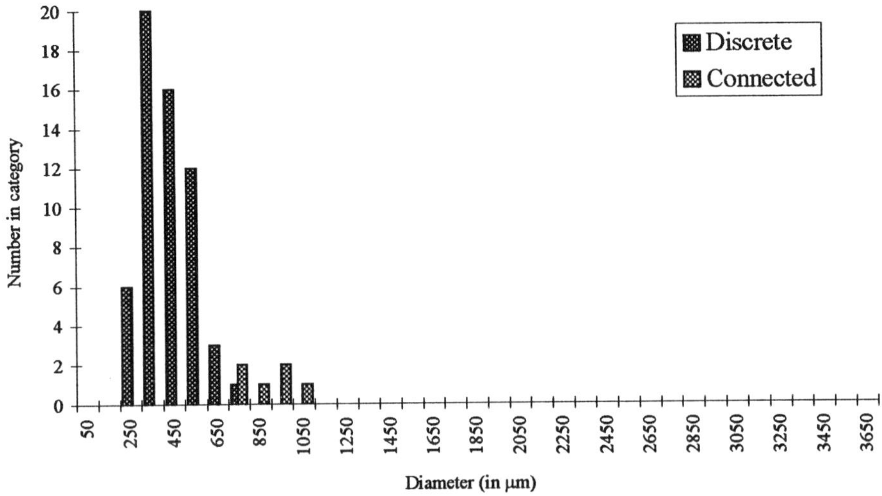

Figure 5. Dendrite core diameters measured in a cube direction (Fig. 3b).

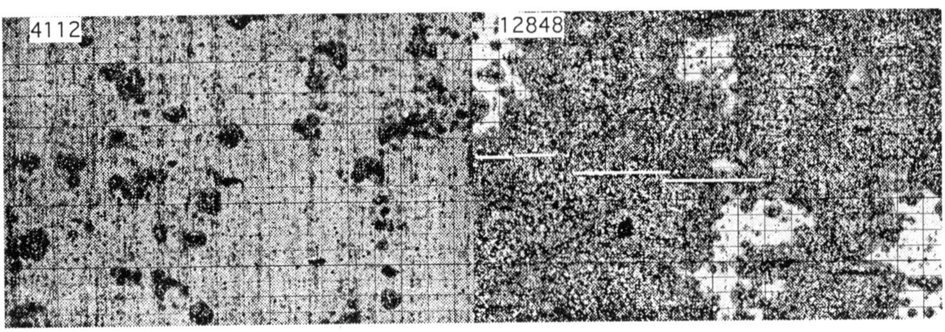

Figure 6. The same 4 mm x 2.6 mm surface area of sample 1 at two cycle numbers (indicated top-left). The white lines indicate the site of the main crack. The grid is 220 μm x 130 μm.

2.3. DISCONTINUOS OXIDE GROWTH & CRACK COALESCENCE EVENTS

As shown in Fig. 6, during the TMF test at a mechanical strain range of 0.7% the sample surface mainly presents copiously oxidised areas (dark contrast) and areas covered with a relatively thin oxide layer. A third component of the surface structure, consisting of mode-I opening cracks, is superimposed.

A small area of the surface of sample 1, corresponding to that shown in Fig. 6, was studied at a pixel resolution of 4.5 μm x 3 μm at intervals of typically 450 cycles during the TMF test. A microstructural parameter based grid ($\lambda_1/2 \times \lambda_2$), also visible in Fig. 6, was overlaid on the images during post-test analysis. Mechanical data collected during the test can be correlated directly with the images of the surface.

The changing stress conditions at the maximum temperature as measured in the course of the test are plotted in Fig. 7. The qualitative changes in the stress response are indicated by means of numbered arrows. Initially the stresses change rapidly from compressive to become increasingly tensile (#1). The whole surface appears to be crossed by slip bands. Temporarily the stress value reaches a constant value (#2), and the earliest 'localised' damage becomes visible at this stage. Both the stress at maximum temperature and the minimum stress start to increase again from #4. Since at the same stage the stress range is decreasing gradually (#3) the sample has undergone an overall softening. This softening appears at an intermediate stage in the test at ~10000 cycles leading to a shift in the stress from the plateau at #4 to a new condition at #5. Surface observations suggest that this mechanical transition is related to the spread of damage originally confined to interdendritic/defective surface material into previously 'sound' surface material associated with the dendrite cores. The sample fails at #6 as the result of fatigue crack growth.

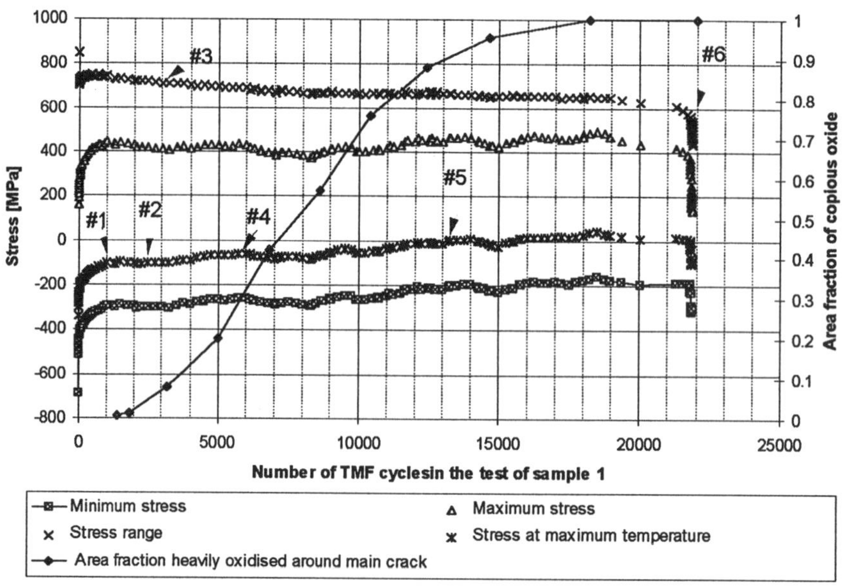

Figure 7. Increase in surface area fraction of regions of dark contrast, generally corresponding to thick oxidation products, simultaneously plotted with the changes in the applied stress during testing.

In order to determine whether important stages in the evolution of a naturally initiated crack population can be related to the microstructure, an initial analysis of some available crack growth statistics was made. As part of a previous project mode-I opening surface crack lengths were measured, from surface images similar to those shown in Fig. 6, for the three tests in Table 1. The data consisted of surface crack lengths, measured perpendicular to the load axis, and a record of the occurrence of coalescence events between crack segments. The data available has been processed, as illustrated in Fig. 8, to give a set of 'coalescence lengths'. The results were separated into coalescence lengths between the first crack segment pairs in a sequence CL2, and all other coalescence events CL3, CL4 and CL5 as defined in Fig. 8. The intention was to arrive at an estimate of the distance between the surface initiation sites of the cracks recorded. It was assumed that cracks would grow symmetrically prior to coalescence. The data set contained 70 coalescence events. The resulting distribution of crack coalescence events is shown in Fig. 9.

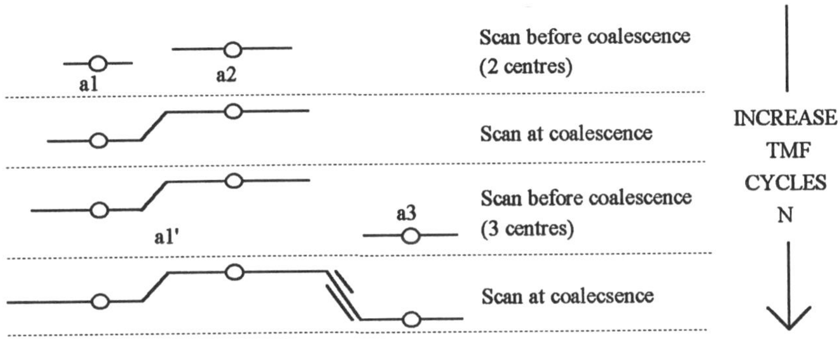

Definition of coalescence length CLn : $CL2 = (a1 + a2) / 2$

$CL3 = (a1' + a3) / 3$ etc.

where a1, a2, a1' and a3 are surface crack lengths measured perpendicular to the load axis.

Figure 8. Definition of coalescence length.

The surface of the test piece is characterised by the appearance of a number of surface structures. A description of the behaviour of some of these features during a TMF test provides help in determining the nature of the damage process. For example some features appear as follows. Small diamond-shaped areas of dark contrast, with diagonals originally ~50 μm long, were developed amidst slip-bands that decorated the whole gauge section early in the test, i.e. before #2 in Fig. 7, when the compressive stresses were still high. Boundaries between these differently contrasting parts of the surface often exhibited a simple geometry, suggesting that the boundaries were associated with the sample crystallography. The test-piece orientation suggests that the boundaries, as they cut through the sample surface, may be based upon {111} slip traces. With further cycling these dark contrasting regions generally expand and change their two dimensional shape. In some cases this expansion is initially isotropic. The expansion often involves short jets directed towards other surface features. In some areas such features are separated in the surface plane typically by less than ~100 μm and may form small clusters. The coalescence process generally upsets the geometrical shape developed earlier, although parts of the geometrical boundaries are maintained until the end of the test. Beyond 3000 cycles the damage zones that appeared first start to coalesce. Coalescence events between damaged surface areas at this stage frequently occur between damage sites spaced apart by approximately $\lambda_1/2$ in directions perpendicular to the load axis, and at approximately λ_2 in directions parallel to the load axis. Large continuous areas of damage are developed, interspersed with relatively un-damaged areas. However, above 10000 cycles the dark contrast spreads across the remaining sound surface areas, sometimes as narrow fingers emanating from cracks, along directions consistent with {111} traces, or as damage extending in similar crystallographic directions but arising from damage areas that are the result of the earlier joining together, along the same direction, of a cluster of closely spaced damage sites.

At the end of the test the dark contrasting areas are found to be heavily oxidised. The relationship between the oxidation and plasticity is clearly shown when slip bands that have developed late in a TMF test, and are still relatively lightly oxidised, have been found to be decorated with numerous titanium ,and oxygen, containing particles.

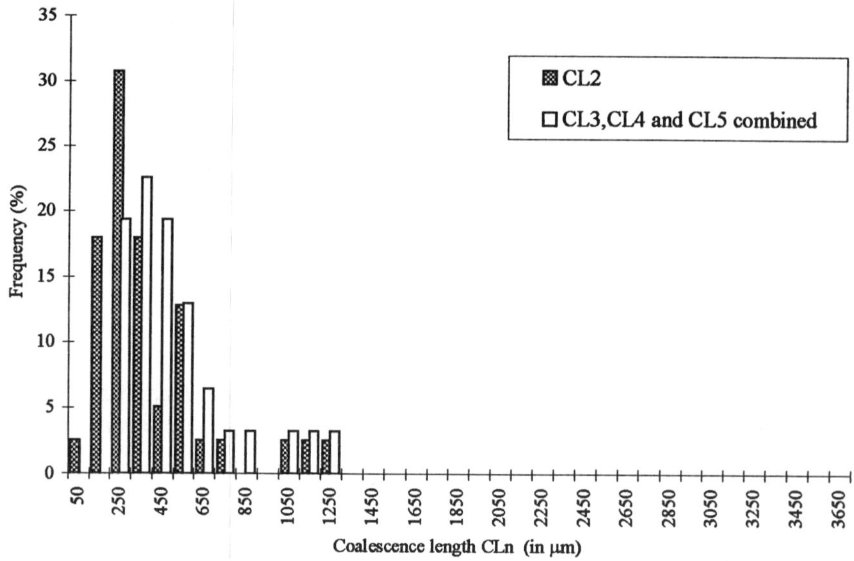

Figure 9. Distribution of coalescence lengths as defined in Fig. 8.

3. Discussion

3.1. CAST MICROSTRUCTURAL PARAMETERS AND THE COALESCENCE OF MODE-I OPENING CRACKS

During the directional solidification of the superalloy SRR99, dendritic γ-Ni separates first, in terms of the local freezing time, followed by any carbide phases that might be present [6]. Shrinkage microporosity will form last. The carbide particles and micropores will be situated within a complexly shaped volume between any adjacent dendrites. The spacings between pores and carbide particles will cover a wide range of values. However, the positions of these features that provide weak interfaces and voids will show a periodicity related to the periodic dendrite structure. When cracks are formed at such weakly bonded features or stress raisers situated at or near the surface of the test piece then the initiation sites of surface cracks might be expected to show modal spacings at the most common defect spacings, which will be related to the general periodicity of the cast structure. When loaded as in the TMF tests described here, one length of interest for describing coalescence event between mode-I cracks, will be λ_1.

In Fig. 9 the distribution of distances over which a crack will travel prior to meeting another crack shows a statistical mode below the mean average dendrite spacing λ_1 value of 440 μm. When both sets of data in Fig. 9 are combined, the mean coalescence length is 430 μm. The most frequent coalescence lengths however are in the range 200 μm-300 μm, and can correspond to either the dimensions within an interdendritic space, or the spacing between interdendritic material separated by a secondary dendrite arm.

It may be informative to compare the distribution of coalescence lengths in Fig. 9 with the distribution of spacings of one type of defect, for example the carbide colony spacings, in Fig. 4. Although the volume fraction of the carbide phases in SRR99 is relatively small, in fact the separation of the carbide from the liquid is undetected by thermal analysis [6], the presence of sometimes coarse and locally concentrated clusters of what are generally considered as an embrittling constituent cannot be ignored. This is especially the case if one is considering sources of scatter in test lifetimes. Crack initiation at carbide particles in superalloys is well known and documented [7]. In the type of tests described here the crack initiation sites are not always positively identified with any particular feature during fractography, principally due to the heavy oxidation of the crack surfaces. Comparing the carbide colony spacings in Fig. 4 with the distribution in Fig. 9, it can be seen that there is a poor agreement between the two sets of data at the lowest categories of spacings. This is in part due to the adoption of a 'crack initiation length' below which a crack is not defined, such that some of the smallest segments in a crack coalescence event sequence will be missing. At the same time a definition of discrete colonies of carbide particles also needs clarification. There are stereological problems, for example, such that some of the colonies recorded in the statistics appear in section as individual platelets which may or may not have been associated with a 'colony'. However, it does appear that there are fewer cracks with crack spacings of the size of the smallest recorded carbide structure spacings then one would expect if cracks initiated at random in the carbide population. However, if the coalescence between pre-crack damage of the type observed on the surface of the test piece is considered then closely spaced carbide particles and microvoid clusters can account for the smallest, sub-$\lambda_1/2$, damage coalescence spacings. During the monitoring of the spread of damage in the area of Fig. 6 it was found that damage coalescence events in general occur at a factor of more than 100 times more frequently than crack coalescence events. The significance of this observation is discussed in section 3.2 below.

A comparison can also be made between the coalescence length distribution in Fig. 9, and the dendrite core diameters in Fig. 5. By inspection there are similarities between the two distributions. The existence of 'interconnected' dendrites can explain the occurrence of some long range coalescence events, and suggest that crack free paths of ~1 mm can exist. The presence of this tail in the distribution supports the proposition that the crack initiation sites show a tendency for clustering.

3.2. SURFACE DAMAGE INITIATION

In the TMF tested sample 1 there are relatively few cracks which are initiated at the surface compared to the total numbers of defects that reside at or near the surface. The defects which are the most severe stress raisers are candidates for providing crack initiation events at the highest rates. However, the inhomogeneous distribution of damage, on the scale of the sound and defective areas of the surface, and the ranges of inter-damage-site spacings, often of the order of $\lambda_1/2$ or λ_2, suggest a correspondence between individual defects and the discrete damage structures visible at the surface early in the TMF test. The defects will be situated at and below the surface of the material, the sub-surface defects capable of mechanically disturbing the surface will indirectly be revealed.

Some damage initiation involved localised sites of intense plastic damage at a time when the bulk of the sample is not undergoing macroscopic cyclic yielding. Identification of the plastic zones at the surface, perhaps bound to sub-surface voids or other defects, allows some pre-crack damage behaviour to be better understood, and to help clarify some aspects of 'small crack behaviour'. Surface observations suggest that some cracks may also be developed more directly from crack-like defects by a time independent process, whereas other cracks develop via a pre-crack damage state. Mixed crack initiation processes, and their correspondingly mixed kinetics, are to be expected.

3.3. DAMAGE INITIATION AND CAST MICROSTRUCTURAL PARAMETERS

A partial interpretation of the development of the surface features described in section 2.3 during the TMF cycle applied is as follows. Individual defects act as local stress raisers. The superalloy yields locally, for example between a sub-surface void and the surface. Repetitive yielding disrupts the surface oxide products and various transport mechanisms allow further oxidation to take place. The locally thickened oxide, having formed heterogeneously by the interaction of localised plasticity with the environment, grows discontinuously via the movement of the interface between the thickened oxide and the non-yielding parts of the surface. A high strain gradient exists across the interface which may consist of active persistent slip bands (PSBs). The traces of the active slip planes define the geometrical boundaries that characterise the observed damage process in the earliest stages as described in section 2.3.

The shortest distances between the damage sites that emerge as the test progresses, are comparable with the smallest inter-carbide and inter-void spacings. Coalescence between the damage on the smallest scale seen creates structures that are often within areas of the order of $\lambda_1/2 \times \lambda_2$ in surface area. The observed damage actually occurs in a volume at the surface, and one would expect the depth of this volume to be load dependent. However, surface observations can be used to determine the crack initiation probabilities within individual, visibly damaged, cells of dimensions $\lambda_1/2 \times \lambda_2$ by defining a local population of damaged cells. By determining the probabilities as a function of time, empirical crack initiation laws can be established for both defective and sound surface material, from the data available.

It is very convenient to consider crack initiation as a fracture event within a microstructurally defined surface unit. A good initial choice for such a cell has dimensions of $\lambda_1/2 \times \lambda_2$ in the situation discussed here. The cell will either fail of not fail. This hypothesis can be the basis for a probabilistic crack initiation model. Examination of Fig. 9 suggests that 'cracks' initiating with mean spacings below $\lambda_1/2$ are infrequent, and this fact supports the proposed use of the aforementioned limiting cell size in which to define either defect state to crack state, or damage state to crack state, transformations.

3.4. STRESS RESPONSE AND DAMAGE IN THE TEST OF SAMPLE 1

The decoration of PSBs by oxidation products allows PSB movement to be traced. When associated with either sub-surface pores or perhaps with undeformable particles embedded in the sample surface, i.e. carbide particles or their oxidation products, associated PSB activity is inferred from the visible geometrical structures developed and the crystallographically directed manner in which many damage sites are seen to link up. From 1000 to about 10000 cycles, sample 1 appears to have reached a saturation state involving only microplastic phenomena. Following 10000 cycles the sample is progressively macroscopically yielded as is revealed by the appearance of PSBs on a larger scale than were present earlier in the test. The change in plastic behaviour appears to take place either because the surface cracks are large enough to drive PSBs across the dendritic parts of the structure and beyond, across the whole sample, or as the result of the oxide coverage having reached a critical state such that stress relaxation at the surface becomes difficult. A build-up of a resistance to the applied compressive stress, by the oxide, may be enough to marginally raise the tensile stresses in the TMF cycle. Some PSBs formed in the pre-crack damage process can be driven from the defective and damaged surface regions into the remaining sound material.

4. Conclusions

The distances over which crack segments grow prior to coalescence appear to be related to the primary dendrite spacing.

Both cracks and localised heavy oxidation are produced in the interdendritic regions that are bisected by the sample surface. The two phenomena appear to be related to localised plasticity and to the presence of defects at or near the sample surface.

The decoration of persistent slip bands by oxidation can be used to help follow the spread of plastic damage. From observations of the spread of damage from interdendritic to dendritic regions in the test of one single crystal it appears that a simple cyclic saturation state is not achieved.

Acknowledgements - Thanks are due to R. De Cat for performing TMF tests and making crack length measurements, D.Arrell for invaluable help in the setting-up of a facility for viewing digitised images, M.Arana and J.L.Vallés for assistance in the preparation of the paper, and to both M.Arana and K.Ostalaza for useful discussions. PKJ is grateful to the European Commission for provision of financial support under the Human Capital and Mobility Program.

References

1. McLean, M., *Directionally Solidified Materials For High Temperature Service*, The Metals Society, London; (1983), pp. 200-201.
2. Weiss, J. and Pineau, A., Microstructurally Based Simulation of Multiaxial Low-Cycle Fatigue Damage Of 316L Stainless Steel in terms of the Behaviour of a crack population, in K.-T. Rie (chief ed.), Proc.Conf., *Low Cycle Fatigue and Elasto-Plastic Behaviour of Materials -3* ,Berlin 7-11 Sept 1992, Elsevier, pp.82-87.
3. Kitagawa, H., Nakasone,Y. and Miyashita ,S., Measurement of Fatigue Damage by Randomly Distributed Small Cracks Data, in *Fatigue Mechanisms: Advances in Quantitative Measurement of Physical Damage*, ASTM STP 811, Lankford et al. (eds.) ASTM, (1983), pp.233-263.
4. Bressers J., Estevas-Guilmain, J., De Cat, R., Knight, S. and Ridge, J., An Automated Computer Vision System for Monitoring the Initiation and Growth of Microcracks, in Einsworth, R.A., and Skelton, R.P. (eds.) *Behaviour of Defects at High Temperature*, ESIS 15, Mechanical Engineering Publications, London, (1993), pp.85-98.
5. Arana, M., Martinez, J.M. and Bressers, J.(1995), this volume.
6. Wills, V.A. and McCartney, D.G., A comparative study of solidification features in nickel-base superalloys: microstructural evolution and microsegregation, *Materials Science and Engineering*, **A145** (1991), 223-232
7. Gell,M. and Leverant, G.R., The Fatigue of the Nickel-base Superalloy, Mar-M200 in Single Crystal and Columnar-Grained Forms at Room Temperature, *Transactions of the Metallurgical Society of AIME*, **242**, (1968),1869-1879.

THERMAL-MECHANICAL FATIGUE BEHAVIOUR OF NiCr22Co12Mo9

B. KLEINPASS, K.-H. LANG, D. LÖHE and E. MACHERAUCH
Institute of Material Science and Engineering I,
University of Karlsruhe,
D-76128 Karlsruhe, FRG

1. Introduction

Every start up and shut down of a system operating at high temperatures causes in the components transient temperature gradients through which complex strain and stress fields occur and local damage may be produced. For example, the lifetime of combustion chambers in gas turbines is often limited by damage resulting from start-stop-cycles and changes of the operating temperature [1,2]. Frequently, data from isothermal strain controlled fatigue tests are used to predict the deformation behaviour and the lifetime of components exposed to thermal-mechanical fatigue (TMF) [3]. This procedure, however, is connected with large uncertainties, especially if the cyclic stress-strain response and the microstructural changes are different under isothermal and thermal-mechanical fatigue. In the present study, the TMF behaviour of the solid solution and carbide precipitation hardened nickel base superalloy NiCr22Co12Mo9, which is commonly used as sheet material in gas turbines, is investigated. The TMF life determined is compared to the lifetime in isothermal fatigue tests.

2. Material

The chemical composition of the investigated nickel-base alloy NiCr22Co12Mo9 (trade names Nicrofer 5520 Co and IN 617) was 0.069 C, 21.9 Cr, 11.65 Co, 8.75 Mo, 1.15 Al, 0.74 Fe, 0.46 Ti, balance Ni (all quantities in wt.-%). Round bars with a diameter of 20 mm were annealed at 1475K and water quenched. Specimens were machined with a cylindrical gauge length of 10 mm, a gauge diameter of 7 mm and conical gripping heads. The microstructure of the material consisted of firmly twinned grains with a low dislocation density, incoherent M_6C carbides and Ti(C,N) carbonitrides [4]. The mean grain size was 160 μm.

J. Bressers and L. Rémy (eds.), Fatique under Thermal and Mechanical Loading, 327–337.

3. Experimental Details

The mechanical loading was applied by a closed loop servohydraulic testing machine with 63kN capacity. Strain measurements were performed with a capacitive extensometer. For heating, an inductor generator was used. Cooling was achieved by thermal conduction into the specimen grips and by blowing compressed air at the specimens by means of a proportionally controlled valve. The temperature was measured with a Ni-CrNi thermocouple spot welded in the middle of the gauge length.

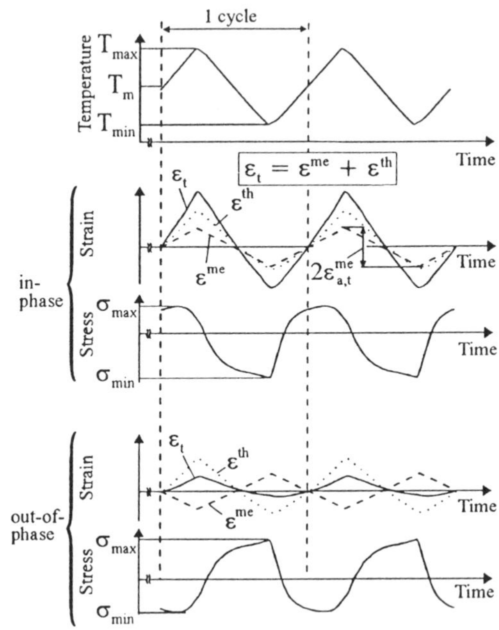

Figure 1. Temperature, strain and stress vs. time during inphase and out-out-phase loading.

Fig. 1 illustrates the triangular temperature-time as well as strain-time courses chosen and the resulting stress-time course. The specimens are allowed to expand and contract freely at zero stress during the initial heating from ambient temperature to the mean temperature T_m of the particular thermal cycle and during the first few cycles, the closed loop control operating in stress control. After this period the controller is switched over from stress to total strain control at T_m, and the first thermal-mechanical cycle starts with heating up the specimen to T_{max}. The closed loop system allows for the thermal strain-time course $\varepsilon^{th}(t)$, which results from cyclic expansion and contraction and which is accurately known from the first pure thermal cycles, superimposing it with a mechanical strain-time course $\varepsilon^{me}(t)$. Hence, the total strain $\varepsilon_t = \varepsilon^{me} + \varepsilon^{th}$ is the signal which is controlled by the testing machine. In-phase (middle part of Fig. 1) and out-of-phase (lower part of Fig. 1) relations between temperature and mechanical strain ε^{me} were realized. The stress measured is the material response to thermal and mechanical loading. During in-phase TMF, tensile stresses occur at high and compressive stresses at low temperatures whereas during out-of-phase TMF stresses occur in the opposite way.

The minimum cycle temperature T_{min} was 473K in all tests. T_{max} was varied between 873K and 1473K. Two amplitudes of mechanical strain, 0.25% and 0.625%, were chosen. The temperature rate, dT/dt, was 14K/s resulting in cycle periods ranging from 60 s to 145 s and mechanical strain rates ranging from $6.9 \cdot 10^{-5}\ s^{-1}$ to $4.2 \cdot 10^{-4}\ s^{-1}$

4. Experimental Results and Discussion

4.1. CYCLIC DEFORMATION BEHAVIOUR

Fig. 2 shows stress-temperature (left) and stress-strain (right) hysteresis loops determined at $N = N_f/2$ from in-phase (top) and out-of-phase (bottom) TMF tests at T_{max} = 873K and 1323K. The mechanical strain amplitude was 0.625%. As the materials resistance against cyclic plastic deformation decreases with increasing temperature, the maximum and minimum stresses measured during TMF at T_{max} = 1323K are much lower than those at T_{max} = 873K.

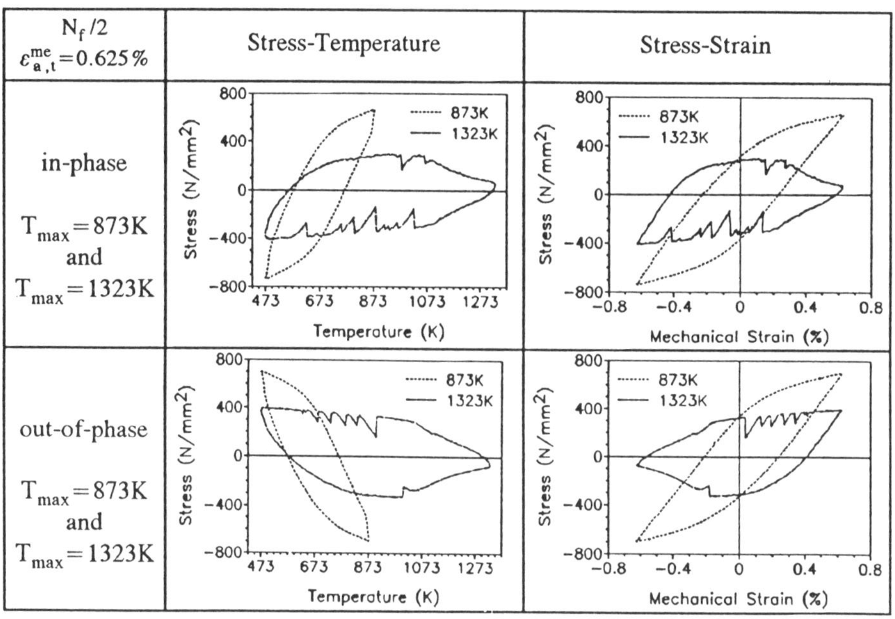

Figure 2. Stress-temperature and stress-strain hysteresis loops developing during in-phase and out-of-phase TMF at T_{max} = 873K and 1323K.

During TMF at T_{max} = 873K always smooth hysteresis loops are observed. Contrarily, the development of the stress-strain loops during TMF at T_{max} = 1323K is characterized by rapid stress drops in distinct temperature ranges and by stress relaxation at very high temperatures. The stress discontinuities which are consistent with repeated tests are produced by dynamic strain ageing effects resulting from interaction between moving dislocations and diffusing alloying atoms. Similar phenomena were observed during isothermal quasistatic and cyclic tests with the same material at similar strain rates and at temperatures ranging from 450K to 950K

[5-7]. In the case of TMF, the range in which stress drops occur is shifted by about 150K to higher temperatures [8]. However, as already stated, a sufficiently high maximum cycle temperature is a precondition for the appearance of discontinuous yielding, the latter being not observed during TMF at T_{max} = 873K although the above mentioned temperature range is partly reached. The rapid stress drops occur mainly during cooling from T_{max} to T_{min}, where compressive and tensile stresses prevail in the case of in-phase and out-of-phase TMF, respectively. Furthermore, during cooling the discontinuous yielding appears down to lower temperatures in contrast to heating. This indicates that at a given temperature the number of diffusing alloying atoms is smaller during the heating phase than during the cooling phase. This effect may be caused by a shift between the temperature ranges in which dissolution of small carbides during heating up and their reprecipitation during cooling down of the specimens occur, respectively. As shown by [8], a sufficiently high T_{max} is a prerequisite for this cyclic dissolution and reprecipitation process. Therefore, no discontinuous yielding is observed during TMF at T_{max} = 873K.

Fig. 3 shows the stress amplitudes σ_a with number of cycles during in-phase TMF. At $T_{max} \leq$ 1223K cyclic hardening is observed, which is the more pronounced the lower T_{max} is. For example, at T_{max} = 873K, the maximum tensile stress amounts to 690 N/mm^2 after approximately 700 cycles and thus exceeds at 873K the ultimate tensile strength of 590N/mm^2 at 873K of the material in the as received state [5]. During TMF at T_{max}= 1323K and 1473K the values measured hardly change. The maximum stress amplitudes monotonously decrease with increasing T_{max}. The drop of σ_a at the end of TMF life is combined with the onset of macro crack propagation.

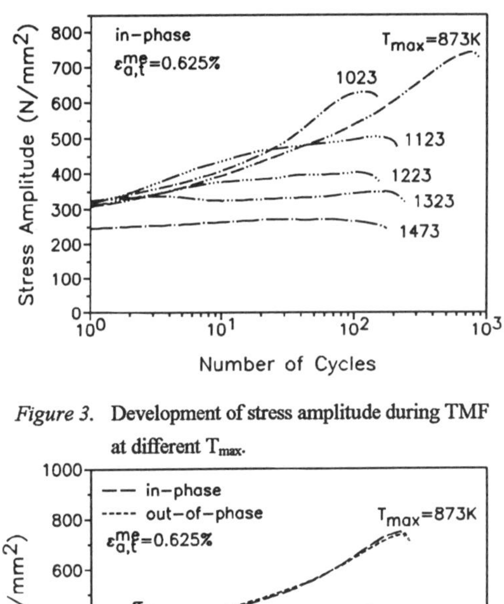

Figure 3. Development of stress amplitude during TMF at different T_{max}.

In Fig. 4 a comparison is made between the development of the stress amplitude σ_a and the mean stress σ_m during in-phase and out-of-phase TMF, respectively, at three different T_{max}. The variation of the stress amplitudes are very similar for both phase relations, whereas the curves of the mean stress markedly

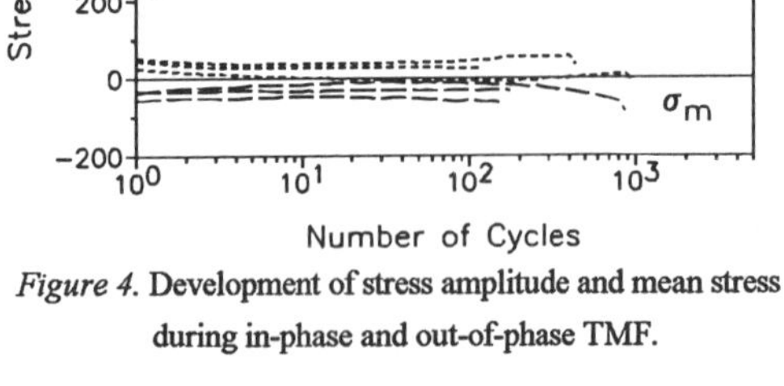

Figure 4. Development of stress amplitude and mean stress during in-phase and out-of-phase TMF.

differ from each other. Compressive mean stresses appear during in-phase loading and tensile mean stresses during out-of-phase loading. The temperature-dependent variation of the material strength in a single TMF cycle results in $\sigma_{max}<|\sigma_{min}|$ and $\sigma_m<0$ during in-phase TMF and $\sigma_{max}>|\sigma_{min}|$and $\sigma_m>0$ during out-of-phase TMF. In all cases the values of the mean stresses remain relatively small and only slightly alter during TMF life.

4.2. LIFETIME

Failure of the specimens was identified with a 10% drop of the maximum stress relative to the σ_{max},lgN-course extrapolated from the cyclic deformation behaviour prior to macro crack initiation. In Fig. 5, a comparison is made between TMF lives evaluated by in-phase tests and out-of-phase tests, respectively, at different T_{max} applying mechanical strain amplitudes of 0.625% (upper part) and 0.25% (lower part). For all loading conditions applied, the highest lifetimes are measured at T_{max} = 873K. In the range of maximum temperatures between 1023K and 1223K (at $\varepsilon_{a,t}^{me}$ = 0.625%) and 1123K and 1323K (at $\varepsilon_{a,t}^{me}$ = 0.25%), respectively, the number of cycles to failure is significantly lower but much less influenced by T_{max}. In these temperature ranges the lifetimes markedly depend on the phase relation between temperature and mechanical strain. Out-of-phase TMF results in significantly higher fatigue lives than in-phase TMF. At the highest T_{max} investigated, the influence of the phase relation on TMF life is again very small.

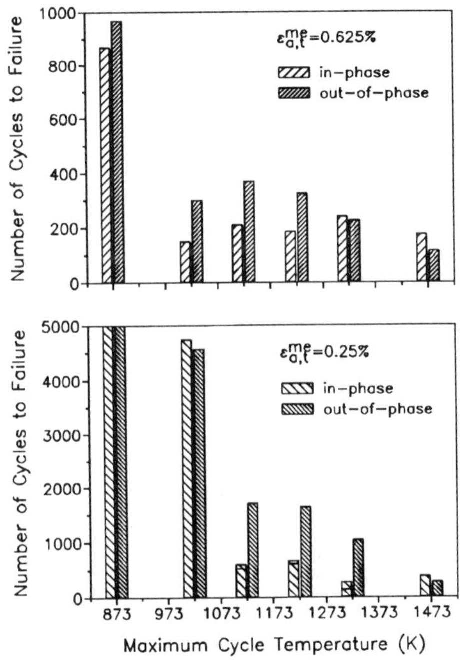

Figure 5. Influence of phase relation on TMF life at different T_{max}.

4.3. MICROSTRUCTURE

The development of the microstructure during TMF is mainly determined by the maximum cycle temperature. Figs. 6 to 8 show some examples of TEM-micrographs obtained from specimens thermal-mechanically fatigued to failure at different T_{max} and $\varepsilon_{a,t}^{me}$ = 0.625%. Fig. 6 demonstrates that at T_{max} = 1023K the microstructures observed after in-phase and out-of-phase cycling are very similar. In both cases a high density and a uniform distribution of dislocations are identified. The electron diffraction pattern and the corresponding bright and dark field images in the lower

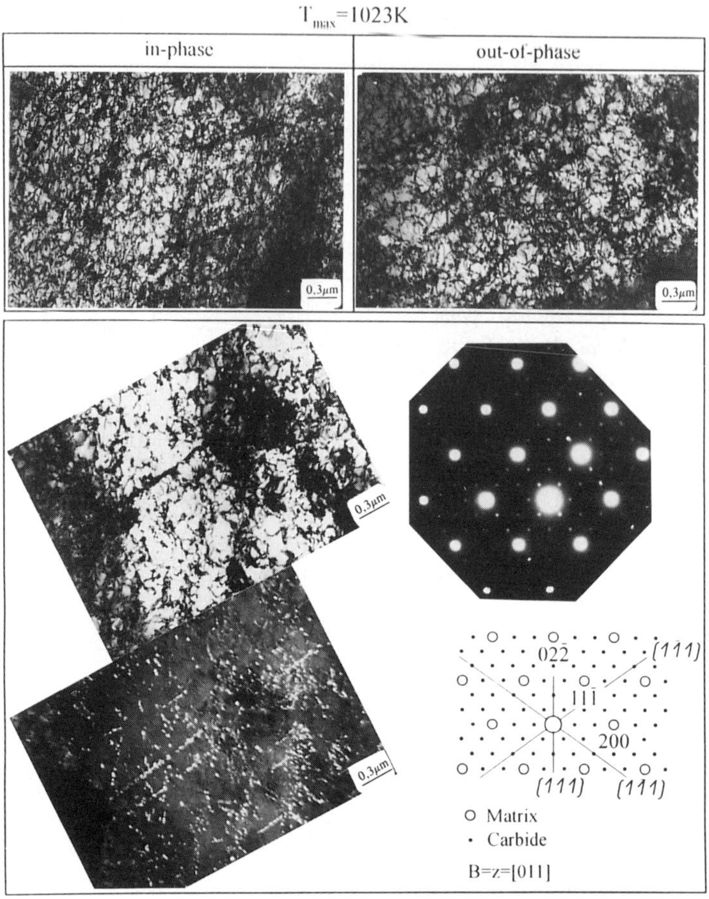

Figure 6. **TEM micrographs of specimens after failure due to in-phase and out-of-phase TMF at T_{max} = 1023K and $\varepsilon_{a,t}^{me}$ = 0.625%.**

part of the figure prove the existence of numerous coherent and very fine carbides which are preferably precipitated at the slip lines, where carbide nucleation is promoted by dislocation pipe diffusion of carbon and substitutional atoms. The small particles and the high dislocation density effectively impede the planar slip of dislocations and yield the pronounced cyclic hardening observed at maximum temperatures up to 1023K.

As no significant influence of the phase relation on the development of the microstructure during TMF at comparable loading conditions could be found, the following descriptions are valid for both test types. Fig. 7 reveals that locally very different dislocation distributions develop in one specimen thermal-mechanically fatigued at T_{max} = 1123K. The planar dislocation structure visible in the left part of the figure is the predominating one. There are coherent carbides precipitated mainly at the intersections of slip planes, which are larger than those observed after TMF at

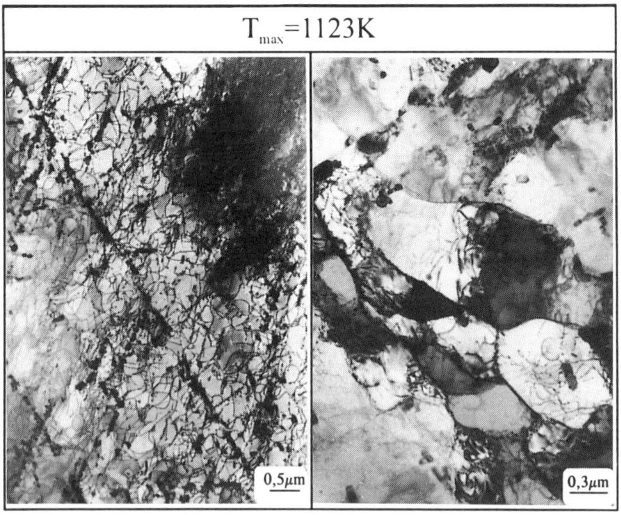

Figure 7. TEM micrographs after failure due to out-of-phase TMF at T_{max} = 1123K and $\varepsilon_{a,t}^{me}$ = 0.625%.

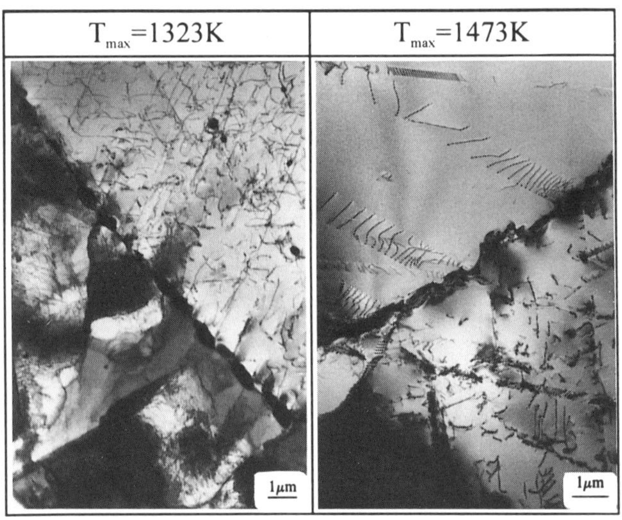

Figure 8. TEM micrographs after failure due to out-of-phase TMF at $\varepsilon_{a,t}^{me}$ = 0.625% and T_{max} = 1323 K and 1473K, respectively.

T_{max} = 1023K. However, as shown in the right part of the figure in some parts of the specimen also subgrain structures developed. They are typical of the influence of recovery processes increasing with increasing T_{max}.

The left side of Fig. 8 shows a characteristic subgrain structure which developed during TMF at T_{max} = 1323K. There are also secondary carbides which were precipitated in the interior of the grains as well as at grain boundaries and twin boundaries. Compared to those observed after TMF at T_{max} = 1123K, they are relatively coarse and less numerous. Besides such subgrain structures, occasionally planar dislocation patterns can be found. A further increase of T_{max} to 1473K results in a microstructure like that shown in the right part of the figure. The dislocation density is very low and comparable to that of specimens before TMF. Almost no secondary carbides can be found. This microstructure reflects a balance between hardening and recovery processes, resulting in almost constant stress amplitudes during TMF at T_{max} = 1323K and 1473K (Fig. 3).

4.4. DEVELOPMENT OF DAMAGE

Along the longitudinal axes of the specimens metallographic sections were prepared to estimate the influence of the TMF loading conditions on the development of damage at the surface as well as in the interior of the specimens. The micrographs in Fig. 9 show some typical surface microcracks observed after failure at $\varepsilon_{a,t}^{me}$ = 0.625%. TMF at T_{max} = 1123K causes intergranular cracks at in-phase and transgranular cracks at out-of-phase loading. Cycling at T_{max} = 1323K produces partial recrystallisation leading to a smaller grain size particularly in the vicinity of the surface. During in-phase cycling intergranular crack growth along the original grain boundaries is observed. After out-of-phase TMF transgranular microcracks appear which are very short and blunt. Oxidation of the surface layer is the more severe the higher T_{max} is.

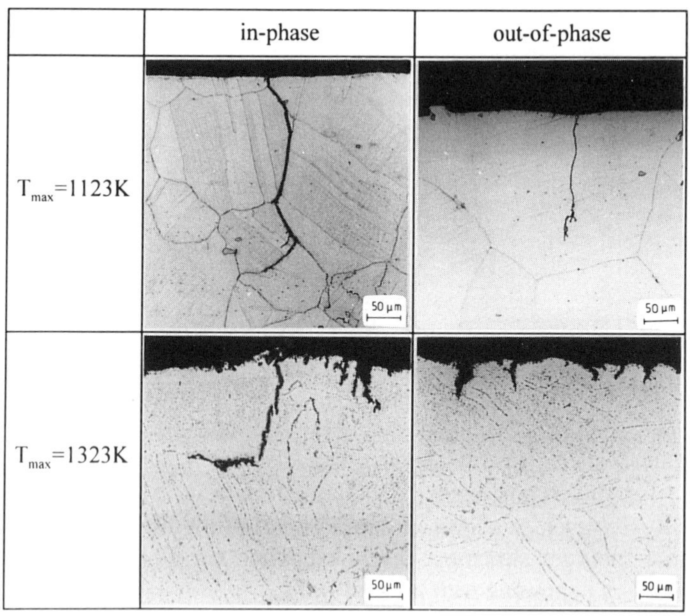

Figure 9. Surface cracks after failure due to in-phase and out-of-phase TMF at $\varepsilon_{a,t}^{me}$ = 0.625% and T_{max} = 1123K and 1323K, respectively.

To determine the damage developing during TMF in the interior of the specimen metallographic sections were checked for cavities and intergranular cracks. To evaluate a measure of total damage at grain boundaries, an average size of 5 μm was assumed for every cavity which could clearly be identified. Besides that the length of every intergranular crack was measured separately and added to the "damage length" obtained from the sum of the cavities. In Fig. 10 the results of these estimations of the total damage at grain boundaries are summarized. The specimens investigated were fatigued at T_{max} ranging from 1023K to 1323K and at $\varepsilon_{a,t}^{me} = 0.625\%$ (top) and 0.25% (bottom), respectively. With the exception of in-phase TMF at $T_{max} = 1323K$ and $\varepsilon_{a,t}^{me} = 0.625\%$, the total grain boundary damage increases with increasing T_{max}. In-phase TMF produces a much larger intergranular damage than out-of-phase TMF. This is an immediate consequence of the phase relation between temperature and mechanical strain. During in-phase testing, tensile stresses occur at high temperatures favouring the nucleation and growth of cavities and intergranular cracks. It is worth noticing that the intergranular damage after TMF at $\varepsilon_{a,t}^{me} = 0.25\%$ is roughly five times higher than at $\varepsilon_{a,t}^{me} = 0.625\%$. Thus, the longer lifetime occuring during TMF at $\varepsilon_{a,t}^{me} = 0.25\%$ is more important for the development of intergranular damage than the high stresses and plastic strains acting during TMF at $\varepsilon_{a,t}^{me} = 0.625\%$.

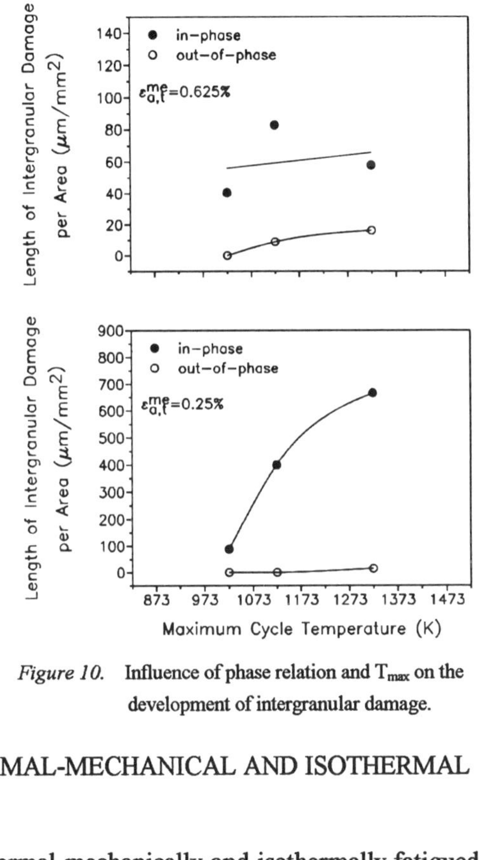

Figure 10. Influence of phase relation and T_{max} on the development of intergranular damage.

4.5. COMPARISON BETWEEN THERMAL-MECHANICAL AND ISOTHERMAL FATIGUE

In Fig. 11 the lifetimes of specimens thermal-mechanically and isothermally fatigued at $\varepsilon_{a,t}^{me} = 0.25\%$ are plotted versus the maximum cycle temperature or the test temperature, respectively. During isothermal testing the number of cycles to failure for all temperatures investigated lies between the in-phase and out-of-phase data. Also the intergranular damage observed after isothermal failure is higher than after out-of-phase, but lower than after in-phase loading. Similar effects of the phase relation in TMF were observed in numerous studies for different superalloys [9-11]. The differences in the lifetimes are caused by the stress-temperature relations. During isothermal cycling tensile and compressive stresses occur at the same temperature, so that small cavities which had nucleated due to tensile stresses can close if compressive stresses become effective [12]. For in-phase tests the low temperature during the compressive part of the cycle inhibits the closure of cavities. The high number of cycles to failure for out-of-phase cycling again demonstrates that compressive stresses at high temperature produce limited intergranular damage within the investigated temperature range.

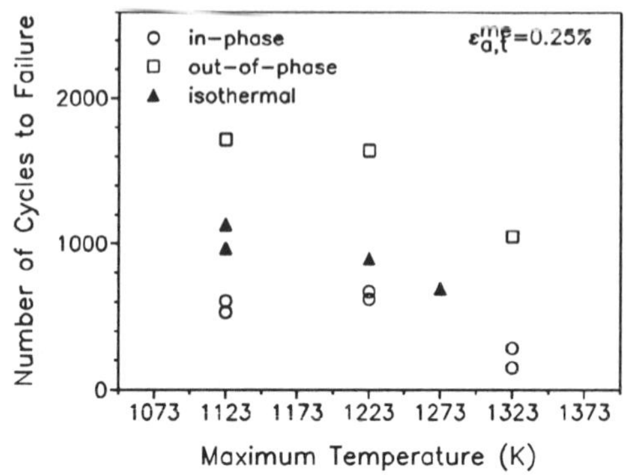

Figure 11. Comparison of fatigue life measured after isothermal and thermal-mechanical fatigue.

5. Summary

The cyclic stress-strain response and the development of the microstructure of the nickel-base superalloy NiCr22Co12Mo9 during thermal-mechanical fatigue were investigated for maximum cycle temperatures between T_{max} = 873K and 1473K and mechanical strain amplitudes of 0.625% and 0.25%, respectively. For lower values of T_{max} the stress-temperature- and stress-strain-hysteresis loops and the cyclic deformation curves prove cyclic hardening of the material during the whole fatigue life, which is the more pronounced the lower the maximum cycle temperature is. The microstructure which develops during TMF is mainly influenced by T_{max}. At lower T_{max} the increase of dislocation density and the precipitation of small carbides are the dominating hardening processes. The higher T_{max} is, the more important recovery processes become. The cyclic stress-strain response changes into neutral behaviour, and in the microstructure supplementary subgrain structures appear.

The development of surface microcracks and the damage in the interior of the specimens depend on the phase relation between temperature and mechanical strain. In-phase loading produces more intergranular damage than out-of-phase loading resulting in smaller lifetimes within a wide range of maximum cycle temperatures. Supplementary isothermal fatigue tests prove that combined thermal-mechanical loading may lead to lower lifetimes, even if compared with isothermal loading having the same strain amplitude at the maximum temperature of the thermal cycle.

Acknowledgements - The investigations were carried out within the frame of the Sonderforschungsbereich 167 which was initiated and financially supported by the Deutsche Forschungs-gemeinschaft (DFG). This is gratefully acknowledged.

References

1. Coles, A., Material Considerations for Gas Turbine Engines, in *Proc. 3rd Int. Conf. Mech. Behaviour of Materials (ICM3)*, 1, Pergamon Press, Oxford, (1979), pp. 3-11.
2. Skelton, R.P., Crack Initiation and Growth in Simple Metal Components during Thermal Cycling, in *Fatigue at High Temperature*, (1983), pp. 1-62.

3. Vogel, W.H., Soderquist, R.W., Schlein, B.C., Prediction of Crack Initiation in Combustion Chamber Liners, *J. Aircraft*, **14** (1977), 1076-1080.
4. Richter, F., Thermophysikalische Eigenschaften des hochwarmfesten Werkstoffes NiCr22Co12Mo (INCONEL 617), *Mat.-wiss. u. Werkstofftech.* **19** (1988), 5-61.
5. Viereck, D., *Das Zug- und Relaxationsverhalten von Hochtemperaturblechwerkstoffen im Temperaturbereich 78K ≤ T ≤ 1473K*, Fortschrittsberichte VDI, Reihe 5, Nr. 202, VDI Verlag Düsseldorf, (1990).
6. Lang, K.-H., Eifler, D., Macherauch, E., Fatigue Behaviour of Ni-base Alloys up to 1273K, in *Proc. of 8th Int. Conf. Strength of Metals and Alloys (ICSMA 8)*, Tampere, Pergamon Press, (1988), pp. 701-706.
7. Merckling, G., Lang, K.-H., Macherauch, E., Das Ermüdungsverhalten der Nickelbasis-Superlegierung NiCr22Co12Mo9 im Temperaturbereich 295K ≤ T ≤ 1473K, *Z. Metallkd.* **84** (1993), 844-853.
8. Pan, Y., Lang, K.-H., Löhe, D., Macherauch, E., Cyclic Deformation and Precipitation Behaviour of NiCr22Co12Mo9 during Thermal Fatigue, *Phys. Stat. Sol. (a)* **138** (1993), 133-145.
9. Marchand, N.J., Pelloux, R.M., Ilschner, B., A Fracture Mechanics Criterion for Thermal-Mechanical Fatigue Crack Growth of Gas Turbine Materials, *Eng. Fract. Mech.*, **31** (1988), 535-551.
10. Kuwabara, K., Nitta, A., Thermal-Mechanical Low-Cycle Fatigue under Creep-Fatigue Interaction on Type 304 Stainless Steels, *Fatigue Eng. Mater. Struct.*, **2** (1979), 293-304.
11. Boismier, D.A., Sehitoglu, H., Thermo-Mechanical Fatigue of MAR-M247: Part 1/2-Experiments, *J. Eng. Mater. Technol.*, **112** (1990), 68-89.
12. Baumgärtner, T., Bothe, K., Hurta, S., Gerold, V., Thermomechanical Fatigue of Nimonic 80A and Alloy 800H, in *High Temperature Materials For Power Engineering*, Proc. of Conf. in Liège, Belgium, Kluwer Academic Publischers, (1990), pp. 1087-1096.

MICROSTRUCTURAL ASPECTS OF DAMAGE OCCURRING DURING THERMO-MECHANICAL AND LOW CYCLE FATIGUE TESTING OF AN OXIDE DISPERSION STRENGTHENED ALLOY

J. KOMENDA, L. LINDÉ and P.J. HENDERSON
Swedish Institute for Metals Research
Drottning Kristinas väg 48
114 28 Stockholm Sweden

1. Introduction

Oxide Dispersion Strengthened (ODS) nickel-base alloys have good strength and corrosion resistance at high temperatures and are produced from their constituent elements in the form of mechanically alloyed powders which are subsequently hot extruded and directionally recrystallised. The ODS alloy MA 754, the subject of this study, consists of a single phase solid solution Ni-Cr matrix containing 1.0 vol.% finely distributed particles of yttria. These stable oxide particles give the alloy strength up to a relatively high proportion of its melting point and the elongated grain structure with a preferred <001> orientation leads to good thermal fatigue resistance.

Unfortunately, the mechanical properties of materials often depend on inhomogeneities in the microstructure which can lead to premature failure. In the case of ODS alloys it has been found that small grains produced by incomplete recrystallisation in the manufacturing process can give rise to additional cracking at grain boundaries, [1,2]. The results of thermo-mechanical fatigue (TMF) and some isothermal low cycle fatigue (LCF) tests on MA 754 have been published previously by the authors, [3]; in this paper more emphasis has been given to the role of the microstructure and the type of damage occurring has been quantified and related to the shape of the hysteresis loops using the plastic strain-range partitioning method, [7].

2. Experimental

2.1. MATERIAL AND MECHANICAL TESTING

The MA 754 was manufactured by Inco Alloys Ltd., in the form of extruded and solution annealed 20 mm diameter bars. The chemical composition is given in Table 1. The average grain size in the longitudinal direction was 1.1 mm and in the short transverse direction 0.1 mm. All specimens were oriented parallel to the

J. Bressers and L. Rémy (eds.), Fatique under Thermal and Mechanical Loading, 339–347.

longitudinal grain direction. TMF specimens were hollow, with an 8 mm outer diameter, 4 mm inner diameter and a 17 mm parallel gauge length. The LCF specimens were solid, but otherwise of the same design.

TABLE 1. Chemical composition of the MA 754 in weight %.

Ni	Cr	C	Fe	S	Al	Yi	N	O	Y_2O_3
bal	19.79	0.047	0.36	<0.001	0.31	0.45	0.073	0.38	0.53

The testing was performed in a servo-hydraulic machine with a 12.5 mm gauge length side entry extensometer, induction heating and compressed air cooling. The LCF tests were performed at 850 and 1100°C in symmetrical (R = -1) total strain control with a triangular waveform and equal compression- and tension-going ramp rates which corresponded to a strain rate of 5 x 10^{-3} s^{-1}. The TMF tests were run in strain control with the sum of the thermal strain + total mechanical strain as the controlling parameter. Equal ramp rates were also used giving a total mechanical strain rate of between 0.5 and 2.0 x $10^{-4}s^{-1}$. The temperature ramp rate was not allowed to rise above 6°C s^{-1} and in most cases was about 3°C s^{-1}. The minimum and maximum temperatures were 600 and 1100°C.

Two types of TMF cycles were used : in-phase (IP) cycles with the maximum tensile strain at the maximum temperature and out-of-phase (OP) cycles with the maximum compressive strain at the maximum temperature.

Before the start of each TMF test the specimen was thermally cycled under zero load and the thermal strain versus temperature was recorded. During testing the temperature-strain response was recorded directly after each stress-strain hysteresis loop recorded. The stress-mechanical strain hysteresis loops were computed from these three curves and an expression for the temperature dependence of Young's Modulus.

During testing the peak tensile load decreased linearly and slowly with increasing cycle number before the onset of final cracking. N_{90}, the cycle number corresponding to a 10% drop in load from linear behaviour was used for graph plotting. More details of the mechanical testing and evaluation procedure are given in Ref. 6.

2.2. METALLOGRAPHY AND MICROSCOPY

The Electron Back Scattering Pattern (EBSP) technique, a diffraction technique, was used to identify recrystallisation defects by measuring local changes in orientation. EBSPs are similar to Kikuchi lines observed in a transmission electron microscope and result from the diffraction of inelastically scattered electrons produced below the surface of the specimen at a depth of up to 10 nm. The patterns can be produced in a scanning electron microscope (SEM) using a stationary or scanning beam which is incident on the specimen surface at an acute or glancing angle (usually 20°) and are imaged on a phosphor screen placed in front of the specimen. An EBSP consists of parallel lines, one pair for each set of atomic planes, separated by a bright band. The bands intersect to form poles which can be identified by considering the number and symmetry of the bands at the intersection. Thus the orientation can be determined.

The resolution is about 100nm. More details of the technique and imaging are given in Refs. 4 and 5.

After mechanical testing specimens for EBSP examination were cut, polished and left unmounted. The last stage of the mechanical polishing was carried out with 0.25 μm diamond paste followed by electro-polishing in a mixture of 10% perchloric acid and 90% ethanol to ensure a deformation-free surface. The temperature of the electrolyte was not allowed to rise above 5°C. The specimens were examined in a JEOL 6400 SEM using a LINK EBSP system with an operating voltage of 15kV and a beam current of 1×10^{-8} amps. The experimental set-up is given in Ref. 5.

Specimens for microscopic examination were polished and etched in a solution of 1.5%H_2O_2, 30% HCl and 68.5% H_2O and examined with light optical microscopy (LOM) and in a SEM. Quantitative metallography of cracks was performed with a Kontron IBAS 2000 image analysis system on specimens magnified 400 times by LOM.

3. Results

3.1. MECHANICAL TESTING

In order to compare LCF and TMF results the thermal strains have been subtracted and only mechanical strain values are shown. The total mechanical strain range versus N_{90} is shown in Fig. 1. The fatigue lives of the 1100°C-LCF, OP-TMF and low strain range 850°C-LCF tests were broadly similar. The fatigue lives of the IP-TMF tests were considerably shorter and there was also a larger amount of scatter in the IP data. The scatter in the IP data could not be correlated with differences in strain rate. Hysteresis loops are displayed in Fig. 2.

3.2. METALLOGRAPHY

Cracks produced by LCF and TMF could be classified into four types which are shown schematically in Fig. 3 and defined below as :-

P-ext Intergranular cracks *parallel* to the stress axis (longitudinal grain direction) which have initiated on the fracture surface,

P-int Intergranular cracks *parallel* to the stress which are formed *internally*, remote from the surfaces

N-ext Edge cracks *normal* to the stress growing in from the *external* surfaces. These cracks were mainly intergranular, but some transgranular cracks were detected and

N-int Intergranular cracks *normal* to the stress which are formed *internally*, remote from the surfaces. They were mainly associated with smaller equiaxed grains.

The EBSP technique was used to confirm that the edge cracks, type *N-ext*, at 1100°C were predominantly intergranular, although some small transgranular cracks were observed. Areas containing *N-int* cracks were associated with smaller, equiaxed grains. The orientations of grains in the vicinity of *N-int* cracks were measured by EBSP and it was found that many of the grains had orientations far removed from the expected <001> texture. Fig. 4 shows an example of cracking and misoriented grains.

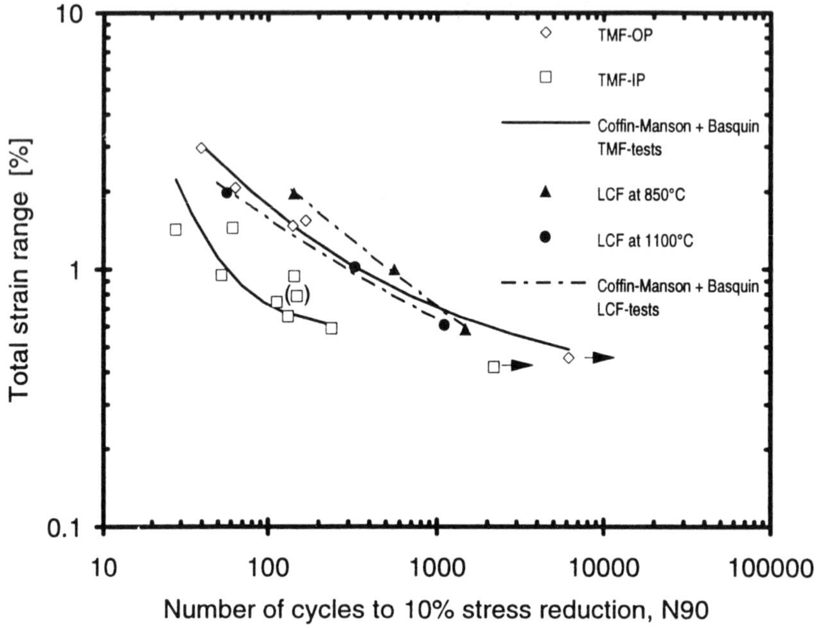

Figure 1. Total mechanical strain range versus number of cycles to a 10% stress reduction for MA 754. The thermal strains have been removed. The points with the arrows are tests interrupted before any sign of load reduction occurred. A test with R<-1 is shown in parentheses.

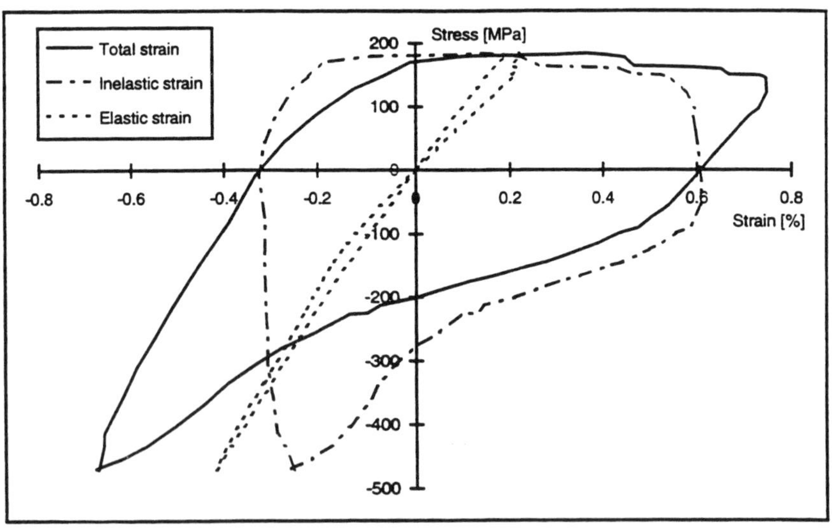

Figure 2. Stress-mechanical strain hysteresis loops for an in-phase test, total strain range, ($\Delta\varepsilon_t$) = 1.43%, inelastic strain range ($\Delta\varepsilon_{in}$) = 0.94%, N_{90} = 28 cycles.

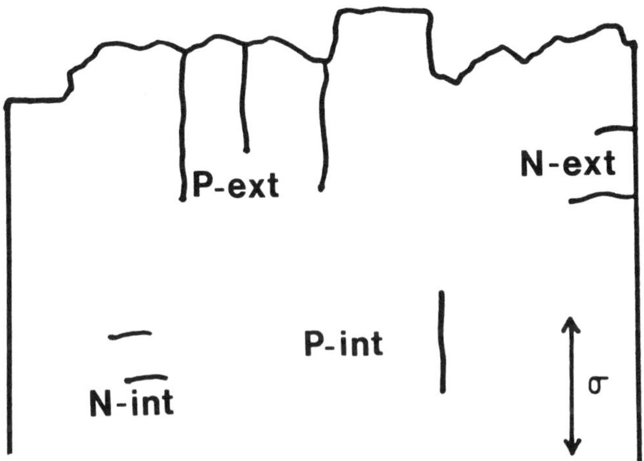

Figure 3. Schematic diagram showing the type of cracks seen on longitudinally sectioned specimens.

TABLE 2. Results of quantitative metallography on longitudinally sectioned fatigue tested specimens showing the numbers of different types of cracks and the results of strain-range partitioning (SRP) for the same specimens.

Test type	$\Delta\varepsilon_t$ (%)	N_{90} (cycles)	*P-ext* (no. mm^{-1})	*P-int* (no. mm^{-2})	*N-ext* (no. mm^{-1})	*N-int* (no. mm^{-2})	$\Delta\varepsilon_{pp}$ (%)	De_{cc} (%)	$\Delta\varepsilon_{pc}$ (%)	$\Delta\varepsilon_{cp}$ (%)
IP	0.42	2200 →	-	-	0.06	0.47	0.07	-	-	-
IP	0.59	240	1.75	0.03	0.12	5.50	0.13	-	-	0.12
IP	0.75	114	6.50	0.62	0.28	14.72	0.12	-	-	0.19
IP	1.43	28	6.50	0.97	0.53	8.19	0.26	-	-	0.67
IP	1.45	62	1.50	0.28	0.75	5.84	0.32	0.12	-	0.35
OP	0.45	6184 →	-	-	-	-	0.17	-	-	-
OP	1.55	169	0.50	-	0.09	-	0.42	0.26	0.39	-
LCF 1100°C	0.61	1112	3.37	0.47	-	0.16	0.25	0.18	-	-
LCF 1100°C	1.97	57	6.37	0.17	0.37	0.86	0.43	1.39	-	-
LCF 850°C	0.59	1483	0.87	0.06	0.56	0.20	0.24	0.14	-	-
LCF 850°C	1.96	144	13.60	-	0.25	-	0.61	1.15	-	-

→ indicates that the test was interrupted before a drop in peak load could be detected.

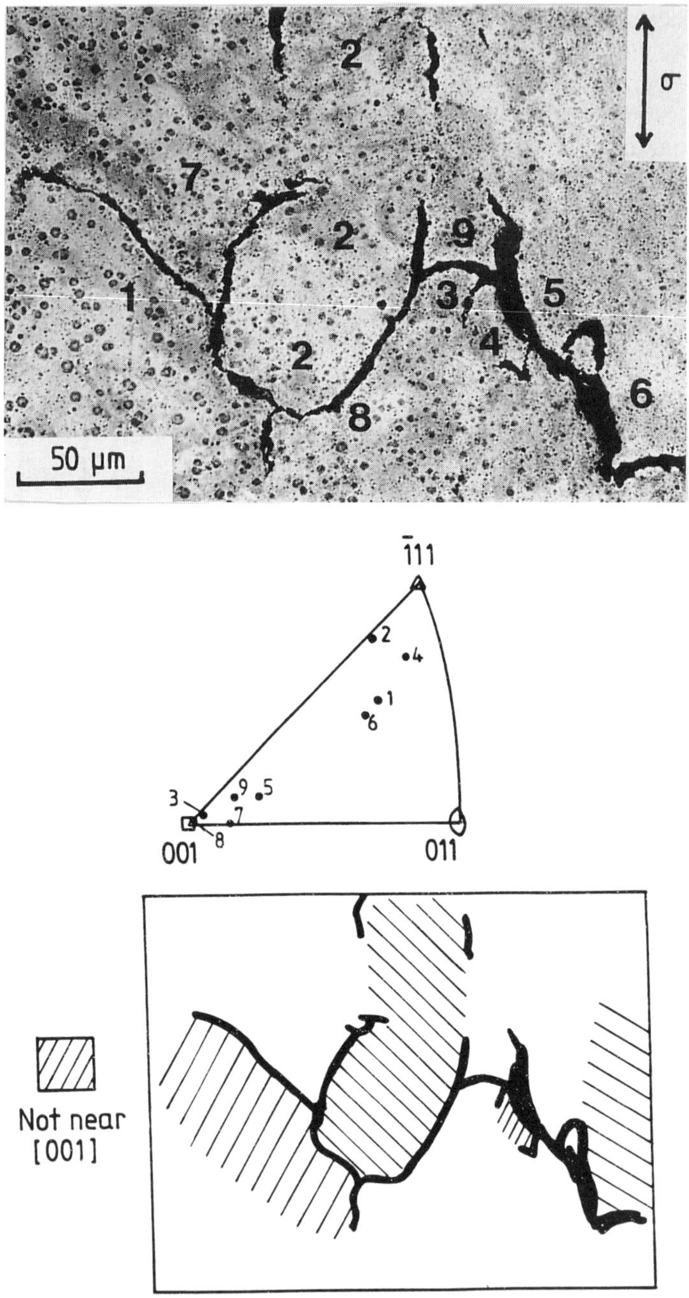

Figure 4. SEM micrograph of a cracked region after LCF at 1100°C, $\Delta\varepsilon_t$ = 1.97%, $\Delta\varepsilon_{in}$ = 1.73% and N_{90} = 144 cycles, stereographic triangle showing the orientations of the grains with respect to the stress axis [001] and a schematic diagram showing the grains which lie more than 25° from [001].

The results of crack counting in tested specimens are given in Table 2. (Strain-range partitioning results are dealt with under section 3.3.). Crack types are shown in Fig. 3. The area examined was in each case the specimen diameter by 8 mm along the gauge length from the fracture surface. Because the TMF specimens are hollow and the LCF specimens are solid, the number of externally initiated cracks has been normalised with respect to the length of the surfaces and the number of internally initiated cracks has been normalised with respect to the area of material from which measurements were taken. The length of the cracks was also measured and shows the same trends. A scan of Table 2 with regard to cracking reveals the following.-

In the OP specimen interrupted before a load-drop was detected, no cracks of any type were seen in the section examined.

In the IP specimen interrupted before a load-drop was detected there were both internal and external cracks present, growing normal to the stress axis.

In OP specimens generally, no internal cracking was seen.

Most *P* type cracks (parallel to stress) were initiated at the fracture surface, but most *N* type (normal to stress) were initiated internally.

The densities of *P-ext*, *P-int* and *N-ext* cracks were approximately similar in IP and LCF specimens, but IP specimens contained much larger numbers of *N-int* cracks.

An increase in the number of *N-int* cracks corresponded to a decrease in the fatigue life, when comparing similar strain ranges.

3.3. STRAIN-RANGE PARTITIONING

The shape of a hysteresis loop depends on the stress, temperature and to a certain extent the strain rate. From Fig. 2, an IP loop, it can been seen that there is a large creep component (horizontal part of curve) in the tension part of the cycle. The reverse is true of an OP cycle, *i.e.* there is a creep component in the compression part of the cycle. The different types of deformation occurring in a fatigue cycle can be defined and quantified using the strain-range partitioning method, [7]. A closed hysteresis loop can be partitioned into a maximum of three of the following four strain-range components:-

$\Delta\varepsilon_{pp}$ - plastic strain in tension reversed by plastic strain in compression
$\Delta\varepsilon_{cc}$ - creep strain in tension reversed by creep strain in compression
$\Delta\varepsilon_{pc}$ - plastic strain in tension reversed by creep strain in compression
$\Delta\varepsilon_{cp}$ - creep strain in tension reversed by plastic strain in compression

The hysteresis loops at $\sim N_{90}/2$ associated with the metallographic specimens were partitioned according to the method in Ref. 7. For partitioning of TMF loops, the inelastic strain loops were used, since the elastic modulus varies with temperature. The results are also given in Table 2.

Table 2 shows that high temperature tensile creep reversed by lower temperature plasticity in compression, ($\Delta\varepsilon_{cp}$), as seen in IP cycles, gives rise to large numbers of *N-int* cracks, whereas tensile plasticity reversed by high temperature compressive creep causes little or no damage in this alloy. Considering the IP tests in more detail it can be seen that the highest value of $\Delta\varepsilon_{cp}$ is associated with the shortest fatigue life.

Progressively decreasing values of $\Delta\varepsilon_{cp}$ lead to increasing fatigue lives. It would therefore appear that $\Delta\varepsilon_{cp}$ is the most damaging type of strain occurring in the IP tests and has the greatest effect on the cyclic lifetimes.

In LCF a large $\Delta\varepsilon_{cc}$ component is considerably more damaging at 1100°C than 850°C.

4. Discussion

MA 754 normally contains elongated grains with a preferred <001> orientation, but grains of orientations other than <001> were associated with internal grain boundary cracking, as shown in Fig. 4. These misoriented grains are processing defects caused by incomplete recrystallisation and are usually smaller and less elongated than the <001> grains. The shape of the grains is often defined by the grain aspect ratio (GAR) which is the ratio of the longitudinal to transverse sections. Research on creep of ODS alloys has shown that a reduced GAR (caused by the presence of these processing defects) leads to a reduction in ductility and creep life of the specimen, [1,2]. An explanation of why these defect grains are more susceptible to cracking is given below.

It has previously been shown that, during high temperature deformation of MA 754, grains oriented close to [001] before deformation rotated towards this pole, but those oriented nearer [011] or [$\bar{1}$11] rotated towards [$\bar{1}$11], [5]. This is consistent with deformation occurring by {111}<112> slip, [8].

When the defect grains deform plastically the slip direction will rotate and a rotation of 30-40° may be necessary before the [$\bar{1}$11] pole is reached. When the elongated grains deform plastically the rotation occurs in the *opposite* direction and smaller rotations are needed to bring the slip direction to a duplex slip boundary and thence to the [001] pole. It is known that the tensile ductility of MA 754 is dramatically reduced at temperatures above 900°C [9], and so it is likely that the difference in the amount and direction of rotation between the elongated and defect grains leads to internal stresses at the grain boundaries which easily induces cracking at high temperatures.

It seems that high temperature (>850°C) tensile creep reversed by lower temperature compressive plasticity, $\Delta\varepsilon_{cp}$, as seen in IP cycles, is the most damaging type of strain, giving rise to large amounts of *N-int* cracks which are most likely to occur around the defect grains. This gives rise to a reduced fatigue life and could also account for some of the scatter in the data.

5. Conclusions

TMF testing between 600-1100°C on MA 754 gave much shorter fatigue lives and more scatter in the data for IP tests than for OP testing. Isothermal LCF lives at 1100°C were similar to those at 850°C at low strain ranges, but reduced at high strain ranges.

Plastic strain-range partitioning of TMF hysteresis loops was found to yield useful quantitative information on the types of strain occurring during cycling.

The numbers and types of cracks on fatigue tested specimens were related to different types of strain obtained from strain-range partitioning of hysteresis loops. It was found that a large amount of fully reversed creep strain ($\Delta\varepsilon_{cc}$), seen in LCF tests, was more damaging at 1100°C than at 850°C and that the most damaging type of strain was high temperature (> 850°C) creep in tension reversed by lower temperature plastic strain in compression , ($\Delta\varepsilon_{cp}$), as seen in IP tests.

$\Delta\varepsilon_{cp}$ gave rise to large amounts of internal cracking, normal to the stress, *N-int* cracking. The EBSP technique was used to show that these *N-int* cracks were associated with the boundaries of grains that did not have the expected <001> texture of the bulk material, (*i.e.* they were processing defects) and it is thought that the presence of *N-int* cracks causes a reduction in the fatigue life.

Acknowledgements - We are grateful for the help given to us by Anders Oscarsson. This work was funded by an internal *Fri Forskning* project and by the companies forming the Advanced Materials group of Sweden's High Temperature Materials Committee (*Varmhållfasthetskommitté*). These companies are ABB STAL AB, Avesta Sheffield AB, Celsius Materialteknik, Kanthal AB, AB Sandvik Steel, Sydkraft Konsult, Vattenfall Energisystem and Volvo Aero Corporation. The financial support is gratefully acknowledged.

References

1. Arzt, E. and Singer, R.F., The effect of grain shape on stress rupture of the oxide dispersion strengthened superalloy Inconel MA 6000, in M. Gell *et al* (eds.), *Superalloys '84*, AIME-TMS, (1984), pp. 367-376.
2. Zeizinger, H. and Arzt, E., The role of grain boundaries in the high temperature creep fracture of an oxide dispersion strengthened superalloy, *Z. Metallkde.*, **79** (1988), 774-781.
3. Henderson, P.J. and Lindé, L., High temperature thermo-mechanical and low cycle fatigue of the oxide dispersion strengthened alloy MA 754, in Y. Hosoi *et al* (eds.), *Aspects of High Temperature Deformation and Fracture in Crystalline Materials*, Japan Institute of Metals, Sendai, Japan, (1993), pp 527-534.
4. Quested, P.N., Henderson, P.J. and McLean, M., Observations of deformation and fracture heterogeneities in a nickel-base superalloy using electron back scattering patterns, *Acta Metall.* **36** (1988), 2743-2752.
5. Komenda, J. and Henderson, P.J., Deformation and cracking studies using electron back Scattering patterns, Swedish Institute for Metals Research Report IM-3075 , 1994.
6. Lindé, L. and Henderson, P.J., Thermo-mechanical and Low Cycle Fatigue Behaviour of the ODS Alloy MA 754. Swedish Institute of Metals Research Report IM-3114, 1994.
7. Manson, S.S. and Halford, G.R., Hirschberg, M.H., Creep-fatigue analysis by strain-range partitioning. *Symposium on Design for Elevated Temperatures,* ASME, New York, (1971), pp 12-24.
8. McLean M., *Directionally Solidified Materials for High Temperature Service.* Chapter 5. The Metals Society, London, 1983.
9. *Inco Mechanically Alloyed Products* , Inconel alloy MA 754. Inco Alloys Ltd, 1983.

MATERIAL BEHAVIOUR AND DEVELOPMENT OF MICROSTRUCTURE DURING THERMAL-MECHANICAL FATIGUE OF A 12% CHROMIUM STEEL

G. PITZ[1], T. SÄUBERLICH[2], D. LÖHE[1]
[1] *Institut für Werkstoffkunde I, Universität Karlsruhe (TH) Kaiserstr. 12, D-76128 Karlsruhe, FRG*
[2] *Universität-GH Paderborn, FB 10, Fachgebiet Werkstoffwissenschaften, D-33095 Paderborn, FRG*

1. Introduction

Ferritic-carbidic steels with chromium contents ranging from 8 to 12% are important structural materials for thermally and mechanically heavily stressed components which operate at temperatures up to 550°C and are used e.g. in power generating plants or in the chemical industry. The complex thermal-mechanical loading of these components usually comprises creep, high cycle fatigue and low cycle fatigue thermally induced by start-ups, load changes and shut-downs, producing instationary temperature gradients and hence strain as well as stress fields. In comparison with austenitic steels, high chromium ferritic-carbidic steels have a relatively low thermal expansion and a relatively high thermal conductivity . Therefore, it is to be expected that thermally induced stresses are generally lower than in austenitic steels. Up to now research work is mainly focused on the creep properties of these steels and the degradation of the microstructure during creep [1]. However, some investigations were carried out concerning the relationship between microstructure and thermal-mechanical fatigue behaviour, indicating that continuous cyclic softening occurs during thermal-mechanical fatigue (TMF) life [2, 3] and that it is combined with microstructural changes, which are important because they may change the material's response to operating stresses.

In the present study, the cyclic deformation behaviour and the development of the microstructure of a 12% chromium steel were investigated during TMF life. A triangular temperature-time course was chosen which was out-of-phase with the mechanical strain-time course. This corresponds to the conditions in fast heating parts of a component, the thermal expansion of which is hindered by colder parts.

J. Bressers and L. Rémy (eds.), Fatigue under Thermal and Mechanical Loading, 349–360.

2. Material

The material investigated was a 12% chromium steel (German grade X 22 CrMoV 12 1) with the chemical composition (wt.-%) 0.22 C, 11.86 Cr, 0.93 Mo, 0.57 Ni, 0.25 V, 0.31 Si, 0.49 Mn, 0.021 P, 0.06 S. During austenitization for 30 minutes at 1030°C, most of the carbides were dissolved. By quenching in oil, a martensitic structure was produced, which was finally tempered for 2 hours at 710°C. During tempering, the martensitic structure was transformed into a structure of elongated subgrains with the majority of carbides at the subgrain boundaries (Fig. 1). Larger carbides are found at the boundaries of the former austenite grains. The quantitative evaluation of the microstructure after the heat treatment yields the data given in Table 1.

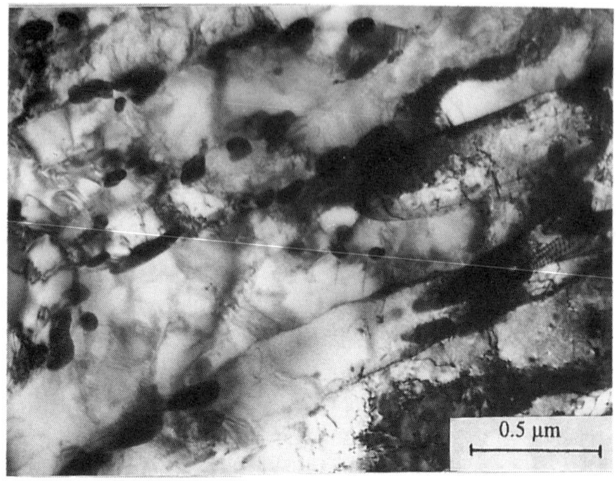

Figure 1. TEM-microstructure of the steel X 22 CrMoV 12 1 before TMF cycling.

TABLE 1. Microstructure of X 22 CrMoV 12 1 before TMF.

Microhardness (Vickers Hardness)	306 HV 0.3
Subgrain size (longitudinal direction) s_l	1.37 μm
Subgrain size (transversal direction) s_t	0.39 μm
Aspect ratio s_l/s_2	3.56
Fraction of carbides at the subgrain boundaries	65 %
Fraction of carbides in the interior of the subgrains	35 %
Mean carbide size	38nm

3. Experimental Procedure

In the present investigation the experiments were run with a triangular temperature-time course. The minimum temperature T_{min} was 150°C, the maximum temperature T_{max} 600°C, and the heating as well as the cooling rates were 10°C s^{-1}. The internal constraint existing in components was simulated by external constraint produced by strain controlled loading. During heating from ambient temperature to the minimum cycle temperature T_{min}, however, the specimens were allowed to expand freely using stress control. Then, the closed loop control was switched over from stress to total strain control, and no change of total strain was allowed during subsequent thermal cycling. By definition, the total strain ε_t at $T = T_{min}$, was set to zero, and therefore

$$\varepsilon_t = \varepsilon_t^{me} + \varepsilon^{th} = \varepsilon_e^{me} + \varepsilon_p^{me} + \varepsilon^{th} = 0 \quad (1)$$

or

$$\varepsilon_t^{me} = \varepsilon_e^{me} + \varepsilon_p^{me} = -\varepsilon^{th} \quad (2)$$

Hence, during subsequent thermal cycling between T_{min} and T_{max}, cyclic mechanical strains develop which are out-of-phase with the triangular temperature-time course chosen, at a strain-rate of about $1.3 \cdot 10^{-4}$ s^{-1}. The TMF tests were stopped at predefined numbers of cycles and after fracture, respectively. The solid specimens used had a gauge length of 15 mm and a gauge diameter of 7 mm. For heating, a 5 kW inductor was used. Cooling was achieved by thermal conduction into the specimen grips and by additionally blowing compressed air on the specimens using a proportionally controlled valve.

The Microhardness of the specimens was measured according to Vickers with a load of 3 N (HV 0.3) and a loading time of 30 s. The development of the subgrain structure was investigated using a 200 kV transmission electron microscope (TEM). TEM-micrographs magnified by a factor of 22.000:1 were evaluated by the intercepted-segment method. Measurement was performed parallel and perpendicular to the longitudinal axis of the elongated subgrains. Size distribution, shape and location of the carbides were analysed using TEM-micrographs magnified to 50.000:1. Similarly to the evaluation of the subgrains, the longitudinal and the transverse dimensions of the elongated carbides were determined.

4. Cyclic Stress-Strain Response

From Eq. 2, the mechanical and the plastic strain generated by the total suppression of thermal strains were determined. Fig. 2 shows the stress-strain response developing during TMF at different numbers of cycles. The evaluation of these hysteresis loops yields the development of the maximum stress, the minimum stress, the mean stress, the plastic strain amplitude and the plastic mean strain during TMF given in Fig. 3.

During the first heating from T_{min} compressive stresses develop and pronounced plastic deformation starts as mechanical strain approaches -0.24%. The maximum compressive stress amounts to 530 N/mm^2 at 390°C. During further heating, stress relaxation occurs even though the mechanical strain still increases in absolute value until T_{max} is reached. Therefore, in the first cycles, the minimum stress σ_{min} is not correlated to T_{max}. During cooling from T_{max}, tensile stresses develop which amount to 580 N/mm^2 at T_{min} . Thus, as a result of cyclic relaxation of compressive stresses, already after the first cycle tensile mean stresses occur which increase until the second cycle is reached. The minimum plastic strain amounts to -0.41% after the first heating up to T_{max} and to -0.27% after the first cooling down to T_{min}, resulting in a (plastic) mean strain of -0.34%. As shown by the data determined at N = 5, the

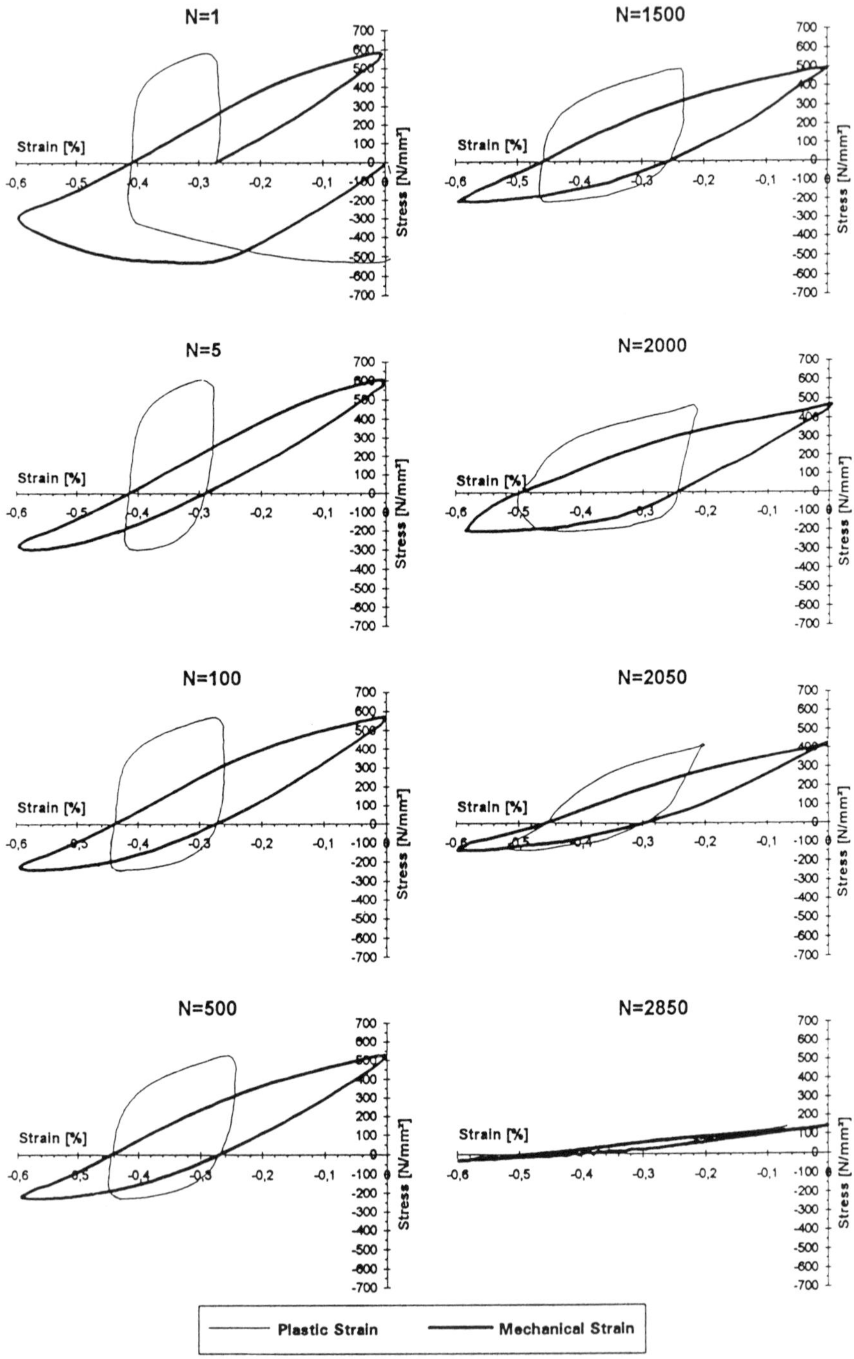

Figure 2. Stress-strain- and stress-plastic strain-hysteresis loops at different cycle numbers.

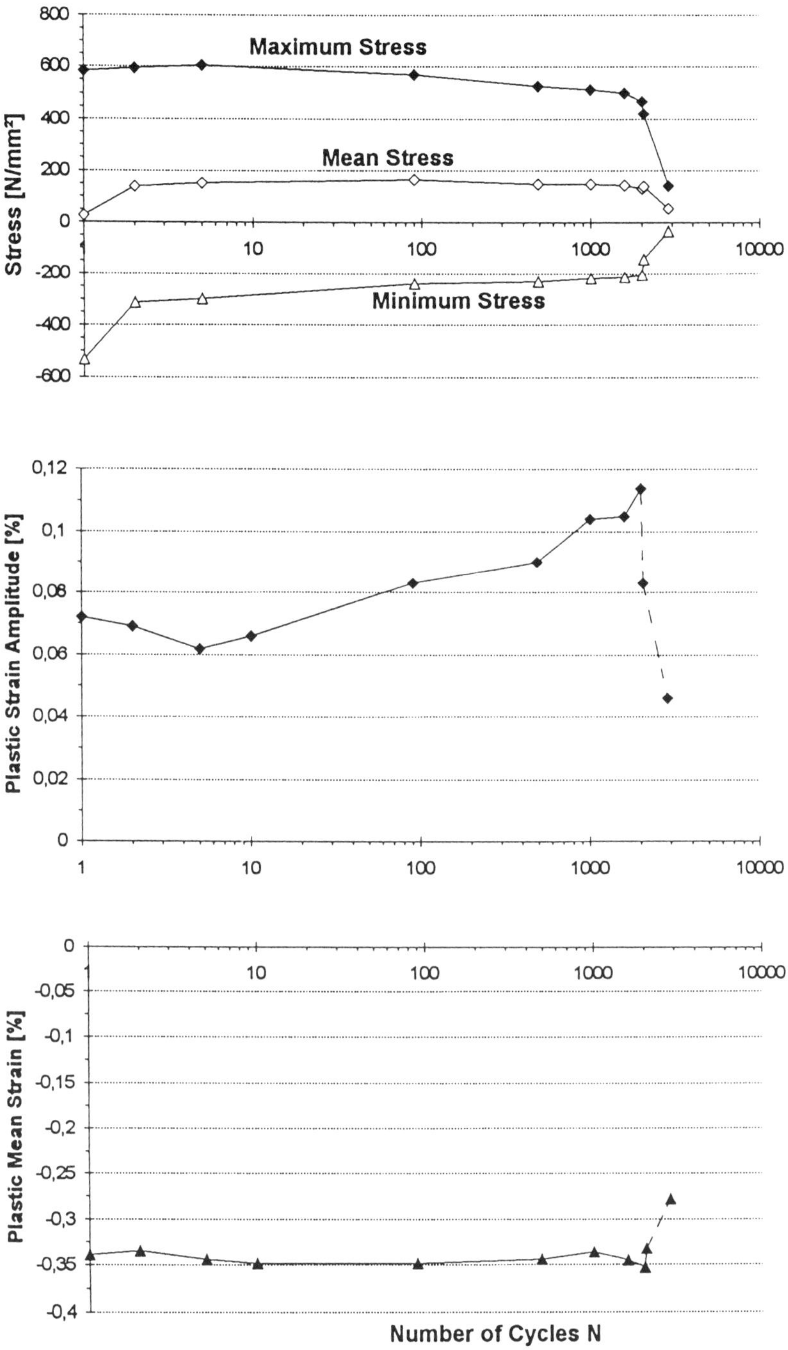

Figure 3. Development of the maximum, the minimum and the mean stress as well as the plastic strain amplitude and the plastic mean strain during TMF at different cycle numbers.

hysteresis loops are markedly shifted towards the tensile range during the first cycles, resulting in a mean stress of $\sigma_m = 150$ N/mm^2, a weak increase of the maximum stress up to 605 N/mm^2 and a decrease of the minimum stress down to 310 N/mm^2. From N = 5 to N = 2000 the mean stress remains almost constant, but the decrease of the stress range $\Delta\sigma = \sigma_{max} - \sigma_{min}$ from 904 to 675 N/mm^2 and the increase of the plastic strain amplitude from 0.062% to 0.114% indicate cyclic softening. The strong decrease of the stress amplitude and the plastic strain amplitude as well as the increase of the plastic mean strain after 2000 cycles result from the propagation of a macrocrack, which strongly changes the shape of the stress-plastic strain hysteresis loop.

5. Change of Microstructure During TMF

The cyclic softening which is clearly visible in the cyclic stress-strain response goes hand in hand with microstructural changes, which in turn change the microhardness of the material. However, as shown by Fig. 4, there is no direct correlation between the cyclic stress-strain response and the microhardness. The latter increases monotonously during the first 100 cycles, where the stress-strain response already indicates cyclic softening. During the remaining TMF life, the microhardness decreases below the value measured after the heat treatment.

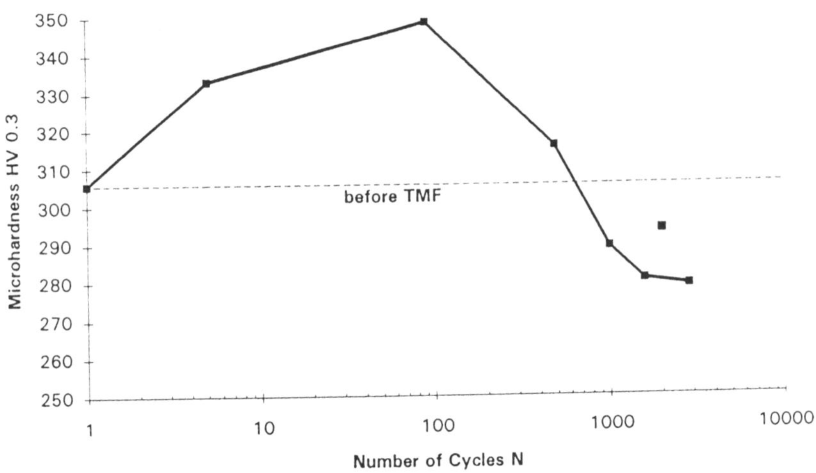

Figure 4. Microhardness HV 0.3 before TMF and after TMF at different cycle numbers.

The evolution of the microstructure during TMF is rather heterogeneous. As an example, Fig. 5 shows TEM-micrographs obtained from different locations in one specimen tested up to 1500 TMF cycles. The upper part of the figure shows almost the same subgrain structure as before TMF; the elongated subgrain structure. In the lower part of the figure a different structure is visible. As compared to the material state before TMF, s_l has decreased to 0.98 μm, but s_t has increased to 0.69 μm. Hence, the subgrains are spheroidized and the aspect ration s_l / s_t has decreased to 1.42. The development of the microstructure during TMF life is shown in Fig. 6 by three TEM-micrographs corresponding to N = 100, N = 1500 ≈ $N_f/2$ and N = 2850 = N_f. During the first 100 cycles, the subgrain size decreases by about 30% in the longitudinal direction, whereas it has hardly changed in the transverse direction. The relationship between carbides and neighbour subgrain boundaries has strongly changed. Only 41% (65% before TMF, see Table 1) of the carbides are still located at the subgrain boundaries, whereas 59% (35% before TMF) are situated in the interior of the subgrains. After 1000 cycles, a minimum subgrain size appears in the longitudinal direction. During further thermal cycling, the subgrains grow both in the longitudinal and in the transverse direction. The three micrographs clearly show the increasing spheroidization of the subgrains during TMF. The results of the microstructural observations are summarised in Figs. 7 to 10. Fig. 7 shows the development of the subgrain structure and Fig. 8 the change of the neighbour relationship between carbides and subgrain boundaries, which is the strongest during the first 100 cycles. The development of the carbide structure is illustrated in Figs. 9 and 10. Fig. 9 shows the evolution of the carbide size classified in four size classes. Coarsening of carbides sets in after 500 cycles. This is also evident from Fig. 10, where the carbide size both in the longitudinal and transverse direction, as well as the mean carbide size are shown as a function of the number of cycles. The growth of the carbides in transverse direction is faster than in longitudinal direction, resulting in a spheroidization of the carbides. This is combined with decreasing aspect ratios, the scatter of which, however, is relatively large.

Figure 5. TEM-micrographs of different locations after 1500 TMF cycles (see text).

6. Discussion

The cyclic stress-strain response and the TMF life of the material investigated is mainly determined by the initial microstructure produced by quenching and tempering and by its development during TMF. During quenching from the austenitization temperature, a martensitic structure with a very high dislocation density in the interior and at the boundaries of the martensite laths is produced. Furthermore, fine ε-carbides and primary Nb-V-carbonitrides are already present. During tempering at 710°C, the martensitic structure is transformed into a subgrain structure, most of the lath boundaries becoming subgrain boun-daries [4, 5, 6]. During this process, the dislocation distribution becomes heterogeneous, resulting in a relatively low dislocation density in the interior of the subgrains and a relatively high dislocation density at the subgrain boundaries. During tempering, the majority of carbides existing after the heat treatment are precipitated. Most of them are located at subgrain boundaries, where nucleation is facilitated (Figs. 1 and 8). Stress relaxation tests at different tempera-

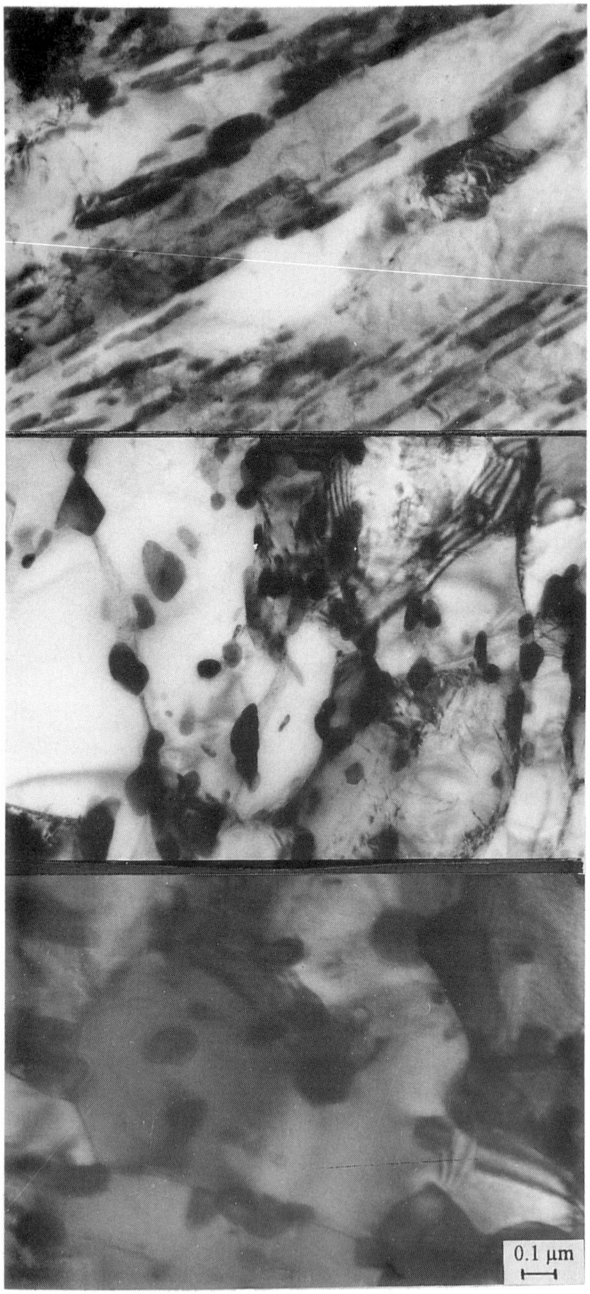

Figure 6. TEM-micrograph at N=100, N=1500 (≈N_f/2) and N=2850 (=N_f).

tures and various initial stresses and plastic strains, respectively, prove that the relaxation rate becomes rather high as the temperature exceeds 450°C and that stress and plastic strain assume values determined during first heating from T_{min} (Fig. 2, N = 1) [7]. Therefore, after passing the minimum stress σ_{min}, the stress increases during further heating to T_{max} because the influence of workhardening is overcompensated by stress relaxation due to the increase of total strain. This results

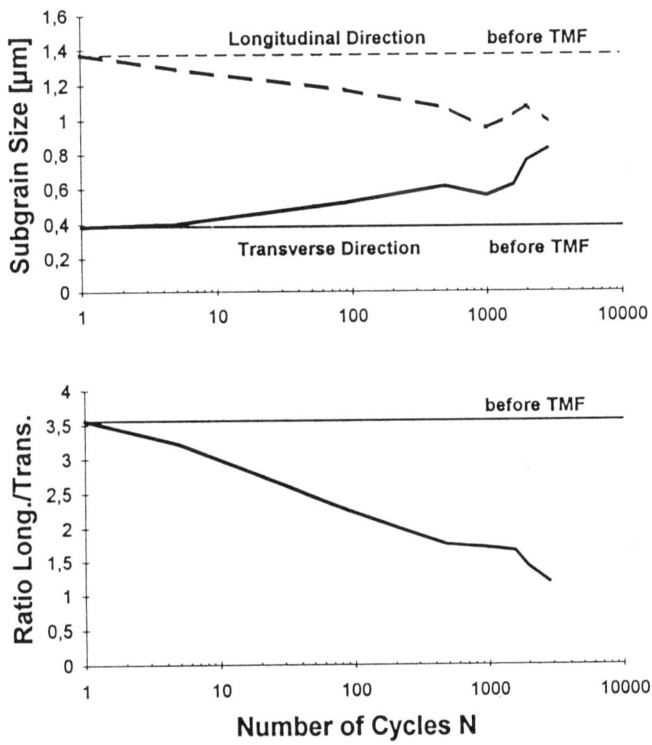

Figure 7. Subgrain size and subgrain aspect ration before and after TMF at different cycle numbers.

in a shift of the hysteresis loop towards the tensile range, in a strong reduction of the amount of the minimum stress and in a formation of tensile mean stresses, which is almost completed after two cycles (Fig. 3). The development of larger tensile mean stresses by cyclic relaxation of compressive stresses is limited by the reduction of the compressive stress itself, resulting in a strong drop of the relaxation rate, and by cyclic plastic deformation occurring during cooling to T_{min}. Hence, up to 100 cycles, the cyclic deformation stage, during which the mean stress and the plastic mean strain remain almost unchanged, is relatively stable. However, the stress range $\Delta\sigma = \sigma_{max} - \sigma_{min}$ weakly decreases and the plastic strain amplitude significantly increases, indicating cyclic softening. This is combined with a marked change of the subgrain structure. The subgrains become spheroidized (lower part Fig. 7) and the neighbour relationships between subgrain boundaries and carbides are strongly changed (Fig. 8). since the loca-

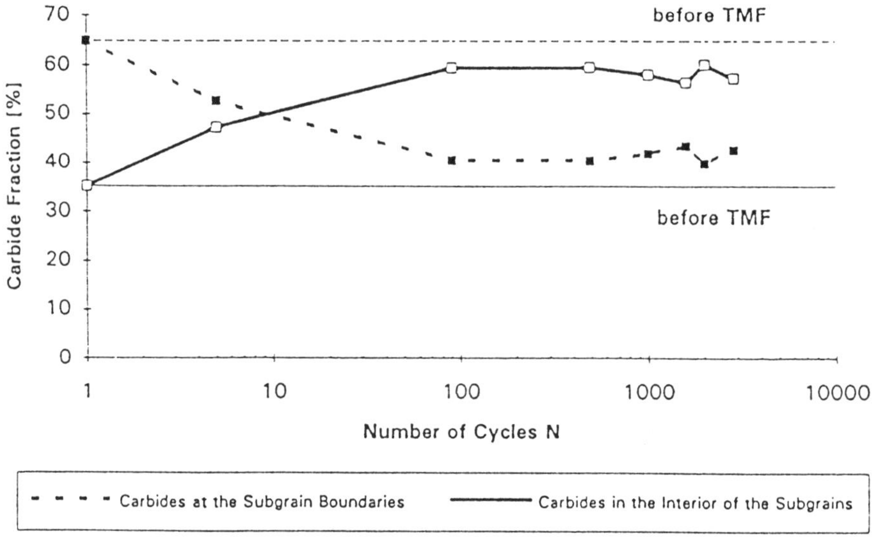

Figure 8. Carbide fractions in the interior of the subgrains and the subgrain boundaries before and after TMF at different cycle numbers.

tion of carbides cannot change during the relative short period of 100 TMF cycles,it becomes evident that many subgrain boundaries move or are dissolved during TMF. In Fig. 6 the former location of subgrain boundaries is clearly indicated by carbides arranged like strings in the interior of subgrains. It is an interesting finding that the change of the subgrain structure, which obviously promotes cyclic plastic deformation, results in an increase of microhardness (Fig. 4). This is assumed to be caused by an increase of the dislocation density in the interior of the subgrains, but has still to be evaluated quantitatively.

During further thermal cycling the spheroidization of subgrains continues, but the neighbour relationships between subgrain boundaries and carbides remain almost unchanged. After failure, the dislocation density in the interior of the subgrains decreases to $\rho_t = 2.7 \cdot 10^{10}$ cm^{-2}, which is somewhat below the value before TMF ($\rho_t = 3{,}5 \cdot 10^{10}$ cm^{-2}). After 500 cycles, significant coarsening and spheroidization of carbides sets in and continues until failure (Figs. 9 and 10). As a result of these processes, microhardness drops to 280 HV 0.3, which is 70 HV 0.3 below the maximum value at N = 100 and 27 HV 0.3 below the value before TMF. However, these rather marked microstructural changes do not result in an acceleration of the cyclic softening process, which continues uniformly without saturation until the propagation of a macrocrack after 2000 cycles (Fig. 3). Cyclic softening in this TMF stage causes a stronger decrease of the amount of maximum stress in comparison with the decrease of minimum stress, resulting in a decrease of the mean stress. Crack propagation strongly changes the shape of the hysteresis loops (Fig. 2, N ≥ 1500) and

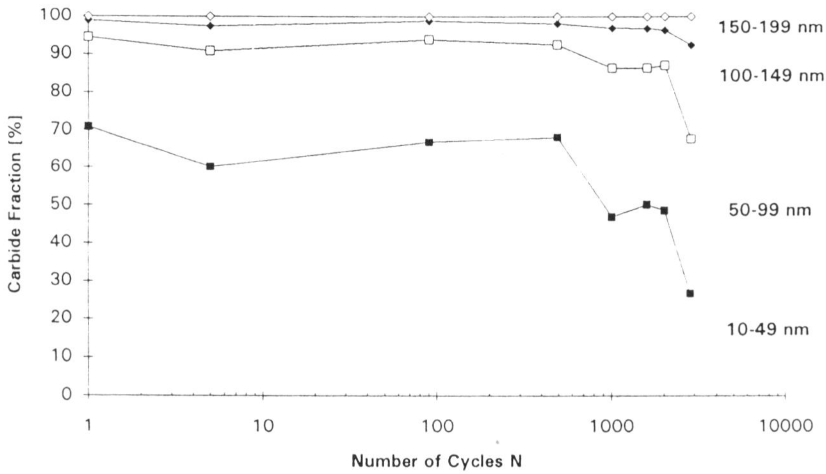

Figure 9. Change of the carbide size distribution during TMF.

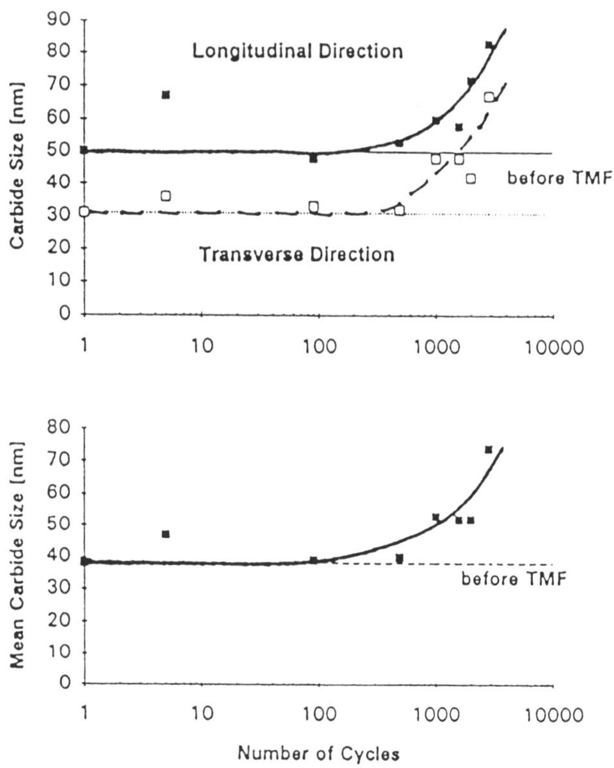

Figure 10. Evolution of the carbide size during TMF life.

the cyclic stress-strain response (Fig. 3). The numbers of cycles to crack initiation and to failure, respectively, determined in TMF tests in the present investigation are lower than those determined in total strain controlled fatigue tests carried out isothermally at 600°C with the same total strain amplitude. This is mainly an effect of the tensile mean stress being generated during out-of-phase TMF and promoting crack initiation and crack propagation. The frequency of the isothermal experiment being more than two orders of magnitude higher, further investigations are made to determine the influence of the frequency [8].

Acknowledgements - The support of the investigation by the Deutsche Forschungsgemeinschaft (DFG) is gratefully acknowledged.

References

1. Straub, S., Polcik, P., Blum, W., Development of Microstructure and Deformation Resistance of the Martensitic Steel X 22 CrMoV 12 1 during Long-Term Creep, in H. Oikawa, K. Maruyama, S. Takeuchi and M. Yamaguchi (eds.), *Proc. of the 10th International Conference on the Strength of Material (ICSMA 10),* Sendai Japan, (1994), 623-626.
2. Alvarez-Armas, I., Armas, A.F., Petersen, C., Thermal fatigue of a 12% Chromium Steel. *Fatigue Fract. Engng. Mater. Struct.* **17** (1994), 671-681.
3. Swindeman, R.W., Cyclic Stress-Strain-Time Response of a 9Cr-1Mo-V-Nb Pressure Vessel Steel at High Temperature. Low Cycle Fatigue, *ASTM STP 942,* H.D. Solomon, G.R. Halford, L.R. Kaisand, B.N. Leis, American Society for Testing and Materials, Philadelphia, (1988), 107-122.
4. Brühl, F., Haarman, K., Kalwa, G., Weber, H., Zschau, M., Verhalten des 9%-Chromstahles X 10 CrMoVNb 9 1 im Kurz- und Langzeitversuch. Teil 1: Grundwerkstoff. *VGB Kraftwerkstechnik* **69,** (1989), 1214-1224.
5. Brühl, F., Cerjak, H., Schwaab, P., Weber, H., Metallkundliche Untersuchungen am Grundwerkstoff und an Schweißverbindungen des 9%-Chromstahles X 10 CrMoVNb 9 1. *Steel Research* **62** (1991), 75-82.
6. Eggeler, G., Nilsvang, N., Ilschner, B., Microstructural changes in a 12% chromium steel during creep. *Steel Research* **58** (1987), 97-103.
7 Säuberlich, T., Löhe, D., to be published.
8. Säuberlich, T.,Pitz, G., Löhe, D., to be published.

THERMAL-MECHANICAL FATIGUE OF THE CAST ALUMINIUM ALLOY GK-AlSi10Mg wa

B. FLAIG, K.-H. LANG, D. LÖHE and E. MACHERAUCH
Institut für Werkstoffkunde I,
Universität Karlsruhe (TH)
Kaiserstr. 12, D–76128 Karlsruhe, FRG

1. Introduction

Cast aluminium alloys represent an important group of materials for practical use [1,2]. Their outstanding physical, mechanical and technological properties have well established them as common materials for important components of internal combustion engines in automotive engineering, e. g. engine blocks, cylinder heads and pistons [3]. The continuous efforts to increase power and to reduce exhaust emission, fuel consumption as well as weight lead to steadily increasing stresses in these components. Furthermore, they are subjected to thermal–mechanical fatigue (TMF) caused by transient processes like startups and shutdowns which result in thermally induced stresses and strains. Due to complex anisothermal loading the design and dimensioning of these components is a very demanding task. Commonly, for this purpose, material data from isothermal tests are employed. However, recently performed investigations show that lifetime predictions of thermal-mechanically loaded specimens based on data from isothermal experiments may be non-conservative to a large extent [4–9]. On the other hand, little information is available concerning the TMF-behaviour of cast aluminium alloys. The investigations carried out so far are limited to the determination of the number of cycles to crack initiation and to failure, respectively [10–16].

In the present work, the influence of the maximum cycle temperature on TMF-life and the development of the cyclic stress-strain response during TMF of a cast aluminium alloy (German grade GK-AlSi10Mg wa) is analyzed. The tests were run with the temperature- time course being out-of-phase with the mechanical strain-time course, thus simulating the conditions existing in a fast heating part of a component, the thermal expansion of which is hindered by colder parts.

2. Material

The material investigated was a below eutectic cast aluminium alloy, German grade GK-AlSi10Mg wa, the chemical composition (in weight %) was 9.53 Si, 0.41 Mg, and 0.13 Fe. Cylindrical blanks with a diameter of 22 mm and a length of 120 mm

J. Bressers and L. Rémy (eds.), Fatique under Thermal and Mechanical Loading, 361–370.

were chill-casted by the manufacturer (Alcan Industries, Nürnberg, FRG) and heat treated to achieve maximum hardness by a T6 procedure consisting of solution annealing for 5h at 520°C, quenching in water of 20°C and finally ageing 5h at 170°C.

Fig. 1 shows the microstructure of the heat-treated state which is composed of Al-rich solid solution phase (α) and eutectic consisting of α and Si. The globular shape of the Si-crystals was produced by treating the melt with Na. During the heat treatment, the Mg_2 Si-phase existing in the chill-cast state is dissolved, and very fine precipitations are formed during final ageing, which are of course not visible in the micrograph due to the magnification chosen.

From the blanks solid round specimens with a cylidrical gauge length of 10 mm and a gauge diameter of 7 mm were machined.

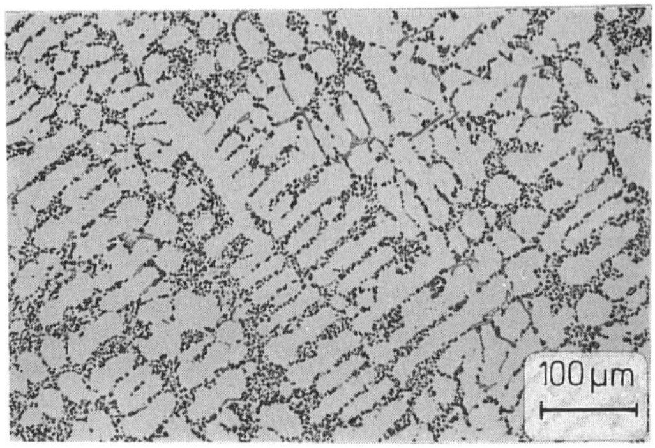

Figure 1. Light microscopic cross section of the cast aluminium alloy GK-AlSi10Mg wa.

3. Experimental Procedure

TMF tests were run on a closed-loop controlled electro-mechanical testing machine. The specimens were fixed with hydraulic grips. For heating, a 5 kW inductor generator was used. Cooling was achieved by thermal conduction into the watercooled grips and by blowing compressed air on the specimens using a proportionally

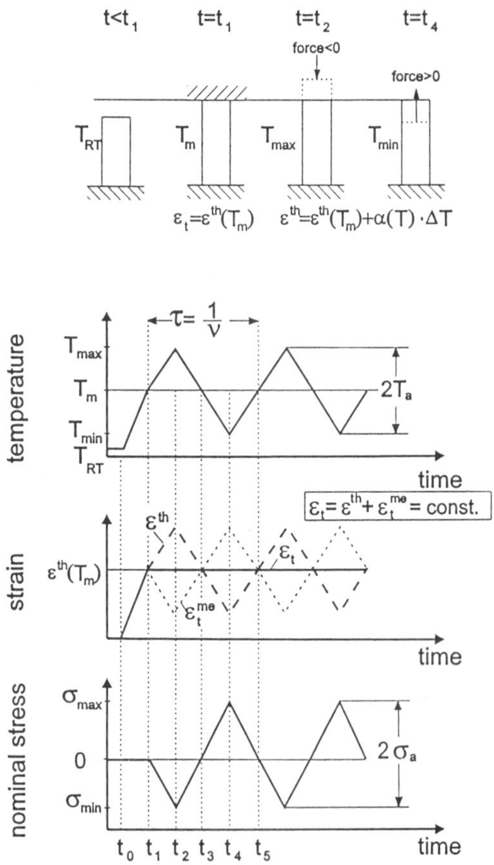

Figure 2. Experimental procedure.

controlled valve. Temperature was measured with a NiCr-Ni thermo- couple which has a diameter of 1 mm and was fixed into a bore very near to the gauge length. Strain measurements were performed with a capacitive strain gauge system [17].

In TMF tests, the internal constraint caused in components by instationary temperature gradients is replaced by external constraint resulting in partial or total suppression of thermal strains. As shown in Fig. 2, in the present investigation the specimens were allowed to expand freely at zero tress during the initial heating from ambient temperature to the mean cycle temperature T_m. After reaching T_m the closed-loop control was switched over from stress to total strain control, and the sum of the thermal strain ε^{th} and the mechanical strain ε_t^{me} was kept at a constant value. Due to $\Delta\varepsilon_t^{me} = -\Delta\varepsilon^{th}$, the appearing cyclic mechanical strains are out-of-phase with the triangular temperature- time course chosen. Thus, compressive stresses develop at high temperatures and tensile stresses at low temperatures of the thermal cycle. T_{min} was always 50°C and T_{max} varied between 150°C and 350°C. The period of a cycle was always one minute, so that the strain rate $\dot{\varepsilon}$ (temperature rate T) varies between $8 \cdot 10^{-5}$ 1/s and $2 \cdot 10^{-4}$ 1/s (3°C/s and 10°C/s). The tests were finished after failure of the specimens or after 10^4 cycles. This ultimate number of cycles led to a maximum test duration of one week.

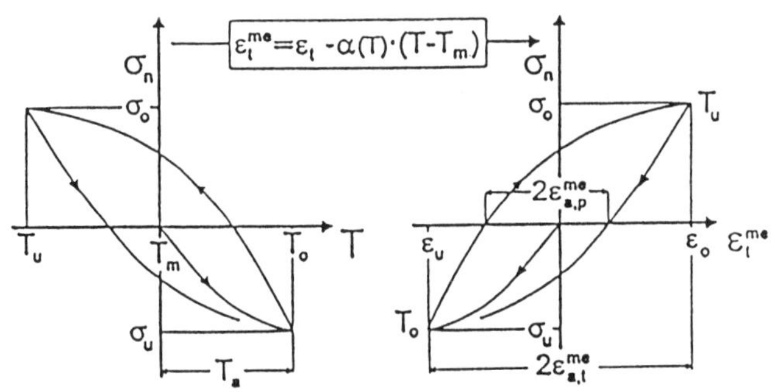

Figure 3. Evaluation of the thermal-mechanical tests.

During TMF nominal stress is measured as a function of temperature, and stress-temperature hysteresis loops are evaluated. The left-hand side of Fig. 3 schematically shows the σ-T-hysteresis loop developing during the first thermal cycle. The width of the loop depends on the amount of plastic deformation. As mentioned above, the thermal strains $\Delta\varepsilon^{th}(T) = \alpha(T) \cdot (T-T_m)$ (α = thermal expansion coefficient) are completely compensated by the mechanical strain ε_t^{me}. Therefore a stress-strain hysteresis loop as shown at the right-hand side of the figure may be determined. From this hysteresis loop, the total strain amplitude $\varepsilon_{a,t}^{me}$ and the plastic strain amplitude $\varepsilon_{a,p}^{me}$ can be drawn out.

4. Cyclic Deformation Behaviour and TMF-Life

As an example, the upper part of Fig. 4 shows stress-temperature hysteresis loops developing during the 10th cycle, the 100th cycle and the cycle preceeding failure in a TMF test with T_{max} = 325°C. During heating from T_m in the 10th cycle, compressive stress develops, and the onset of pronounced plastic deformation is observed as temperature exceeds 170°C. During further heating to T_{max}, cyclic stress relaxation starts, and the amounts of compressive stress decrease even though the mechanical strain still increases until T_{max} is reached. Consequently, the minimum stress is not correlated with T_{max}.

During cooling from T_{max} to T_{min}, tensile stresses develop, which are much larger than the amounts of compressive stress. With increasing number of cycles, the hysteresis loops become wider and the stress range σ_{max} - σ_{min} decreases, indicating cyclic softening. In the lower part of the figure, stress-strain hysteresis loops obtained by transformation of the stress-temperature hysteresis loops are shown. Again, the existence of a tensile mean stress and cyclic sofening is clearly visible.

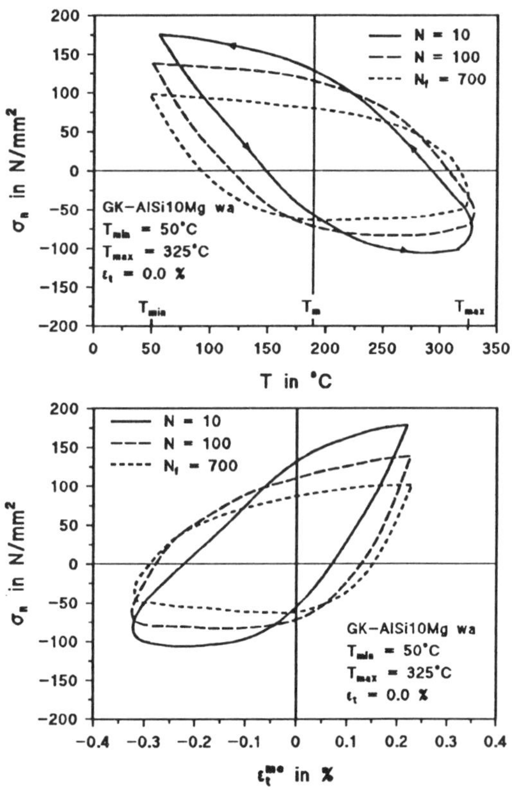

Figure 4. Stress temperature hysteresis loop (above) and stress strain hysteresis loop (below).

From stress-strain hysteresis loops, the development of stress amplitude σ_a, mean stress σ_m and plastic strain amplitude $\varepsilon_{a,p}^{me}$ during TMF-life as shown in Fig. 5 was evaluated. During TMF atT_{max} = 200°C, σ_a remains almost constant, and no macrocrack formation is observed within the ultimate number of cycles. Also during TMF at T_{max} = 225°C, the stress amplitude hardly changes until crack initiation and crack propagation occur, resulting in a marked drop of the measured values. During TMF at T_{max} = 250°C, the stress amplitude decreases after 100 cycles, indicating cyclic softening. At still higher T_{max}, a decrease of σ_a is observed from the very beginning of the test, the initial slope of the curves being the steeper the higher T_{max} is. In the range 200°C ≤ T_{max} ≤ 275°C, the initial stress amplitudes increase with increasing T_{max}. However, in the range 275°C ≤ T_{max} ≤ 350°C at a given number of cycles the stress amplitudes decrease with increasing T_{max} throughout TMF life because of lower material resistance at higher temperatures.

Due to cyclic relaxation of compressive stresses at sufficiently high temperatures the mean stress σ_m shown in the middle part of Fig. 5 at first increases during TMF at 200°C ≤ T_{max} ≤ 325°C. The initial gradient $d\sigma_m/d\log N$ increases with rising T_{max}. Whereas σ_m monotonously increases during TMF at T_{max} = 200°C and 225°C, maxima of σ_m occur in the range 250°C ≤ T_{max} ≤ 325°C. The value of the maximum mean stress climbs as T_{max} rises from 250°C to 275°C and hardly changes at still higher T_{max}. The number of cycles corresponding to the maxima is the lower the higher T_{max}

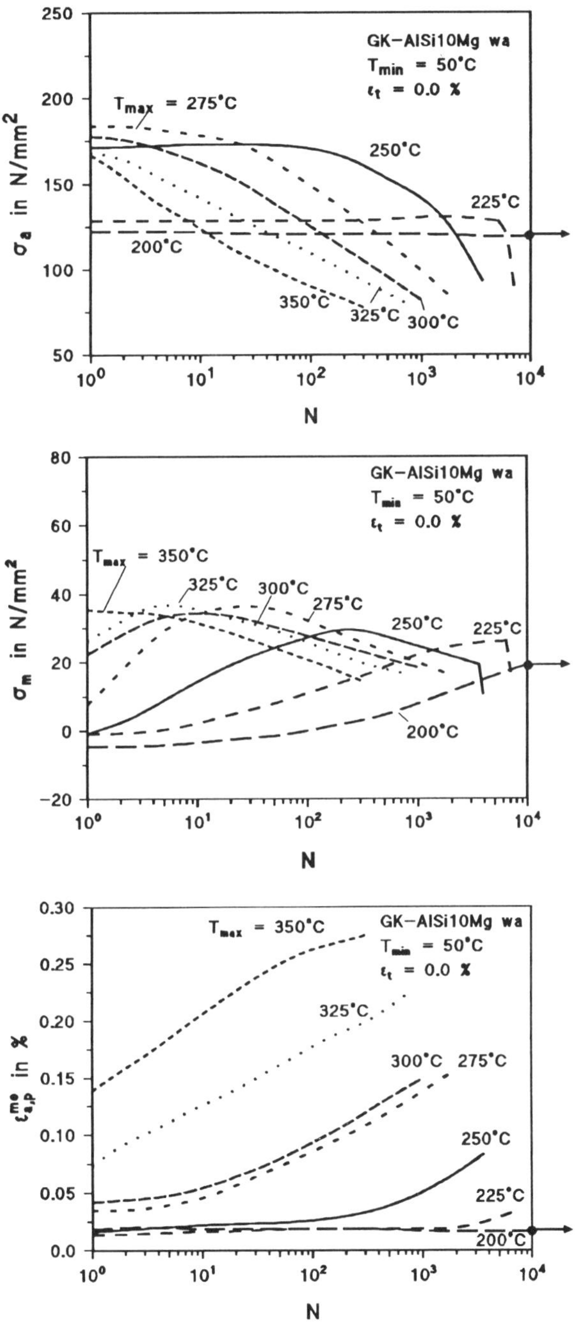

Figure 5. Stress amplitude, mean stress and plastic strain versus number of cycles.

is with the exception of the test at T_{max} = 350°C, where a relatively high mean stress exists already in the first cycle and continuously decreases during TMF life.

The development of the plastic strain amplitude is illustrated in the lower part of Fig. 5. During TMF at T_{max} = 200°C and 225°C, the values measured are very low, and only after 1500 cycles at T_{max} = 225°C a weak increase of $\varepsilon_{a,p}^{me}$ is observed. Low initial values are also measured a T_{max} = 250°C. However, after an incubation period of 100 cycles, the plastic strain amplitude increases markedly. A similar observation, however at higher $\varepsilon_{a,p}^{me}$-values and shorter incubation periods, is made at T_{max} = 275°C and 300°C. At even higher T_{max}, the initial plastic strain amplitudes are rather high and increase from the very beginning of the test, indicating strong cyclic softening.

The upper part of Fig. 6 shows the relationship between the maximum temperature and the temperature amplitude versus the number of cycles to failure. The data points of the semilogartihmic plot are well fitted by a straight line. TMF at T_{max} = 150°C and 200°C does not result in the formation of a macrocrack within the ultimate number of cycles. An increase of maximum temperature from 200°C to 350°C reduces TMF life from 10^4 to $3 \cdot 10^2$ cycles. At maximum temperatures exceeding 300°C, failure of the specimens is not caused by crack propagation, but rather by pronounced barrelling of the gauge length.

In the lower part of the Fig. 6, the relationship between total strain amplitude and number of cycles to failure is given. Again, the measured values are fitted quite reasonably by a straight line in the semilogarithmic graph. The decrease of TMF life as mentioned above is connected with an increase of the total strain amplitude from 0.14 to 0.3%.

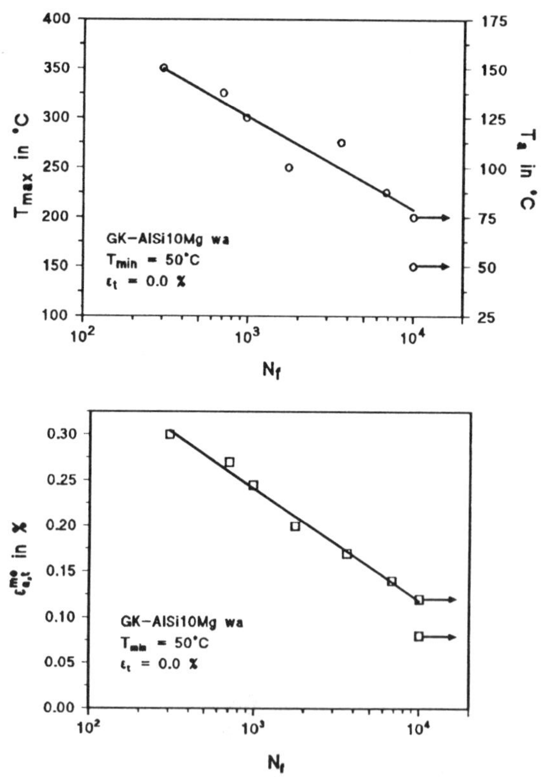

Figure 6. Temperature amplitude and transformed total strain versus cycles to failure.

5. Microscopic Observations

Crack initiation was investigated by optical microscopy of the cross section of specimens thermal-mechanically fatigued at T_{max} = 250°C to different number of cycles and to failure, respectively. The results obtained are typical of TMF tests at T_{max} up to 300°C.

Already after ten cycles, cross- shaped microcracks with a length of about 5 μm are observed in the α- matrix close to Si-particles. After 100 cycles, the length of these microcracks is almost the same, and their number is hardly increased. After 1000 cycles (corresponding to approximately 0.2 N_f) microcracks initiating at the surface are observed. Their length amounts to about 30 μm. The cracks mainly propagate through the α-matrix, occasionally cutting broken Si- particles especially near the surface. After specimen failure, several secondary cracks are observed around the fractured surface. Again, the crack propagation path mainly follows the α-matrix, rarely cutting through damaged Si-particles.

In TMF tests at T_{max} > 300°C, specimen failure did not occur by rupture, but rather by large inhomogeneous deformations resulting in a barrelling of the gauge length. In these cases, the tests were terminated after a 50% drop of the maximum stress σ_{max}. The formation of some 150 μm long, widely opened cracks, which initiated at the surface, is combined with barrelling.

TEM-micrographs of specimens tested after 100 cycles, 1000 cycles and after failure, respectively, under the same TMF loading conditions firstly show a homogeneous dislocation distribution after 100 cycles. Upon further cycling, the dislocation structure becomes heterogeneous, and regions with high and low dislocation density, respectively, are formed. After failure, a pronounced cell and subgrain structure is observed. Simultaneously, overaging occurs resulting in coarse precipitates. The grain size increases during TMF at high T_{max}, the value measured after failure being the larger the higher T_{max} is.

6. Discussion

The out-of-phase TMF-tests carried out in the present investigation simulate the loading conditions in the hot part of a cooled component (e.g. the combustion side of a watercooled cylinder head). Increasing temperature ranges T_{max} - T_{min} increase the temperature gradient and consequently also the mechanical strain. Due to the constant minimum temperatures T_{min} in all TMF tests, the total mechanical strain increases with rising T_{max}. On the other hand the material's resistance against plastic deformations decreases with increasing temperatures. As a result, during heating from T_m to T_{max} cyclic plastic deformation as well as cyclic relaxation of compressive stresses increase with increasing T_{max} due to both decreasing strength and increasing total mechanical strain (see lower part of Fig. 5).

The stress-temperature hysteresis loops are cyclically shifted towards tensile stresses by cyclic relaxation of compressive stresses during heating, resulting in the development of tensile mean stresses (see middle part of Fig. 5). The initial slopes of the σ_m-log N curves prove that this effect is the more pronounced the higher T_{max} is, as was to be expected. However, only at T_{max} = 200°C and 225°C, the shift is more or less continuous during TMF life. During TMF at higher T_{max}, the development of larger tensile mean stresses is limited by plastic deformation occurring during cooling to T_{min}. The occurrence of maximum mean stresses during TMF at 250°C ≤ T_{max} ≤ 325°C is combined with cyclic softening, which lowers the material's resistance against plastic deformation. Consequently, cyclic plastic deformation during cooling to T_{min} increases, resulting in decreasing σ_m, and cyclic softening is the more pronounced the higher T_{max} is. Therefore, as T_{max} increases, the maximum of the mean stresses is shifted to smaller numbers of cycles. During TMF at 350°C, marked softening already occurs during the first cycle, resulting in continuously decreasing mean stresses.

Cyclic softening becomes clearly evident from the decrease of the stress amplitude and the increase of the plastic strain amplitude (see upper part and lower parts of Fig. 5, respectively). As T_{max} exceeds 275°C, σ_a decreases although the total mechanical strain amplitude increases. Obviously, at high T_{max} pronounced softening already occurring after the first cycle, as mentioned above, outweighs the influence of the mechanical strain amplitude on the stress amplitude. Cyclic softening is mainly a result of the overageing of the finely distributed partly coherent β'-precipitation phase, which is transformed into the incoherent Mg_2Si-phase during TMF at high T_{max}. The removal of effective obstacles to dislocation motion and the enhancement of thermal activation of dislocation movement promote recovery processes and diffusion controlled plastic deformation, resulting in a significant alteration of the shape of stress–temperature or stress–strain hysteresis loops during heating to T_{max} as it becomes evident from the sequence of loops shown in Fig. 4. These processes finally result in the formation of a cell and/or subgrain structure during later stages of TMF at high T_{max}. An interesting observation is the formation of non-propagating cracks in the eutectic areas of the microstructure. As the thermal expansion of silicon particles differs from that of the α-matrix, additional cyclic stresses are generated in microscopically small areas, which superimpose on the stresses produced by the overall suppression of thermal expansion. Consequently, microscopically short cracks are initiated rather early during TMF life close to the silicon particles. However, these microcracks hardly propagate into the surrounding α-matrix, because the additional stresses and hence the driving force necessary for crack propagation strongly drop as the crack tip leaves the vicinity of the silicon particles.

Several proposals have been made in the literature to evaluate damage accumulating during cyclic loading and to account for the influence of loading conditions on fatigue life. One of the best known is the Manson–Coffin–relationship (MCR), which is almost entirely used for the assessment of isothermal low cycle fatigue behaviour. Nevertheless, the original motive, from which MCR was derived, was to evaluate damage produced by thermal fatigue [6, 18–19]. Fig. 7 shows a plot of the plastic strain amplitude at N = N_f /2 versus log N_f according to MCR. The straight line fitted to the datapoints yields

$$\varepsilon_{a,p} = 54.7 \cdot N_f^{-0,87} \qquad (1)$$

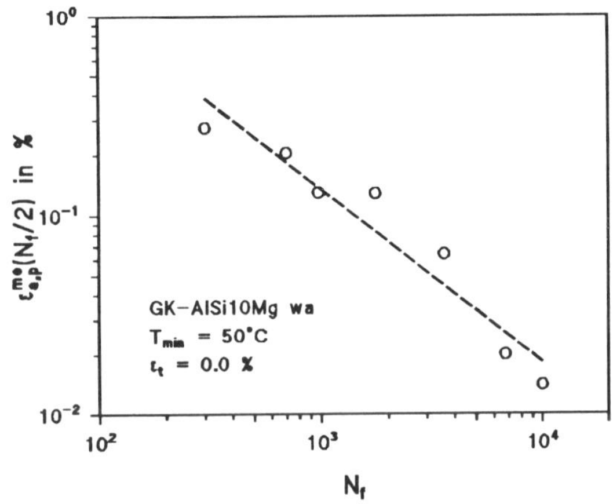

Figure 7. Manson-Coffin-relationship.

However, the way in which the experimental values fit MCR is at best acceptable. This may be the result of the mean stress being different during TMF at different T_{max}, which is not explicitly accounted for by MCR. However other damage parameters like the Smith–Watson–Topper parameter or the Ostergren parameter, which take into account the maximum stress $\sigma_{max} = \sigma_a + \sigma_m$ describe the experimental results even more inaccurately. Therefore, it is assumed that relations like that shown in Fig. 7 which are commonly used for the assessment of isothermal fatigue behaviour are strongly influenced by pronounced cyclic softening occurring during TMF at high T_{max}.

References

1. Aluminium: High-Tech-Werkstoff mit Zukunft, Ingenieur-Werkstoffe 3,(1993), Nr. 6, 10 - 14.
2. Taschner, W., Mit Aluminium in die Industrie, Ingenieur-Werkstoffe 4, (1993), Nr. 12, 52 - 54.
3. Köhler, E., Maßgeschneiderte Werkstoffe für schadstoffoptimierte Motoren, Aluminium, (1993), 7, 627 - 628.
4. Taira, S., Relationship Between Thermal Fatigue and Low-cycle Fatigue at Elevated Temperatures, *Fatigue at Elevated Temperature*, ASTM STP 520, 1972, 80 ff.
5. Manson, S.S., Thermal Stress and Low Cycle Fatigue, McGraw-Hill Book Company, New York, S. Francisco, Toronto, London, Sydney, (1966), Kap.6.
6. Halford, G.R., Saltsman, J.F., Calculation of Thermomechanical Fatigue Life Based on Isothermal Behaviour, Thermal Stress, Material Deformation and Thermo-mechanical Fatigue, Special Pub. PVP - 123, ASME, New York, (1987), 9 - 21.
7. Hallstein, R., Das Verhalten von Gußeisenwerkstoffen unter isothermer, thermischer und thermisch-mechanischer Wechselbeanspruchung, Dissertation, Universität KLarlsruhe, (1991).
8. François, M., Rémy, L., Thermal-mechanical Fatigue of MAR-M 509 Superalloy. Comparison with Low-cycle Fatigue Behaviour, *Fatigue Fract. Engng. Mater. Struct.*, **14**, No. 1, (1991), 115 - 129.
9. Weronski, A., Hejwowski, T., *Thermal Fatigue of Metals*, Marcel Dekker Inc., New York, Basel, Hong Kong, 1991.
10. Röhrle, M., Rißneigung von Leichtmetall-Kolben-Legierungen durch periodische Wärmebeanspruchung, Dissertation, Universität Stuttgart, (1968).
11. M. Röhrle: Rißneigung von Leichtmetall–Kolben–Legierungen durch periodische Wärmebeanspruchung, MTZ 30, (1969), 344 - 353 und 472 - 479.
12. Zimmermann, P., Gobrecht, J., Warmfeste Aluminium-Magnesium-Silizium-Gußlegierungen für temperaturwechselbeanspruchte Gußstücke, Aluminium **56**, (1980), 323 - 328.

13. Babes, F., Gobrecht, J., Legierungsoptimierung bei luftgekühlten Aluminium-Zylinderköpfen mittels technologischer Prüfverfahren, Berichstband 7, Int. Leichtmetalltagung Leoben, Vereinigte Metallwerke Ranshofen Berndorf AG, (1981), 208 - 209.
14. Essig, G., Mielke, S., Henning, W., Faserverstärkte Serienpressgußkolben, Ingenieur Werkstoffe, **3**, (1991), Nr. 6, 48 - 50.
15. Mielke, S., Bucher, G., Eschenweck, D., Faserverstärktes Aluminium für den Einsatz bei hohen Betriebstemperaturen, Konf.: *3. Int. Symposium Aluminium + Automobil,* 03. - 04.02.1988 in Düsseldorf, 176 - 178.
16. Ohuchi, H., Tsuda, M., The Influence of the Primary Si-Crystal Shape on the Thermal Shock Properties of the Hyper–Eutectic Al-Si Casting Alloys, Aluminium **59**, (1983), 463 - 465.
17. Lang, K-H., Dehnungsmessung bei Dauerschwingversuchen, Bad Nauheim, DVM, 1993
18. Manson, S.S., Behaviour of Materials under Conditions of Thermal Stress, ASTM STP1170, (1954), 1 - 34.
19. Kuwabara, K., Nitta, A., Thermal-Mechanical Low Cycle Fatigue under Creep Fatigue Interaction on Type 304 Stainless Steel, ICM 3, Vol 2, (1979), 69 - 78.

THE EFFECT OF PARTICLE SIZE ON THERMO-MECHANICAL FATIGUE OF Al/SiC METAL MATRIX COMPOSITES

H. SEHITOGLU
Department of Mechanical and Industrial Engineering
University of Illinois
Urbana, Illinois 61801

1. Introduction

Numerous research papers have focused on elevated temperature behaviour of metal matrix composites with aluminium matrices and silicon carbide reinforcements in recent years [1-6]. These materials exhibit superior elevated temperature strength, higher elastic modulus and lower coefficient of thermal expansion compared to matrix alloys. There is abundant information on experimental, numerical and analytical studies of volume fraction effects on mechanical properties of the materials. There is far less information on the influence of particle size on the strengthening and flow behaviour [7-9]. Preliminary experimental results confirmed that the particle size effect is significant and could even surpass the volume fraction effect [8-9]. The reasons for the increased strengthening associated with the particle size have been linked to grain refinement [10-11], geometrically necessary dislocations [12], Orowan [13] looping mechanism, and dislocation pileup or blockage of plastic flow at the particles. It has been forwarded that smaller particle sizes result in smaller grain sizes of the matrix resulting in strengthening through a Hall-Petch type of relationship. Simple analyses for geometrically necessary dislocations have been reported by Ashby [12] resulting in a dependence of the strength on particle size to one-half power. The Orowan looping mechanism could be operative especially at small particle sizes but its contribution to strengthening becomes small as the particle size increases and nears the grain size. The dislocation pileup at particles could be an effective mechanism for blocking slip. We note that the local shear stress distribution near particles is non uniform, therefore, the strengthening is expected to differ depending on the type of loading.

This paper deals with the deformation and fatigue behaviour of Al/SiC metal matrix composites under isothermal and thermo-mechanical loading. The composite consisted of 10%, and 30% volume fraction of silicon carbide particles with particle sizes of 2 and 30 μm in diameter. Experimental results indicate that the decrease in particle size resulted in longer fatigue lives. Furthermore, decreasing the particle size resulted in higher strengthening in thermo-mechanical fatigue (100°C-300°C) loading conditions compered to isothermal fatigue (20°C, 200°C and 300°C). The

J. Bressers and L. Rémy (eds.), Fatique under Thermal and Mechanical Loading, 371–380.

local shear stress distribution in the vicinity of particles was determined for isothermal and thermo-mechanical loading. A dislocation particle interaction model was derived which explains the increase in strengthening with decrease in particle size, and more interestingly, the enhanced strengthening in the thermo-mechanical loading case. A relationship between strength and particle size based on pileup of dislocations against particles in established. Thermo-mechanical loading is conducive to higher strengthening compared to the isothermal case in view of the different shear stress in the vicinity of inhomogeneities.

2. Experiments and Material

Two kinds of experiments were performed. In the isothermal cease, the specimens were deformed in monotonic tension at temperatures of 20°C, 200°C and 300°C. In the thermo-mechanical loading case the strain and temperature were varied either in-phase or out-of -phase as shown in Fig. 1. This figure defines the thermo-mechanical out-of-phase (OP), thermo-mechanical in-phase (IP) and isothermal fatigue (IF) loadings. The net strain, e_{net}, is the sum of thermal and mechanical strain components. The thermal strain is defined as the product of thermal expansion coefficient, α and the temperature range where T_o is the reference temperature (=200°C in this study). The e_{net} and e_{mech} represent overall, and not internal, strain components on the monolithic and composite materials. The mechanical strain is the sum of elastic and inelastic strain components. The procedure for conducting TMF experiments has been outlined in early work [2-4, 5-6, 8-9].

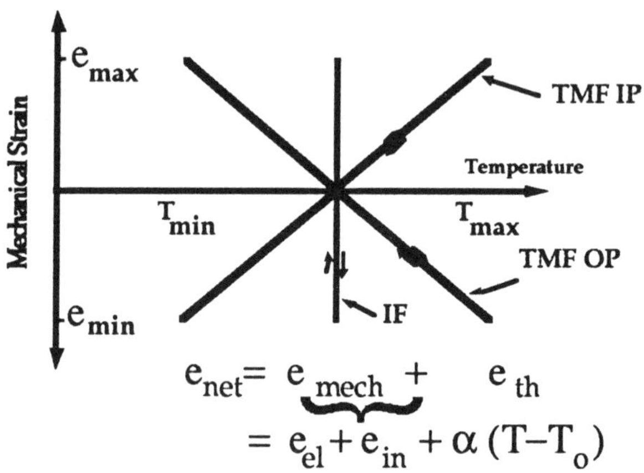

Figure 1. Schematic of strain-temperature variation in isothermal and thermo-mechanical fatigue loading.

The experiments were conducted on solid cylindrical specimens of Al 2024 reinforced with 10% and 30% volume fraction, V_f, of silicon carbide particles. The material is a powder metallurgy alloy with a matrix chemical composition of 4.35% Cu, 1.54% Mg, 0.59% Mn, 0.19% Fe, 0.13 Si, 0.032% Zr, 0.03% Cr produced by Kobe Steel of Japan. The microstructure of the V_f=30% composite with two different particle sizes is illustrated in Fig. 2.

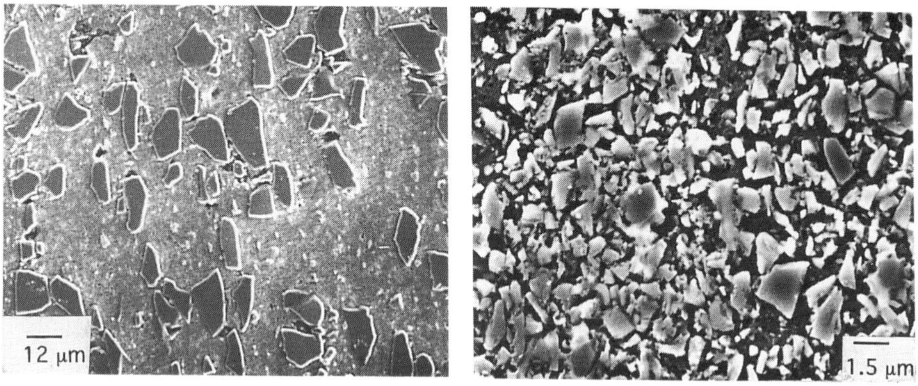

Figure 2. Microstructure showing the aluminum reinforced with 30% volume fraction particles for the (a) 30 micron and (b) 2 micron particle size case.

3. Summary of Experimental Results

The stress-strain behaviour for the thermo-mechanical fatigue experiments is shown in Figs. 3(a) -3(b). The strain rate for the thermo-mechanical experiments was 1. 10^{-4} 1/s corresponding to a cycle period of 60-120 s. In Fig. 3a the results for TMF OP and in Fig. 3(b) the TMF IP experiments are shown.

In all figures the thinner line denotes the hysteresis loops at first cycle (N=1) and the thicker line the behaviour at half line($N=N_f/2$). The results for the unreinforced material are shown on the far left, the 30% SiC material with 30 μm reinforcement in the middle and the 30% SiC material with 2 μm reinforcement on the far right. The stress range associated with both reinforced materials is higher compared to the monolithic alloy.

TMF OP experiments were conducted on two MMCs (V_f = 30% with d = 2 μm and 30 μm) and the unreinforced material at a strain rate of approximately 10^{-4} s^{-1}. Fig. 4 summerizes the TMF OP experiments which show the life, in cycles, as a function of stress range observed during the first cycle of each strain controlled experiment. The thermo-mechanical fatigue lives were considerably longer for the 30% volume fraction 2 μm material. The lives of the monolithic aluminium alloy were lower than the two composite microstructures.

In the course of the research, we noted that the strengthening was a strong function of temperature of the experiment, the particle size and the type of loading, ie. isothermal versus thermo-mechanical. Specifically, the strengthening of the composite relative to the matrix was considerable at room temperature and at 200°C and this strengthening decreased as 300°C was approached. The most noteworthy result was that the strengthening in thermo-mechanical fatigue experiments (all of them with T_{min}=100°C and T_{max}=300°C) was considerably higher than that in the isothermal case. To illustrate this point, in Fig. 5, the strengthening (Yield stress -

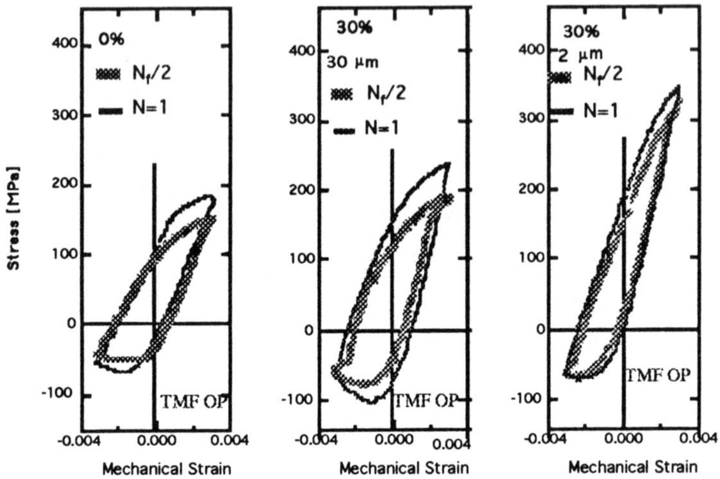

Figure 3a. Stress-mechanical strain behaviour during out-of-phase experiments with T_{min}=100°C and T_{max}=300°C and a mechanical strain range of 0.006.

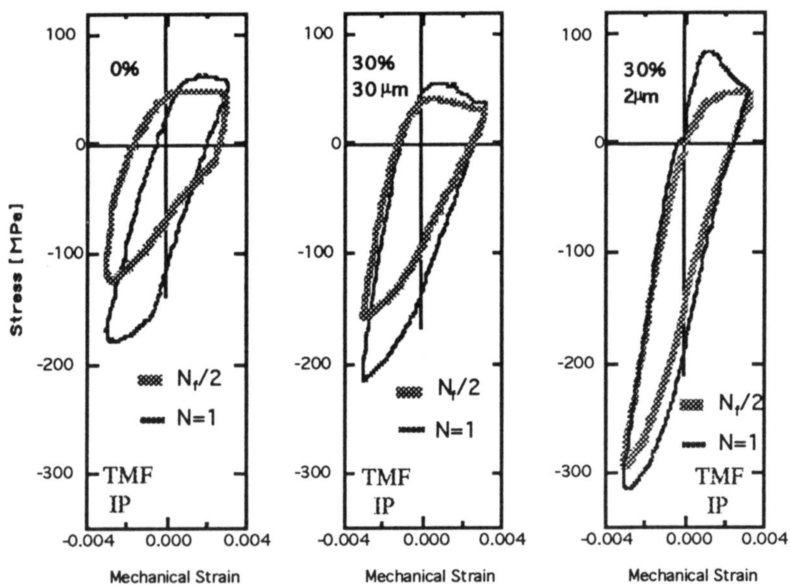

Figure 3b. Stress-mechanical strain behaviour during in-phase experiments with T_{min} 100°C and T_{max}=300°C and a mechanical strain range of 0.006.

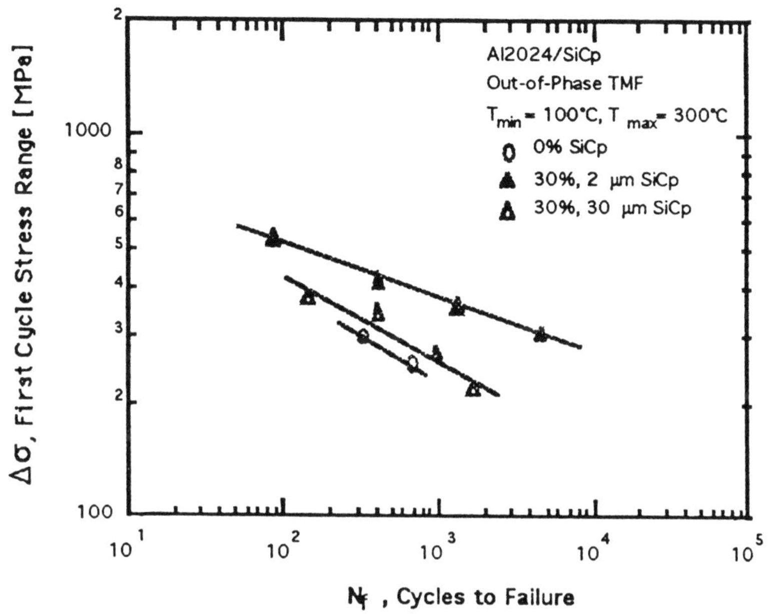

Figure 4. Comparison of thermo-mechanical fatigue lives of three materials under out-of-phase thermo-mechanical fatigue loading.

Matrix Yield Stress) is plotted versus $1/d^{½}$ where d is the particle size. The yield stress represents the composite material strength and the matrix yield stress was obtained form the monolithic material under the same loading conditions. The volume fractions for these experiments were 30%. The $1/d^{½}$ = 0.18 corresponds to the 30 μm particle case and the $1/d^{½}$ = 0.7 corresponds to the 2 μm particle size case. The results from 25 experiments are summarized in Fig. 5. The figure contains several monotonic experiments at room temperature on samples that have been exposed to 300°C for a period of 10 and 100 hours. The data points "o" represent the strengthening in the thermo-mechanical fatigue case (both TMF OP and IP) and we note that this strengthening increment is as high as 250 MPa. This is considerably higher than strengthening obtained under all other conditions. The remainder of the paper is devoted to explanation of this strengthening effect in thermo-mechanical loading.

A photo obtained from transmission electron microscopy of a composite specimen is shown in Fig. 6. The matrix microstructure and the particle/matrix interface have been examined with a Philips EM420 transmission electron microscope operated at 120 KeV. These photographs confirm that blockage of slip by the particles is a contributing factor to strengthening in metal matrix composites. If this is the case, the dislocation pile-up behaviour is expected to be different in isothermal versus thermo-mechanical loading because the local shear stresses are different for these two cases.

In the next sections, the local shear distributions in the vicinity of an inhomogeneity are determined under thermo-mechanical and isothermal loading conditions.

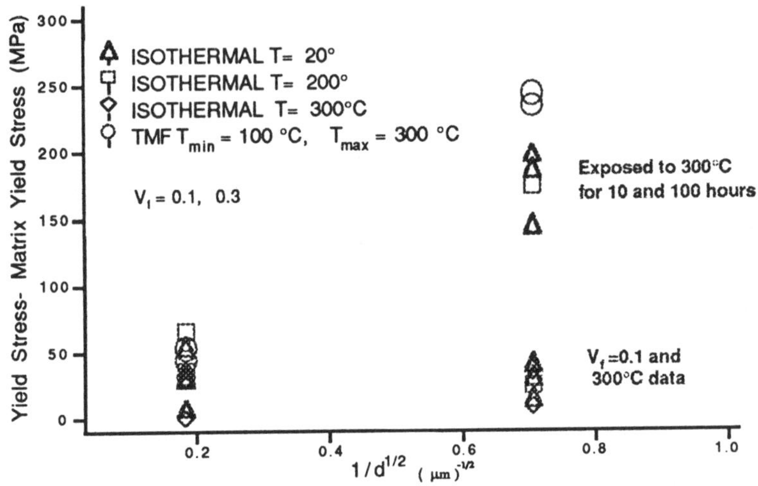

Figure 5. Summary of results showing strengthening as a function of particle diameter (25 experiments).

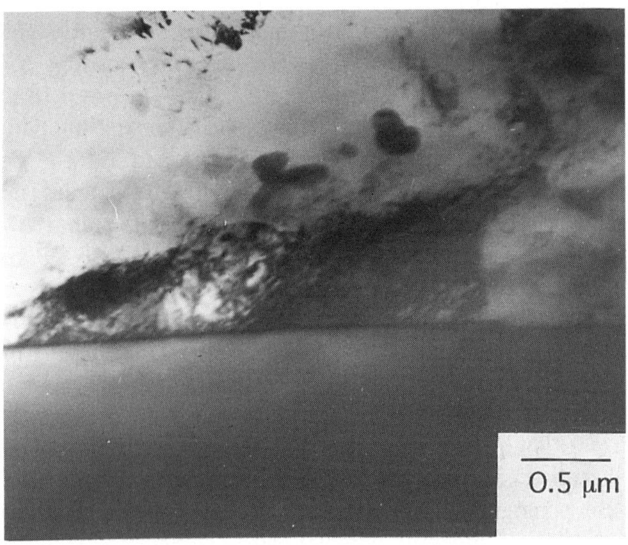

Figure 6. Dislocation pileup and density in the vicinity of particle matrix interface.

4. Determination of Shear Stress Distribution near an Inhomogeneity under Isothermal and Thermo-mechanical Loading

Eshelby [14] reported stresses for the case where the inhomogeneity (which has different elastic constants than the rest of the medium) is embedded in an infinite elastic body. Since its deformation is constrained by the surrounding matrix, a perturbed strain field results. The problem, then, is to find the local (matrix) strain fields due to the existence of the inhomogeneity. The Eshelby's equivalent inclusion theorem is applied to the case of a cylindrical inclusion with radius a. Some preliminary work on this geometry has been published by Mikata an Taya [15]. The details of the current analysis are reported by Sehitoglu [6].

4.1. THERMAL LOADING CASE

Using the Eshelby's equivalent inclusion theorem, the local shear strain, e_{12}, normalized by the thermal mismatch strain in Fig. 7 where e^{th} is the thermal mismatch strain ($e^{th} = (\alpha_2 - \alpha_1)\Delta T$), α_2 an α_1 represent the coefficient of thermal expansion of the particle and the matrix respectively and ΔT is the temperature change. The shear strain is evaluated on the plane $\frac{y}{a} = 1$ because on the $\frac{y}{a} = 0$ plane the shear stresses are zero. The z (33) is the long direction of the cylinder. For the Al/SiC system the G_2/G_1 ratio of 5 and v=0.3 was used where G_2/G_1 is the shear modulus ratio of particle to matrix and v is the Poisson's ratio of the matrix.

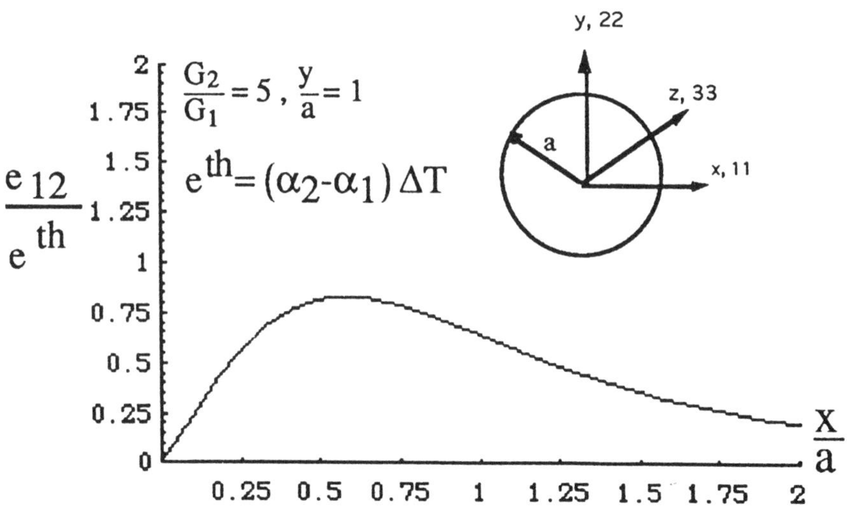

Figure 7. The variation of shear strain under thermal loading as a function of normalized distance from the inhomogeneity ($y/a = 1$, $0 \le x/a \le 2$)

4.2. ISOTHERMAL CASE

The local shear strain field under an applied strain field of e_{12}^{A} under isothermal loading is plotted as a function of $\frac{x}{a}$ $(1 \le \frac{x}{a} \le 2)$ for the plane $\frac{y}{a}=0$ in Fig. 8.

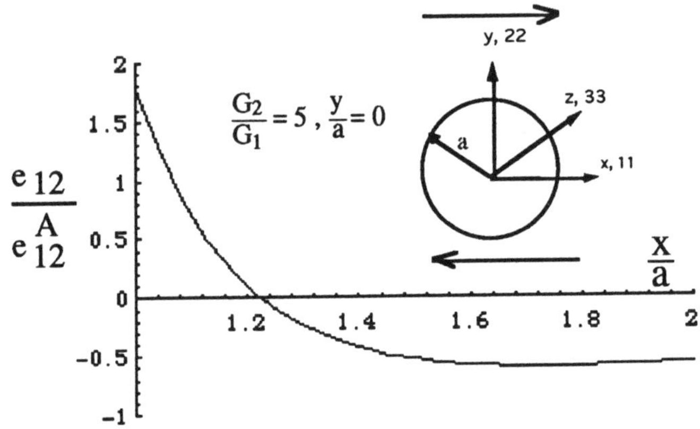

Figure 8. The variation of shear strain under isothermal loading as a function of shear moduli ratio and normalized distance from the inhomogeneity ($y/a = 0, 1 \le x/a \le 2$).

Two major observations can by made from these plots: (i). The maximum local shear strain under isothermal loading is higher relative to that under thermo-mechanical loading when e^{th} and e_{12}^{A} are similar, (ii) the polynomials that depict the shear stress for the isothermal case and thermo-mechanical case are rather different.

5. Analysis of Continuously Distributed Dislocations in the Presence of a Nonuniform Shear Stress Field

We seek to determine the dislocation density distribution, force on the lead dislocation under an arbitrary polynomial distribution of shear stress. The background information can be found in the books by Hirth and Lothe [15], Mura [16] and in the work of Bilby and Eshelby [17]. The shear stress distribution is given as $\tau(q) = j + kq + \ell q^2$ where $q = 0$ is the midpoint between two particles and j, k, ℓ are constants. We consider the pileup between two particles with an edge tot edge spacing of 2m.

$$\text{Flow Stress} \sim \frac{1}{\sqrt{m}\{2j+k+\ell\}}$$

We note that the local shear stress distribution appears in the denominator of the above equation. This means that when the flow stress is plotted versus $\frac{1}{\sqrt{m}}$ the slope will change as a function of local shear stress distribution.

The above approach explains two results observed experimentally in Fig. 5. The particle size affects the yield strength of these composites and the dependence on the particle size can be obtained from the above equation. Secondly, there is a dependence of strength on the local shear stress field through the $\{2j + k + l\}$ term and we note that the strengthening in thermo-mechanical fatigue is expected to be higher since the magnitude of $\{2j + k + l\}$ is lower for the TMF case compared to the isothermal case. The detailed results will be available in Ref. [6]. Further refinements in the results can be obtained by utilizing (i) anisotropic elasticity when the particle size is smaller than the grain size, (ii) accounting for the interaction of dislocation stress field and the inhomogeneity [18]. However, these modifications are not expected to alter the conclusions reached in this study.

6. Conclusions

(1) There is a strong effect of particulate size on the cyclic stress-strain response of MMCs. The yield strength and fatigue strength of the composite with 2 μm size particles is significantly higher than the composite with 30 μm particles for the same volume fraction of 30%.
(2) The stress range and strengthening in thermo-mechanical loading exceed that in isothermal loading. The particle size dependence under TMF IP and TMF OP loading is highest among the cases considered. Under isothermal loading at 300°C, very little increase in either flow stress or work hardening is realized by the addition of particles.
(3) Accounting for local strains the flow stress particle spacing relationship is modified. This explains the increased strengthening in the thermo-mechanical case.

Acknowledgements - The work was partially supported by Mitsubishi Motors Corporation, Japan and a fellowship from the Center for Advanced Study, University of Illinois, Urbana. The Mathematica software was used for some of the calculations.

References

1. Pickard, S.M. and Derby, B., The Deformation of Particulate Reinforced Metal Matrix Composites During Temperature Cycling, *Acta Metall.*, **38**, (1990), 2537-52.
2. Karayaka, M. and Sehitoglu, H., Thermo-mechanical Fatigue of Particulate-Reinforced Aluminium 2xxx-T4", *Metall. Trans.*, **22A**, (1991), 697-707.
3. Karayaka, M. and Sehitoglu, H., Thermo-mechanical Cyclic Deformation of Metal Matrix Composites, *ASTM STP* 1157, (1992), pp. 101-120.

4. Karayaka, M. and Sehitoglu, H., Thermo-mechanical Cyclic Deformation Modeling of A12xxx-T4 Composites, *Acta Metallurgica* **41**:1, (1993), 175-189.
5. Sehitoglu, H., Karayaka, M., Prediction of Thermo-mechanical Fatigue Lives in Metal Matrix Composites, *Metall. Trans.*, **23A**, (1992), 2029-2038.
6. Sehitoglu, H., The Influence of Particle size on Strengthening and Fatigue at High Temperatures, Manuscript to be submitted for publication, (1995).
7. Kamat, S.V., Hirth, J.P. and Mehrabian, R., Mechanical Properties of Particulate Reinforced Aluminum Matrix Composites, *Acta. Metall.*, **37**, (1989), 2395-2402.
8. Van Arsdell, W., The Effect of Particle Size on the Thermo-mechanical Fatigue of Metal Matrix Composites, M.S. Thesis, Department of Mechanical and Industrial Engineering, University of Illinois, 1993.
9. Van Arsdell, W., Sehitogly, H. and Mushiake, M., The Effect of Particle Size on the Thermo-mechanical Fatigue Behaviour of Metal Matrix Composites, Fatigue 93, 1993.
10. Miller, W.S. and Humpreys, F.J., Fundamental Relationship Between Microstructure and Mechanical Properties of Metal Matrix Composites in P.K. Liaw and M. Gungor (eds.),TMS, 1990, pp. 517-41.
11. Miller, W.S. and Humpreys, F.J., Strengthening Mechanisms in Particulate Metal Matrix Composites," *Scripta Metall.*, **25**, (1991). 33-38.
12. Ashby, M.F., The Deformation of Plastically Non-Homogeneous Alloys in A. Kelly and R.B. Nickelson (eds.), Elsevier, Amsterdam, (1971), 137-193.
13. Orowan, E., Symposium on Internal Stresses in Metals and Alloys, Institute of Metals, (1948), p. 451.
14. Eshelby, J.D., Chapter 3: *Elastic Inclusions and Inhomogeneities, Progress in Solid Mechanics*, **3**, (1961), pp. 87-140.
15. Hirth, J.H. and Lothe, J., *Theory of Dislocations*, Chapter 21, 2nd Edition, Krieger Publishing, Malabar, Florida 1992.
16. Mikata, Y., and Taya, M., Stress Field in a Coated Continuos Fiber Composite Subjected tot Thermo-mechanical Loadings, *J. Composite Materials*, (1985), 554-578.
17. Mura, T., *Micromechanics of Defects in Solids*, Martinus Nijhoff Publishers, (1982).
18. Bilby, B.A. and Eshelby, J.D., Dislocations and the Theory of Fracture, *Fracture, 1*, H. Liebowits (ed.), Academic Press, New York, (1968), pp. 99-182.
19. Dundurs, J. and Mura, T., Interaction Between an Edge Dislocation and a Circular Inclusion, *J. Mech. Phys. Solids*, **12**, (1964), 177-189.

MODELLING OF THE THERMAL-MECHANICAL FATIGUE BEHAVIOUR OF COATED AND BARE NICKEL-BASE SUPERALLOY SINGLE CRYSTAL AM1

E. CHATAIGNER[1], E. FLEURY[2] and L. RÉMY[1]
[1] *Centre des Matériaux P.M. Fourt, Ecole des Mines de Paris URA CNRS 866, BP 87, F-91003 EVRY Cedex, France*
[2] *Institute for Advanced Materials, Joint Research Centre, P.O. Box 2, NL-1755 ZG Petten, The Netherlands*

1. Introduction

Thermal Fatigue (TF) tests have been widely used as a standard test to study the damage mechanisms under thermal transient conditions. Thermal-Mechanical Fatigue (TMF) tests were then introduced to complement the thermal fatigue tests and to simulate the volume element behaviour of critical areas of components subjected to non-isothermal conditions. This volume element test, where strain, stress and temperature are known throughout the test, can thus be used to evaluate the validity of constitutive equations and damage models developed to predict the stress-strain behaviour and lifetime, respectively.

Lifetime prediction for jet engine components is still a matter of controversy and most authors attempt to describe the fatigue under high temperature by creep or oxidation interaction mechanisms. Engineering methods have been proposed which claim to account of there interactions, but most of them use parametric equations with a limited physical basis.

The introduction of directionally solidified single crystals as turbine blades has led to an increase of the performance of advanced jet engines. Considerable evidence has been produced that such components are submitted to thermal and mechanical loading, and that oxidation is a major cause of damage. Lifetime prediction for turbine blades, as for all components submitted to non-isothermal fatigue, is still an actual challenge for metallurgists and mechanical engineers. Most proposed models are essentially the same as for high temperature low cycle fatigue (LCF). But for the last few years the trend has been to integrate observable physical phenomena in the modelling. A crystallographic approach has been developed in several models to describe the mechanical behaviour of single crystals [1, 2]. In parallel, a metallographic approach has been introduced in models to achieve more reliable life prediction [3, 4]

The purpose of this paper is to report on the evaluation of predictive models able to describe the mechanical and lifetime behaviour of nickel-base superalloy single crystals under thermal-mechanical fatigue conditions. The application of a

J. Bressers and L. Rémy (eds.), Fatique under Thermal and Mechanical Loading, 381–392.

crystallographic model to simulate the mechanical behaviour of anisotropic materials is first described, followed by the predictions obtained from two lifetime physics based models. The present work includes the influence of the crystallographic orientation of bare and coated AM1 single crystals. A "C1A" coating was applied by a chromising-aluminising deposition process on samples with a [001] crystallographic orientation.

2. Thermal-Mechanical Fatigue Tests

The objective of the TMF tests is to simulate the behaviour of critical parts of components. Many authors have studied the effect of the synergy between the temperature cycle and the mechanical loading [5, 6]. In-phase and out-of-phase cycles have been often used as baseline experiments with the aim to evaluate the potential of different materials for turbine blades [7]. In our approach, thermal-mechanical fatigue tests were performed using a specific cycle that simulates thermal loading conditions experienced in service (Fig. 1). A mechanical strain (ε_m) - temperature (T) loop was used ranging from 600°C to 1100°C (873 to 1373K), with peak strains at intermediate temperatures: 950°C (1223K) in compression on heating and 700°C (973K) in tension on cooling. Both the temperature and the mechanical cycle are synchronously generated by a microcomputer in a period of 210 seconds.

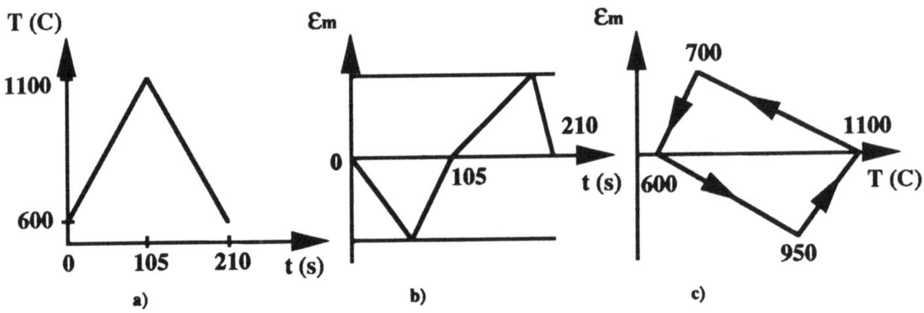

Figure 1. TMF cycle: a) temperature *vs* time, b) mechanical strain *vs* time, and c) mechanical-strain *vs* temperature.

2.1. CYCLIC STRESS-STRAIN CURVES FOR THE [001] CRYSTALLOGRAPHIC ORIENTATION

Figure 2 shows the variation of the stress, σ, as a function of the mechanical strain, ε_m. With our TMF cycle, the inelastic strain is primarily created during the heating phase in compression while an elastic behaviour is observed in tension at 700°C. The variation of the mechanical properties of the materials (Young's modulus, flow stress, and cyclic hardening) with the temperature lead to an unbalanced stress cycle although the imposed mechanical strain is fully reversed. For the [001] crystallographic orientation, the maximum compressive stress is reached at a temperature below 950°C, and the inelastic strain minimum at about 1040°C.

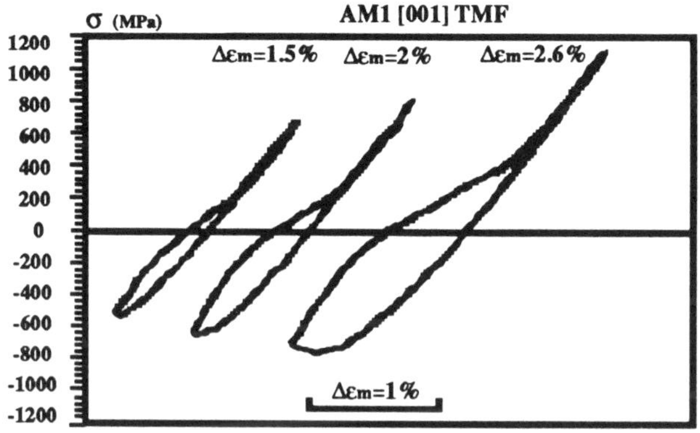

Figure 2. Stress-mechanical strain hysteresis loops of bare material with a [001] crystallographic orientation.

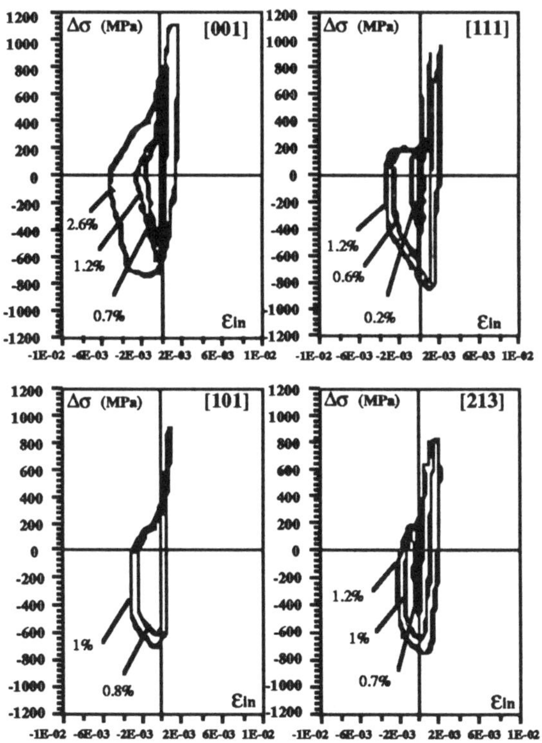

Figure 3. Comparison of the cyclic stress-strain loops for the [001], [111], [101] and [213] crystallographic orientations.

2.2. CRYSTALLOGRAPHIC ORIENTATION EFFECTS ON THE CYCLIC STRESS-STRAIN BEHAVIOUR

The temperature dependence of the mechanical properties varies with the crystallographic orientation. For example, the stress corresponding to the compressive peak strain is very different from the minimum stress on a specimen with the [111] crystallographic orientation because of the important decrease with the temperature of the flow stress and the low hardening (Fig. 3). Specimens with the [101] and [213] crystallographic orientations, present a behaviour intermediate between the results obtained for [001] and [111].

Inelastic deformation of single crystals is accommodated in localised slip bands which are parallel to either cubic or octahedral planes. Determination of the slip systems has led to the identification of multiple octahedral slip for samples near the [001] orientation, and multiple cube slip occurring for samples near [101] and [111] orientations, while near the [213] crystallographic orientation only a single cube slip plane was activated [8].

3. Modelling of the Mechanical Behaviour

Another advantage of TMF is to give the possibility to test constitutive equations under non-isothermal conditions. For this purpose, we have applied a crystallographic model to simulate the mechanical behaviour of our anisotropic material under non-isothermal conditions. This model, proposed by G. Cailletaud [2], used a micro-mechanical approach of the stress-strain behaviour which is as close as possible to the 'internal behaviour' of the material. For the single crystal, the deformation is analysed through the identification of the active slip systems.

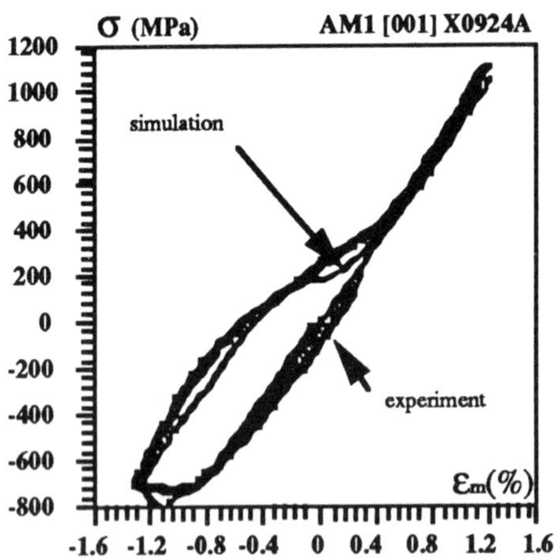

Figure 4. Comparison between the cyclic stress-mechanical strain loops computed by the model and the experimental data for a specimen with a [001] crystallographic orientation tested with our TMF cycle 600° ⇔1100°C cycle.

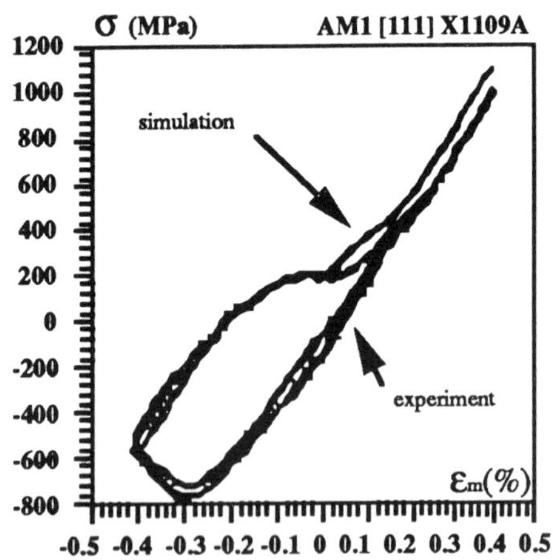

Figure 5. Comparison between the cyclic stress-mechanical strain loops computed by the model and the experimental data for a specimen with a [111] crystallographic orientation tested under TMF condition.

The macroscopic strain is partitioned into an elastic and a viscoplastic part. But whereas the relations describing the elastic strain are written at the macroscopic level, the viscoplastic constitutive equations are written for each slip system. The identification of the model parameters has been achieved in an earlier study and the results obtained in isothermal conditions have been presented in [8, 9]. The parameters, deduced from isothermal tests by a fully automated procedure, were used to simulate mechanical behaviour of the AM1 superalloy tested with our TMF cycle. The model is able to describe the shape of the cyclic stress-strain loops under anisothermal conditions whatever the crystallographic orientations (Figs. 4 and 5) and can predict the activated slip systems [10].

4. Thermal-Mechanical Fatigue life

4.1. INFLUENCE OF CRYSTALLOGRAPHIC ORIENTATION

The TMF life to 1 mm crack depth, N_f, *vs.* mechanical strain range, $\Delta\varepsilon_m$, is presented in Fig. 6. As it has been reported for LCF [11, 12], the TMF life is strongly dependent on the crystallographic orientation. The higher endurance is exhibited by samples near the [001] and [010] orientation, while lower TMF lives were observed for the other crystallographic orientations.

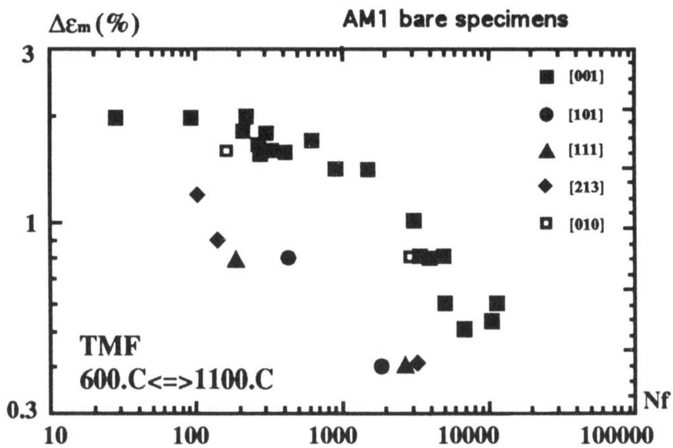

Figure 6. Variation of total lifetime (N_f) with mechanical strain range ($\Delta\varepsilon_m$) for bare AM1 specimens of different orientations tested under TMF conditions.

Depending on the mechanical strain range, cracks initiated from sub-surface micropores or from oxidised areas at the surface. Only specimens with the [213] crystallographic orientations tested at high mechanical strain ranges exhibited a different crack initiation and crack propagation mechanism with nucleation along the unique active slip band. The crack initiation period, studied from observations of plastic replicas, represents around 5 pct of the total life, which means that most of the life is spent in the propagation of a small crack. In the strain range studied, these small cracks grow in stage II mode (*i.e.*, mode I opening) from the initiation site to a size approximately equal to 0.4 mm (for a wall thickness of 1 mm). If crack initiation under TMF may result from interaction between fatigue and oxidation mechanisms, oxidation plays also an important role in crack propagation. An oxide layer is covering the crack surface up to the crack tip, and a zone depleted of γ' precipitates is formed ahead of the crack tip [13].

4.2. INFLUENCE OF THE COATING

The comparison of the TMF lives of coated and uncoated AM1 obtained for specimens with the [001] crystallographic orientation is presented in Fig. 7. The endurance of coated specimens is almost the same at that of bare specimens for the TMF cycle used, as observed in LCF at high temperature [14]. The damage is manifested as a major crack initiating mainly from casting micropores located within the sub-surface area. Cracks observed behave as initiating early from the surface within a negligible initiation period. Strong localised oxidation occurs at the coating-substrate interface which can lead to local delamination of the coating. After failure of the coating the oxidation is playing an important role in the propagation of the crack.

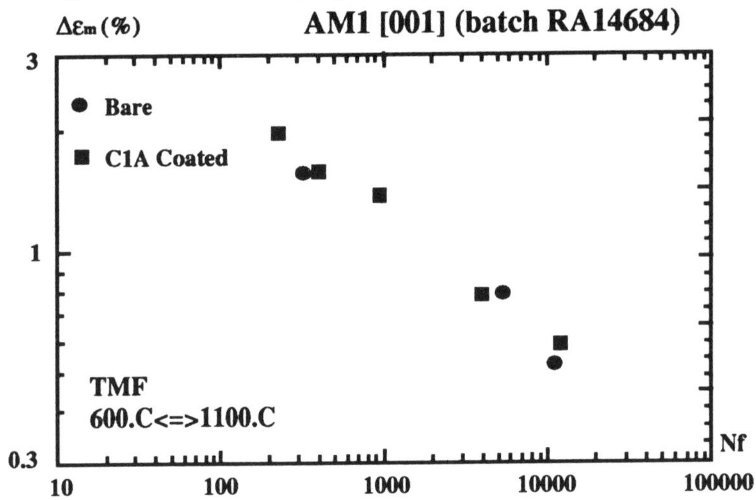

Figure 7. Variation of TMF total lifetime with mechanical strain range for the [001] orientation: comparison between bare and coated specimens.

5. Lifetime Prediction

The observations made on the AM1 superalloy single crystals have shown a strong synergy between oxidation and fatigue damage. Thus two physically-based models were introduced. The first one, named oxidation embrittlement model, was proposed to describe the oxidation-fatigue interaction in conventionally cast superalloys [4, 15]. This approach uses the process zone concept of Mc Clintock [16], who considered a process of repeated crack nucleation ahead of a crack tip. The second one, named oxide cracking model and proposed by Reuchet and Rémy [3], is simply a summation of both types of damage resulting from fatigue and from oxidation. Equations are derived assuming that the elementary crack propagation is the result of the advance due to the crack opening under fatigue and of an additional contribution due to oxidation at the crack tip. We will give more details of the two different models in this paragraph and discuss the predictions obtained in each case.

5.1. MODEL I: OXIDATION EMBRITTLEMENT MODEL

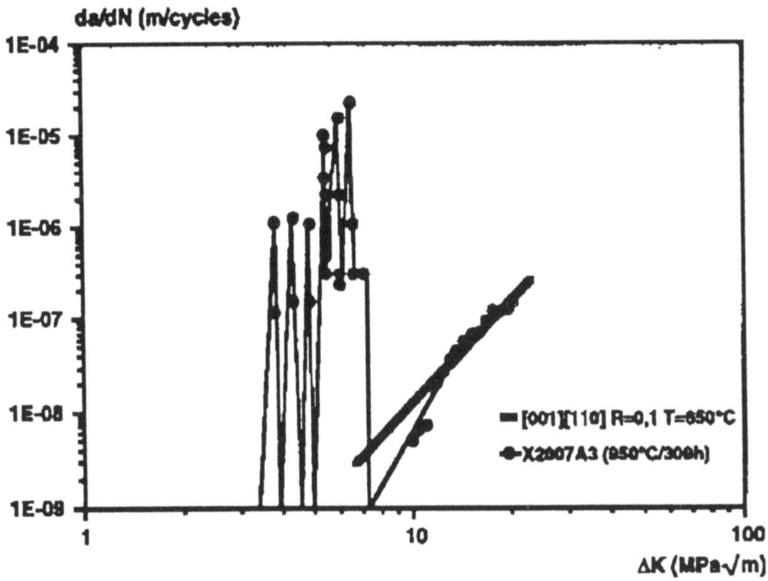

Figure 8. Variation of FCGR (da/dN) as a function of stress intensity range (ΔK) at 650°C for a non-oxidised specimen which has been oxidesed at 950°C.

The TMF damage is considered as a microcracking process and the exposure to high temperature is assumed to embrittle the material at the crack tip. An empirical equation using local stresses at the crack tip has thus been proposed, and the crack grows by rupture of a microstructural element ahead of the crack tip when a local fracture creterion is fulfilled. The damage equation is deduced form isothermal fatigue crack propagation tests made on CT specimens without environment effect. Among the equations proposed to describe the fatigue crack growth rate, Formans's equation has been shown to be in good agreement with experimental data. This equation was used to analyse our experiments with a local approach, which necessitated to deduce the local stresses at the crack tip from finite element computation [17]. The equation can be written as follows:

$$\frac{da}{dN} = \frac{C.\Delta\sigma_{yy}^{m}}{[(1-R).(\sigma_c - \sigma_{yy})]^{\beta}} \quad (1)$$

where C, m and β are constats at a given temperature, and the stress ratio is defined as: $R=\langle 1-\Delta\sigma_{yy}/\sigma_{yy}\rangle$, i.e. $R=(1-\Delta\sigma_{yy}/\sigma_{yy})$ when $\Delta\sigma_{yy}<\sigma_{yy}$ and R=0 when $\Delta\sigma_{yy}>\sigma_{yy}$. The normal stress range $\Delta\sigma_{yy}$, normal stress σ_{yy}, and critical fracture stress σ_c are calculated at a distance λ ahead of the crack tip. Equation (1) was fitted to data obtained in fatigue crack growth tests for two load ratios R=0.1 and R=0.7, and to LCF data for R=-1. The β parameter of equation (1) is adjusted to superimpose the curves obtained for different stress ratios. The critical fracture stress (σ_c) is influenced by the loading condition (temperature, environment): σ_c describes the oxidation effect in our damage equation, and is derived from experiments on oxidised CT specimens. Oxidation gives rise to an embrittlement of the material ahead of the crack tip, visible as a bright zone on the fracture surface. This embrittlement leads to an acceleration of the crack growth rate for the low ΔK values, while no propagation was observed in non-oxidized specimens. The size of the embrittelement zone (Δa), which has been determined at two temperatures (Fig. 9), is varying as:

$$\Delta a = \alpha_{ox}(T).t^{1/4} \tag{2}$$

where $\alpha_{ox}(T)$ is the oxidation constant of the material.

As shown previously [4], this embrittled zone is induced by the oxygen diffusion in front of the oxide and leads to the reduction of the local fracture toughness which increases the fatigue crack growth rate. When the crack grows farther from the oxidised region, the critical stress σ_c increases more rapidly until it approaches values typical of the virgin material. The evolution of the critical fracture stress in the embrittled zone is shown in Fig. 10 and has been described by the equation:

$$\sigma_c=\sigma_{co}.[1-u+u.\exp(m.x/l_{ox})] \tag{3}$$

where u and m are two temperature dependent constants.

Progress of the crack Δa (µm)

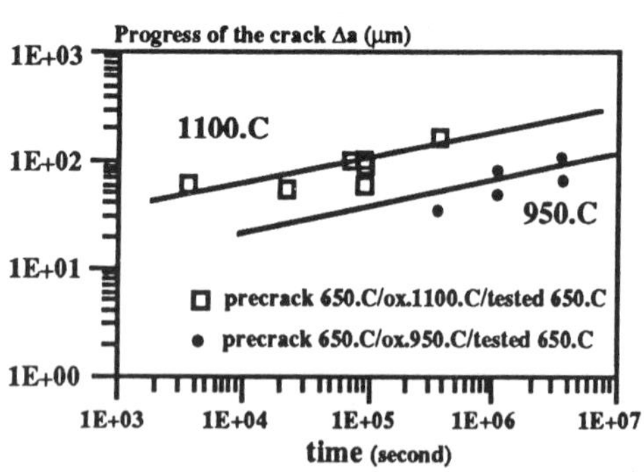

Figure 9. Variation of the crack progress as a function of the exposure time at 950°C and 1100°C.

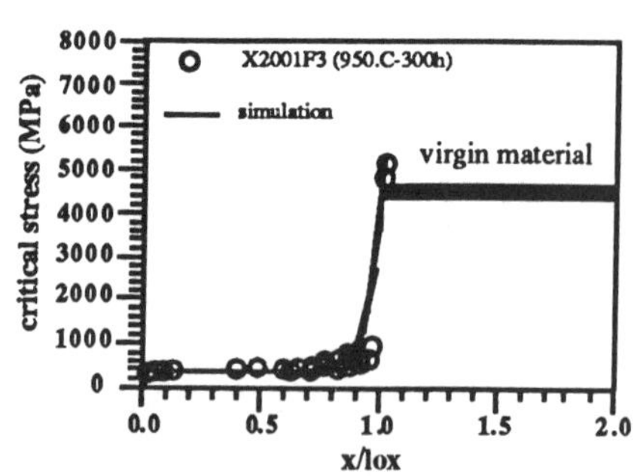

Figure 10. Variation of the critical stress to fracture (σ_c) with the position of the crack in the oxidised zone on a CT specimen oxidised at 950°C.

Under anisothermal conditions, the oxidation constant $\alpha_{ox}(T)$ is calculated by integration between the minimum and maximum temperatures of the cycle:

$$\left(\overline{\alpha_{ox}}\right)^4 = \frac{1}{\Delta t} \cdot \int_0^{\Delta t} [\alpha_{ox}(T)]^4 . dt \tag{4}$$

The number of cycles, $N(\lambda)$, to achieve a crack growth increment of size λ, is given by the following condition:

$$\int_0^{N(\lambda)} dD = 1 \tag{5}$$

and TMF life is obtained by the harmonic average of the number of cycles at the minimum an maximum stress.

5.2. MODEL II: OXIDE CRACKING MODEL

As it can be observed in Fig. 11, when plotting the TMF data versus maximum tensile stress, the crystallographic orientation dependence of the TMF life is much reduced. A damage equation, proposed to describe the isothermal LCF life at high temperature of conventional cast superalloys [3], has been used. Fatigue is considered as a micro-cracking process, and the model takes account of the interaction between fatigue and oxidation. The damage equation is simply written as the summation of two terms: one due to the crack opening under fatigue and a second one due to the oxidation of the crack tip:

$$da/dN = (da/dN)_{fat} + (da/dN)_{ox} \tag{6}$$

The fatigue contribution to the crack advance is estimated by the model proposed by Tomkins [18] assuming that the crack is opened only by a tensile stress:

$$(da/dN)_{fat} = B.a \qquad \text{where } B = \Delta\varepsilon_{in} \cdot \left[\frac{1}{\cos\left(\frac{\pi}{2} \cdot \frac{\sigma_{max}}{\sigma_u}\right)} - 1 \right] \tag{7}$$

and $\Delta\varepsilon_{in}$ is the inelastic strain range, σ_{max} the maximum cyclic tensile stress and σ_u the ultimate tensile strength in montonic tension.

The contribution due to oxidation was derived from the oxidation kinetics which is enhanced by cyclic straining, and is a function of the period and the parameters of the fatigue cycle. For a given period and mechanical strain range, this contribution to crack advance is a constant length, $\Delta 1_{ox}$:

$$(da/dN)_{ox} = \Delta 1_{ox} \tag{8}$$

Consequently, the oxidation-fatigue interaction damage can be written as:

$$da/dN = B.a + \Delta 1_{ox} \tag{9}$$

By solving this equation for the crack depth, a can be expressed as a function of the number of cycles, N:

$$a = \Delta l_{ox}. (\exp B.N-1)/B \tag{10}$$

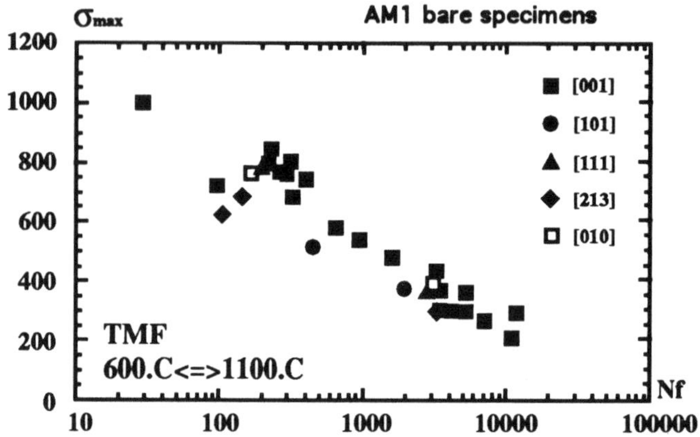

Figure 11. Variation of the lifetime (Nf) of bare AM1 specimens of different orientations tested in TMF with peak tensile stress (σ_{max}).

From interrupted tests we can estimate the oxide length formed at the crack tip in each cycle using Eq. (10). As it has been already shown by oxide depth measurements in various superalloys [4, 15], the depth of oxide formed in every cycle increases with the maximum stress according to a power law:

$$\Delta l_{ox} = \alpha(\sigma_{max}). \Delta t^{1/4} \tag{11}$$

The parameters were thus indentified from experimental crack growth data using only [001] bare specimens tested in LCF at 650°C, 950°C and 1100°C at a frequency of 0.05Hz. To predict the endurance under TMF loading, the oxidation term is obtained by integration of the oxidation constant over the whole temperature-time cycle and the average oxidation constant, $\alpha(\sigma_{max})$, is thus given by:

$$\overline{\alpha}^4(\sigma_{max}) = \int_0^{\Delta t} \alpha^4[\sigma_{max}(t), T(t)].dt \tag{12}$$

5.3. TMF LIFETIME PREDICTION

TMF lifetimes have been computed and predictions are compared with the actual life to achieve a crack depth of 0.1 mm, which is more demanding than predicting the total life to 1 mm crack depth. Fig. 12 presents the predictions obtained with the model I on bare [001] specimens and Fig. 13 the results achieved using the model II for coated [001] specimens as well as bare specimens of different crystallographic orientations. In both cases, these physically based models give reliable predictions of lifetime for bare and coated specimens.

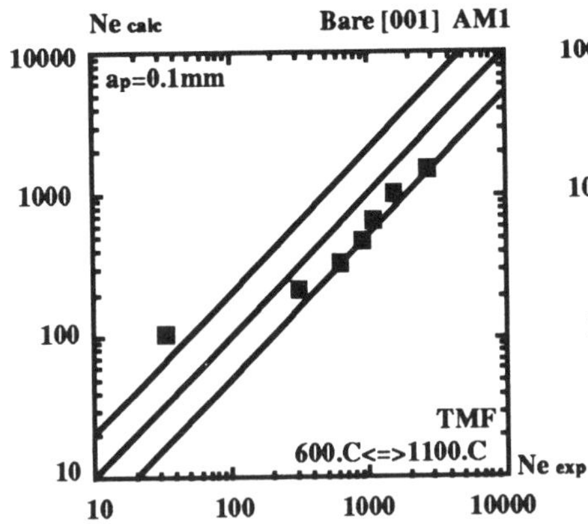

Figure 12. Model I: Comparison between predicted (Ne calc) and experimental lifetime to 0.1 mm crack depth (Ne exp) of bare specimens with [001] crystallographic orientation tested under TMF conditions.

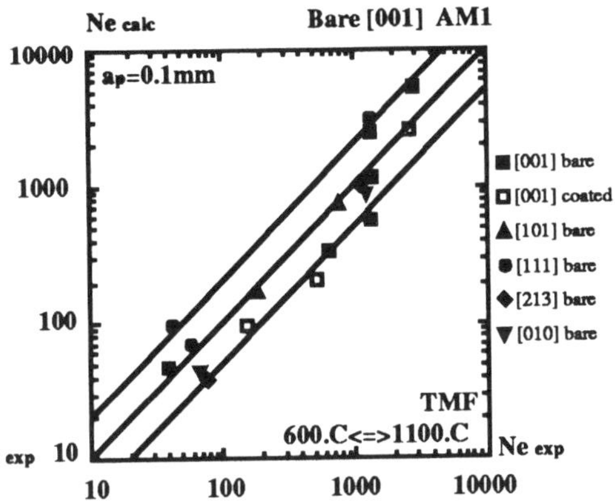

Figure 13. Model II: Comparison between predicted (Ne calc) and experimental lifetime to 0.1 mm crack depth (Ne exp) of bare specimens with different orientation and C1A coated [001] specimens tested under TMF conditions.

6. Conclusions

The thermal-mechanical fatigue test is a very useful tool which enables to study the damage mechanisms in realistic conditions and to test predictive models developed either for the strain-stress behaviour or lifetime. A crystallographic model using phenomenological viscoplastic equations, and Schmid's law gave a good agreement with experimental stress-strain behaviour under non-isothermal conditions. It has been shown that models which describe oxidation-fatigue interactions give a fairly good description of the thermal-mechanical fatigue life in the anistropic material studied here.

Acknowledgements - This study was financially supported by the French engine manufacture SNECMA, and by the Direction de Recherches, Etudes et Techniques (DRET) of the French Ministry of Defence, which the authors acknowledge gratefully. The crystallographic visco-plastic model has been used in collaboration with F. Hanriot and G. Caillitaud of the Centre de Matériaux of Ecole des Mines de Paris (France).

References

1. Pierce, D., Asaro, R and Needleman, A., *Acta Metall.*, **31** (1983), pp. 1951-1976.
2. Cailletaud, G., (1987), Thesis, Ecole Centrale Paris.
3. Reuchet, J. and Rémy, L., *Metall. Trans. A.*, **14A** (1983), pp. 141-149.
4. Rezai-Aria, F. and Rémy, L., *Eng. Fracture Mechanics*, **34** (1989), pp. 283-294.
5. Manson, S.S. and Halford, G.R., Thermal Fatigue of materials and Components, ASTM STP 612, D.A. Spera and D.F. Mowbrays Eds. Amercan society for Testing Materials, (1976), pp. 239-254.

6. Nichols, T., Heil, M. and Haritos, G.K., *International Journal of Fracture*, **41** (1989), pp. 1769-1783.
7. Marchand, N., Thermal-Mechanical Fatigue Behaviour of Nickel-Base Superalloys, Sci. D. Thesis, Massachusetts Institute of Technology, 1986.
8. Hanriot, F., Cailletaud, G. and Rémy, L., High Temperature Constitutive Modeling Theory and Applications, A.D. Freed and K.P. Walker Eds. *Am. Soc. of Mech. Engn.*, New York, USA, MD-**26** and **121**, pp. 139-150.
9. Hanriot, F., Thesis Ecole Nationale Supérieure des Mines de Paris, (1993).
10. Espié, L, Hanriot, F., Cailletaud, G. and Rémy, L., in Proceedings ICM, 1995, Den Haag.
11. Gabb. T.B., Gayda, J. and Miner, R.V, *Metall. Trans. A.*,**17** (1986), 497-505.
12. Fleury, E. and Rémy, L., *Materials Science and Engineering*, **A167** (1993), 23-30.
13. Fleury, E. and Rémy, L., *Metall. and Mat. Trans. A.*, **25** January, (1994), pp. 99-109.
14. Chataigner, E. and Rémy, L., presented in the second Symposium "Thermomechanical Fatigue Behaviour of Materials". ASTM, Phoenix, Arizona, 14-15 November 1994.
15. Rémy, L., Bernard, H., Malpertu, J.-L. and Rezai-Aria, F., in Proceeding of the Symposium Thermo-Mechanical Fatigue Behaviour of Materials, San Diego, Oct.16, 1991, H. Sehitoglu, Ed, ASTM, Philadelphia, P.A., STP 1186, pp. 3-16.
16. Mc Clintock, F.A., Fracture of Solids, D.C. Drucker and J.J. Gilman Eds., Inter Science, New York, (1963), pp. 65-102.
17. Fleury, E., Thesis Ecole Nationale Supérieure des Mines Paris, 1991.
18. Tomkins, B., Philosophical Magazine, **18** (1968), pp. 1041-1066.

CRACK PROPAGATION AND LIFE PREDICTION IN A NICKEL-BASED SUPERALLOY UNDER TMF CONDITIONS

M. ARANA[1, 3], J.M. MARTÍNEZ-ESNAOLA[1] and J. BRESSERS[2]
[1] *Centro de Estudios e Investigaciones Técnicas de Gipuzkoa (CEIT)*
Paseo Manuel de Lardizábal, 15
20009 San Sebastián-Donostia Spain
[2] *Institute for Advanced Materials- JRC Petten*
P.O. Box 2, 1755 ZG Petten
The Netherlands
[3] *Currently at IAM-JRC Petten*

1. Introduction

Single crystal nickel-based alloys are widely used as blade materials in aero-gas-turbines because of their excellent resistance to high temperature deformation. Coatings are applied in order to provide the blades with adequate protection against environmental degradation and loss of mechanical performance. The major cause of failure in current single crystal blades of aero-gas-turbines is thermally induced stresses, which result from thermal strains over the blade thickness caused by temperature gradients during heating and cooling cycles.

Actual blade behaviour is closely simulated by means of thermo-mechanical fatigue (TMF) tests, which are designed to reproduce the temperature and strain cycles seen by critical volume elements of the blade. As opposed to isothermal fatigue testing, during TMF cycling the test specimen is repeatedly cycled through a wide range of temperatures, which gives rise to different damage and failure mechanisms. The synergy between the different damage mechanisms makes it difficult to estimate the life of a component by using traditional low cycle fatigue life prediction models. During the past few years some suitable models for the prediction of the lifetime of uncoated and coated specimens subjected to TMF have been proposed [1-3].

The aim of this contribution is to represent the TMF life of the single crystal superalloy SRR99 as a function of two damage mechanisms: oxidation and fatigue. These have been identified as the operating mechanisms under TMF conditions in this superalloy. The model is a consequence of the analysis of crack growth evolution during the TMF tests and of observation of the fracture surface of *post mortem* TMF samples.

J. Bressers and L. Rémy (eds.), Fatique under Thermal and Mechanical Loading, 393–402.

2. Experimental Results

Strain controlled out-of-phase TMF tests with a lag of -135° between temperature and strain were carried out using the single crystal nickel-based superalloy SRR99. Some of the samples tested had been subjected to a standard pack aluminising process covering their surface with a nickel-aluminide coating approximately 30 μm thick. The temperature cycle is between 1323 and 573 K with a period of 90 s, 30 s for heating up and 60 s for cooling down.

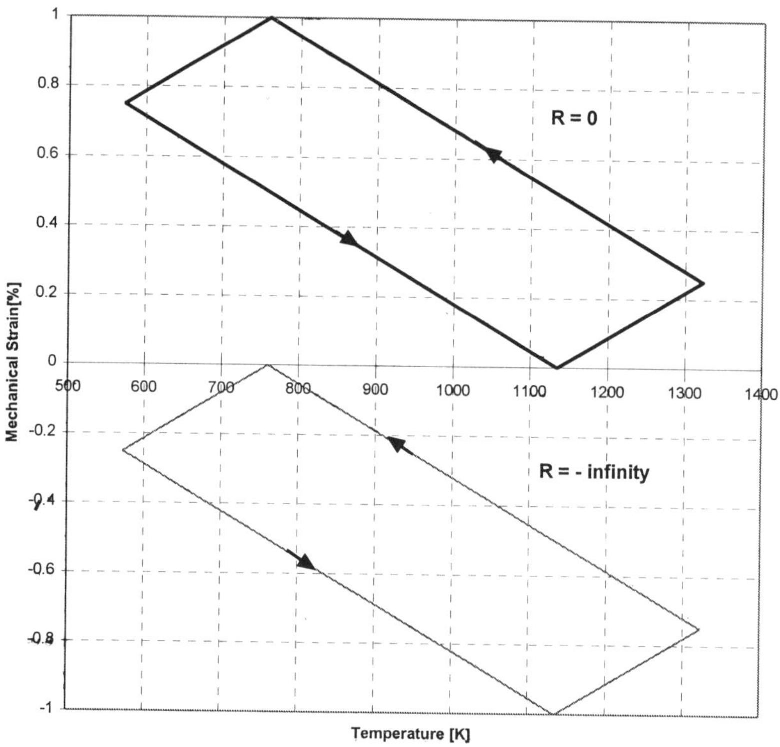

Figure 1. Strain-Temperature cycle for two -135° out-of-phase TMF cycles.

The mechanical strain ranges were varied between 0.5 and 1%. For each strain range two types of cycle are considered corresponding to two different minimum to maximum strain ratios, i.e. $R = 0$ and $R = -\infty$. The resulting mechanical strain-temperature cycles corresponding to a mechanical strain range of 1% are shown in Fig.1. Mechanical strain rates during the heating and cooling parts of the cycle are approximately 3×10^{-5} and 1.5×10^{-5} s^{-1}.

During testing, the surface of the sample is periodically scanned using a video camera system. The recorded images are stored and analysed after testing. This technique provides a large amount of information related to the evolution of damage at the surface and, consequently, the damage parameters that produce failure of the specimen. The analysis of some representative images showed that the failure of uncoated specimens under -135° out-of-phase conditions is due to the multiple initiation of cracks at oxidation spikes on the surface, followed by growth and crack coalescence. The oxidation spikes appear early in the test at preferential sites [4]. The size of such oxidation spikes and of the cracks which grow from them is measured and recorded as a function of the cycle number. The analysis reveals a transition in the crack evolution which suggests a change in the crack growth mechanism. Fig. 2 shows some representative examples of the recorded crack evolution.

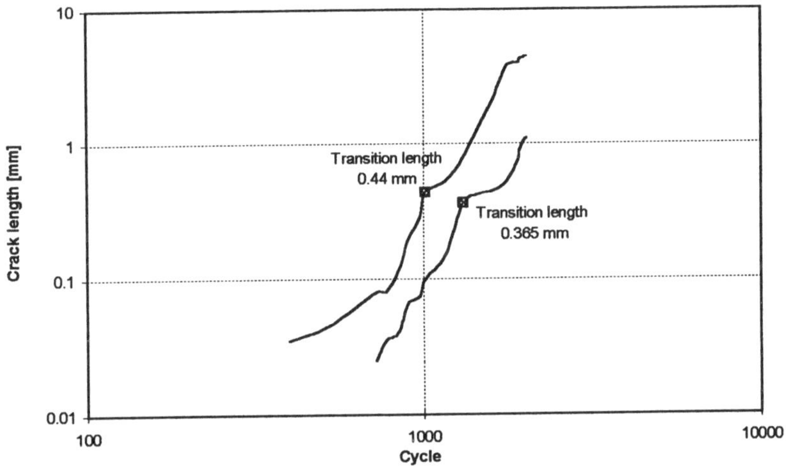

Figure 2. Example of crack evolution in TMF tests.

The analysis of the fracture surface shows that the previously mentioned oxidation spikes are responsible for crack initiation. It is also possible to identify a transition in the surface roughness from a nearly flat fracture surface, when the crack is small, to a hilly and faceted surface once the cracks have reached a specific size. This suggests that, for long cracks, crack growth is governed by local crystallographic failure on the {111} facets, whereas for short cracks the growth mechanism could be associated with oxidation. Finally, it is worth pointing out that the faceted fracture surface is also covered by several oxide layers. This suggests that, even for long cracks, oxidation plays an important role in crack growth, at least in terms of a crack closure effect.

3. Description of the Oxidation-Fatigue Life Prediction Model

This model follows a fracture mechanics approach based on the experimental observations that have been described in the previous section, and assumes that TMF crack growth can be represented by two stages which are controlled by different damage mechanisms: oxidation through the first stage and fatigue during the second. The transition between these two stages is assumed to occur when the crack reaches a specific microstructural size.

The model aims to reproduce the observed behaviour of the TMF cracks through the growth of a unique *equivalent crack* which represents the general evolution of the TMF cracks leading to failure for the different testing conditions.

In the following sections, the formulation of the equations that reproduce the evolution of such a *characteristic crack* for each damage mechanism is described and discussed.

3.1. OXIDATION MECHANISM

Oxidation is assumed to dominate crack growth behaviour during the earlier part of a test. This first stage is considered to occur until the crack reaches a critical length. The crack growth mechanism within this stage is sketched in Fig. 3. At first, there is an initiation site that enhances local oxidation (1). An oxide layer starts to grow (2) until it reaches a critical thickness and it fractures (3). This allows the oxygen to penetrate and form a new oxide layer at a deeper level (4). Again, once a critical thickness is reached, fracture of the oxide layer occurs (5) and the diffusion process restarts (6).

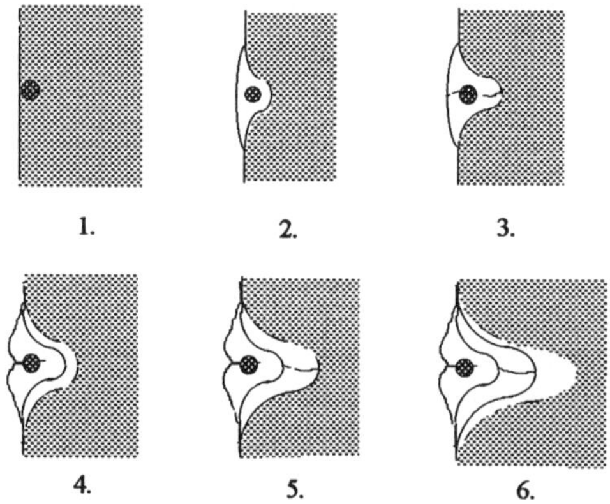

Figure 3. Schematic representation of the successive growth and fracture of oxide layers.

The growth mechanism was chosen to be mathematically represented as follows:

$$a = A \cdot \sigma_{max}{}^{\beta} \cdot N^{\alpha} \tag{1}$$

where a is the crack length, σ_{max} is the maximum tensile stress, N stands for the number of cycles, A and β are constants calculated by fitting the TMF crack growth data of 6 selected tests, and α is assumed to be 1. Eq. 1 represents the growth of an oxide layer under stress based on oxidation kinetics law. This usually is linear ($\alpha = 1$) when the material does not present any impedance to the oxidation, and has been reported to be true during the first oxidation stage in nickel-based superalloys [5]. Note that, according to the schematic in Fig. 3, the oxidation process reinitializes when fracture of the oxide layer occurs, consequently, for this specific type of TMF test the dependence on time remains linear through the whole testing period. The dependence on the peak stress represents the effect of the applied load on the successive cracking of the oxide layer and has been taken into account by a power law relationship with the peak stress. A similar oxidation mechanism has already been observed for TMF conditions[3].

It is important to note that this type of TMF test favours the fracture of the oxide layer since the maximum stresses occur at low temperature (Fig. 1), when the ductility of the oxide layer is low. The described crack growth mechanism, and consequently equation (1), is not appropriate for other type of TMF tests where the maximum temperatures and stresses occur simultaneously.

3.2. FATIGUE MECHANSM

The fractographic observations suggest that the second stage of the TMF lives is governed by local crystallographic failure on {111} planes, which has been identified as the fatigue mechanism for nickel-based single crystal superalloys [6,7]. Consequently, it is assumed that cracks grow according to a fatigue mechanism described by a Paris law

$$\frac{da}{dN} = C \cdot \left(K_{max} - K_{CL}\right)^n \tag{2}$$

where K_{max} is the maximum stress intensity factor and K_{CL} is the stress intensity factor below which the crack is closed, and C and n are constants to be determined. Several authors have proposed life prediction models based on Paris type laws under TMF conditions [1,2,8].

Regarding the formulation of the crack closure effect, the TMF life data showed a dependence of stress intensity factor below which the crack is closed, K_{CL}, on the minimum stress, σ_{min}, as follows

$$K_{CL} = \frac{B}{\left|\sigma_{min}\right|^m} \tag{3}$$

This dependence on the minimum stress can be physically explained in the following terms: the larger the compressive stress the smaller the crack closure effect, since compressive stresses tend to smooth out crack surface defects and, consequently, reduce crack closure. This phenomenon has been observed experimentally [9]. The crack closure effect is considered to be different in coated and uncoated specimens.

4. Model Equations

The evolution of a crack following the growth mechanisms described in the previous section is represented in Fig. 4 in a log a-log N plot. The parallelism between the experimental (Fig. 2) and the theoretical (Fig. 4) behaviour is noticeable.

In order to obtain the constants of the equation that reproduces the oxidation crack growth the actual transition points were recorded for the TMF cracks which had no apparent interaction with the surrounding cracks. A regression analysis between the number of cycles to the transition and the peak stress of each test provides the numerical form of Eq. (1)

$$a = \frac{a_{trans.}}{3.5 \times 10^{10}} \cdot \sigma_{max}^{\ \ 2.58} \cdot N \tag{4}$$

where σ_{max} is expressed in MPa, and a_{trans} and a have the same units. The transition size was calculated as the arithmetic mean of the measured transition sizes, resulting in a value of $a_{trans.} = 0.3$ mm.

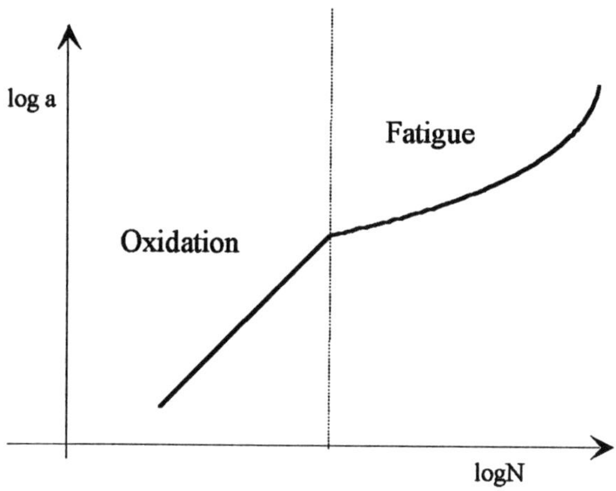

Figure 4. Sketch of the behaviour of a crack following the theoretical oxidation-fatigue behaviour.

Concerning the fatigue equation, the exponent in Eq. (2) was set to 2. This is the mean value of the slope found in the log(da/dN) versus log(ΔK) plots for the TMF crack growth of four representative tests [2]. Both the pre-exponential and crack closure constants in Eq. (2) were calculated by means of a regression analysis of the cycles to failure versus the result of integrating the effective stress intensity factor over the testing time.
The resulting Paris equation is

$$\frac{da}{dN} = 2 \times 10^{-6} \cdot (K_{max} - K_{CL})^2 \tag{5}$$

where the stress intensity factor, K, is in MPa√m and da/dN is in mm/cycle.

The crack closure equations that correctly fit the TMF lives are

$$K_{CL} = \frac{337}{|\sigma_{min}|^{0.7}} \tag{6}$$

for the uncoated specimens, and

$$K_{CL} = \frac{139}{|\sigma_{min}|^{0.64}} \tag{7}$$

for the coated specimens, where the stress intensity factors for the crack closure, K_{CL} are expressed in MPa$\sqrt{m}$ and the stresses in MPa.

5. Results and Discussion

By using the equations described in the previous section it is possible to reproduce the TMF crack length evolution for different test conditions.

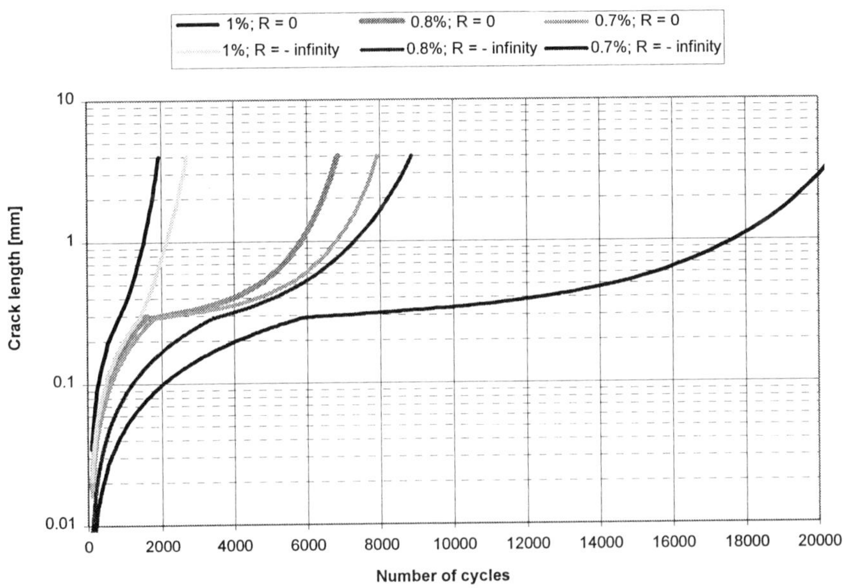

Figure 5. Predictions of crack length evolution for different test conditions.

Note that a specific test condition (in terms of applied strain) would not imply a unique crack growth behaviour since the crack growth equations are expressed as a function of the stress and the tests are not stress controlled. Fig. 5 shows the predicted crack growth behaviour for some tests obtained with Eqs. (1) and (2) by using the stress response. Figs. 6 and 7 compare the predicted crack evolution with the experimental data for two extreme test conditions. The results satisfy the initial aim of reproducing the general trend of TMF cracks.

Finally, Fig. 8 shows the predicted lives versus the actual lives for all the -135° out-of-phase tests, coated and uncoated, based on the assumption that the specimen fails when the *representative crack* covers 30% of the cross sectional area. From Fig. 8 it is possible to observe a good correlation between predicted and actual lives except for two tests which showed complete crystallographic failure and lay outside the error bands. For strain ranges lower than 0.5% with R = 0 and than 0.65% for R = - ∞ approximately, the model predicts crack closure stress intensity factors higher than those applied, which suggests that, for those conditions, the fatigue mechanism is not controlling the general damage process.

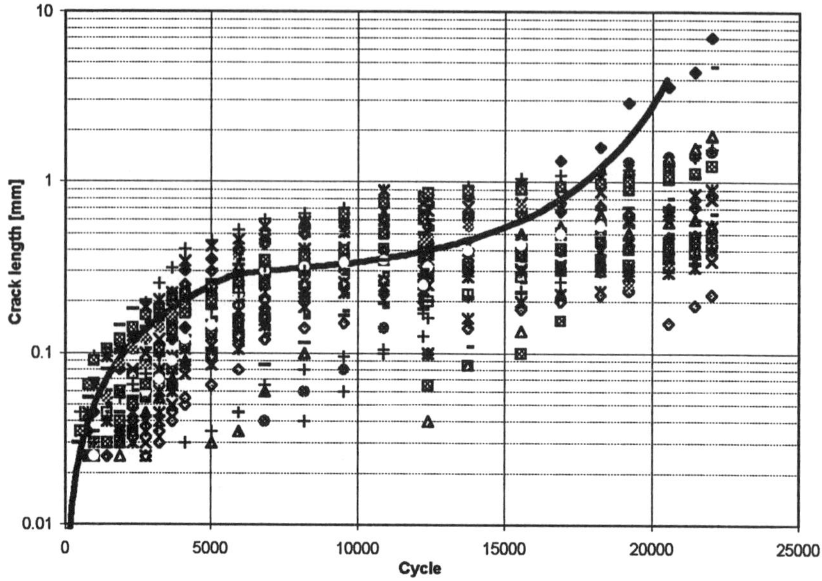

Figure 6. Comparison between real and predicted crack growth ($\Delta\varepsilon_m = 0.7\%$, $R = -\infty$).

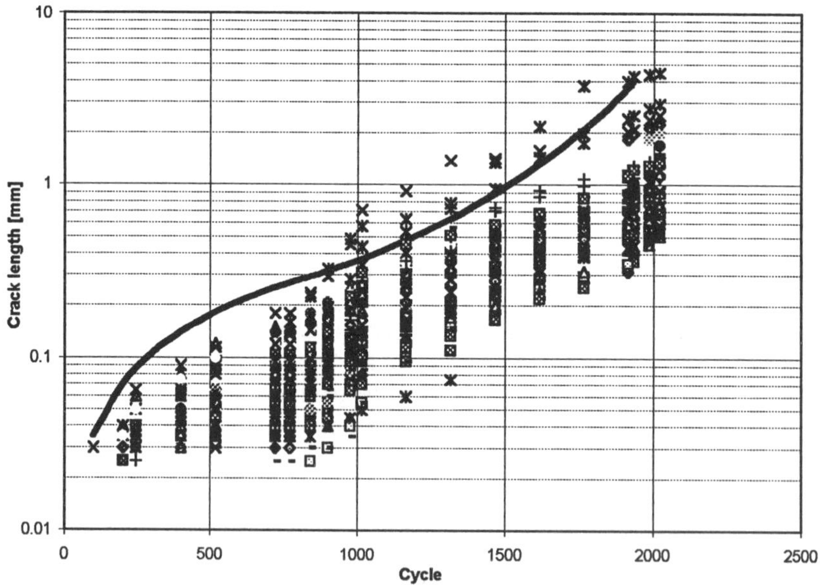

Figure 7. Comparison between real and predicted crack growth ($\Delta\varepsilon_m = 1\%$, $R = 0$).

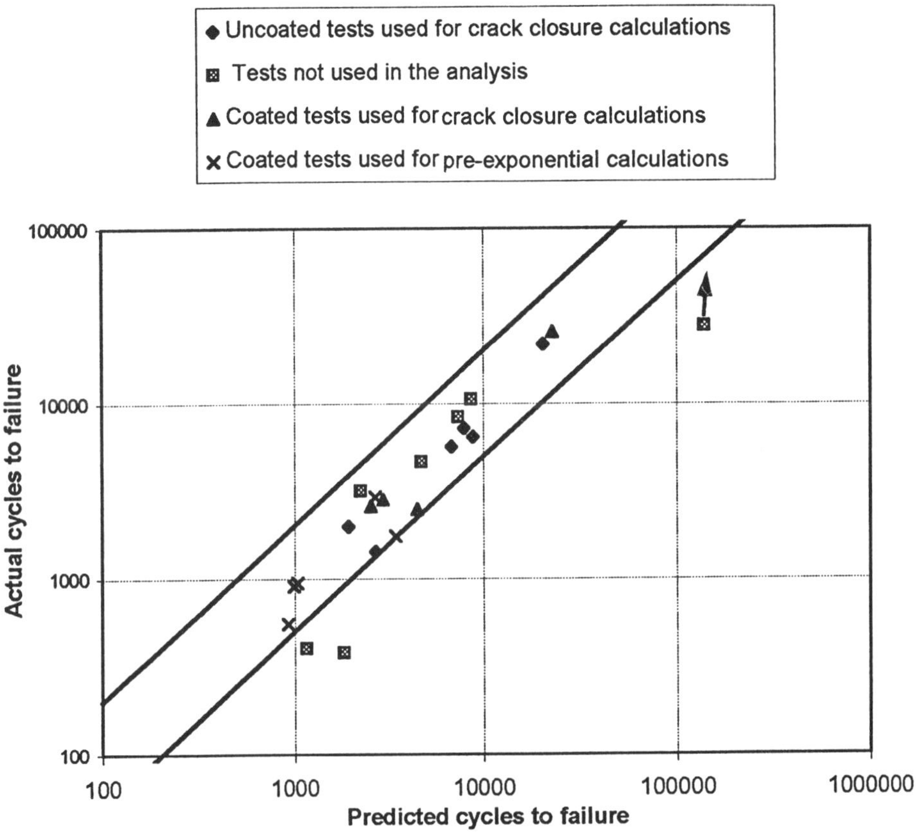

Figure 8. Life prediction for -135° out-of-phase TMF tests in SRR99.

It is interesting to highlight that coated samples showed shorter lives than the uncoated samples [11]. The model explains these differences by considering a lower crack closure effect in coated samples. In these samples, the oxidation effect is smaller than in the uncoated samples and consequently the crack grows faster.

The model, as a first approximation, is able to reproduce the main crack growth behaviour reasonably well by emphasising the role of oxidation as a crack growth mechanism. Although several other factors should be taken into account for a more realistic representation, such as the influence of the microstructure and the interaction among the growing cracks, the described approach is able to fit TMF lives within a factor of two for -135° out-of-phase TMF testing of bare and aluminide coated SRR99.

Acknowledgements - This work is part of the Brite-Euram project BE3338-89 with financial support from the European Commission, coordinated by Rolls-Royce plc (UK) in collaboration with Motoren und Turbinen Union, MTU (Germany), CEIT

(Spain) and IAM-JRC Petten (The Netherlands). Financial support of CEIT and of IAM-JRC by Rolls-Royce plc and MTU is also gratefully acknowledged. One of the authors (M.A.) wants to thank the Departamento de Educación, Universidades e Investigación of the Basque Government for the grant received to carry out her Ph.D.

References

1. Rezai-Aria, F. and Rémy, L., An Oxidation Fatigue Interaction Damage Model for Thermal Fatigue Crack Growth, *Engineering Fracture Mechanics* **34** (1989), 283-294.
2. Miller, M.P., McDowell, D.L. and Oehmke, R.L.T., A Creep-Fatigue-Oxidation Microcrack Propagation Model for Thermomechanical Fatigue, *Journal of Engineering Materials and Technology* **114** (1992), 282-288.
3. Neu, R.W. and Sehitoglu, H., Thermomechanical Fatigue, Oxidation and Creep: Part II. Life Prediction, *Metallurgical Transactions* **20A** (1989), 1769-1783.
4. Johnson, P.K. and Bressers, J., Characteristics of a Population of Naturally Initiated Cracks that Evolve During Thermomechanical Fatigue Testing of Bare and Aluminide Coated SRR99, in *Fatigue Under Thermal and Mechanical Loading*, International Symposium, May 22-24 Petten (N.H.), The Netherlands, 1995.
5. Strangman, T.E., Thermal-Mechanical Fatigue Life Model for Coated Superalloy Turbine Components, in S.D. Antolovich, R.W. Stusrud, R.A. Mackay, D.L. Anton, T. Khan, R.D. Kissinger, and D.L. Klarstrom (eds.), *Superalloys 1992*, The Minerals, Metals & Materials Society, 1992, pp. 795-804.
6. Crompton, J.S. and Martin, J.W., Crack Growth in Single Crystal Superalloy at Elevated Temperature, *Metallurgical Transaction* **15A** (1984), 1711-1719.
7. Telesman, J. and Ghosn, L.J., The Unusual Near-Threshold FCG Behaviour of a Single Crystal Superalloy and the Resolved Shear Stress as the Crack Driving Force, *Engineering Fracture Mechanics* **34** (1989), 1183-1196.
8. Okazaki, M. and Koizumi, T., Crack Propagation of Steels During Low Cycle Thermal-Mechanical and Isothermal Fatigue at Elevated Temperature, *Metallurgical Transactions* **14A** (1993), 1641-1648.
9. Rodríguez Ibabe, J.M., Propagación de grietas por fatiga, Escuela Superior de Ingenieros Industriales de San Sebastián, Universidad de Navarra, 1986.
10. Arana, M., Fatiga termomecánica de una superaleación monocristalina de base níquel (SRR99). Comportamiento mecánico y predicción de vida, Doctoral Thesis, Escuela Superior de Ingenieros Industriales de San Sebastián, Universidad de Navarra, 1995.
11. Bressers, J., Timm, J., Williams, S.J., Bennett, A. and Affeldt, E., Effects of Cycle Type and Coating on the TMF Lives of Single Crystal Based Gas Turbine Blade Alloy, in M.J. Verrilli and M.G. Castelli (eds.), *Thermo-Mechanical Fatigue Behaviour of Materials: 2nd Volume*, ASTM STP 1263, American Society for Testing and Materials, Philadelphia, 1995.

CRACK GROWTH PREDICTION IN FATIGUE AND CREEP WITH ENVIRONMENTAL EFFECTS

S. KRUCH
Division OR/ce
B.P. 72 - 92322 CHATILLON Cédex France
Office National d'Études et de Recherches Aérospatiales (ONERA)

1. Introduction

This paper presents a phenomenological model able to simulate the crack propagation in the powder metallurgical nickel based alloy Astroloy at 650°C under complex cyclic loading conditions. An extensive experimental program in addition with a numerical analysis were necessary to build the model. The model uses the actual stress intensity factor as the loading variable but also several material parameters are introduced in order to describe the history of the various processes that operate near the crack tip during the hold period at high temperature:

- -The mechanical stress redistribution due to viscoplasticity.
- -The creep damage process under tensile (open crack) conditions.
- -The environmental driven embrittlement effect, that increases the fatigue crack propagation rate.
- -The interaction of overloads with both the fatigue and creep crack growth process.

The various state variables that store the post loading events are:

- -A threshold for fatigue crack growth, similar to the opening stress intensity factor. Its evolution is influenced both by fatigue overloads and by the hold time, during which a relaxation does occur.
- -Another threshold to describe the creep crack growth.
- -The size of the zone embrittled by oxidation, from which is deduced an effective local toughness of the material, included in the fatigue crack growth model through a Forman's expression.
- -The size of the plastic zone.

The main features of the model are discussed, together with the procedure for determining the model constants, involving specific creep-fatigue crack growth tests, tests with prior oxidation and test with hold times after peak load. Moreover, a special numerical procedure based on the cyclic viscoplastic analysis and a node release technique were developed for the determination of the purely mechanical effect of the hold time, associated to the stress redistribution at the crack tip.

The last section of this paper presents the model itself with all the constitutive equations and their physical meaning concerning the creep-fatigue-environment

J. Bressers and L. Rémy (eds.), Fatique under Thermal and Mechanical Loading, 403–416.

processes and their interaction. A comparison is made between the experimental results and the model predictions, leading to a discussion about the accuracy of the model and the necessity of further developments.

2. Experimental Analysis

Tests were performed on Single Edge Notch (SEN) specimens previously cracked by electro-erosion over a length of 0.3 mm [1]. The experimental set up is represented in Fig. 1.

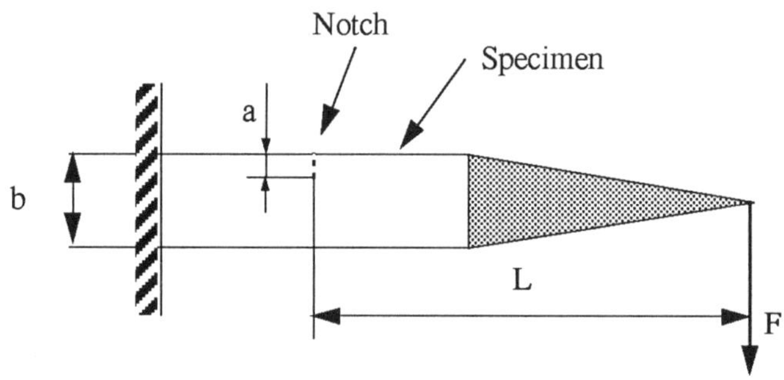

Figure 1. Experimental device.

A cyclic force F is applied at a distance L inducing a couple. Crack propagation is measured using a displacement sensor. Tests were performed at 650°C with complex loads schematically represented in Fig. 2.

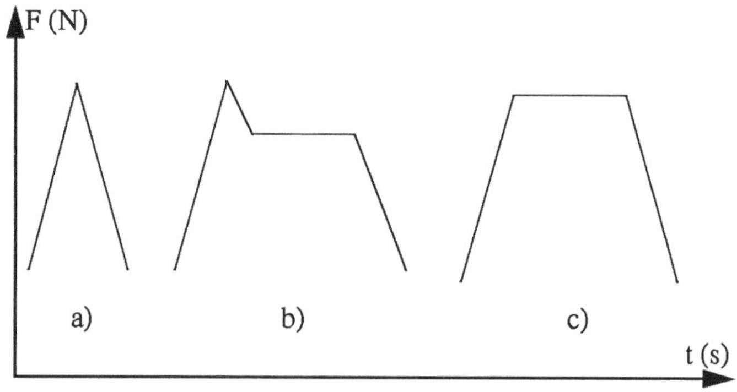

Figure 2. Shape of analysed cycles.

Test a) represents a triangular fatigue test without hold time. This load is able to represent the pure fatigue test if the frequency of the cycle is sufficiently high (over 5 Hz). However, an effect of creep can appear depending on the loading-unloading rate. An overload is introduced in tests b) before the application of a hold time in order to analyse its influence during the subsequent cycles [2, 3, 12]. Test c) represents a cycle with a hold time, that can vary from 0 (see cycle a) to ∞ (for pure creep tests). In fact, we analysed the following values of the hold time: 90s, 300s, 900s and also we performed tests up to 18000s in order to have some data on pure creep crack growth. This is the most damaging test. All these tests with or without a hold time are representative of the loads that the structure (i.e. aircraft engine turbine disks) experiences. The experimental results were recorded on a computer and it was possible to plot:

-The evolution of the crack length, a, as a function of the number of cycles, N.

-The evolution of da/dt (in creep) and da/dN (in fatigue) as a function of the Stress Intensity Factor (SIF), K.

-The variation of the crack length as a function of time.

3. Numerical Analysis

A numerical analysis using the Finite Element Method (FEM) was performed in order to complete the information obtained by tests.

All phenomena concerning the crack opening and closure effects and the stress redistribution at the crack tip induced by plasticity were analysed in detail during the crack propagation. We used an internal code developed at ONERA [4], named EVPCYCL, where sophisticated constitutive equations have been introduced to take account of all the non linear phenomena encountered in the study of viscoplastic materials. These laws involve a series of differential equations, the number of which varies according to the complexity of the selected model [5], [6].

The constitutive equations of the material used in this analysis are reduced to a viscoplastic formulation with only one nonlinear kinematic hardening parameter.

The specimen was discretized in 636 nodes and 287 elements which are 6 nodes triangles with 3 integration points. Taking the symmetries of the structure into account, only half on the specimen was meshed. The line y=0 represents the axis of symmetry (Fig. 3).

The numerical analysis was performed as close as possible to the experimental tests. Two steps were necessary to simulate the crack propagation inside this specimen. The first one consisted in applying triangular fatigue cycles (15s-0s-15s) and to release, each 3 cycles, 2 nodes of the mesh in front of the crack tip. This step was very important in order to have a good estimate of the real "plastic train" during the cycling. Once this step is done (i.e. when the crack reaches a given length), the complex cycles with a hold time (90s or 300s or 900s) can be applied to the specimen. Again, it is necessary to release 2 nodes each 3 cycles in order to simulate the crack propagation and store the correct "plastic train". To stabilize the evolution of viscoplasticity, it was essential to compute 3 full cycles at least, after each released nodes procedure.

As it can be seen in Fig. 3, the mesh was very fine in the region where the crack propagates and even finer in the zone where the complex load was applied. This kind of computation represents the mechanical behaviour of the specimen without any influence of external phenomena, like oxidation. The hypothetical differences which are observed between the numerical and experimental results allow to derive the following information about these phenomena: Their importance, their effects and which mechanical parameters are affected, enabling to take account of the effects of these external phenomena in the theoretical formulation of a crack growth model.

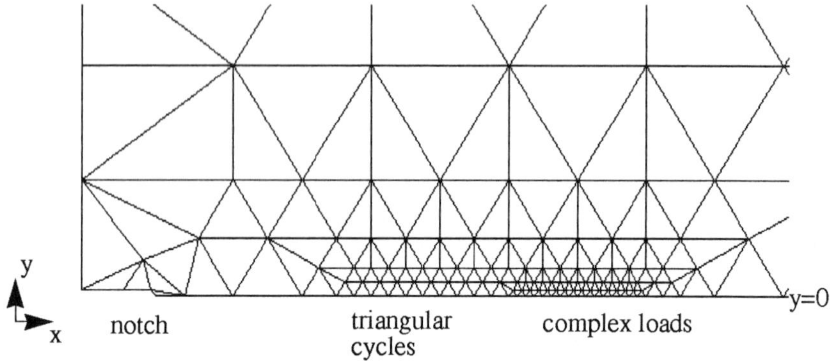

Figure 3. Zoom of the mesh at the crack propagation zone.

The main results of this computation used to calibrate the phenomenological model were:
a) The evolution with cycling of local variables like the viscoplastic strains and the induced stress redistribution at the crack tip as shown in Fig. 4 where the tensile stress is plotted during the 300s of the hold time. The stress relaxation is fast in the first 15s and the stress continues to decrease gradually until the end.

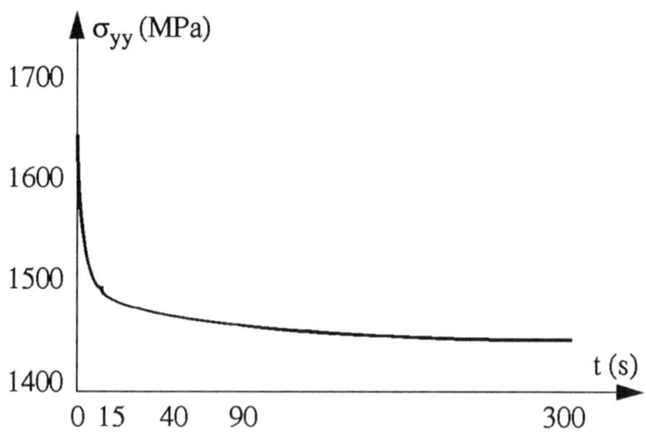

Figure 4. Evolution of the tensile stress with the hold time.

b) The crack opening and closure loads which are directly linked to the fatigue and creep thresholds, corresponding to the load level at the time of contact between the released nodes ($y \neq 0$) and the line of symmetry ($y = 0$) of the specimen. The opening load is determined at the beginning of the cycle and the closure load is determined at the end of the same cycle. Based on the numerical analysis, and in order to be consistent with previous experimental results (not presented here), the value of the initial fatigue load threshold, normalized with respect to the maximum applied load, was chosen as 0.35. It is important to have a good estimate of this initial value because it will govern, indirectly, the evolution of the crack propagation during creep and fatigue.

4. Phenomenological Model

4.1. MAIN ASSUMPTIONS

The proposed model is built up in the framework of classical LEFM and must be able to reproduce the crack growth when the specimen is loaded with the cycles defined in the second section of this paper (Fig. 2). The main points the model must take into account are:

- the fatigue crack propagation,
- the creep crack propagation,
- the SIF relaxation during creep,
- the creep-fatigue interaction,
- the evolution of the plastic zone,
- the environmental effects.

The crack growth rate is classically divided in 2 parts, the first one concerns the fatigue propagation, the second one concerns the creep:

$$da = C_F\left[K_M - K_S\right]^{\eta_F} dN + C_C\left[K_t - K_S\right]^{\eta_C} dt \tag{1}$$

where:

C_F, C_C, η_F, η_C are material parameters,
K_M is the maximum value of the SIF for one cycle,
K_S is the SIF threshold for one cycle,
K_t is the SIF during the hold time.

The evolution of K_S during a cycle with a hold time is the most important concept of this model. The crack growth process and the induced creep-fatigue interaction are based on the term (K_M - K_S) for both fatigue and creep.

As it was pointed out before, the model is calibrated with the hypothesis of an initial threshold of 0.35 in fatigue. This value, combined with experimental results concerning in particular the growth crack rates plotted as a function of the SIF in fatigue (i.e. da/dN=f(K)) and creep (i.e. da/dt=g(K)) for 4 different hold times (0s, 90s, 300s, 900s), gives the basic evolution of the threshold in creep and fatigue as a function of the applied hold time. This evolution is schematically presented in Fig. 5. Underlying the figure are the following experimentally observed facts: The hold time increases the fatigue (transient) growth rate (decreasing the fatigue threshold), but decreases the creep growth rate during the hold time (increasing the creep threshold).

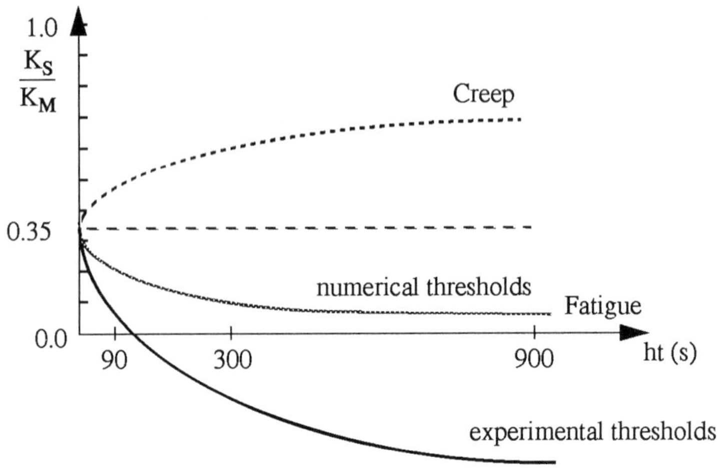

Figure 5. Creep and fatigue threshold.

The calibration of the thresholds based on experimental results as mentioned before gives inconsistent results as it can be seen in Fig. 5, where the "measured" fatigue thresholds are negative for hold times greater than 90s. Now the same calibration based on numerical results performed with hold times of 90s, 300s, 900s gives good results, in fact, the numerical fatigue thresholds decrease continuously towards zero as it is shown in this figure.

4.2. INFLUENCE OF ENVIRONMENT

The differences observed between the experimental and numerical approaches can be explained as follows. The numerical analysis only involves the mechanical phenomena which are developing in the specimen during the crack propagation. On the other hand, the experimental results involve additional external phenomena which can be induced by environment (i.e. oxidation). Tests performed by other researchers [7] on this material in vacuum show a great influence of environment on the crack propagation process. The main environmental effect is the material oxidation at the crack tip leading to a reduction of the mechanical properties. A high level load applied to the specimen increases the crack opening, so, the oxidation phenomenon develops and the material becomes brittle at the crack tip over an extended area (a surface or volume), increasing the fatigue crack growth rate. It is important to introduce these effects in the model in order to simulate accurately all the physical mechanisms induced by oxidation.

To reach this objective, a Forman expression [8] was introduced in the model in order to simulate the evolution of the penetration length, 1, of oxide inside the material as a function of time, that is: $l = \alpha\, t^{1/4}$; where α is a material parameter.

The evolution of the local toughness induced by oxidation is given by:

$$K_C = K_{C_0}\left[1-u+u\,exp\left(\frac{mx}{l}\right)\right] \quad (2)$$

where m, x and u are material parameters, an K_{c_0} is a material parameter representing the minimum value of K_c at the crack tip, i.e. of the completely embrittled material. Fig. 6 represents the evolution of K_c ahead of the crack tip, as deduced from special tests performed at EMP [9] and SNECMA [10] on oxidized specimens. It serves to determine the material parameters of the above relationship between the effective thoughness and the oxide penetration length 1. K_{c_v} is the toughness value for the non-oxidized material.

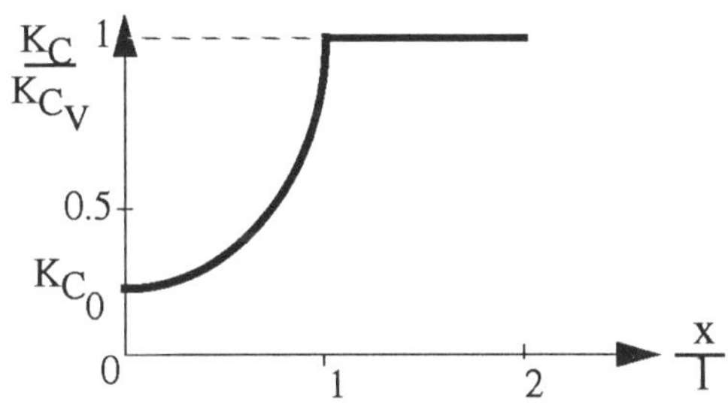

Figure 6. Evolution of K_c at the crack tip.

This new expression of K_c is introduced in the equation of the fatigue crack growth as presented below.

4.3. DESCRIPTION OF THE MODEL

This model is an iterative one, where each iteration represents a cycle (triangular or complex). The flowchart of the model with the constitutive equations is presented in Fig. 7. In this figure, f_1, f_2, f_3 and f_4 represent functions whose expressions are not developed in the paper (see [2] for more detail). At the beginning of each cycle, knowing the applied load, the maximum and the minimum values of the SIF (K_M and K_m respectively) are determined [Eq. a]. Since the value of the fatigue threshold at the end of the previous cycle is known, the crack increment in fatigue, Δa_F, is computed with equation b). This value is used to determine an equivalent SIF (K_{Meq}) proportional, by hypothesis, to K_M by a factor Φ, given by the relation c), which takes account of the actual fatigue threshold, the crack length and the applied load. For cycles without creep, i.e. without a hold time or with a high frequency, Φ=1 and $K_{Meq}=K_M$.

The introduction of K_{Meq} in the model allows to take account of an overload effect indirectly induced by the increased fatigue crack growth during the previous transient. A cycle with a hold time increases the fatigue crack propagation in the next cycle with respect to a cycle without a hold time. Physically, this effect induces an increase of the plastic zone and can be interpreted as an overload with respect to the previous cycle.

Fatigue

$$K_M = f_1\left[f_M ; a\right] \qquad K_m = f_1\left[f_m ; a\right] \qquad \text{a)}$$

$$\Delta a_F = f_2\left[K_M - K_S ; K_C - K_M\right] \qquad \text{b)}$$

$$\Phi = \Phi\left(\Delta a_F ; K_M\right) \qquad \text{c)}$$

$$K_{Meq} = \phi\, K_M \qquad \text{d)}$$

$$\rho = f_3\left[K_{Meq}\right] \qquad \text{e)}$$

$$K_S = g\, K_{Meq} \qquad \text{f)}$$

Creep

$$\Delta a_C = f_4\left[K_{tm} - K_S\right] \qquad \text{g)}$$

$$\frac{d\left(\dfrac{K_S}{K_M}\right)}{dt} = -A\left(\frac{K_S}{K_M}\right)^{-\omega} \qquad \text{h)}$$

Oxidation

$$\frac{dlp}{dt} = \frac{1}{4}\left[c\,\mathrm{B}\frac{F_{tm}}{F_M}\right]^4 l_p^3 \qquad \text{k)}$$

$$K_C = K_{C_0}\left[1 - u + u\exp\left(\frac{mz^*}{lp}\right)\right] \qquad \text{l)}$$

Figure 7. Flowchart of the phenomenological model.

The fatigue threshold, K_S, coming from the classical fatigue crack propagation model [11] is used now to compute the creep crack growth. K_S is assumed to be proportional to K_{Meq} by a factor g (equation f) which is given by:

$$g(Req) = \alpha\, g_1\,(Req) + (1 - \alpha)\, g_2\,(Req) \qquad (3)$$

with $$Req = \frac{Kmeq}{KMeq}$$

g_1 and g_2 are functions introduced in the classical model to have more degrees of freedom in order to analyse more complex loading conditions, and K_{meq} represents an equivalent minimum SIF introduced to take account of the effects of overloads in compression. In this analysis we have:

$$\alpha = 0 \quad \text{and} \quad f_2\,(Req) \;=\; \frac{1}{r_2 - (r_2 - 1)\;Req} \tag{4}$$

where r_2 is a material parameter. Once the value of K_S for creep is known at the beginning of the hold time, it is possible to compute the creep growth Δa_c using equation g) and the relaxation of K_S during the hold time, integrating equation h) step by step in order to take account of the evolution of K_M with the crack length. Once the fatigue and creep crack growth are determined, the following step consists in computing the oxide penetration length using equation k), as well as the new value of K_C using equation 1) as presented before. This value of K_C is introduced in the expression of Δa_F (Eq. b) for the next cycle analysis. Equation 1) is the last one for a given cycle. After that, the process begins again with a new cycle.

The calculation stops when a critical value of the crack length is reached, that means when the specimen breaks. In this analysis, the critical length was equal to 5 mm in accordance with the experimental tests.

4.4. DISCUSSION

A large number of experimental tests were simulated with this model. Some results are summarized in Fig. 8 which presents a comparison of experimental and predicted number of cycles to failure as a function of the shape of the cycle (i.e. from pure fatigue to pure creep, depending on the value of the hold time).

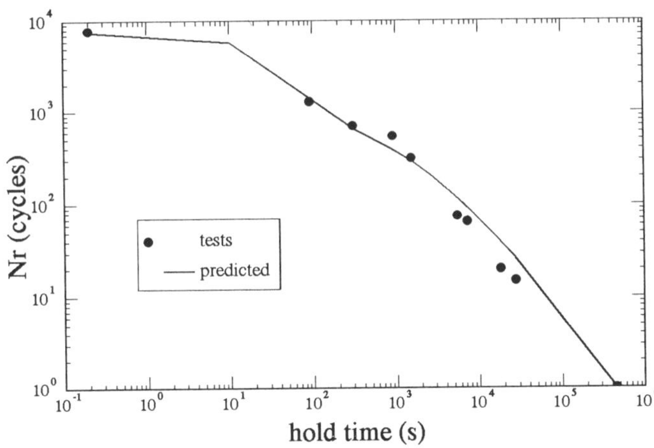

Figure 8. Comparison of experimental and predicted results.

As it can be seen, the results predicted by the model are quite accurate, but they overestimate the experimental results for cycles with a hold time greater than 5000s.

This fact can be explained by the procedure adopted for introducing the environmental effects in the model. As presented before, only the fatigue crack propagation is affected by oxidation. Now, for long hold times, it is obvious that oxidation will take place at the crack tip and will change the material characteristics, increasing the crack propagation and decreasing the lifetime.

4.5. INFLUENCE OF PLASTICITY

The application of an overload during a cyclic loading has several consequences with respect to the crack propagation rate. It has been observed experimentally that the rate increases as long as the overload is applied, but it decreases following the overload until its effect vanishes. At that moment in time, the crack propagates at the same rate it propagated for cycles without overloads. These variations of the crack rate can be related to the evolution of the plastic zone at the crack tip. An overload instantaneously increases the plastic zone size. After that, it will take some cycles to observe once again an increment of the plastic zone. This mechanism is schematically represented in Fig. 9.

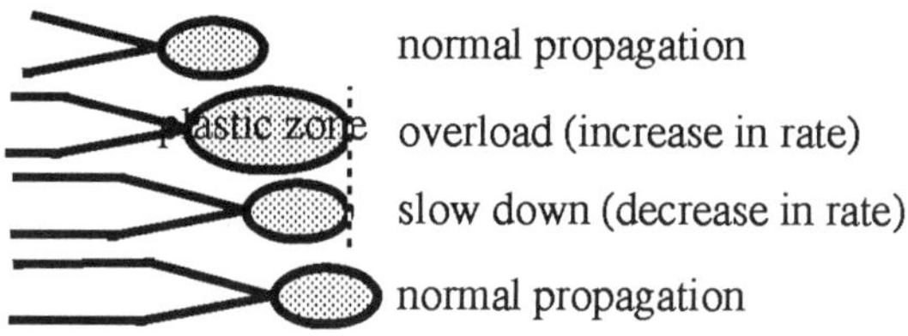

Figure 9. Evolution of the plastic zone during an overload.

In order to take account of these phenomena, an new parameter ρ, representing the radius of the plastic zone, was introduced in the formulation (relation e) in Fig. 7). Using the plane stress hypothesis, ρ can be related to the SIF by the following relation:

$$\rho = \frac{1}{2\pi}\left(\frac{K}{\sigma_l}\right)^2 \tag{5}$$

where K is the SIF and σ_l is the yield limit. In this analysis $K=K_{Meq}$, in order to take account of cycles with hold time. The plastic zone radius is determined as follows:
At cycle N, K_{Meq} (N) is computed using relation d) in Fig. 7.

a) if $K_{Meq}(N)$ is greater than $K_{Meq}(N-1)$, the plastic zone size increases and the new value of ρ is given by relation (5).

b) if $K_{Meq}(N)$ is less than $K_{Meq}(N-1)$ the plastic zone doesn't evolve but ρ is affected by the increment of the fatigue crack. The new value of ρ is then $\rho(N) = \langle \rho(N-1) - \Delta a_F \rangle$, where <x> represents Max(0,x).

The corresponding value of $K_{Meq}(N)$ is determined inverting relation (5).

$$K_{Meq}(N) = \sigma_l \sqrt{2\pi\rho(N)} \ .$$

c) The process continues computing the creep crack growth.

4.5.1. Pure Fatigue

The cyclic load in pure fatigue with an overload is schematically represented in Fig. 10. Cycles of 8daN of amplitude are applied, followed by a deliberately introduced overload of 14daN during 30 cycles in order to clarify how the model works in this case.

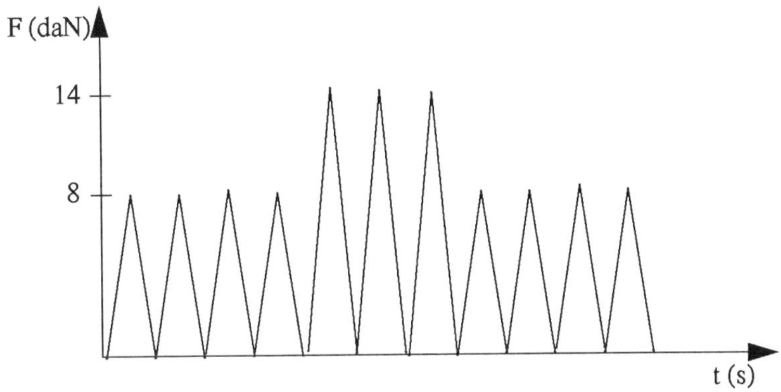

Figure 10. Analysed load in pure fatigue.

In pure fatigue K_M and K_{Meq} are identical and follow the same evolution, even during the application of the overloads. Their evolution changes following the overload, K_M decreases instantaneously but K_{Meq} decreases progressively until it reaches the value of K_M. At that time the effect of the overload is completely annihilated. These evolutions are presented in Fig. 11 where the solid and dotted lines represent the evolution of K_{Meq} and K_M respectively, as a function of the crack length.

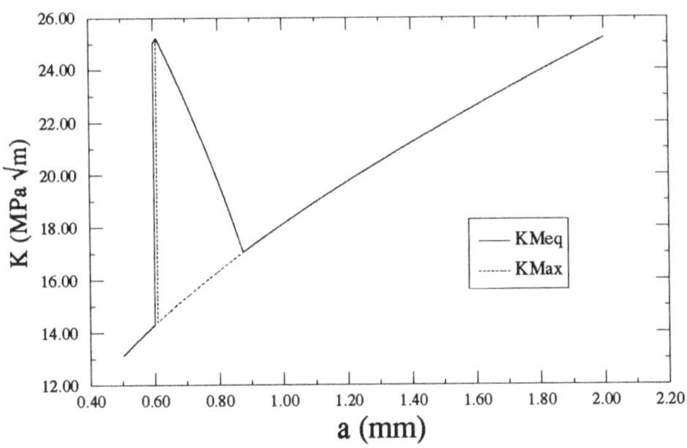

Figure 11. SIF evolutions in pure fatigue during and after the application of an overload.

Fig. 12 presents the consequence of these phenomena on both the evolution of the crack growth (solid line) and the plastic zone (dotted line) during the cyclic loading. The dashed line represents the evolution of the crack growth for a cyclic loading without any overload. The bifurcation point between the solid and dashed line represents the application of the overload. The plastic zone increases as long as the overload is applied, after which it decreases progressively as the crack propagates. The overload induces an acceleration of the crack growth rate (not visible in the figure due to the low number of cycles during which the overload is applied) but following the overload, the rate decreases and the crack propagates slowly until the effect of the overload vanishes.

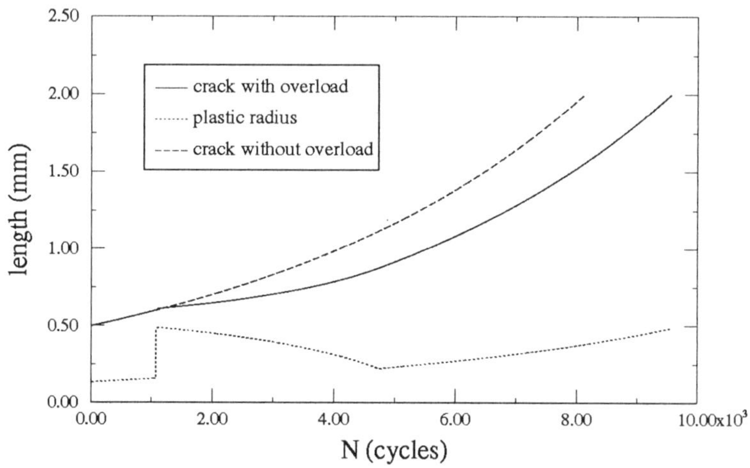

Figure 12. Evolutions of the crack length and plastic radius during pure fatigue with an overload.

Applying an overload during a cyclic loading has a favourable effect on the crack propagation rate and it increases the lifetime.

4.5.2. Creep-Fatigue interaction

In this section we analyse the effects of an overload applied during cycling with a 300s hold time. The load is schematically represented in Fig. 13 and it consist again of cycles of 8daN of amplitude interrupted by 150 cycles of 14daN.

Fig. 14 represents the evolution of K_M (dotted line) and K_{Meq} (solid line) as a function of the crack length, showing the differences induced by the shape of the cycle, the influence of the overload and finally the influence of plasticity at the crack tip after the application of the overload. The dashed line and the dashed-dotted line represent the evolution of K_{Meq} and K_M respectively, for cycles without an overload.

Fig. 15 shows the evolution of the crack length (solid line) and the plastic zone size (dotted line) during the cyclic loading. The increase in crack growth rate during the overload is clearly shown, as well as the increase of the plastic zone. The further evolution of both curves follows the mechanism described in section 5.4.

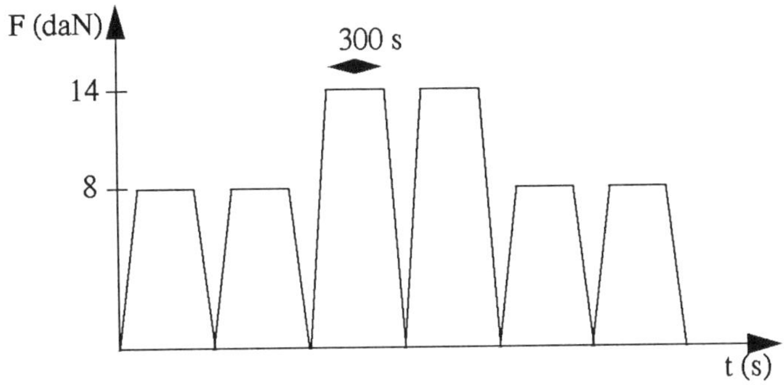

Figure 13. Analysed cycles with a hold time and overload.

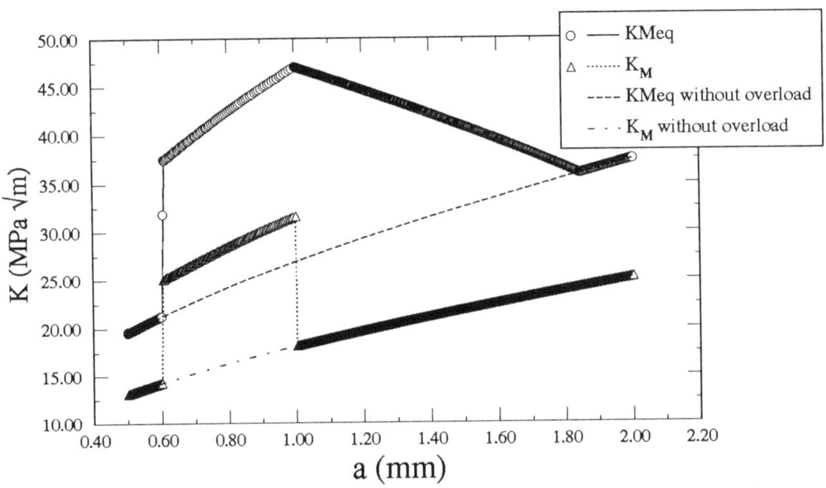

Figure 14. SIF evolutions for cycles with a hold time and overload.

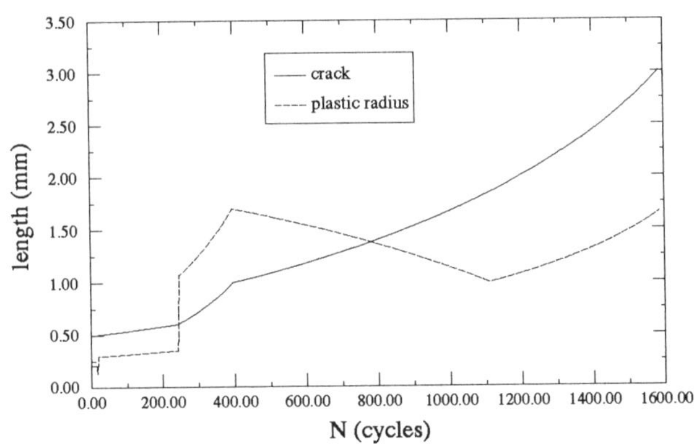

Figure 15. Evolution of crack length and plastic radius during creep-fatigue interaction with overload.

5. Conclusion

An extensive experimental program complemented with a numerical analysis based on FEM, resulted in a crack propagation model for Astroloy loaded in fatigue and creep. The model takes environmental effects into account. The numerical simulation of crack propagation by FEM with cyclic loads including hold times of 90s, 300s or 900s required lengthy and costly computations in order to closely reproduce experimental processes. By means of the numerical analysis, mechanical effects were separated form the environmental effects in the experimental tests, and both effects were introduced in the phenomenological model. The introduction of the evolution of the plastic zone at the crack tip allows to take the effects of overloads on the fatigue crack propagation into account. All these different processes were introduced in a computational code enabling the analysis of several interactions like creep-fatigue, fatigue-environment and fatigue-plasticity, which are often neglected in other models.

The model predictions compared well with the experimental results. However, the model is prone to further improvement with respect to the influence of environmental effects on the creep crack propagation.

It is planned to extend the model to an anisothermal formulation. In this case, material parameters will vary as functions of temperature and new finite element computations will be performed, at several temperatures, in order to analyse the evolution of the new fatigue and creep thresholds which form the basis of the model.

References

1. Baudin, G., Lemaître, J. and Robert, M., Intérêt de l'essai de fissuration en flexion pour la détermination des lois de progression de fissures de fatigue, *La Recherche Aérospatiale,* **6** (1976), 349-358.
2. Prigent, P., Modélisation de la propagation de fissure dans l'Astroloy à 650°C en prenant en compte les interaction fatigue-fluage-environment, PhD thesis of École Nationale des Ponts et Chaussées, Paris, (1993).
3. Prigent, P.,Kruch, S. and Devoucoux, J., Fissuration des alliages obtenus par métallurgie des poudres, *Proc. Journées Hispano-Française sur la Mécanique de la Rupture, Aiguablave, Espagne,* (1992).
4. Chaboche, J.L., EVPCYCL: Un code d'éléments finis en viscoplasticité cyclique, *La Recherche Aérospatiale,* no. 2.(1986).
5. Chaboche, J.L., Constitutive equations for cyclic plasticity and cyclic viscoplasticity, *Int. J. Plast* **5**, no. 3 (1989), 247-302.
6. Chaboche, J.L. and Cailletaud, G., On the calculation of structures in cyclic plasticity or viscoplasticity, *Computer* and *Structure* **23**, no 1 (1986), 23-31.
7. Bernède, P., Fissuration à chaud de l'Astroloy en métallurgie des poudres. PhD thesis of École des Mines de Paris, (1994).
8. Forman, R.G., Kearney, V.E. and Engle, R.M., Numerical analysis of crack propagation in cyclic-loaded structures. Trans. ASME, *Journal of Basic Engineering*, **89**, 459-464.
9. Reger, M. and Rémy, L., Influence de l'oxydation sur le comportement en fatigue à haute température d'un alliage à base de nickel. Journées Internationales de Printemps de ls SFM, Paris (1986), 176-184.
10. Lautridou, J.C., Private communication. SNECMA Corbeil (France).
11. Baudin, G. and Robert, M., Crack growth life-time prediction under aeronautical loading. *Proc. of ECF5,* Lisbonne, (1984), Portugal.
12. Kruch, S., Chaboche, J.L. and Prigent, P., A fracture mechanics besed fatigue-creep-environment crack growth model for high temperature, *CAPE '93, 2nd International Colloquium Ageeing, Life Assessment.* Stellenbosch (South Afrika Republic), (1993).

LIFETIME PREDICTION IN THE CREEP-FATIGUE INTERACTION REGIME APPLYING THE SRM RULE COMBINED WITH AN APPROPRIATE CONSTITUTIVE MODEL

D. RUBEŠA and R. DANZER
Institut für Struktur- und Funktionskeramik,
Montanuniversität Leoben
Magnesitstraße 2, A-8700 Leoben, Austria

1. Introduction

The application of any conventional lifetime prediction method requires some knowledge about the response of a material to the imposed loading. The stress-strain response should preferably be computationally simulated by an appropriate constitutive model of material behaviour, instead of carrying out the corresponding tests.

Extensive research has been carried out to find an effective couple of a lifetime prediction method for the creep-fatigue interaction regime and an adequate constitutive model. Within a Japanese project [1], for instance, various combinations of eight types of lifetime prediction rules and ten representative constitutive models of viscoplasticity were examined on the example of 2¼Cr–1Mo steel at 600°C, but none of them was found to be outstanding. Another combination, which is outlined in the following, was proposed by the authors [2, 3]: the empirical rule of Strain Rate Modified linear accumulation of creep damage (the SRM rule) for lifetime prediction is supported by Chaboche's viscoplastic constitutive model. The accuracy and reliability of lifetime predictions given by this method for various loading conditions in which both creep and fatigue take place were verified on the example of IN 738 LC superalloy at 850°C.

2. The SRM Lifetime Prediction Rule

The applied method of lifetime prediction is based on the Strain Rate Modification of linear accumulation of creep damage. Assuming the linear, i.e. additive, accumulation of damage, in variable loading conditions the damage, D, evolves according to

$$D(t) = \int_0^t \dot{D}(t')\,\mathrm{d}t' . \tag{1}$$

If the failure condition is defined by $D = 1$ and damage is only due to creep, then the damage grows at the rate

J. Bressers and L. Rémy (eds.), Fatique under Thermal and Mechanical Loading, 417–424.

$$\dot{D}_c(t) = \frac{1}{t_{f,c}[\sigma(t), T(t)]}, \qquad (2)$$

where $t_{f,c}$ is the time to failure in a creep test performed under the stress, σ, and temperature, T, actual at time t, held constant [4, 5]. To measure the deviation from the steady-state creep regime Franklin [6, 7] proposed the ratio of the inelastic strain rate, $\dot{\varepsilon}_{in}$, at any instant in time to the minimum creep strain rate, $\dot{\varepsilon}_{c,min}$, in the corresponding stationary loading conditions, i.e. in the creep test under the actual stress and temperature:

$$r(t) = \frac{\dot{\varepsilon}_{in}(t)}{\dot{\varepsilon}_{c,min}[\sigma(t), T(t)]} \qquad (3)$$

This refers to both the processes of deformation and damage evolution. Hence the r-ratio may be used to modify the rate of creep damage (2) and thus take into account the effects of fatigue damage mechanisms and of the possible creep-fatigue damage interactions in variable loading conditions. Following Franklin's suggestion [6, 7] a simple power law was found to be appropriate for this purpose [8–13]:

$$\dot{D}_{SRM}(t) = \dot{D}_c(t)\, |r(t)|^{\nu}, \qquad (4)$$

where ν is an adjustable material parameter. Applying the law of linear accumulation of damage, (Eq. 1), to the Strain Rate Modified damage rate, (Eq. 4), the damage increment in one cycle of a periodically repeated loading can be evaluated as $\delta D = D(\Delta t)$, where Δt is the cycle period. Its reciprocal value, according to the failure condition $D = 1$, gives the prediction of the number of cycles to failure

$$N_{f,SRM} = \left(\int_0^{\Delta t} \frac{1}{t_{f,c}[\sigma(t), T(t)]}\, |r(t)|^{\nu}\, dt \right)^{-1} \qquad (5)$$

As a special case, for $\nu = 0$, the prediction assuming the linear accumulation of creep damage only (Spera's rule [4, 5]), $N_{f,c}$, is obtained.

An additional multiplying factor for compression, c, has to be introduced into the kinetic law of damage evolution (Eq. 4), and accordingly into the SRM life prediction rule (Eq. 5), to take into account that compression may not be as damaging as tension. Specifically, $c = 1$ if compression and tension are equally damaging, $c = 0$ for non-damaging compression, and $c = -1$ if compression is fully damage healing.

The ratio of the cyclic lives predicted by Spera's and the SRM rules, $R_{SRM}=N_{f,c}/N_{f,SRM}$, can serve as a measure of the overall deviation from pure creep conditions and of the influence of fatigue. The highest R-ratio calculated for the cyclic tests used for the determination of the material parameter ν can be regarded as an approximate upper bound of the validity of predictions given by the SRM rule.

3. The Selection of Chaboche's Viscoplastic Constitutive Model

Knowledge of the functional relationships of time to failure and steady-state creep rate to the applied stress and temperature in creep tests, $t_{f,c} = t_{f,c}(\sigma, T)$ and $\dot{\varepsilon}_{c,min} = \dot{\varepsilon}_{c,min}(\sigma, T)$ respectively, is fundamental for the application of the SRM life prediction rule (Eq. 5). Along with this, the history of stress, temperature and inelastic strain rate ($\sigma(t)$, $T(t)$ and $\dot{\varepsilon}_{in}(t)$) over the entire (stabilized) load cycle has to be known. The inelastic strain rate history, however, is not known in advance, nor is the stress history in a strain controlled load cycle typical for HT LCF (high-temperature low-cycle fatigue) tests and common in service conditions. The uniaxial restriction of Chaboche's constitutive model of viscoplasticity [14] was found appropriate to supply the SRM life prediction rule with the needed stress and inelastic strain rate histories, $\sigma(t)$ and $\dot{\varepsilon}_{in}(t)$ respectively, by simulating and thus replacing the corresponding experiments. The model is, however, restricted to isothermal conditions.

Chaboche's viscoplastic model is of the unified type, so that the joint effect of time-independent and time-dependent (creep) inelastic deformations is described by a single state variable, the inelastic strain ε_{in}. Two additional internal state variables, R and X, describe the isotropic and kinematic hardening, respectively, and the accumulated inelastic strain, $p = \int_0^t |\dot{\varepsilon}_{in}| \mathrm{d}t'$, is used as a hardening parameter. The inelastic deformation is driven by the non-equilibrated viscous stress in excess of the current threshold value determined by the internal (equilibrium) stress, X, and the size of the elastic region of half width R. The version of uniaxial Chaboche's model selected to support the SRM rule consists of the following system of differential equations: the rate-type stress-strain relationship

$$\dot{\sigma} = E\left(\dot{\varepsilon} - \dot{\varepsilon}_{in}\right), \tag{6}$$

which involves Hooke's law of linear elasticity, and the kinetic equations of the evolution of internal variables

$$\dot{\varepsilon}_{in} = \left(\frac{\langle |\sigma - X| - R \rangle}{K}\right)^n \cdot \mathrm{sgn}(\sigma - X) \tag{7}$$

$$\dot{p} = |\dot{\varepsilon}_{in}|, \tag{8}$$

$$\dot{R} = b\left(R_\infty - R\right)\dot{p}, \tag{9}$$

$$\dot{X} = C_1\,\dot{\varepsilon}_{in} - C_2\,\varphi(p)\,X\,\dot{p} - C_3\,|X|^\rho\,\mathrm{sgn}\,X, \tag{10}$$

where

$$\varphi(p) = \varphi_\infty + \left(1 - \varphi_\infty\right)e^{-\omega p}, \tag{11}$$

$\langle\cdot\rangle$ denotes the operator

$$\langle x \rangle = \begin{cases} x & \text{for} \quad x > 0 \\ 0 & \text{for} \quad x \leq 0, \end{cases} \tag{12}$$

and E, K, n, b, R_0, R_∞, C_1, C_2, C_3, ρ, φ_∞ and ω are material constants. The initial conditions for a material which has not yet experienced any inelastic deformation are defined as $\varepsilon_{in} = 0$, $p = 0$, $R = R_0$, $X = 0$ for $t = 0$. The loading history can be given by specifying either $\sigma = \sigma(t)$ or $\varepsilon = \varepsilon(t)$. Note that (Eq. 9) is readily integrable, and for small strains, (Eq. 6), is integrable as well [2].

The particular feature of Chaboche's model, advantageous for its use in conjunction with the SRM method, is that it faithfully reproduces the *shapes* of stress-strain hysteresis loops, which ensures a correct prediction of the required inelastic strain rate at any instant in time.

4. Prediction of Cyclic Life of IN 738 LC at 850°C

4.1. CALIBRATION OF THE METHOD

The computer-assisted lifetime prediction by means of the SRM rule supported by the viscoplastic constitutive model of Chaboche was verified on the example of the nickel-base superalloy IN 738 LC in isothermal creep-fatigue interaction conditions at 850°C. The constitutive model was calibrated to IN 738 LC at BAM, Berlin [15]. The behaviour of the material in creep tests was analysed in [16] and the analytical relationships $t_{f,c} = t_{f,c}(\sigma)$ and $\dot{\varepsilon}_{c,min} = \dot{\varepsilon}_{c,min}(\sigma)$ were established. From the metallographic study it was concluded that compression is harmless [17] and, accordingly, $c = 0$ was applied.

The single adjustable parameter of the SRM rule, ν, was determined on the basis of seven fully reversed saw-tooth strain controlled HT LCF tests performed with the high strain rate of $\dot{\varepsilon} = \pm 10^{-2}\ s^{-1}$ and the systematically varied strain amplitudes from $\varepsilon_a = 0.0022$ to $\varepsilon_a = 0.0100$ [7]. The minimization of the logarithmic standard deviation, s, between the predicted and the measured cycles to failure gave the optimal value of $\nu = 0.61$ with the corresponding lowest value of $s = 0.322$.

4.2. EVOLUTION OF DAMAGE IN THE COURSE OF A LOAD CYCLE

The evolution of damage in the stabilized cycle of a fully reversed saw-tooth strain controlled cyclic test with strain rate $\dot{\varepsilon} = \pm 10^{-2}\ s^{-1}$ and strain amplitude $\varepsilon_a = 0.0075$ as given by the applied method is shown, as an example, in Fig. 1. For the given time-variation of strain, $\varepsilon(t)$, the constitutive model, (Eqs. 6–11), predicts the stress response, the evolution of the inelastic strain, and its rate—$\sigma(t)$, $\varepsilon_{in}(t)$, and $\dot{\varepsilon}_{in}(t)$, respectively. The deviation of the latter from the steady-state creep rate $\dot{\varepsilon}_{c,min}(t)$ under the stress actual at the moment determines the damage rate under tension according to (Eq. 4); the unmodified creep damage rate, (Eq. 2), is also shown. By integrating these damage rates over the cycle according to the principle of linear accumulation, (Eq. 1), the damage growth per cycle of $\delta D = 0.0143$ and $\delta D = 0.0024$, respectively, is obtained. This gives the prediction of 70 cycles to failure according to the SRM rule, whereas 415 cycles would be expected if damage were due to creep alone. There were 100 cycles to failure measured in the correspond-

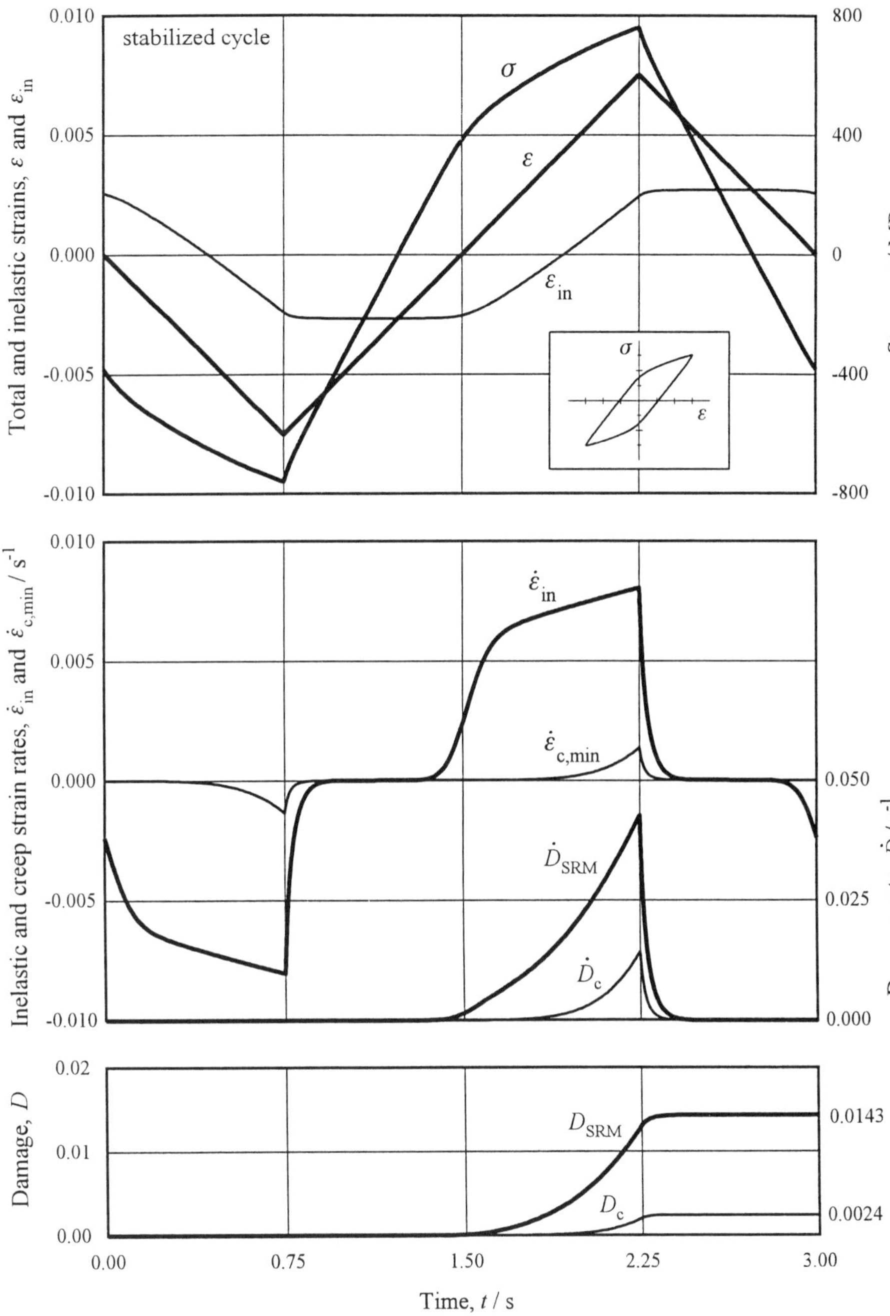

Figure 1. **Damage evolution in the simulated fully reversed saw-tooth strain controlled cyclic test with $\dot{\varepsilon} = \pm 10^{-2}\ s^{-1}$ and $\varepsilon_a = 0.0075$.**

ing test [7]. The $\dot{D}_{SRM}$ curve in Fig. 1 clearly shows which part of the loading cycle is damaging and to what extent.

4.3. RESULTS OF LIFETIME PREDICTION

On the basis of the 7 simplest tests used to calibrate the method, the cyclic lives in a further 52 HT LCF tests were predicted and compared with the experimental values measured in different laboratories. All the tests were strain controlled and fully reversed with constant strain rates in tension and compression, as shown in Table 1 (for the experimental data sources see [2] or [3]). Group 1 refers to the calibration tests. Group 2 contains the tests of the same kind performed in three other laboratories. The tests from groups 3 to 11 were performed in (mostly) quite different cyclic loading conditions. By including fast-fast, slow-slow, slow-fast and fast-slow tests, with cycle frequencies from about $4 \cdot 10^{-4}$ Hz to 1 Hz the whole range of creep-fatigue interaction conditions is considered to have been adequately covered. Note that slow-fast and fast-slow tests are a useful alternative to tests with dwell times [18].

The quality of the given lifetime predictions is shown in Fig. 2. They were all within the range of validity, which in the present example is given by $R_{SRM} < 70$. Solid and open circular symbols in Fig. 2 refer to the tests from groups 1 and 2, respectively, and both show the quality of interpolation. Triangulars represent the extrapolated predictions given for the tests from groups 3 to 11. Under the justified assumption that the logarithmic deviations between the predicted and the measured cycles to failure, $\log N_{f,SRM} - \log N_{f,meas}$, are approximately normally distributed, 95.5% predictions are expected to be within a scatter band of width $4s$ in the logarithmic scale, where s is the logarithmic standard deviation. This corresponds to a factor of $10^{\pm 2s}$ in cyclic life. The logarithmic standard deviation of $s = 0.234$ calculated for tests from groups 1 and 2 together, indicates the quality of interpolation. This is better than could have been expected on the basis of $s = 0.322$ found for the calibration tests, which means that the experimental data for the set of calibration tests were less consistent. The quality of extrapolation is indicated by $s = 0.245$ obtained for tests from groups 3 to 11. The value of $s = 0.239$ for all 59 tests together corresponds to the inherent empirical data scatter which was estimated to lie between 0.20 and 0.30 for all the analysed tests (cf. [9] and [12]). The scatter band of 95.5% confidence as judged from $s = 0.322$ of the calibration tests, and the actual one corresponding to $s = 0.239$, are marked in Fig. 2 with dashed and full lines, respectively.

Extrapolation to different cyclic loading conditions appears to be as reliable as interpolation. There is, however, a clear rationale for this beneficial outcome. The method was intentionally calibrated to the tests where fatigue had the strongest influence taken into account by the adjustable parameter ν. The cycle frequency of all other tests was equal or lower than in the calibration tests, so the conditions were either the same or closer to those of pure creep. Thus all the predictions were made for the conditions of creep-fatigue interaction between the two appropriately modelled extremes: steady-state creep, as a starting point of the SRM lifetime prediction rule, and the conditions of the most pronounced action of fatigue.

TABLE 1. HT LCF tests used in the analysis.

Group #	No. of tests	Strain rate tensile, $\dot{\varepsilon}$ / s^{-1}	Strain rate compr., $-\dot{\varepsilon}$ / s^{-1}	Strain amplitudes, $\varepsilon_a \cdot 100$	Measured cycles to failure, $N_{f,meas}$
1	7	10^{-2}	10^{-2}	0.22 ... 1.00	39 000 ... 45
2	14	10^{-2}	10^{-2}	0.25 ... 1.25	54 000 ... 16
3	2	$1.4 \cdot 10^{-3}$	$1.4 \cdot 10^{-3}$	0.28 and 0.39	4 100 and 720
4	4	10^{-3}	10^{-3}	0.36 ... 0.99	1 200 ... 33
5	6	$3 \cdot 10^{-5}$	$3 \cdot 10^{-5}$	0.28 ... 0.96	750 ... 15
6	3	$2.5 \cdot 10^{-5}$	$2.5 \cdot 10^{-5}$	0.29 ... 0.94	2 500 ... 28
7	4	10^{-5}	10^{-5}	0.24 ... 0.45	440 ... 82
8	1	$3.5 \cdot 10^{-6}$	$3.5 \cdot 10^{-6}$	0.21	580
9	10	10^{-5}	10^{-2}	0.24 ... 0.59	420 ... 33
10	2	10^{-2}	10^{-4}	0.23 and 0.50	4 500 and 190
11	6	10^{-2}	10^{-5}	0.31 ... 0.59	250 ... 32

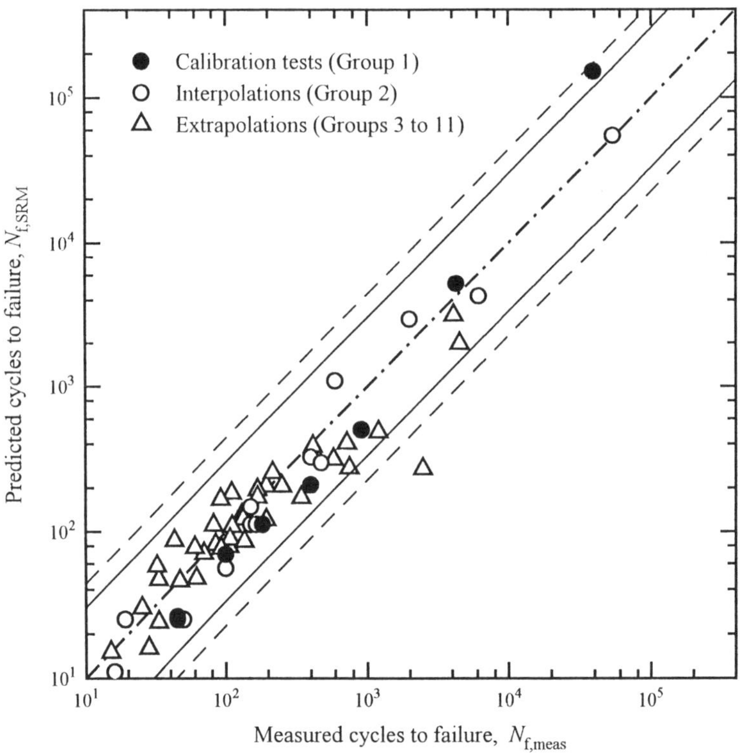

Figure 2. **Predicted vs. measured cycles to failure.**

5. Conclusions

A particular empirical rule of Strain Rate Modified linear accumulation of creep damage (the SRM rule) has been supported by Chaboche's viscoplastic constitutive model to predict lifetime under a non-stationary loading at high temperatures, i.e. in the creep-fatigue interaction regime. The ability of this combination to extrapolate lifetime predictions to various cyclic loading conditions other than those of the few fatigue tests required for the calibration of the SRM rule has been proved for IN 738 LC superalloy under uniaxial cyclic loading at 850°C. An inherent criterion for the validity of a given prediction appears as a particular advantage of the method.

References

1. Inoue, T., Igari, T., Okazaki, M., Sakane, M. and Tokimasa, K., Fatigue-creep life prediction of 2¼Cr–1 Mo steel by inelastic analysis, *Nucl. Eng. and Design* **114** (1989), 311–321.
2. Rubeša, D., Lifetime Prediction for Metallic Materials in the Creep-Fatigue Interaction Regime Based on the Constitutive Modelling of Material Behaviour, Ph. D. Thesis, Montanuniversität Leoben, 1995.
3. Rubeša, D. and Danzer, R., Reevaluation of the SRM lifetime prediction rule by coupling it with a suitable constitutive model, in M.H. Aliabadi *et al.* (eds.), *Localized Damage III : Computer-Aided Assessment and Control*, Computational Mechanics Publications, Southampton–Boston, 1994, pp. 601–613.
4. Spera, D.A., A Linear Creep Damage Theory for Thermal Fatigue of Materials, Ph. D. Thesis, University of Wisconsin, Madison, WI, 1968.
5. Spera, D.A., The Calculation of Elevated-Temperature Cyclic Life Considering Low-Cycle Fatigue and Creep, NASA TN D-5317, Lewis Research Center, Cleveland, OH, 1969.
6. Franklin, C.J., Cyclic creep and fatigue life time prediction, in D. Coutsouradis *et al.* (eds.), *High Temperature Alloys for Gas Turbines*, Applied Science Publishers, London, 1978, pp. 513–547.
7. Franklin, C.J. and Martens, H.J., Zyklisches Kriechen oder die Wechselwirkung zwischen Kriechen und Ermüdung, COST 50, 2. Runde, Proj. CH 4, Schlussbericht, Gebrüder Sulzer AG, Winterthur, 1981.
8. Danzer, R., Buchmayr, B. and Thomas, G.B., The influence of creep on the high temperature cyclic life of IN 738 LC, in R. Brunetaud *et al.* (eds.), *High Temperature Alloys for Gas Turbines 1982*, D. Reidel Publishing Comp., Dordrecht, The Netherlands, 1982, pp. 547–560.
9. Danzer, R., Life Time Prediction of the Gas Turbine Materials PM Astroloy, MAR M 509, Waspaloy, IN 738 LC in the Creep-Fatigue Interaction Regime, COST 50, Round III, Joint Research Report, Inst. f. Metallkunde u. Werkstoffprüfung, Montanuniversität Leoben, Leoben, 1985.
10. Danzer, R. and Bressers, J., A new method to predict the life under high-temperature low cycle fatigue conditions, *Fatigue Fract. Engng Mater. Struct.* **9** (1986), 151–168.
11. Danzer, R., Eine Methode der Lebensdauerprognose für hochfeste metallische Werkstoffe im Bereich der Kriech-Ermüdungs-Wechselwirkung, *Z. Metallkunde* **78** (1987), 19–31.
12. Danzer, R., *Lebensdauerprognose hochfester metallischer Werkstoffe im Bereich hoher Temperaturen*, Gebrüder Borntraeger, Berlin–Stuttgart, 1988.
13. Danzer, R., The influence of creep on the fatigue life of nickel base superalloys, in D.A. Woodford, C.H.A. Townley and M. Ohnami (eds.), *Creep: Characterization, Damage and Life Assessment*, ASM International, Materials Park, OH, 1992, pp. 359–368.
14. Chaboche, J.-L., Constitutive equations for cyclic plasticity and cyclic viscoplasticity, *Int. J. of Plasticity* **5** (1989), 247–302.
15. Sievert, R., Ein Beitrag zur Bestimmung der Materialparameterwerte viskoplastischer Stoffgesetze mittels eines Optimierungsverfahrens, BAM 1.31 92/1, Bundesanstalt f. Materialforschung u. -prüfung, Berlin, 1992.
16. Buchmayr, B., Analyse des Ermüdungs- und Kriechverhaltens der Ni-Basislegierung IN738LC im Hinblick auf den Einsatz als Gasturbinenschaufel, Dissertation, Montanuniversität Leoben, 1983.
17. Day, M.F. and Thomas, G.B., Creep-fatigue interaction in alloy IN 738 LC, in *AGARD Conf. Proc. No. 243, Aalborg, Denmark*, 1978, Ref. II–10.
18. Rodriguez, P. and Bhanu Sankara Rao, K., Nucleation and growth of cracks and cavities under creep-fatigue interaction, *Prog. Mater. Sci.* **37** (1993), 403–480.

THE STRESS-STRAIN BEHAVIOUR OF IN738LC UNDER THERMOMECHANICAL UNI- AND MULTIAXIAL FATIGUE LOADING

J. MEERSMANN, J. ZIEBS, H.-J. KÜHN, R. SIEVERT,
J. OLSCHEWSKI and H. FRENZ
Bundesanstalt für Materialforschung und -prüfung (BAM),
Unter den Eichen 87, D-12205 Berlin,
Germany

1. Introduction

High temperature components such as cooled or uncooled first stage turbine blades are subjected to triaxial stress fields, mainly induced by constrained thermal strain fields during service. The steepest temperature gradients and consequently the largest thermoinduced stresses are generated during start-up and shut-down phases of turbine operation.

The aim of our experiments and numerical simulations is to study the local material behaviour of hot components under thermal conditions as described above.

Furthermore, the applicability of the J_2-theory is verified by the comparison of uni- and multiaxial non-isothermal test results. To investigate life prediction the inelastic work density and the strain rate modified (SRM) time-fraction rule are used.

2. Material

The material studied was the cast nickel base alloy IN738LC. The material was solution treated at 1120°C for two hours, air cooled and aged at 850°C for 24 hours. The material was not hot isostatically pressed (HIPed). The microstructure consisted mainly of cuboidal γ' particles embedded in a γ-matrix. The volume fraction of γ' was approximately 43%. Thin-walled specimens, 200 mm in total length, 25 mm gauge length, 26.5 mm outside diameter and 1.25 mm wall thickness in the gauge section were used. The specimens were supplied as cast-to-size tubes [1].

3. Experimental Set-up

The Thermo-Mechanical-Fatigue (TMF)-tests were conducted on a tension-torsion-internal pressure closed loop test system under computer control. This system is equiped with an induction heating unit and a temperature monitoring device [2].

J. Bressers and L. Rémy (eds.), Fatique under Thermal and Mechanical Loading, 425–434.

4. Thermo-Mechanical Straining Paths

The operating cycles were simulated by strain controlled TMF-tests with several phase angles φ_T between mechanical strain ε_m and temperature T. The controlled total strain ε_{tot} was achieved by adding mechanical strain ε_m and thermal strain ε_{th} (T). ε_m complies with constrained thermal strain. To calculate $\varepsilon_{th}(T)$, a polynominal approximation of second order for the coefficient of thermal expansion $\alpha(T)$ based on experimental data was used. A further step to study the material behaviour under more realistic load conditions is to create a ε_m/T-history representative of the conditions at the leading edge of the first stage of a cooled blade in service [3].

The multiaxial TMF-tests discussed in this paper belong to three groups according to the different ε_m/T-paths: 1. proportional with phase angles φ_T=0° (**In Phase (IP)**) and φ_T =180° (**180°Out of Phase (180OP)**), 2. nonproportional (diamond, sinusoidal) with φ_T =90° (**90OP**) and 3. complex nonproportional ε_m/T- cycles (service cycles).

To compare multiaxial with uniaxial tests an equivalent strain range has to be defined:

$$\Delta\bar{\varepsilon}_m = \sqrt{(\Delta\varepsilon_m)^2 + \frac{1}{3}(\Delta\gamma)^2} \tag{1}$$

The tests were performed at equivalent strain ranges $\Delta\varepsilon_m$=0.6%, 0.8%, 0.9%, 1.2%, equivalent strain rate $\dot{\varepsilon}_m=10^{-4}$ 1/s, phase angle of mechanical values φ_m=0° (proportional) with $\varepsilon_m=\gamma/\sqrt{3}$, and φ_m=90° (nonproportional, sinusoidal) in the temperature range 450°C<T<950°C and with temperature rate T=4 K/s.

5. Stress-Strain Behaviour

5.1. PROPORTIONAL STRAIN-TEMPERATURE PATH, φ_T =0°

Only a brief summary of the tension-torsion stress-strain behaviour will be given in Fig. 1 due to the large variety of TMF cycle shapes which were investigated. The in-phase (φ_T=0°) proportional strain-temperature path results in an asymmetric stress-response (Fig.1a (stress response)). In the range of high temperatures (750°C→950°C) a significant stress relaxation is observed due to the time-dependent inelastic strains. The stress-response for tension and torsion is similar.

5.2. NONPROPORTIONAL STRAIN-TEMPERATURE PATH (DIAMOND), φ_T=90°

The non-proportional diamond (φ_t=90°) strain-temperature path achieves extremes of strains at T = 700°C, Fig. 3b. For temperature T < 700°C mechanical properties are nearly independent of temperature and strain rate [4]. The TMF stress response meets at comparable mechanical strains the responses of isothermal tests in the temperature range 700°C→ 450°C→ 700°C. In the range 850°C→ 950°C→ 850°C softening is observed.

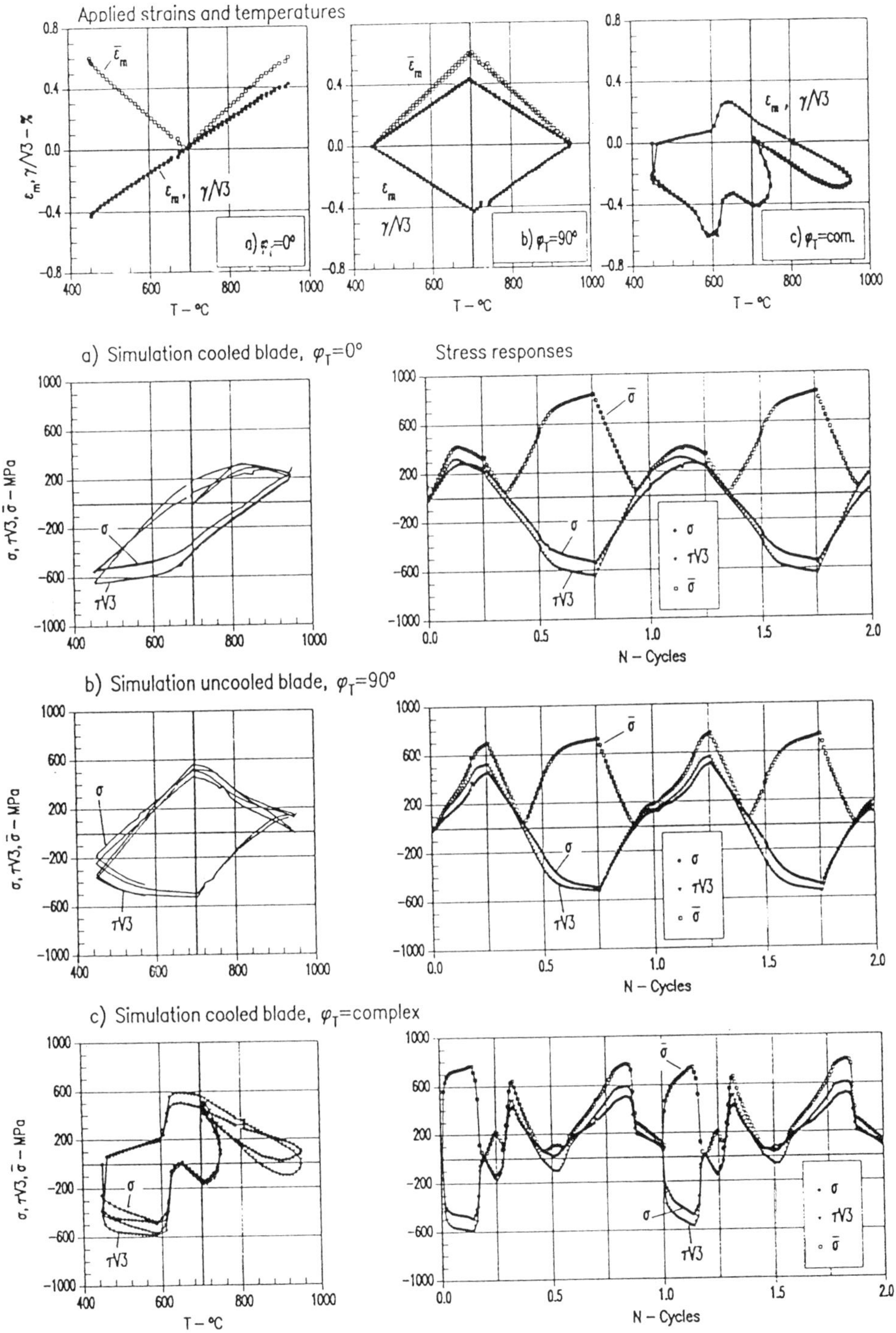

Figure 1. Multiaxial TMF-tests, $\dot{\bar{\varepsilon}}_m = 10^{-4}$ 1/s.

5.3. THE COMPLEX STRAIN-TEMPERATURE PATH, φ_T=COMPLEX

The applied ε_m/T-history (Fig.1c (applied strains)) was developed from a component analysis [3]. As shown in Fig.1c (stress response) compressive stress peaks occur after warm-up, during acceleration and at base load. Tensile stress peaks occur before 'load' and during 'unload'. The largest inelastic strain is achieved after 'warm up'. It is mainly reduced during the unload phase by tensile stress. The influence of a variable temperature on the mechanical properties can be seen in the range 850°C→ 950°C→ 850°C, i.e. the base load temperature range, as described in 5.2.

In all groups it can be observed that the magnitudes of the shear-stresses are generally higher than the magnitudes of the normal stresses.

6. Comparison of Uni- and Multiaxial Stress-Strain Behaviour

6.1. EXPERIMENTAL VERIFICATION OF THE J_2-THEORY

To verify the von Mises-hypothesis, in this case for a plane stress-state, the behaviour of the equivalent stresses and strains in multiaxial and uniaxial tests under equal thermal loads, equivalent strain ranges and strain rates have to be compared. It should be remarked that the curves of equivalent stresses versus equivalent strains lead to a very complex plot because all values are positive. Therefore the curves were changed by signs to get the usual hysteresis loops. The results are shown in Fig.2.

Pure tension-compression- and torsion-tests as well as tension-compression-torsion-tests are compared. Fig.2a represents the IP-path with φ_T =0, Fig.2b the 90OP-path with φ_T =90° and Fig.2c the complex path. In Fig.2c the pure torsional load is left. The equivalent stress values of pure tension-compression-tests are about 100 MPa lower than the others. These deviations are inside the scatterband of this material. The equivalent stresses and strains, calculated by the von Mises relation, of several mechanical loadings are in a good agreement, thus the J_2-theory is altogether applicable to TMF-loadings.

6.2. NUMERICAL SIMULATION OF MULTIAXIAL HARDENING BEHAVIOUR

Simulations of the stress-strain behaviour under TMF-loading (Fig.5) were performed using simple non-isothermal constitutive equations [5] based on the viscoplastic Chaboche model [6]. Material constants for IN738LC have also been given in [5]. Essentially, a finite state function for the isotropic internal stress depending on the accumulated inelastic strain and temperature has been used [7]. Only isothermal uniaxial test data were used to calibrate the model. Further simulations under non-proportional loading are presented in [8].

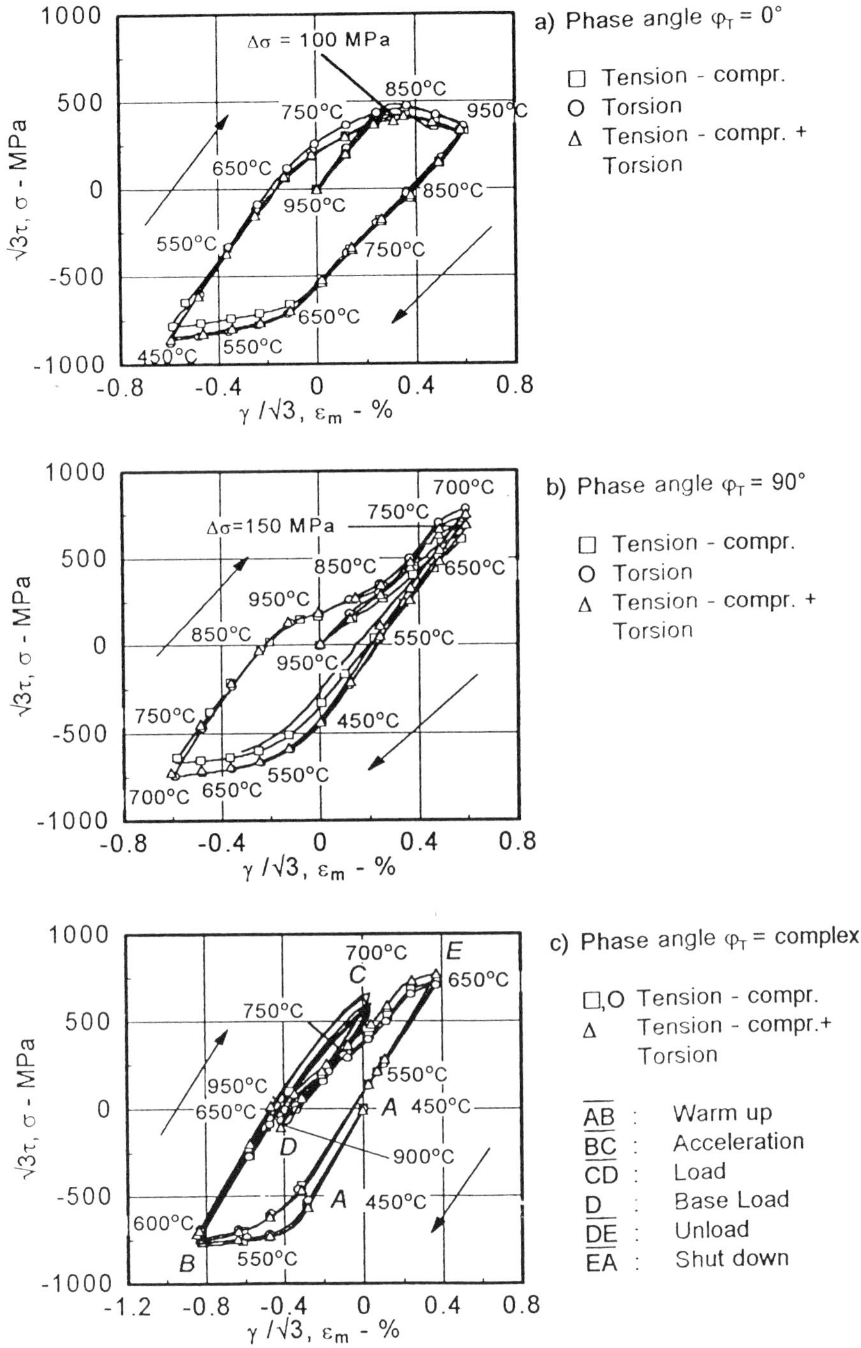

Figure 2. **Comparison of uni- and multiaxial TMF-tests, 450°C<T<950°C, $\dot{\bar{\varepsilon}}_m=10^{-4}$ 1/s, $\Delta\varepsilon_m$=1.24%.**

The described non-isothermal simulations show the applicability of the *constitutive stress-strain model* for the case of TMF-loading of a material element. At this stage, the non-isothermal viscoplastic model can be applied for the computation of the inhomogeneous (temperature, strain, stress etc.) fields in structures. Examples are a turbine blade [9] and also a wedge-type specimen (Ramteke *et al.* [10]).

7. Lifetime Assessment

7.1. THERMO-MECHANICAL FATIGUE

For lifetime prediction the following tests were performed: tension-compression tests with $\varphi_T = 0°$, $\varphi_T = 180°$ and the complex ε_m/T-path, further tension-compression-torsion-tests with the complex path and a sinusoidal path with the phase angle $\varphi_T = -90°$, $\varphi_m = 90°$. The tests were carried out with three comparable equivalent strain ranges (Fig.3).

The cycles to failure N_f were found by a 1%-stress-decrease in σ_{max}/N with respect to decreasing regression straight-lines which include also points behind saturation [11]. The *inelastic work* $\Sigma\sigma_{ij}\Delta\varepsilon_{ij}^{in}$ of the cycle $N_f/2$ was determined and plotted double logarithmic against N_f (Manson-Coffin) [12, 13].

The coefficient A and the exponent m in the correlation

$$\Sigma\sigma_{ij}\Delta\,\varepsilon_{ij}^{in} = A\ N_f^{\,m} \tag{2}$$

are obtained by linear regression for each test group. Fig.4 shows the points of five test groups with their straight-lines of regression. The most outside points limit a scatterband with an N_f width-factor of 2.25 with respect to the scatterband midline.

7.2. HIGH TEMPERATURE FATIGUE

For lifetime prediction the approach of Franklin [14] and Danzer [15] has been combined with Chaboche's viscoplastic stress-strain model [6]. The damage rule can be considered as a *rate equation for the consumed lifetime fraction* that is denoted by D and set 0 at the beginning of loading and 1 at failure:

$$dD = \frac{dt}{t_{f,c}(|\sigma|\ ,T)}\left(\frac{|\dot{\varepsilon}_{in}|}{r_o\dot{\varepsilon}_{c,min}(|\sigma|\ ,T)}\right)^{\nu}\beta(\sigma)\ ,\ \ \beta = \begin{cases} 1\ ,if\ \sigma \geq 0 \\ k_c\ ,\ if\ \sigma < 0 \end{cases} \tag{3}$$

Here, $t_{f,c}$ is the time-to-creep-fracture. β gives the amount k_c of damaging under compression with respect to the damaging under tension (Rubeša and Danzer [16]).

In the strain rate modified time-fraction rule (3) the deviation of the current inelastic strain rate $\dot{\varepsilon}_{in}$ from the minimal creep rate $\dot{\varepsilon}_{c,\,min}$ under the current stress is taken into account. The deviation from monotonous deformation under fatigue is

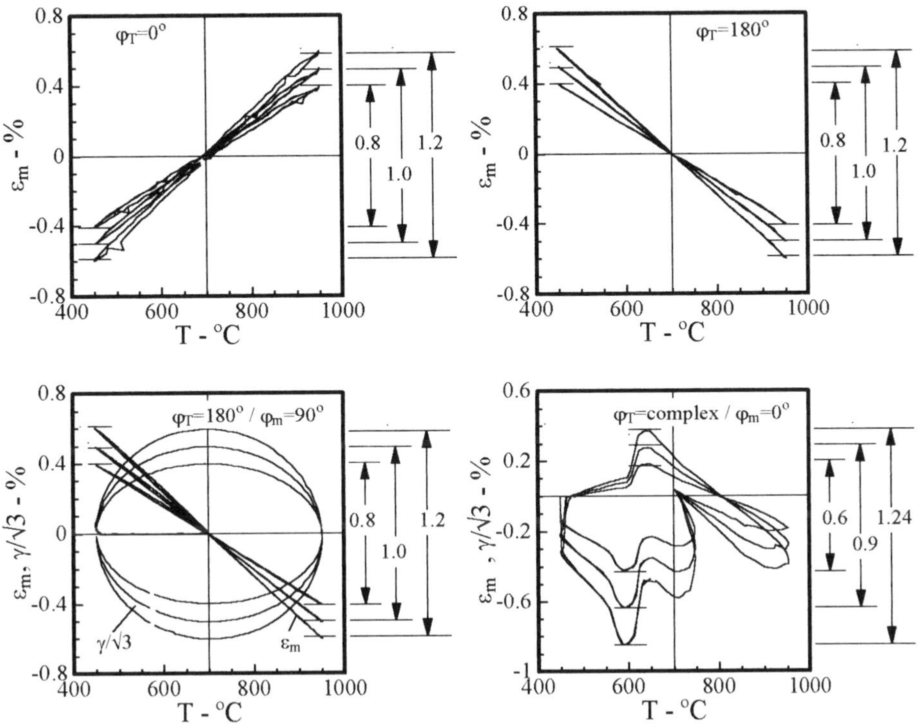

Figure 3. Applied strain-temperature paths to study TMF-lifetime, $\dot{\varepsilon}_m = 10^{-4}$ 1/s.

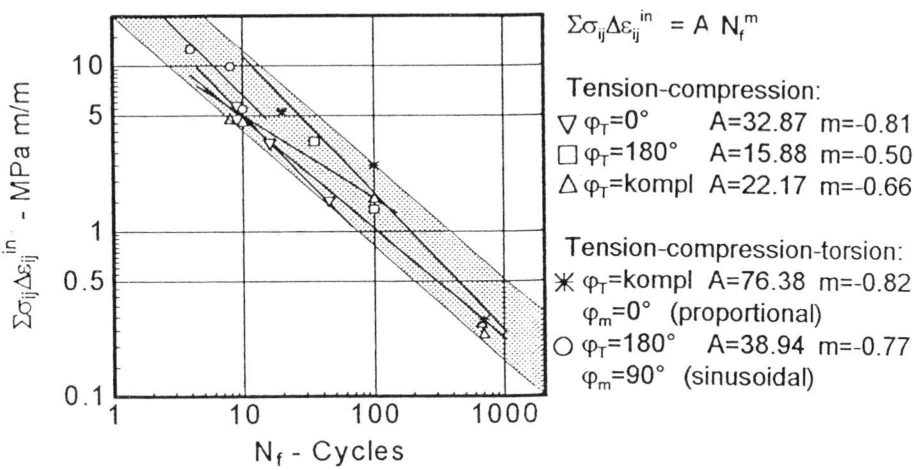

Figure 4. Manson-Coffin-Plot of uniaxial and multiaxial TMF-tests.

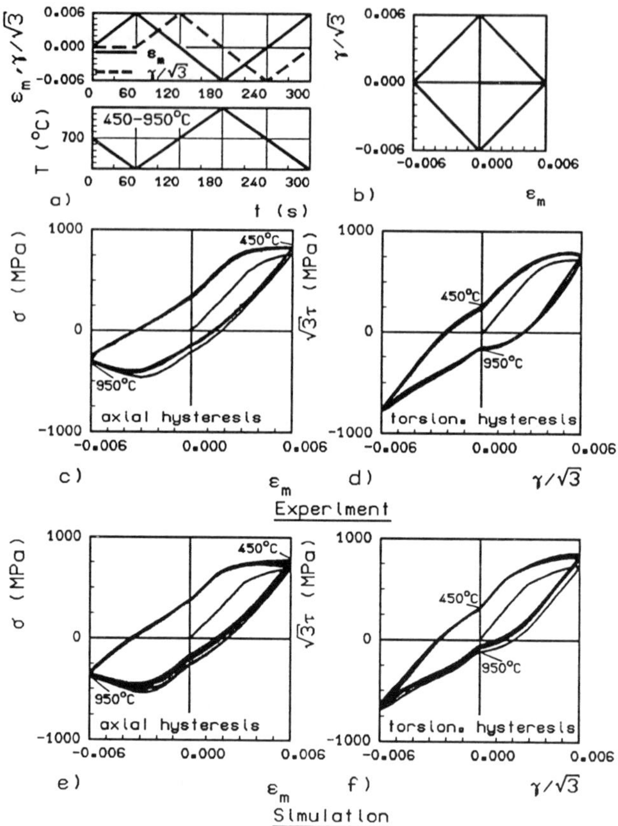

Figure 5. Axial and torsional hysteresis loops in a non-isothermal process: a) loading history, b) mech. loading path, c, d) material response and e, f) simulation.

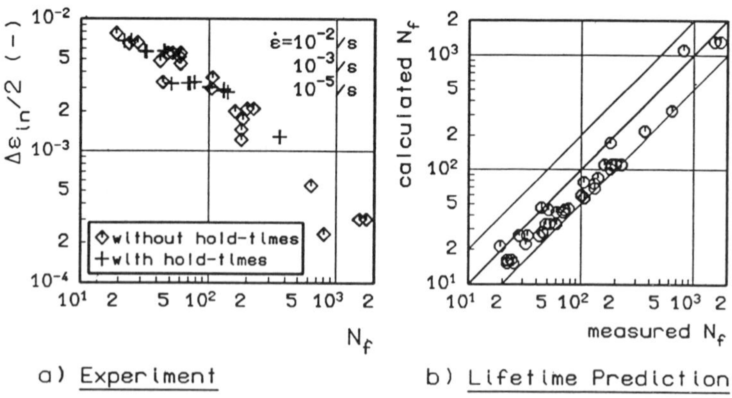

Figure 6. Low-cycle-fatigue of IN 738 LC under uniaxial loading at 850°C: a) Representation of number of cycles to failure N_f vs. inelastic strain range $\Delta\varepsilon_{in}$, hold-times: 30 s and 2 min in tension and/or compression strain amplitudes b) lifetime assessment with the modified time-fraction rule (3) using Chaboche's viscoplastic model to compute the varying inelastic strain rate.

indicated on the microscopic scale by the initiation and propagation of small cracks.

The material constants r_0, v and k_c have been estimated by Danzer [15] for IN738LC in air: r_0=2.6, v=0.71, k_c=0. In the present paper the predictive capability of these fatigue constants, which have been determined without using a viscoplastic stress-strain model, is investigated.

The damage rate equation (3) was applied to high temperature fatigue including relaxation periods, see Fig.6a. The tests performed have been strain-controlled without mean-strain at different strain rates. The predicted lifetimes are given in Fig.6b. They are within a factor of 2. This result of coupling the damage rule with a viscoplastic stress-strain model can be considered as a verification of the procedure given in [17] to approximate fatigue lifetime, because with this procedure the essential fatigue constant v has been determined.

In cyclic loaded structures multiaxial fatigue occurs. An application to this problem is the next step of investigation of the presented lifetime approach. To assess life under TMF-loading (including oxidation effects) [18] an additional thermal-mechanical cycling factor has been proposed by Neu and Sehitoglu [19].

8. Conclusions

The stress-strain behaviour of IN738LC under TMF-loading is nearly independent of temperature for T<700°C [20]. The shear stresses are usually about 10% higher than the normal stresses under the load of equal shear and normal strains.

It was verified by experiments that the von Mises-hypothesis is altogether applicable to the hardening behaviour of IN738LC under TMF-loading.

By using inelastic work, a simple relation to predict lifetime for the investigated kinds of TMF-tests can be proposed. The sinusoidal time histories of applied strains and temperatures generate the highest work and result in the fewest cycles to failure. No significant dependence on the type of the load path and nonproportionality of load components can be observed. The scatterband with a N_f-width-factor of 2.25 is nearly the usual of 2.0.

Constitutive equations based on the viscoplastic Chaboche model describe the non-isothermal multiaxial hardening behaviour in the fatigue tests performed.

The coupling of the Franklin-Danzer damage rate equation with the viscoplastic model is successful for not HIPed IN738LC at 850°C without any redetermination of the material constants.

Acknowledgements - The authors are grateful for the financial support of the Deutsche Forschungsgemeinschaft (DFG) within the Sonderforschungsbereich 339. The contributions of Mr. M. Bendig in performing the numerical work to evaluate the damage rate Eq. 3 is gratefully acknowledged.

References

1. Ziebs, J., Meersmann, J., Kühn, H.-J., Effects of proportional and non-proportional straining sequences on the hardening/softening behaviour of IN738LC at elevated temperatures, *Eur. J. Mech. A/Solids*, **13**, 5 (1994), pp. 605-619.
2. Ziebs, J., Meersmann, J., Kühn, H.-J., Ledworuski, S., Experimental Techniques for Testing Materials under Multi-Axial Loading, *Fourth Int. Conf. on Biaxial/Multiaxial Fatigue*, Vol. I, Societé Française de Metallurgie et de Matériaux (SF2M), 1994, pp. 61-77.
3. Embley, G.T., Russel, E.S., Thermal-mechanical fatigue of gas turbine bucket alloys, *1st Parsons Int. Turbine Conf.*, Parsons Press, Trinity College Dublin, Inst. of Mech. Engineers, London, 1984, pp. 157-164.
4. Ziebs, J., Meersmann, J., Kühn, H.-J., Frenz, H., High Temperature Inelastic Deformation of IN 738 LC Under Uniaxial and Multiaxial Loading. Proc. *8th CIMTEX-World Congress and Forum on New Materials*, Florenz, P. Vencenini (ed.), Techna Publ., 1994.
5. Bertram, A., Olschewski, J., Sievert, R., *Arch. Mech.* **46**, 3 (1994), 413-429.
6. Chaboche, J.-L., Viscoplastic Constitutive Equations for the Description of Cyclic and Anisotropic Behaviour of Metals. *Bulletin L'Academie Sciences* **XXV**, 1, Serie Sciences Techniques (1977), 33-42.
7. Benallal, A., Ben Cheikh, A., *Constitutive Laws for Engineering Materials*, C.S. Desai et al. (eds.),ASME Press, 1987, 667-674.
8. Frenz, H., Meersmann, J., Ziebs, J., Kühn, H.-J., Olschewski, J., Sievert, R., Book of Abstracts *7th Int. Conf. Mechanical Behaviour of Materials* (ICM7), The Hague, A. Bakker (ed.), European Structural Integrity Society (ESIS), Delft University Press, The Netherlands, 1995, 59-60.
9. Olschewski, J., Haftaoglu, C., Noack, H.-D., Thermo-Mechanical 3D-FE-Analysis of a Cooled Turbine Blade Using an Unified Constitutive Model. *Inelastic Analysis, Fatigue, Fracture and Life Prediction,* Seminar Post-SMiRT(12) No.**5**, Paris, Ministry for Research and Space, France, G. Baylac (Chairman),Electricité de France (EDF), Paris, 1993.
10. Ramteke, A.L., Meyer-Olbersleben, F., Rézaï-Aria, F., *Low Cycle Fatigue and Elasto-Plastic Behaviour ofMaterials-3*, K.-T. Rie (ed.), Elsevier Applied Science, 1992, 281-286.
11. Tomas, G.B., Hales, R., Ramsdale, J., Suhr, R.W., Sumner, G., *Fatigue Fract. Eng. Mater. Struct.*, **12**, 2 (1989), 135-153.
12. Ellyin, F., Xia, Z., A General Fatigue Theory and Its Application to Out-of-Phase Cyclic Loading, *J. Eng. Mat. Technology* **115**, (1993), 411-416.
13. Engler-Pinto Jr., C.C., Härkegård, G., Ilschner, B., Nazmy, M.Y., Noseda, C., Rézaï-Aria, F., *Materials for Advanced Power Engineering*, Part I, D. Coutsouradis et al. (eds.), Kluwer Academic Publ., 1994, 853-862.
14. Franklin, C.J., *High Temperature Alloys For Gas Turbines*, D. Coutsouradis *et al.* (eds.), Applied Science Publ., 1978, 513 -547.
15. Danzer, R., Lebensdauerprognose hochfester metallischer Werkstoffe im Bereich hoher Temperaturen, Gebrüder Borntraeger, Berlin, Stuttgart, 1988.
16. Rubeša, D., Danzer, R., Reevaluation of the SRM lifetime prediction rule by coupling it with a suitable constitutive model. *Localized Damage III, Computer Aided Assessment and Control*, M.H. Aliabadi *et al.* (eds.), Computational Mech. Publ., Southampton, Boston, 1994, 601-613.
17. Danzer, R., Bressers, J., A new method to predict life under high temperature low cycle fatigue conditions, *Fatigue Fract. Engng. Mater. Struct.* **9**, 3 (1986), 151-168.
18. Rézaï-Aria, F., Dambrine, B., Rémy, L., *Fatigue Fract. Engng. Mat. Struct.* **11**, 4 (1988), 291-302.
19. Neu, R.W., Sehitoglu, H., Thermo-Mechanical Fatigue, Oxidation and Creep, Part II - Life Prediction, *Metallurgical Trans. A*, **20A**, (1989), 1769-1783.
20. Ziebs, J., Meersmann, J., Kühn, H.-J., Hülsmann, N., Olschewski, J., Multiaxial Thermomechanical Behaviour of IN738LC, *Fourth Int. Conf. on Biaxial/Multiaxial Fatigue*, Vol. II, Societé Française de Metallurgie et de Matériaux (SF2M), 1994, 247 - 259.

THERMO-MECHANICAL BEHAVIOUR AND MODELLING OF AN AUSTENITIC STAINLESS STEEL UNDER ANISOTHERMAL CYCLIC LOADINGS

A. BOUCHOU and P. DELOBELLE
Laboratoire de Mécanique Appliquée R. Chaléat
URA 04 CNRS
UFR Sciences et Techniques
La Bouloie - Route de Gray
25030 Besancon Cedex, France

1. Introduction

The industrial materials used for the construction of certain components of modern thermal machines are often submitted to severe thermomechanical loadings and the precise knowledge of the behaviour and the development of the isothermal and anisothermal laws proves to be indispensable for an eventual prediction of the lifetime of these structures. Toward this end and with the considerable development of internal variable models which lead to a precise modelling of the mechanical behaviour under isothermal conditions, many recent studies attempt to enlarge their domain of application to the case of anisothermal loadings [1-16].

These works demonstrate the real predictive possibilities of this type of strategy given a correct mathematical formulation for the transition of the isothermal case to the anisothermal case [2-5, 7-16]. Meanwhile, certain materials possess temperature history effects meaning that a one to one relation between mechanical response and temperature does not exist [1, 3, 6, 17]. In this context, in addition to the correct expression of the kinetic equations, it is necessary to understand and to analyse these history effects in order to model them with a view to their integration into the general model framework.

In the present case an austenitic stainless steel is studied between 20 and 650°C. This steel is mainly used in nuclear industry, namely in the construction of heat exchangers for the fast breeder reactors. An adequate knowledge of its behaviour under isothermal conditions constitutes an essential preliminary step before the study of anisothermal loadings. The results of this analysis were given by Delobelle [18]. This article presents a precise study of the response of this steel to anisothermal loadings in order to demonstrate an eventual temperature history effect.

J. Bressers and L. Rémy (eds.), Fatique under Thermal and Mechanical Loading, 435–444.

2. Experimental Techniques

The test specimens are obtained from slices removed from 30 mm thick sheets and annealed at 1200°C. The weight composition of this low carbon steel is given in Table 1. Several machines have been used to perform the different experimental sequences, namely a screw tensile machine controlled by strain and monitored by a microcomputer for the stepwise isothermal tests, an electrodynamic torsion machine which allows the strain to be a function of the temperature signal for the real anisothermal tests, and finally a hydraulic tensile-torsion machine, used for the biaxial tests (in and out of phase) at isothermal levels.

TABLE 1. Weight composition of the steel in mass %.

C ≤ 0.03 | S ≤ 0.001 | P ≤ 0.021 | Si 0.44 | Mn 1.084 | Ni 12.3 | Cr 17.54 | Mo 2.47 | N 0.075 | B 0.001 | Co 0.15 | Cu 0.175 | Ti < 0.005 | Nb 0.015 | Al 0.100 | Fe bal.

3. Experimental Results and Analysis

The best way to approach and analyse potential temperature history effect is by performing thermal cycles at different isothermal levels and successively increasing and then decreasing the temperature stepwise, or vice versa, while having the possibility of changing the values of the cycle limits. This approach is realized for monotonic, uniaxial and biaxial cyclic loadings (in and out of phase).

3.1. THE MONOTONIC TESTS

Fig. 1 shows an example of a monotonic test with successively increasing then decreasing isothermal steps. If the monotonic curves obtained for each isotherm are superimposed, it is shown that within experimental error due to the presence of the yield point return phenomena, the strain levels are identical. This means that there is no noticeable temperature history effect, an observation which is in agreement with the results of Niitsu and Ikegami [19] obtained for a 304 stainless steel.

3.2. UNIAXIAL CYCLIC TESTS

The peak values of the stress are shown in Fig. 2 for a cyclic strain test and for three cycles of increasing and decreasing temperature steps (20 ↔ 200 ↔ 400 ↔ 560 ↔ 600°C). The test characteristics are: $\dot{\varepsilon}^T_{zz} = 6.6 \times 10^{-4}\ s^{-1}, \Delta\varepsilon^T_{zz}/2 = \pm 4 \times 10^{-3}$, 50 mechanical cycles for each step and a zero force control at temperatures of approximately one hour at each step. If the average of the peak stress values $\Delta\sigma_m/2$ is plotted as a function of the temperature T, for the three thermal cycles at the end of each step, it is shown in Fig. 3 that the relation $\Delta\sigma_m/2 = f(T)$ is not one to one, which means that there is a temperature history effect.

Indeed, whereas the stress levels correspond to those obtained under isothermal conditions for the first half thermal cycle, this is not true for the remaining thermal cycles, the stress levels being larger than those of the isothermal tests. For 20°C, a hardening of the order of 80 MPa is obtained. In addition, it can be remarked that, on the one hand, the stress levels are nearly identical for the thermal cycles two and three, which means that the temperature history has been established from the first thermal half cycle, and on the other hand, the paths for the increasing temperature are slightly different than that of decreasing temperatures. This experiment tends to show that the material memorises the maximum temperature reached during the thermal cycling (T_{max} = 600°C) and that the hardening amplitude is a function of T_{max} and of the amplitude of the domain where $n^* = (\partial(\Delta\sigma_m / 2) / \partial T) > 0$. These two facts are confirmed by the experiments reported in Figs. 4 and 5. Indeed, if two thermal cycles are performed starting from T_{max} = 600°C, it is shown in Fig. 4 that in this case a trajectory higher than in the previous experiment is directly obtained, corresponding to the material in its memorisation state for T_{max} = 600°C. In the experiment shown in Fig. 5, six thermal cycles are performed between 100°C and T_{max} by progressively increasing the upper limit (T_{max} = 200 (point (1)), 300 (2), 400 (3), 525 (4) and 600°C (points (5) (6)). For T_{max} = 200°C and belonging to the zone $n^* < 0$, the cycle is reversible and no supplementary hardening appears ; for T_{max} = 300°C and corresponding approximately to $n^* = 0$, only a very slight hardening is noticeable, whereas for T_{max} = 400, 525 and 600°C corresponding to $n^* > 0$, a significant hardening is obtained as an increasing function of T_{max}. The material thus memorises T_{max} when $n^* = (\partial(\Delta\sigma_m / 2) / \partial T) > 0$.

The hardening amplitude can be evaluated by plotting a path parallel to the branch $n^* < 0$, starting from the maximum temperature reached and obtained from isothermal tests. It can be shown that if the same type of experiment is performed starting from 500°C and with decreasing T_{max}, then the hardening is obtained immediately after the first stage at this temperature and its amplitude corresponds closely to cycle four of the previous experiment. There again the material has memorised T_{max} = 500°C. It can also be noted that past 550-600°C, n^* is negative again and that in this zone the material no longer memorises T_{max}. The highest memorisation temperature is situated in the neighborhood of 600°C, which corresponds to $n^* = 0$ and $(\Delta\sigma_m/2)$ maximum.

In conclusion, it can be affirmed that the memorisation of T_{max} is made in the domain found between the two zeros of the curve $\Delta\sigma_m/2 = f(T)$ and when its derivative is positive. Only by maintaining a high temperature (T ≥ 650°C for a few hours) can the hardening resulting from the temperature history be progressively erased by the recovery effect of the structure.

These different observations agree globally with those reported by Murakami *et al.* [17] and Ohno *et al.* [8], respectively on 316 and 304 stainless steel.

3.3. CYCLIC BIAXIAL TESTS IN TENSILE-TORSION

The effect of multiaxial stresses on the phenomena of temperature history memorisation is studied using cyclic tension-torsion tests, both in phase (ϕ = 0°) and out of phase (ϕ = 90°) along different stepwise isothermal loading trajectories. It is well known that this steel possesses a supplementary cyclic hardening which is a function of the non-radiality of the loading and has a maximum for a phase shift of 90° and a ratio between the maximum strain amplitudes equal to $2/\sqrt{3}$ [20, 21]. However, for radial loadings (ϕ = 0°) no supplementary hardening appears. Since the objective is to analyse the consequences of the temperature history effect for non-radial loadings, the same stepwise isothermal loadings are adopted as those shown in Figs. 3 and 4, with ϕ = 0° and ϕ = 90°.

The results mentioned in Figs. 3 and 4 are found again, both qualitatively and quantitatively, for radial loadings, indicating that the same conclusions as before can be drawn.

However, in the case of out of phase loadings (ϕ = 90°), the stress levels $\Delta\overline{\sigma}_{max}/2$ (equivalent von Mises stress) are higher than those for the radial tests. This result is in agreement with the cyclic out of phase properties of this steel under isothermal conditions. The supplementary hardening is a decreasing function of temperature [21,22]. Hence, during the first thermal half-cycle in Fig. 6, the levels of $\Delta\overline{\sigma}_{max}/2$ are in agreement with those of the isothermal tests while for the rest of the temperature sequence, as for the uniaxial tests, the material memorises its passage through 600°C (Figs. 6 and 7). At room temperature, this new supplementary hardening is of the order of 70 MPa, a value which is identical to that obtained for uniaxial loadings. It should be noted that these observations are in disagreement with those of Murakami et al. [22], who report only a very small temperature history effect for ϕ = 90°.

In summary, it has been demonstrated that the conclusions resulting from uniaxial experiments and relating to the temperature history effects remain applicable to radial and non-radial multiaxial loadings. Two types of supplementary hardening have been clearly described at room temperature, one pertaining to the non-radiality of the mechanical loading and the other one to a temperature history effect.

3.4. ANISOTHERMAL TESTS

Several real anisothermal tests have been performed in order to confirm the results presented above and to compare them to the numerical predictions given by the behaviour model [16,18]. The problems of thermal dilatation are avoided by performing anisothermal torsion tests with uninhibited axial strain. The shear strain is controlled by the temperature signal via an extensometer and a microcomputer. Tests are performed in phase, in ¼ phase, in phase opposition, in ¾ phase and in double phase (two mechanical cycles for a thermal cycle), for three successively increasing strain levels (around 40 cycles at each level). The experimental conditions are such that $\dot{T}$ = 7°C min^{-1} and ΔT is between 600 and 200°C. The results of the in 1/4 phase and double phase tests are shown in Figs. 8 and 9 for the last cycles at each strain level. If the stress levels ($\Delta\sigma_{max}/2$) are compared at the characteristic points of the cycle (inversion and singular points) where the strain and temperature are known for the same conditions under isothermal loading, then it can be shown that a difference of about 70 MPa exists for the anisothermal loadings. These results confirm the conclusions drawn for stepwise isothermal experiments.

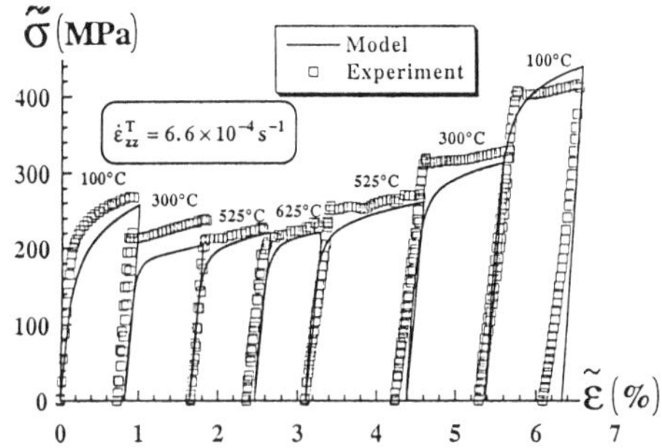

Figure 1. Stepwise isothermal monotonic tests. Experimental results and model.

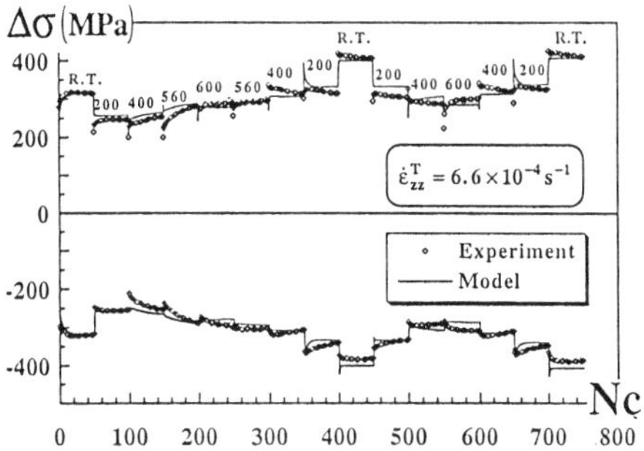

Figure 2. Stepwise isothermal cyclic tests. Experimental results and model.

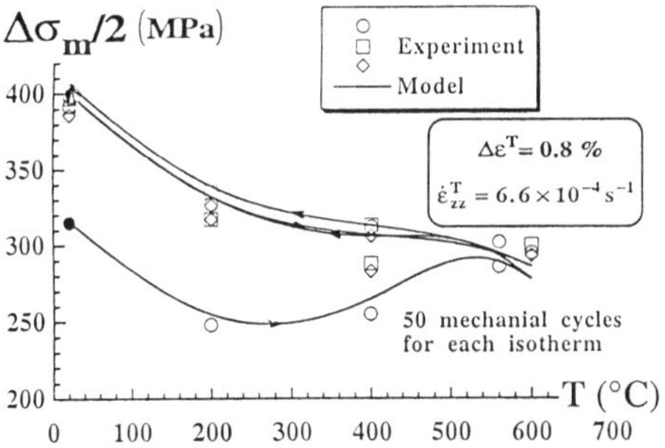

Figure 3. Evolution of the average peak values as a function of the temperature. Experimental results and model.

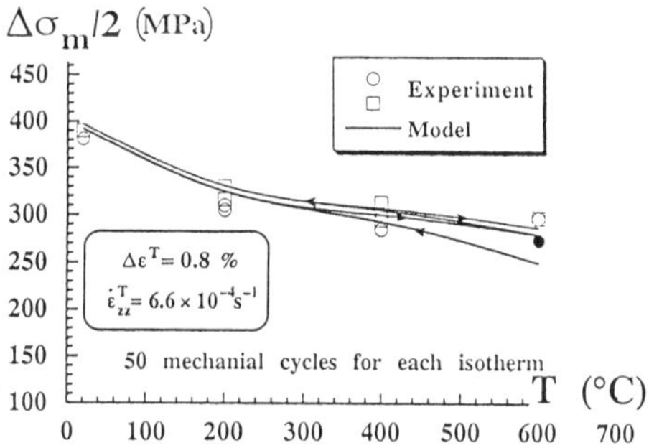

Figure 4. Idem Fig. 3 but starting from T=600°C. Experimental results and model.

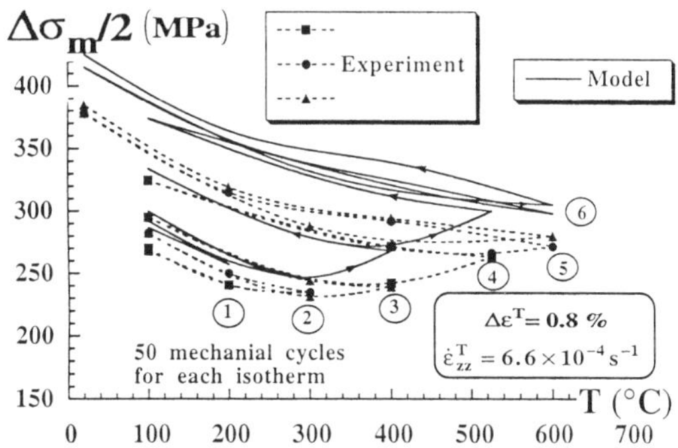

Figure 5. Idem Fig. 3 demonstration of the memorisation of T_{max} when: $\partial(\Delta\sigma_m / 2)/\partial T > 0$ Experimental results and model.

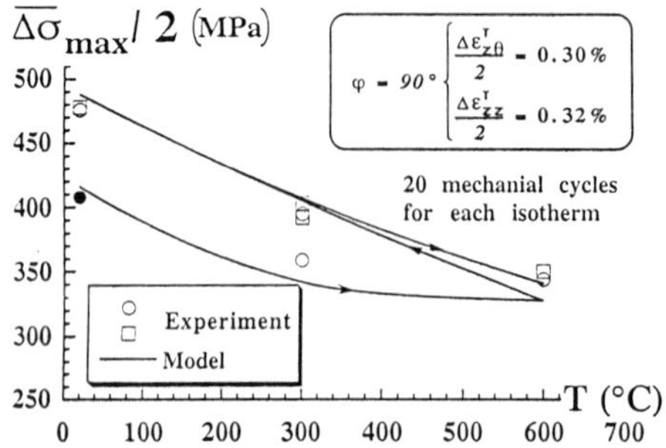

Figure 6. **Stepwise tests with cyclic biaxial out of phase loadings.**

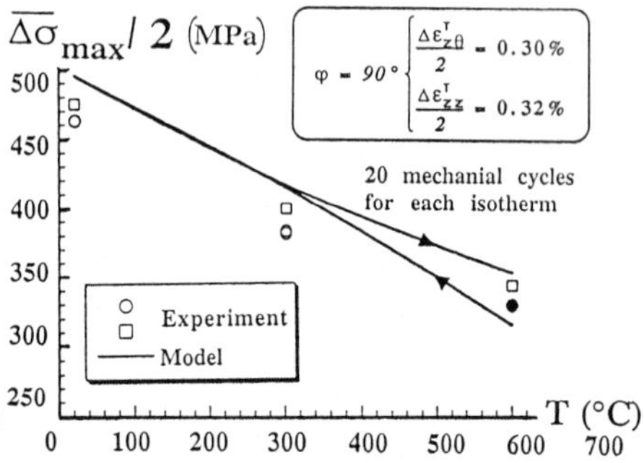

Figure 7. **Idem Fig. 6 but starting from T_{max} = 600°C. Experimental results and modelling.**

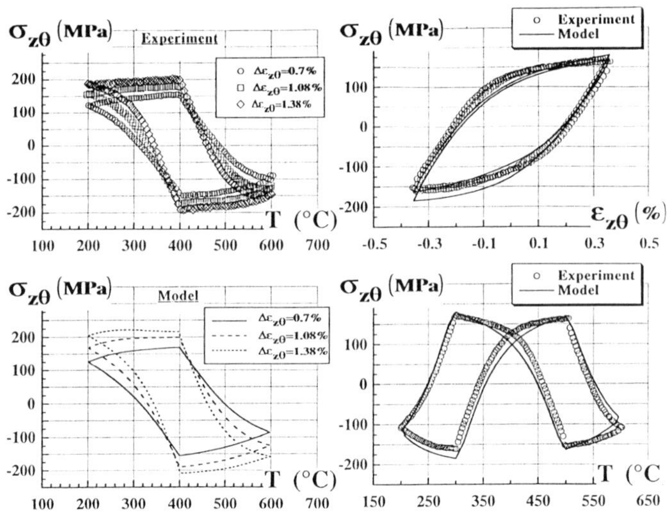

Figures 8-9. **Anisothermal in 1/4 of phase and in double phase. Experimental results and modelling.**

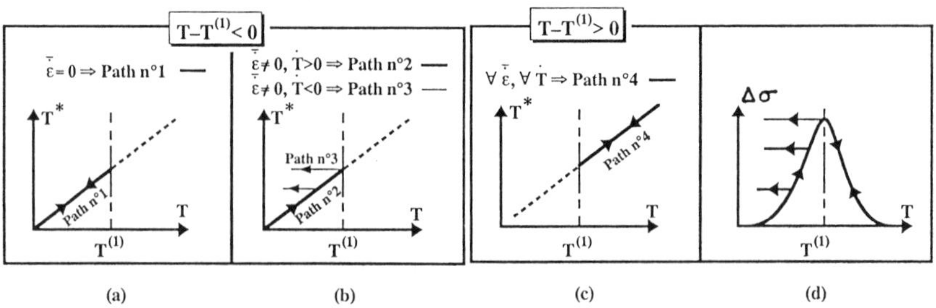

Figure 10. **Implications of Eq. (4), flow chart of the functioning.**

4. Discussion on the Temperature History Effect

The existence of a temperature history effect has been demonstrated and its characteristics described. However, its physical nature remains to be explained. This effect is directly related to the existence of a maximum in the curve $\Delta\sigma_m / 2 = f(T)$, creating a non admissible domain where $\partial(\Delta\sigma_m / 2)/\partial T$ is positive. In other words, the physical nature of the temperature memory phenomena is related to physics of the hardening mechanism associated to this maximum.

Several studies [23-26] clearly show that between 200 and 600°C there exist very strong interactions between dislocations and point defect configurations in these steels (dynamic strain aging) causing this hardening at intermediate temperature. These interaction phenomena are very sensitive to the loading rate and can lead to a negative coefficient of the stress sensitivity with respect to the rate, a fact which has already been observed for this steel [18]. The amplitude of the maximum is thus a function of the strain rate, and in all logic the effect of the temperature history should also depend on the rate, in the sense that the maximum is more pronounced for smaller rates. This point has been verified experimentally. Since the interactions occur with the dislocations present in the material, thus independently of the activated slip systems which are a function of the non-radiality of the loading, it would seem logical to find this history effect in uniaxial as well as biaxial tests. This mechanism, which is intrinsic to solid solutions, explains the memorisation of the maximum temperature. Note that this type of hypothesis has already been suggested by several authors [3, 22].

5. Modelling

Many recent studies show the current predictive possibilities of internal variable models for both isothermal and anisothermal thermomechanical loadings [1-16, 27]. The methodology which is generally accepted and which allows access to an anisothermal modelling can be summarised as follows :

a) Study of the material behaviour under isothermal conditions.
b) Modelling of the isothermal behaviour and correlation between calculations and experiments for each isotherm.
c) Introduction of smoothing functions to describe the evolution of the different model parameters which depend on the temperature and correlation between calculations and experiments.
d) Study of the behaviour under anisothermal conditions in order to detect a possible temperature history effect related to a microstructural instability phenomenon.
e) A correct mathematical formulation for the transition from the isothermal case to the anisothermal case based on thermodynamical considerations, namely taking the temperature as a state variable in the free energy

expression. This leads to the presence of terms incorporating the temperature rate ($\dot{T}$) and allows the absence of a temperature history effect, i.e., the reversibility of the mechanical cycle response as a function of the temperature to be described.

f) Modelling of a possible temperature history effect as well as its introduction in the anisothermal model without a temperature history effect. The objective of this step is to take into account the phenomenon of microstructural instability.

g) Comparisons between the model predictions and experimental results of true anisothermal tests.

Compared to this methodology, it can be said that the phases (a) to (d) are admitted without change [18], and that the steps (e) (f) and (g) need to be performed.

The limited scope of this article does not allow the details of the modelling to be described. We only present the general principles of the procedures e) and f).

5.1. TRANSITION FROM THE ISOTHERMAL CASE TO THE ANISOTHERMAL CASE WITHOUT TAKING INTO ACCOUNT THE TEMPERATURE HISTORY

As mentioned earlier, the methodology consists in modelling initially the absence of a temperature history and then introducing a temperature history module in accordance with experimental observations. The first step can be analyzed using the general thermodynamical considerations concerning internal variable models [15, 28-29]. The knowledge of the thermodynamical potential ψ (free energy) as a function of the state variables allows the definition of the corresponding thermodynamical forces. For example, if we consider the inelastic part of the free energy, $\psi^{v}(X_{ij}^{(k)}, T)$, we have :

$$\alpha_{ij}^{(k)} = \frac{\partial \psi^{v}}{\partial X_{ij}^{(k)}} = f(X_{ij}^{(k)}, X_{ij}^{(l)}, T) \tag{1}$$

The thermodynamical flux $\dot{X}_{ij}^{(k)}$ is defined from the dissipation potential $\Omega^{v}(\alpha_{ij}^{(k)}, T)$, which is given by :

$$\dot{X}_{ij}^{(k)} = \frac{-\partial \Omega^{v}}{\partial \alpha_{ij}^{(k)}} = g(\alpha_{ij}^{(k)}, \alpha_{ij}^{(l)}, T) \tag{2}$$

Thus, in order to obtain the constitutive laws $\dot{\alpha}_{ij}^{(k)} = F(\alpha_{ij}^{(k)}, \alpha_{ij}^{(l)}, T)$, Eq. (1) must be differentiated and then combined with Eq. (2), whence the appearance of terms in $\dot{T}$ due to the temperature dependence of certain parameters $a^{(m)}$ in the kinetics of the evolution laws. This leads to :

$$\dot{\alpha}_{ij}^{(k)} = \dot{\alpha}_{ij}^{(k)}(\dot{T}=0) + \alpha_{ij}^{(k)} \sum_{m} \frac{1}{a^{(m)}} \frac{\partial a^{(m)}}{\partial T} \dot{T} \tag{3}$$

So, these equations are divided into an isothermal part ($\dot{T}$ = 0) and a part depending of the temperature rate. This procedure is applied to all the internal variables of the model.

5.2. FORMULATION OF THE TEMPERATURE HISTORY MODULE AND INCORPORATION INTO THE MODEL

In the first part of this article, it is shown that this material memorises the maximum value of the peak cyclic stress reached when the derivative of the cyclic stress as a function of temperature is positive, and that this phenomenom is certainly due to dynamic strain aging. In order to describe this property mathematically, the notion of a "fictitious temperature" T* is defined by Eq. (4) :

$$\dot{T}^* = \left[\left(1 - H(T - T^{(1)})\right)\left(1 - H(T_{Max} - T)\right) + H(T - T^{(1)})\right]\dot{T}, \tag{4}$$

$$\text{with } T_{Max} = \text{Max}(0, T.H(\dot{\bar{\varepsilon}}))$$

or, in condensed form : $\dot{T}^* = H\,\dot{T}$, H being the temperature history module. It is recalled that H(.) is the Heaviside function. This fictitious temperature is then introduced in the Eq. (5) for the saturation value $Y_{C\infty}^{(1)}$ of the variable $Y_C^{(1)}$ which describes the hardening corresponding to dynamic strain aging.

$$Y_{C\infty}^{(1)} = Y_{C\infty}^{(1)\,sat} \exp -\left(\frac{T^* - T^{(1)}}{A^{(1)}}\right)^2 \tag{5}$$

The implications of the Eq. (4) can be made explicit with the aid of the flowchart in Fig. (10). The Eq. (4) combined with a transformation by means of Gaussian laws (Eq. 5) allows the temperature history effect observed at the stabilized cycles to be described.

6. Comparisons Between Predictions and Experiments

In Fig. 1 we report the predictions for monotonic temperature step tests. In agreement with experiments, there is only a weak temperature history effect. This can be explained by the fact that in this kind of test, the accumulated plastic strain is too small to significatively activate the isotropic variable. The numerical simulations of the principal cyclic experiments are reported in Figs. 2 to 7. The agreement of maximum stress levels at the stabilized cycles is globally satisfactory, which means that the principal effects of temperature history are accounted for by this model.

For real anisothermal loadings, the simulations are reported in Figs. 8 and 9 and we show that they are fairly close to the experimental results.

7. Conclusions

With the aid of uniaxial and biaxial ($\phi = 0$ and 90°) stepwise isothermal tests, it is shown that this austenitic steel possesses a temperature history phenomenon which is independent of the type of loading and is intrinsically related to a solid solution effect. The material memorises the maximum temperature attained in the domain where the derivative of the maximal cyclic stress with respect to the temperature is positive. Real anisothermal tests, where mechanical and thermal loadings evolve simultaneously, have confirmed this observation.

A simple mathematical formulation is proposed which, after being integrated into a unified viscoplastic model developed and identified elsewhere [30], leads to an acceptable phenomenological representation of the totality of the performed thermomechanical experiments. While the introduction of the temperature history module is absolutely necessary to describe the anisothermal behaviour of this steel, it can be shown that the introduction of terms in T into the evolutionary laws of the hardening variables only slightly improves the numerical predictions.

References

1. Cailletaud, G., Chaboche, J.L., In K.J. Miller and J.L. Smith (Eds), ICM3, Cambridge, England, **3**, (1979), 23-32.
2. Walker, K.P., Research and Development Programme for Non-Linear Structural Modeling with Advanced Time-Temperature Dependent Constitutive Relationships, NASA Report n° CR 16 55 33, (1981), 1-181.
3. Robinson, D.N., Bartolotta, P.A., Viscoplastic Constitutive Relationship with Dependence on Thermomechanical History, NASA 174 836, (1985), 1-39.
4. Moreno, V., Jordan, E., *Int. J. Plast.*, **2**, (1986), 223-245.
5. Benallal, A., Ben Cheikh, A., Proc. *2nd Int. Conf. on Constitutive Laws and Short Course : Theory and Application*, Tucson, Arizona, USA, (1987), 667-674.
6. Chan, K.S., Page, R.A., *Metall. Trans.*, **19A**, (1988), 2477-2486.
7. Freed, A.D., Thermoviscoplastic Model with Application to Copper, NASA, Report n° 2845 (1988).
8. Ohno, N., Takahashi, Y., Kuwabara, K., *J. Eng. Mat. Techn.*, **111**, (1989), 106-114.
9. Chan, K.S., Lindholm, U.S, *J. Eng. Mat. Techn.*, **112**, (1990), 15-25.
10. Ghoneim, H., *J. Appl. Mech.*, **57**, (1990), 828-835.
11. Ramaswamy, V.G., Stouffer, D.C., Laflen, J.H., *J. Eng. Mat. Techn.*, **112**, (1990), 280-286.
12. Ohno, N., Wang, J., Nucl. Eng., *Design*, **133**, (1991), 369-381.
13. Moosbrugger, J.C., *J. Eng. Mat. Techn.*, **114**, (1992), 354-361.
14. McDowell, D.L., *J. Eng. Mat. Techn.*, **114**, (1992), 297-303.
15. McDowell, D.L., *Int. J. Plast.*, **8**, (1992), 695-728.
16. Belhedi, B., Delobelle, P., *La Rech. Aérosp.*, **6**, (1992), 43-63.
17. Murakami, S., Kawai, M., Ohmi, Y., *J. Eng. Mat. Techn.*, **111**, (1989), 32-39.
18. Delobelle, P., *Int. J. Plast.* Part I and II, **9**, (1993), 65-85, 87-118.
19. Niitsu, Y.,d Ikegami, K., Bull. of J.S.M.E., **28**, (1985), 2853-2858.
20. Benallal, A., Le Gallo, P., Marquis, D., *Nucl. Eng. Design*, **114**, (1989), 345-353.
21. Delobelle, P., Lachat, R., *La Rech. Aerosp.*, **1**, (1991), 21-37.
22. Murakami, S., Kawai, M., Aoki, K., Ohmi, Y., *J. Eng. Mat. Techn.*, **111**, (1989), 278-285.
23. Rose, K.S.B. and Glover, S.G., *Acta Metall.* **14**, (1966), 1505-1516.
24. Schmidt, C.G., Miller, A.K., *Res. Mech.* **3**, (1981), 109-129.
25. Blanc, D., Thesis, ENSMP, Paris, France, (1986).
26. Billa, R., Delobelle, P., *Mem. Etu. Sci. Rev. Metall.*, Mai, (1990), 311-320.
27. Olschewski, J., Sievert, W., Qi, W., Bertram, A., SMIRT XII, Kussmaul, K. (Ed), Univ. Stuttgart, Germany, paper L01/B, (1993), 17-22.
28. Germain, P., Nguyen, Q.S., Suquet, P., *J. Appl. Mech.*, **50**, (1983), 1010-1020.
29. Freed, A.D., Walker, K.P., Chaboche, J.L., *Acta Mech.*, **90**, (1991), 155-174.
30. Bouchou, A., Thesis, University Franche-Comté, Besançon, (1993).

A MARTENSITIC 10.6% Cr - STEEL UNDER MECHANICAL AND THERMAL FATIGUE

G. DEGALLAIX[1], J. RECH[1,2], Y. DESPLANQUES[1], C. PETERSEN[2] and F. WOLTER[2]

[1] *Ecole Centrale de Lille, Laboratoire de Mécanique de Lille (URA CNRS 1441),BP 48, F-59651 Villeneuve d'Ascq Cedex, France*

[2] *Forschungszentrum Karlsruhe, Institut für Materialforschung II, P.O. Box 3640, D-76021 Karlsruhe, Germany*

1. Introduction

Structural components of an ITER- or DEMO-blanket are subjected during service to alternating thermal and mechanical stresses as a consequence of the high heat fluxes and pulsed reactor operation, as well as coolant pressure, magnetic loads and temperature gradients. The material of the structure must be dimensionally stable and retain adequate mechanical properties during exposure to the environmental conditions, if the required performance and prolonged endurance of the structure are to be achieved. Of particular concern is the fatigue endurance of martensitic steels under cyclic strains and stresses produced by the temperature changes. Along with radiation damage, this is currently considered as the most detrimental lifetime phenomenon for the above structure [1,2].

In order to design such structures operating under combined mechanical and thermal cycling, fatigue life has to be calculated with reasonable accuracy. Currently, fatigue life prediction analysis is based on isothermal fatigue data obtained at a chosen (often maximum operation) temperature. It has been shown that in some cases this approach is non-conservative. The generation of test data by thermal-mechanical fatigue experiments, simulating more accurately the service conditions, has thus become necessary. However, these experiments are very expensive ; they are neither standardized nor applied in design codes. Therefore, thermal fatigue resistance still has to be predicted from isothermal fatigue data [3,4].

The aim of the present work is to analyse a large set of existing isothermal data on the MANET I steel, in a wide range of temperatures, strains and strain rates, in order to extend a modified Manson-Coffin damage law. In a further study, this model can be applied to the thermal-mechanical results, as an isothermal-result-based life prediction model. The MANET I steel (MArtensitic steel for the Next European Torus) had been selected by the EC Fusion Programme as one of the prime candidate materials for applications as a first wall of the blanket structure.

J. Bressers and L. Rémy (eds.), Fatique under Thermal and Mechanical Loading, 445–455.

2. Material and Experimental Procedure

The studied alloy is a ferritic-martensitic 10.6% Cr stainless steel, in tempered condition. Its chemical composition is given in Table 1.

TABLE 1. Chemical composition of MANET I (wt %).

Type DIN 1.4914	C	Cr	Ni	Mo	V	Nb	Si	Mn	B	N
Heat 53645	0.13	10.6	0.87	0.77	0.22	0.16	0.37	0.82	0.08	0.02

Mechanical and thermal fatigue tests have been performed in air on hourglass specimens (Fig. 1), solid in case of isothermal fatigue, hollow in case of thermal fatigue. After machining, the samples were vacuum heat-treated as follows: homogenization (2 h at 960°C, air cooled), austenitization (30 min at 1075°C, air cooled), tempering (2 h at 750°C, air cooled).

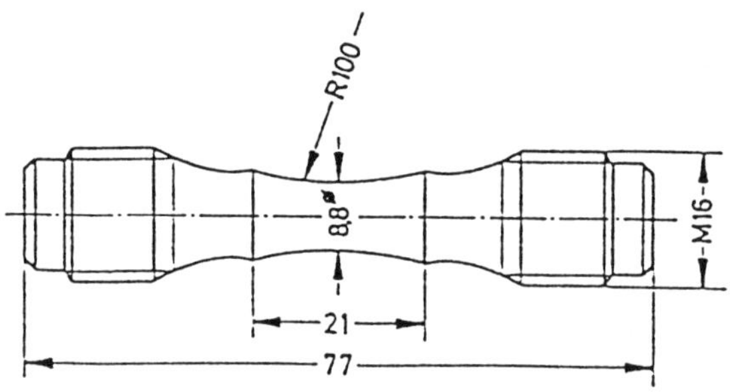

Figure 1. Solid hourglass specimen configuration (dimensions in mm).

The fatigue experiments were conducted at Forschungszentrum Karlsruhe (FZK). For isothermal low-cycle fatigue (LCF), the test facility was a servohydraulic MTS testing machine equipped with a radiant furnace. The fatigue tests were performed under axial displacement control with a triangular fully-reversed waveshape, using a longitudinal extensometer placed on the ridges of the specimens. Numerous tests were carried out at various temperatures from 20 to 550°C, with various average total strain ranges , $\Delta\varepsilon_t$, from 0.4 to 1.5% and with average total strain rates, $\dot{\varepsilon}_t$ from 0.003 to 0.3%.s^{-1}. The fatigue life N_R is defined as the number of cycles up to complete rupture or to a drop of 10% in the tensile stress in relation to the half-life stabilized stress.

3. Result Analysis

The shape of the hourglass specimens was chosen in order to slightly induce the concentration of the strain in the central part of the specimen and, therefore, to initiate and propagate the fatigue cracks there. The average strain in-between the ridges of the specimen was controlled and the load was recorded. This results in an inhomogeneous distribution of the strain along the gauge length of the specimen, which is even amplified by the softening behaviour at various strain amplitudes.

Therefore, a detailed analysis is necessary to obtain the relation between the applied average strain range and the local strain range at the half-life of the specimen, which is a decisive factor for the life evaluation. We have performed this analysis, incorporating the following steps:

3.1. STEP 1

Complementary isothermal LCF tests were performed on cylindrical specimens with constant cross-section on the whole gauge length at six different temperatures in the range 20 to 550°C. During these tests, all the hysteresis stress (σ) - strain (ε_t) loops were recorded.

3.2. STEP 2

The Ramberg-Osgood relation was fitted to the second tensile hysteresis half-loop plotted in relative coordinates (with origin at peak compression), and the parameters E, K and n in the relation

$$\varepsilon_t = \frac{\sigma}{E} + \frac{1}{K}\sigma^{\frac{1}{n}} \tag{1}$$

were evaluated from the experimental data at each temperature and strain range. Since the form of the hysteresis loop in cyclic straining changes only slowly, we can assume that Eq. (1) remains valid during cycling, keeping the Young modulus E constant, and fitting the parameters K and n to obtain the load measured experimentally. This procedure was performed cycle by cycle for the tests at all temperatures. Fig. 2 shows good agreement between the measured hysteresis loops and calculated loops using this procedure for two different cycles.

3.3. STEP 3

The hourglass specimen was separated into thin discs in which stress and strain are constant. The stress-strain relation found in step (2) was applied to the deformation of each disc, imposing the condition of equal load and the condition of applied average strain, derived from the displacement of the ridges of the specimen. Fig. 3 shows one example of the distribution of the local strain along the specimen longitudinal axis for three imposed average strains at 450°C. The result of this procedure was compared with the result of the numerical simulation of the strain distribution, using the finite element method (ABAQUS program) with the identical stress-strain law. Both methods yield fairly similar results which shows that the thin disc approximation describes the distribution of the strain very well. Fig. 4 shows again an example of the isostrain curves in the specimen deformed to the average strain range $\Delta l/l = 10^{-2}$ at temperature of 450°C.

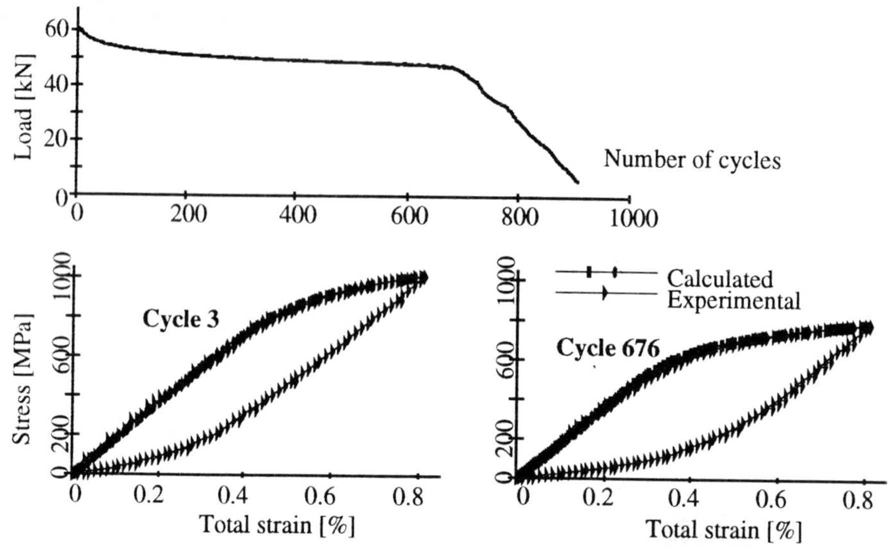

Figure 2. **Comparison between measured hysteresis loops and calculated tensile half-loops, at the beginning and at the end of a test (test conditions , cylindrical specimen, T = 450°C, $\dot{\varepsilon}_t$ = 0.133% s^{-1}). The evolution of the load range during this test is given above.**

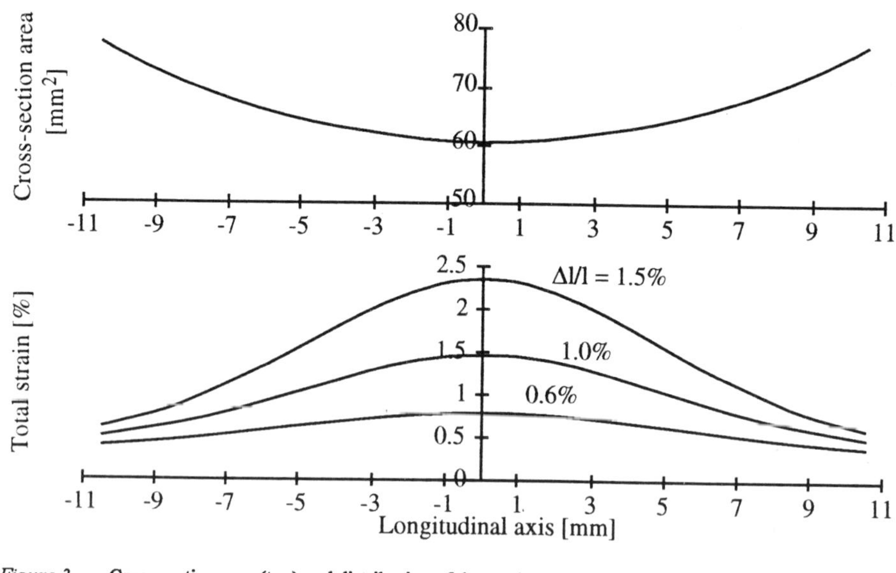

Figure 3. **Cross-section area (top) and distribution of the total strain (bottom) along the gauge length of the hourglass specimen (test conditions , T = 450°C, average strain rate = 0.133% s^{-1}).**

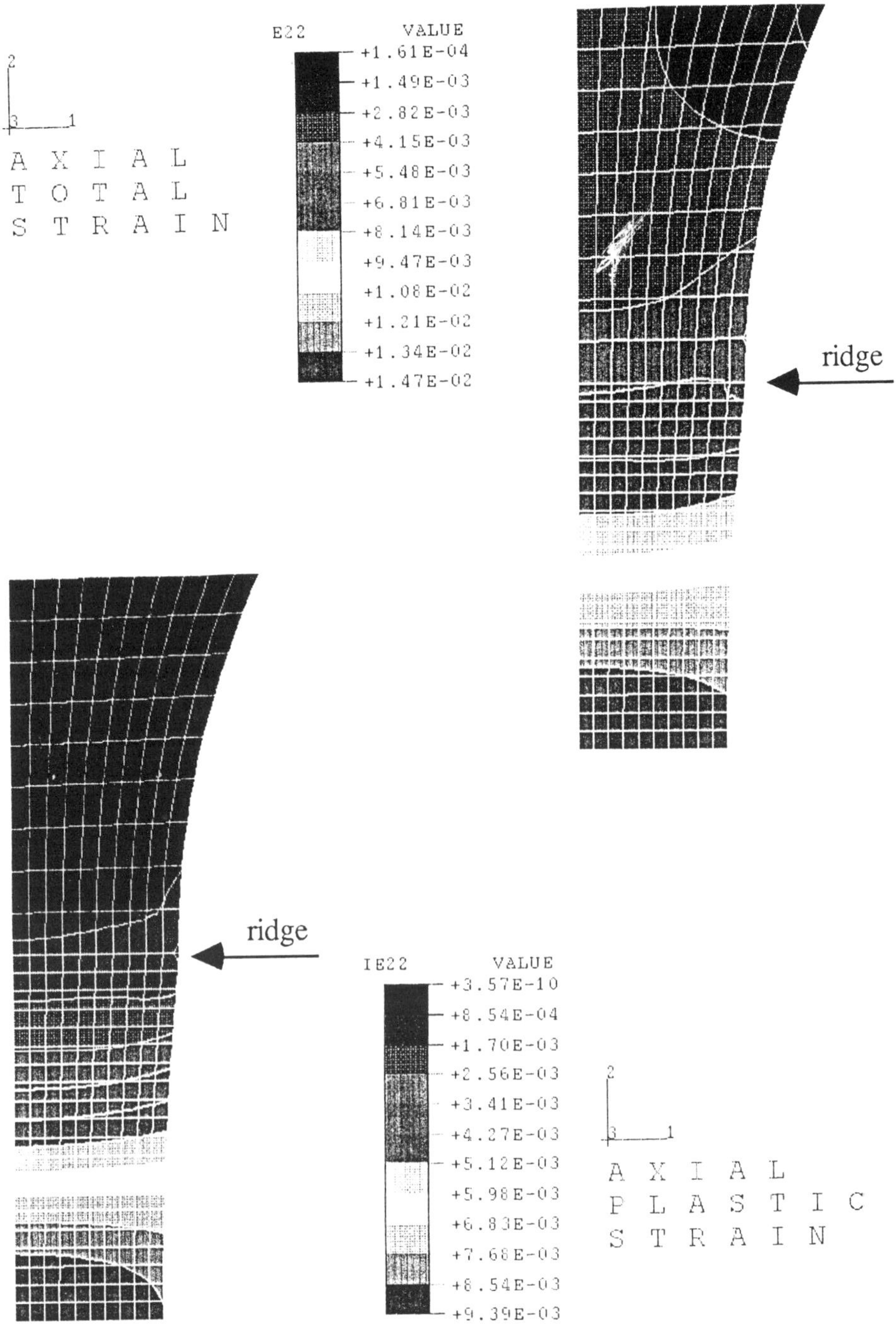

Figure 4. Isostrain curves in the hourglass specimen under the conditions T = 450°C, average strain range $\Delta l/l = 10^{-2}$. Axial total strain (top) ; Axial plastic strain (bottom).

3.4. STEP 4

The evolution of the hysteresis loops during cycling obtained in step (2) for cylindrical specimens was then inserted into the evaluation procedure for hourglass specimens described in step (3). It results in an iterative numerical procedure which adjusts cycle by cycle the parameters of the hysteresis loops to the load measured in the experiments. In first approximation, using the same evolution of the stress-strain law in all discs, the total and the plastic strain ranges in the central part of the specimen can be found, for each average strain range and each temperature. Table 2 gives these strain ranges at half-life, as well as the strain rates in the centre of the specimen assuming the triangular shape of the local strain.

TABEL 2. Isothermal low cycle fatigue results.

Number	T (°C)	$\Delta\varepsilon_t$ (%)	$\Delta\varepsilon_p$ (%)	Strain rate (% s^{-1})	N_R
FE225	20	0.71	0.18	0.3535	16200
FE203	20	1.32	0.72	0.3945	2900
FE210	20	2.09	1.45	0.4184	960
FE181	250	0.72	0.21	0.3575	10480
FE176	250	1.33	0.78	0.3998	2110
FE188	250	2.11	1.50	0.4221	1000
FE141	450	0.61	0.17	0.3654	17500
FE174	450	0.77	0.30	0.3855	7820
FE163	450	1.45	0.92	0.4362	1460
FE155	450	2.35	1.79	0.4695	700
FE190	550	0.49	0.12	0.3678	33100
FE191	550	0.67	0.28	0.4026	11100
FE193	550	0.85	0.43	0.4245	4250
FE16	550	1.64	1.19	0.4923	1520
FE144	550	2.61	2.12	0.5228	620
FE212	20	0.70	0.17	0.03521	12590
FE208	20	1.32	0.72	0.03945	2300
FE211	20	2.09	1.44	0.04181	950
FE246	250	0.72	0.19	0.03595	10400
FE263	250	1.32	0.74	0.03963	2260
FE262	250	2.10	1.48	0.04202	850
FE171	450	0.77	0.30	0.03861	6300
FE278	450	1.46	0.93	0.04371	1360
FE221	450	2.35	1.79	0.04691	580
FE363	550	0.70	0.34	0.04218	7800
FE255	550	0.89	0.51	0.04445	3960
FE280	550	1.68	1.27	0.05052	1260
FE269	550	2.68	2.24	0.03566	540
FE353	20	2.11	1.49	0.004218	820
FE359	250	2.10	1.48	0.004198	770
FE183	450	1.48	0.99	0.004446	1020
FE220	450	2.37	1.83	0.004731	550
FE361	550	1.69	1.37	0.005076	1010
FE277	550	2.65	2.30	0.005294	500

4. Development of the Damage Model

Previous studies [5,6] showed that the complete high temperature LCF behaviour, in a wide temperature range, can be described by considering that the damage process is controlled simultaneously by two kinds of mechanisms, depending upon the plastic strain range, $\Delta\varepsilon_p$, athermal or thermally-activated. Having noted that, at room temperature, there is no significant thermally-activated part in the damage per cycle, the proposed model was written as:

$$\frac{1}{N_R} = \left(\frac{\Delta\varepsilon_p}{C}\right)^{\frac{1}{D}} \cdot \left[1 + A.\Delta\varepsilon_p{}^B.exp\left(-\frac{Q}{RT}\right)\right] \tag{2}$$

where C and D are the well-known constants of the Manson-Coffin law at 20°C, and where A, B and the apparent activation energy Q (in $kcal.mol^{-1}$ and with $R = 2 \cdot 10^{-3}$ $kcal.mol^{-1}.K^{-1}$), are material constants, independent of $\Delta\varepsilon_p$ and T, in the whole domain of strains and temperatures.

This damage model expresses a detrimental effect of a higher temperature on the fatigue life. Exponent B can take on a positive value, a negative value, or can also be equal to zero. These three cases correspond to three forms, typically observed, of the Manson-Coffin curves depending on temperature , divergent, convergent or parallel. Moreover, this model presents notable advantages to the designer. As it corresponds to a single and continuous "fatigue strength surface", it enables a reliable interpolation to be made throughout the studied domain of strains and temperatures, and allows for a reasonable extrapolation out of this domain, provided that no different metallurgical phenomena occur.

This model is also interesting in that it appears as a generalisation of models proposed elsewhere for describing high temperature LCF behaviour [7] (Table 3). In the damage per cycle, if the contribution of thermally-activated mechanisms is negligible, the model becomes the Manson-Coffin law. On the contrary, if such mechanisms are more prevalent than athermal ones, the expressions proposed, in chronological order, by Taira [8], Udogochi and Wada [9], Antolovich [10], Hong and Rie [11], can be found again. It must however be emphasised that our own work on various stainless or engineering steels, as well as that of Korn [12] and of Shi [13], have shown that, in the case of these materials, the competition - in the damage process - between both thermally-activated and athermal mechanisms had to be taken into account.

Eq. 2 does not explicitly bring out a strain rate effect or a frequency effect (nor a hold time effect). To study this issue, it is necessary to have experimental data with sufficiently different strain rate values.

In the present work, we have started from the Coffin approach [14] (frequency modified fatigue life, Table 3) and proposed the following expression:

$$\frac{1}{N_R} = \left(\frac{1}{\overset{o}{\varepsilon}_t}\right)^{\alpha} \cdot \left(\frac{\Delta\varepsilon_p}{C}\right)^{\frac{1}{D}} \cdot \left[1 + A.\Delta\varepsilon_p{}^B.exp\left(-\frac{Q}{RT}\right)\right] \tag{3}$$

TABLE 3. Isothermal low cycle fatigue models [7].

Present model	[5,6, this paper]
$$\frac{1}{N_R}=\left(\frac{1}{\dot{\varepsilon}_t}\right)^{\alpha}\cdot\left(\frac{\Delta\varepsilon_p}{C}\right)^{\frac{1}{D}}\cdot\left[1+A.\Delta\varepsilon_p{}^{B}.exp\left(-\frac{Q}{RT}\right)\right]$$	
	models with T as an explicit variable :
Manson-Coffin (1954) $\frac{1}{N_R}=C_0.\Delta\varepsilon_p{}^{n}$	Taira (1963) [8] $\frac{1}{N_R}=C_1.\lambda(T).\Delta\varepsilon_p{}^{n}$
Coffin (1976) [14] $\frac{1}{N_R}=C_4.\left(\frac{1}{\nu}+t_h\right)^{1-k}.\Delta\varepsilon_p{}^{n}$	Udoguchi and Wada (1969) [9] $\frac{1}{N_R}=C_2.exp\left(-\frac{Q}{RT}\right).\Delta\varepsilon_p{}^{n}$
	Antolovich and al (1981) [10] $\frac{1}{N_R}=C_3.\left(\frac{1}{\nu}+t_h\right).exp\left(-\frac{Q}{RT}\right).\Delta\varepsilon_p{}^{8n'}$
	Hong and Rie (1985) [11] $\frac{1}{N_R}=C_5.\left[\frac{exp\left(-\frac{Q}{RT}\right)\int_0^{t_h}\sigma(t)dt}{T}\right]^{\frac{2}{5}}.\Delta\varepsilon_p{}^{\frac{3}{5}}$

In this expression, the strain rate effect acts simultaneously and in similar fashion on the thermally-activated and the athermal terms.

Note that it would be very simple, depending on the materials and/or the experimental conditions, to express a different strain rate effect on both terms, or more particularly, a strain rate effect limited to the thermally-activated term.

This approach was applied to the whole set of experimental data in Table 2. Since this study has a practical aim, the numerical values of the model constants were determined according to a purely numerical criterion, i.e. those leading to the best fitting of the experimental data (minimising the scatter band).

For the MANET I, the following expression was thus obtained, with $\Delta\varepsilon_p$ in %, $\dot{\varepsilon}_t$ in $\%s^{-1}$ and T in K:

$$\frac{1}{N_R}=\left(\frac{1}{\dot{\varepsilon}_t}\right)^{0.0314}\cdot\left(\frac{\Delta\varepsilon_p}{249.1}\right)^{\frac{1}{0.743}}\cdot\left[1+11.2exp\left(-\frac{7.7}{RT}\right)\right] \tag{4}$$

Fig. 5 shows the very good correlation between the experimental values and the calculated values. All the points fall within a scatter band with a factor of 1.34 (with 72% of the results with a factor equal to or less than 1.2).

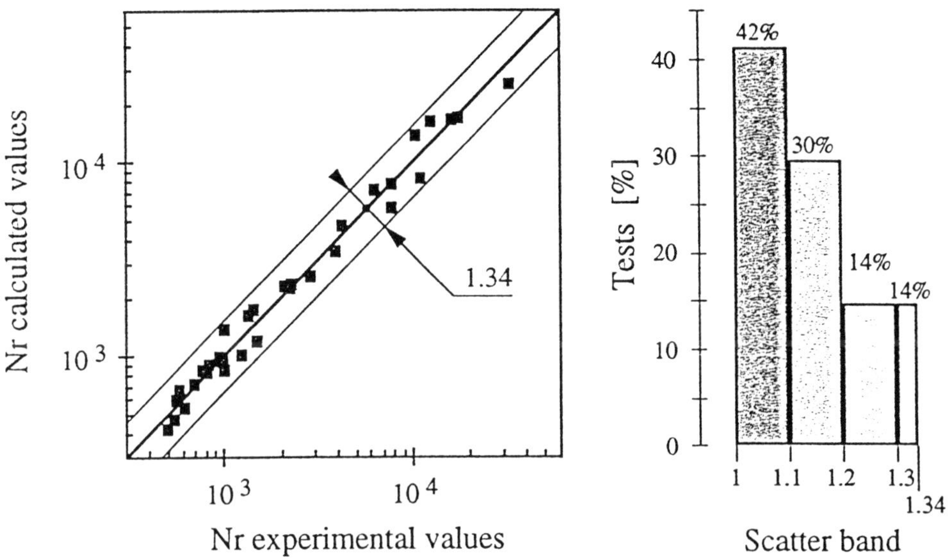

Figure 5. Data correlation using the proposed model, scatter band (left) and distribution of the predicted values (right).

The different coefficients of the model were obtained relatively easily. Experimental data at 20°C, for the highest strain rate, lead to the C and D Manson-Coffin law coefficients. In this case, it was sufficient to use the room temperature data to determine, by a least-squares method, the value of coefficient α. Finally, an iterative computer code, using a non-linear least-squares method, was used to obtain the values of A and Q. Coefficient B was taken equal to zero, expressing that the Manson-Coffin curves are roughly parallel to each other for the various temperatures. It can be noted that the relative small value (7.7 kcal. mol^{-1}) of the apparent activation energy indicates that the contribution from thermally-activated mechanisms to the damage per cycle remains relatively weak, even at the highest temperatures, and thus, that the temperature dependence of LCF behaviour of the MANET I steel is moderate. This was confirmed by the impossibility of correlating the experimental data, using a damage model with only a thermally-activated term. With the Antolovich form [10], for instance, no scatter band with a factor less than 14 can be found.

In conclusion, this paper proposes a description of the isothermal low-cycle fatigue of the MANET I steel using a single damage model including plastic strain, temperature and strain rate as variables.

We intend, in a further study, to evaluate whether the proposed model can be applied to the thermal-mechanical tests already conducted at the FZK [1] and presented in Fig. 6. Based on the Taira equivalence concept [8], the calculation of the damage will be done on the stabilized hysteresis loop, using a linear damage rule.

The integration of the elementary increment of damage (assumed to be isothermal, and connected at each time with the increments of plastic strain and temperature) on the thermal-mechanical cycle will give an estimation of the fatigue life. This method yielded a number of highly acceptable results in other studies [15,16].

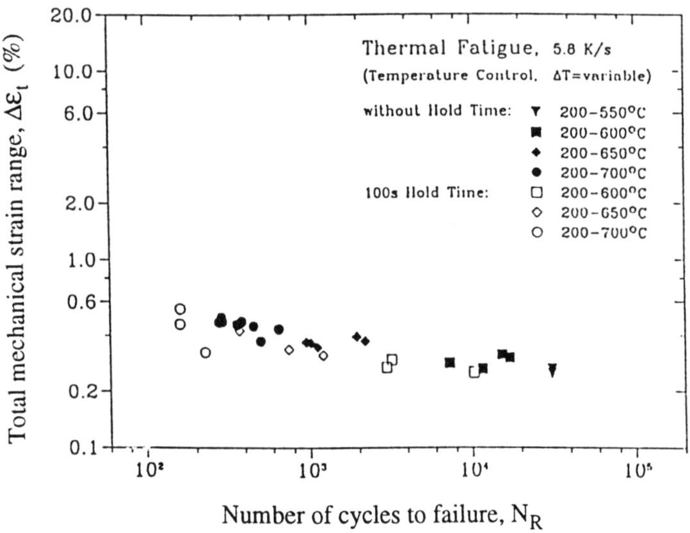

Figure 6. Thermal-mechanical fatigue data of the MANET I steel [1].

Acknowledgements - The isothermal LCF experiments have been conducted by R. Schmitt and M. Pfeiffenroth from FZK.

References

1. Petersen, C., Alvarez-Armas, I., Armas, A.F., Thermal fatigue on steels for ITER and advanced fusion reactors, *Plasma Devices and Operations* **3** (1994), 317-329.
2. Alvarez-Armas, I., Armas, A.F., Petersen, C., Thermal fatigue of a 12% Chromium martensitic stainless steel, *Fatigue Fract. Engng. Mater. Struct.* **17** (1994), 671-681.
3. Petersen, C., Wolter, F., Prediction of thermomechanical fatigue on MANET I, Proceed. 18th SOFT, Karlsruhe (Germany) (September 1994), paper PB - 128, 1-3.
4. Wolter, F., Petersen, C., Thermal and thermomechanical fatigue of a martensitic 10.6% Cr-steel, Proceed. *Ann. Meeting on Nuclear Technology*, Stuttgart (Germany), (Mai 1994), 493-496.
5. Degallaix, G., Degallaix, S., Foct, J., A damage law for predicting the elevated temperature low cycle fatigue life of a martensitic stainless steel, *Mat. Sci. Engng.* **58** (1983), 55-62.
6. Degallaix, G., Lesne, P.M., Low cycle fatigue of engineering materials (20 to 700°C) , Extension of the Manson-Coffin law by a thermally-activated term, Proceed. *ECF 6, Fracture Control of Engineering Structures*, Amsterdam (The Netherlands), T.U. Delft Ed., (1986), 44-49.
7. Degallaix, G., Prédiction de la durée de vie en fatigue thermomécanique, *MAT-TEC 90, Technologie, comportement et traitement des matériaux*, Bathias, C. and al Eds, IITT-Int. Publ. Paris (France), (1990), 37-45.
8. Taira, S., Relationship between thermal fatigue and low cycle fatigue at elevated temperature, ASTM STP 520 (1973), 80-101.

9. Udoguchi, T., Wada, T., Thermal effect on low cycle fatigue strength of steels, *Thermal Stresses and Thermal Fatigue,* Littler, D.J. Ed, Butterworths (1969), 109-124.
10. Antolovich, S.D., Liu, S., Baur, R., Low cycle fatigue behavior of René-80 at elevated temeperature. *Met. Trans.* **12A** (1981), 473-481.
11. Hong, J.W., Nam, S.W., Rie, K.T., A model for life prediction in low cycle fatigue with hold time, *J. Mat. Sc.* **20** (1985), 3763-3779.
12. Korn, C., Aspects expérimentaux et modélisation de la durée de vie en fatigue thermomécanique des aciers Cr-Mo-V et 316L et de l'alliage TZM, *Thesis,* University of Metz (France), (dec 1988).
13. Shi, H.J., Etude en fatigue thermomécanique d'aciers inoxydables. Réalisation d'essais et prédiction de la durée de vie, *Thesis,* University of Metz (France), (mars 1992).
14. Coffin, L.F. Jr, The concept of frequency separation in life prediction for time-dependent fatigue, *Creep-Fatigue Interaction* ASME-MPC-3 Symp., New York (USA), (1976), 349-364.
15. Degallaix, G., Korn, C., Pluvinage, G., Lifetime prediction on Cr-Mo-V and 316L steels under thermal and mechanical cycling, *Fatigue Fract. Engng. Mater. Struct.* **13** (1990), 473-485.
16. Shi, H.J., Pluvinage, G., Thermal-mechanical fatigue life prediction on engineering materials, Proceed. *ICF 8, Int. Conf. on Fracture,* Panasyuk, V.V. Ed., Kiev (Ukraine), (1993), Paper 488, 1-8.

MODELLING OF THE STRESS-STRAIN BEHAVIOUR OF METAL MOULDS

J. OKRAJNI
Silesian University of Technology
ul. Krasinskigo 8, 40-019 Katowice, Poland

1. Introduction

The prediction of the safe and failure-free life of technical devices is one of the most important issues in industrial practice and, hence, the designers have to apply an appropriate methodology to the selection of suitable materials. In this case, it is necessary to apply theoretical and experimental methods from different fields of material science, and particularly to be able to simulate the behaviour of complex mechanical systems and their elements [1, 2]. Numerical methods for evaluating temperature, stress and strain fields in machine equipment parts have to be used and carrying out simulated tests of materials is also necessary.

In this paper, durability prediction methodology is presented for technical structures subjected to thermal-mechanical fatigue. Taking as an example some selected devices, the work concentrates upon the problem of durability of metallic moulds which are commonly used in the metallurgical industry. Three types of moulds have been taken into account i.e. thick-walled vessels filled with liquid metal, the pipe centrifugal casting moulds and permanent moulds for fire-bar casting. Vessels are assumed to be thick-walled tubes. The modelling of moulds from the point of view of solid mechanics is discussed. The paper also presents results of temperature, stress and strain fields evaluation. An attempt has been made to calculate the fatigue life of metal moulds and results of the calculation are shown here.

2. Evaluation of Temperature, Stress and Strain Fields in Metal Moulds

2.1. PROBLEM DEFINITION

In order to discuss the way of selecting an appropriate criterion for thermal-mechanical fatigue life prediction of moulds under mechanical and thermal loading a model approach to heat transfer phenomena was applied. The axially-symmetrical case of thick-walled vessels filled with liquid metal was taken as an example. The temperature, stress and strain variations in thick-walled vessels filled with liquid metal were analysed. An intense heat exchange on the inner surface was assumed

J. Bressers and L. Rémy (eds.), Fatique under Thermal and Mechanical Loading, 457–466.

with the heat transfer coefficient α_l = 5 400 W/m^{2}K, and a much less intensive heat exchange at the outer surface, α_0 = 100 W/m^{2}K. These parameters were to reflect the conditions close to those typical for metal mould surfaces exposed on their interior to liquid metal impact, on their exterior to free air flow. It has been assumed that the heat conduction in the mould material is different from that of the liquid metal inside moulds. Changes in the thermal conductivity and thermal diffusivity with cooling in time have been taken into account. Thermal conductivity of mould materials and of casting materials in solid state is $\lambda_m = \lambda_s$ = 29 W/mK. Thermal conductivity of a liquid metal is λ_l = 23 W/mK. Thermal diffusivity of mould materials and of casting materials in solid state is $a_m = a_s = 6.88\times10^{-6}$ m^2/s. Thermal diffusivity of a liquid metal is $a_l = 3.97\times10^{-6}$ m^2/s. To discuss the way of predicting a fatigue life of the chosen objects, examples of pipe centrifugal casting mould and moulds for fire-bar casting were analysed. In the case of pipe centrifugal casting moulds it has been assumed: α_l = 15 399 W/m^{2}K, α_0 = 500 W/m^{2}K, λ_m = 34.9 W/mK, λ_s = 29 W/mK, λ_l = 23 W/mK, $a_m = 7.72\times10^{-6}$m^2/s, $a_s = 6.88\times10^{-6}$m^2/s, $a_l = 3.97\times10^{-6}$m^2/s.

The temperature field was determined by the method of finite differences (FDM) while the finite element method (FEM) was used in evaluating the time-dependent stress and strain fields. The elasto-plastic material model with exponential hardening was used, and temperature effect on material properties was taken into consideration. The material model is assumed as:

$$\sigma = R_H(T) + m(T)k\left(\frac{\varepsilon_p}{\varepsilon_H}\right)^n \tag{1}$$

where

- $R_H(T)$ elastic limit as a temperature function,
- ε_p plastic strain,
- ε_H strain corresponds to elastic limit,
- k,n material constants,

$$m(T) = \begin{cases} 1 & \text{for} T \le T_0 \\ \frac{1}{2}\left[1+\cos\left(\pi\frac{T-T_r}{T-T_0}\right)\right] & \text{for} T > T_0 \end{cases} \tag{2}$$

$$R_H(T) = R_H(T_o)\,, \tag{3}$$

and T_o, T_r are coefficients of interpolation.

The yield surface is expressed by

$$F_y = f_y(\sigma_{ij}) - f_0(\varepsilon_{int}, T) = 0, \tag{4}$$

where σ_{ij} are stress tensor components, the function f_o depends on the material model and type of hardening,

$$f_y(\sigma_{ij}) = s_{ij}s_{ij} \,, \tag{5}$$

and

$$\varepsilon_{int} = \frac{\sqrt{2}}{3} \times \sqrt{(\varepsilon_{p1} - \varepsilon_{p2})^2 - (\varepsilon_{p2} - \varepsilon_{p3})^2 - (\varepsilon_{p3} - \varepsilon_{p1})^2} \tag{6}$$

with ε_{p1}, ε_{p2}, ε_{p3} as the principal values of plastic strains and s_{ij} as the deviatoric stress.

The von Mises type of material and isotropic work-hardening have been analysed.

2.2. SOLUTION PROCEDURES

The finite differences method was used to determine the temperature field in all moulds mentioned in cases of uniaxial and multiaxial problems. This method was useful in problems of heat exchange in which boundary conditions have been changed with time and they were mutually dependent.

In the case of an axially symmetric problem the heat equation has been expressed as:

$$\frac{\partial^2 T}{\partial r^2} + \frac{1}{r}\frac{\partial T}{\partial r} = \frac{1}{a}\frac{\partial T}{\partial t} \tag{7}$$

This equation was used to determine the temperature field in thick-walled vessels filled with liquid metal and in the pipe centrifugal casting moulds.

The following equations present the boundary conditions in the problem of heat transfer in thick-walled vessels filled with liquid metal:

- inner surface of a mould (outer surface of a casting):

$$h(T_m - T_l) = \left(\frac{\partial T}{\partial r}\right)_m \,, \tag{8}$$

$$\lambda_l \left(\frac{\partial T}{\partial r}\right)_l = \lambda_m \left(\frac{\partial T}{\partial r}\right)_m \tag{9}$$

- outer surface of a mould:

$$h_o(T_e - T) = \frac{\partial T}{\partial r} \tag{10}$$

The boundary conditions in the problem of heat transfer in the pipe centrifugal casting mould are as follows:

- inner surface of a casting:

$$h_i(T_e - T) = \frac{\partial T}{\partial r} \tag{11}$$

- inner surface of a mould (outer surface of a casting): Eqs. (8) and (9)

- outer surface of a mould: Eq. (10),

where the subscripts m and l correspond to a mould (m) or to a casting (l), T_e is the temperature of an environment, subscripts o and i correspond to the outer surface (o) or to the inner surface (i),

$h_0 = \frac{\alpha_0}{\lambda_m}$, $h = \frac{\alpha_1}{\lambda_m}$, $h_i = \frac{\alpha_0}{\lambda_S}$ (casting in a solid state), $h_i = \frac{\alpha_0}{\lambda_l}$ (casting in a liquid state).

A polynomial approximation [3] was applied in order to present the temperature distribution.

The conception of an elasto-viscoplastic material model [4] was applied in the analysis of the elasto-plastic problem. Fictitious time and constants of viscosity were introduced. The calculations were continued until a steady-state (convergence of the solution) was obtained on each step of the iteration.

2.3. RESULTS

Fig. 1 shows the calculated profiles of temperature evolution with time at the internal end external vessel surfaces made of cast iron. These calculations were carried out for a selected vessel wall thickness g = 40 mm. Two different inner radius (r_i) /wall thickness (g) ratios were taken into account. Fig. 2 presents the maximum and minimum equivalent plastic strain ε_{int} during heating and cooling as functions of the radius. The equivalent plastic strain was determined by the Eq. (6).

The diagrams show that the size of plastic zone near the inner mould surface depends on the ratios mentioned above. The temperature-stress characteristics in this plastic zone were calculated. Fig. 3 presents relations between the axial stress and the temperature at the inner surface of thick-walled vessels. An effect of the inner radius value upon the course of the temperature changes and the thermal-mechanical fatigue characteristics is apparent in Figs. 1, 2 and 3.

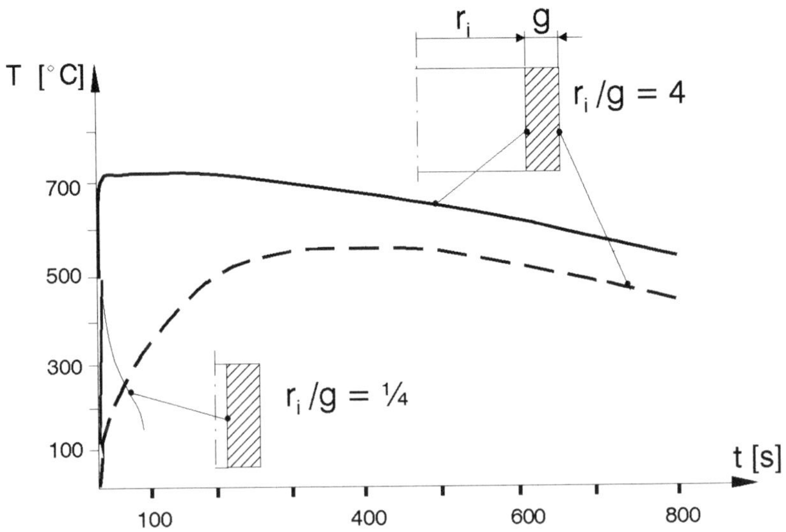

Figure 1. Time-temperature profiles at the inner and outer surfaces of thick-walled vessels filled with liquid metal.

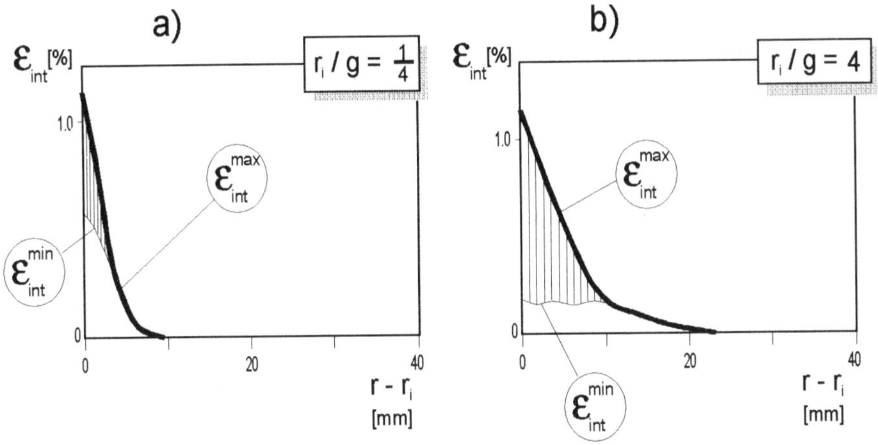

Figure 2. Equivalent plastic strain as a function of the tube radius exposed to cyclic heating and cooling.

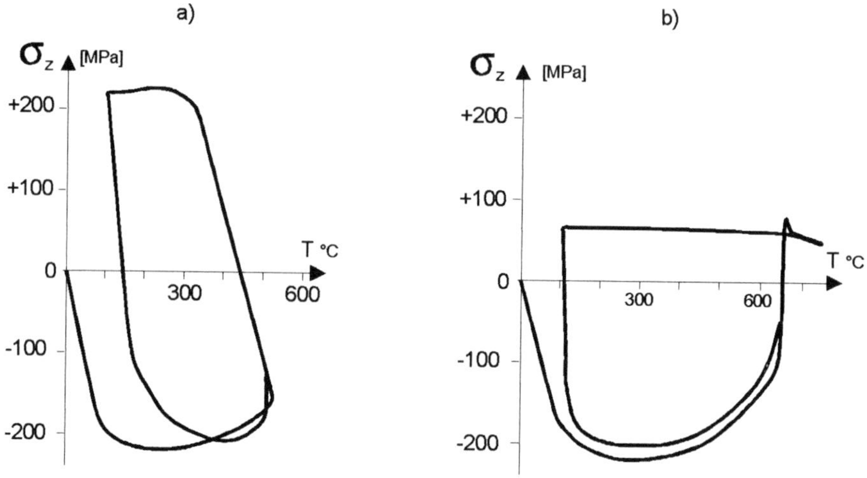

Figure 3. The axial stress course as a function of temperature at the inner surface of the thick-walled vessel: r_i / g = 1/4 (a); r_i / g = 4 (b).

Fields of temperature, axial σ_z and tangential σ_0 stresses, axial ε_z and tangential ε_0 strains in the pipe centrifugal casting mould made of 35 HM steel (34CrMo4 according to DIN) were also analysed (Figs. 4 and 5).

The characteristics of calculated stress-strain behaviour were compared with those measured in operating conditions as reported by [5]. The comparison of these characteristics determined by means of numerical simulation with their equivalents measured in operating conditions shows that it is possible to obtain a considerable convergence (Fig. 5).

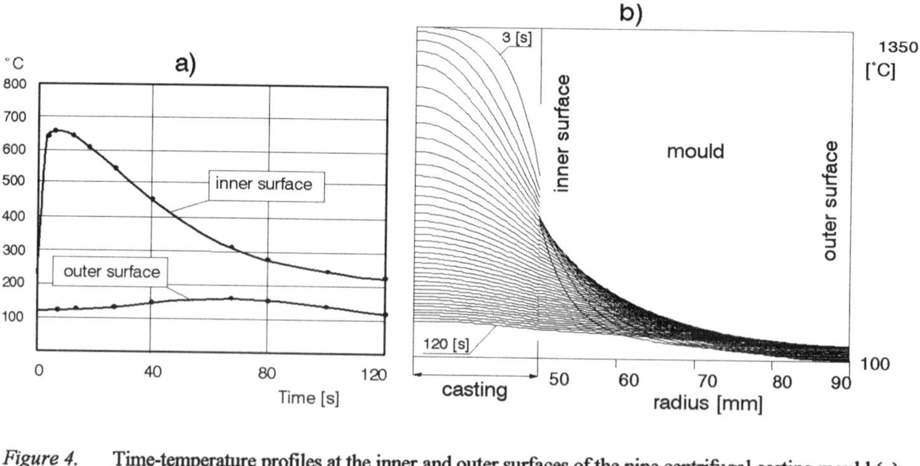

Figure 4. Time-temperature profiles at the inner and outer surfaces of the pipe centrifugal casting mould (a). Temperature as the function of the radius of the casting and mould (b).

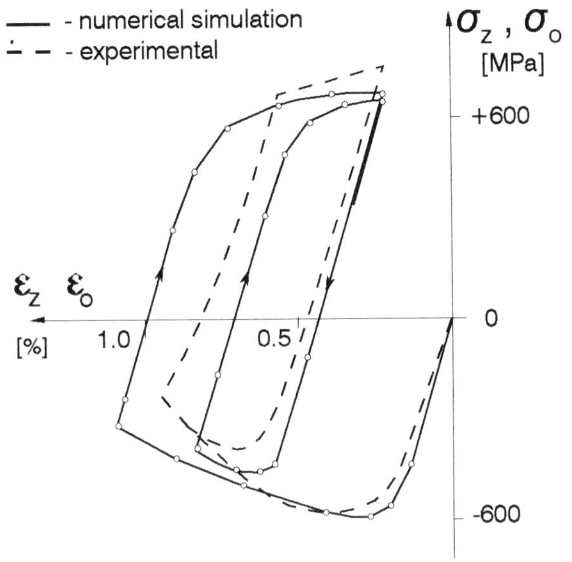

Figure 5. The stress courses in function of strains on the inner surface of the pipe centrifugal casting mould.

3. Fatigue Life Prediction

Crack formation in machine parts subjected to fatigue occurs at the surface, and is connected with the cumulative effect of cyclic plastic strains. Thus it has been assumed that the intensity of the material damage process is conditioned by the size of the range of shear plastic strains at the surface of the above mentioned parts. For the plane stress condition, which prevails at the metal mould surface, the equivalent strain range has been determined [2] as

$$\Delta\varepsilon_{pz} = 0.67\left[(\Delta\varepsilon_{p1} - \Delta\varepsilon_{p3})T_{min} - (\Delta\varepsilon_{p1} - \Delta\varepsilon_{p3})T_{max}\right] \tag{12}$$

where:

ε_{p1}, ε_{p3} -principal values of plastic normal strains

T_{min}, T_{max} -minimum and maximum temperature of cycle

$\Delta\varepsilon_{pz}$ -equivalent strain range.

The Manson - Coffin formulation was used to predict fatigue lives:

$$N = C \times \left(\Delta\varepsilon_{pz}\right)^{n} \tag{13}$$

The fatigue diagram which corresponds to this formulation is shown in Fig. 6. For the value of the strain range $\Delta\varepsilon_{pz} = 0.0028$ the 35HM steel (34CrMo4 according to DIN) centrifugal casting mould would have a durability of $N_f = 257$ cycles.

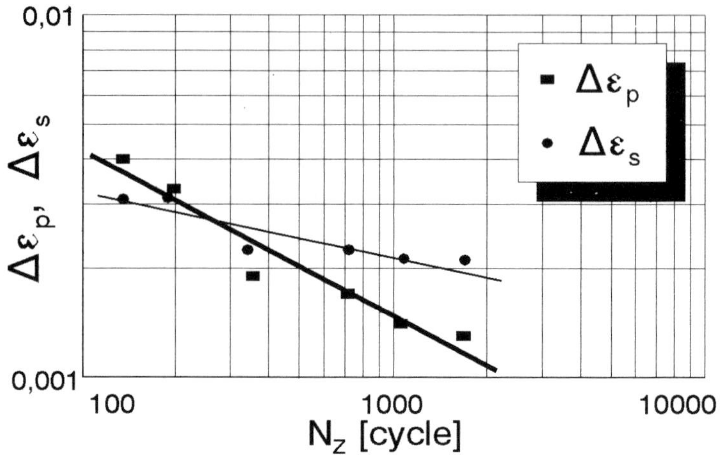

Figure 6. The fatigue diagrams of 35 HM steel (34CrMo4 according to DIN) determined in laboratory conditions; minimum temperature of the cycle T_{min} = 180°C, maximum temperature of the cycle T_{max} = 650°C. $\Delta\varepsilon_p$ is a plastic strain range, $\Delta\varepsilon_s$ is an elastic strain range.

The next example is a grey cast iron permanent mould [2, 6]. The temperature changes with time on the surface of the gate assembly and the mould cavity measured in operating conditions, are shown in Fig. 7. These diagrams have been used to define boundary conditions, when temperature, stresses and strains were calculated (Fig. 8).

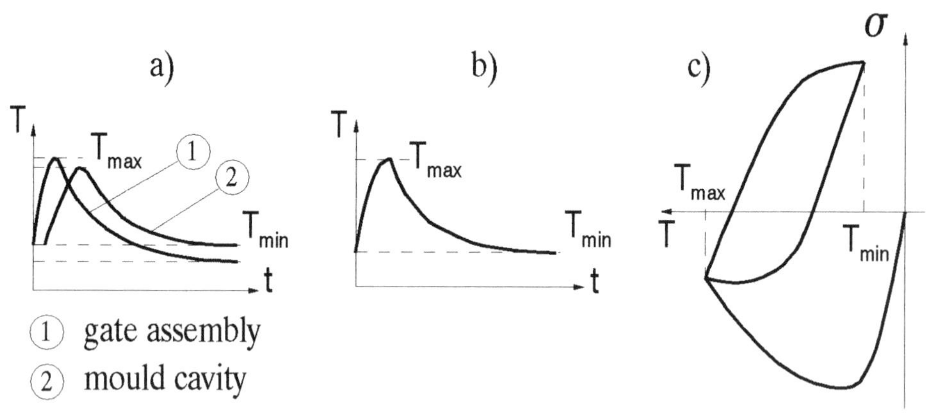

Figure 7. Temperature changes with time in operating conditions at the surface of the gate assembly and the mould cavity (a), temperature changes with time in a specimen for thermal-mechanical fatigue (b), stress/temperature diagram of a specimen (c).

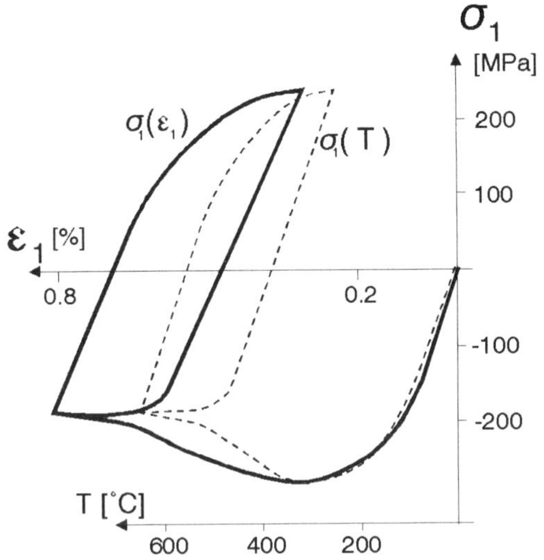

Figure 8. The characteristics of the stress- strain and stress-temperature behaviour determined for the surface of the gate assembly by means of numerical simulation, where σ_1 and ε_1 are the principal values of stress and strain.

The data made it possible to define the thermal-mechanical fatigue parameters for simulation tests of material (grey cast iron). As a result of these tests, characteristics of fatigue life as a function of plastic strain range have been determined [2, 6, 7]. For the value of strain range $\Delta\varepsilon_{pz} = 0.0016$ the grey cast iron permanent mould would last for $N_f = 36$ cycles. The calculated results can be interpreted only in a qualitative sense. However, the difference in the calculated durability of pipe centrifugal casting moulds and the permanent moulds is in line with experimental data. Cracks in the pipe centrifugal casting moulds are formed, as it is shown by [8], after several hundred cycles of pouring. The number of thermal load cycles of the cast iron permanent moulds to the moment of the start of the first cracks is several dozen cycles [6]. Hence the calculated number of cycles for crack formation is consistent with the experimentally observed number of cycles to crack formation of real moulds. That number, however, can not be identified with the object durability in many cases. In pipe centrifugal casting moulds, cracks formed after several hundred cycles of pouring are removed by mechanical treatment, after which the moulds are re-used. Crack formation on the fire-bar casting permanent mould made of cast iron, described in [6], does not disqualify further use of the mould because of the limited requirements for the surface quality. The number of cycles calculated numerically should be taken to be the number of cycles to the first crack formation. It should not be understood as the time of failure free and effective working of the mould, which, because of the possibility of regeneration by mechanical treatment or working, is usually higher than the values calculated above.

4. Conclusions

The lifetime of moulds under thermal-mechanical fatigue loads can be estimated on the basis of parameters which describe the course of the cyclic elasto-plastic deformation, in combination with life curves determined by means of simulation tests on mould materials. An equivalent range of plastic strain can be used as one of these parameters. In this way the life-time of a mould can be estimated but without taking the influences of material structure on the fatigue life into consideration.

The value of the inner radius of the mould has a significant effect upon the course of temperature, strain and stress changes in thick-walled vessels filled with liquid metal, which can be closely related to the influence of the casting size. As far as large castings are concerned, the important factor affecting the life-time is prolonged high temperature exposure. On the other hand, for small castings a crucial role can be attributed to thermal-mechanical elasto-plastic deformations.

The investigations conducted and the attempt to use their results for the behaviour analysis of selected machine parts showed its usefulness in problems of metal mould fatigue life prediction.

Acknowledgements - This work was performed under the Contract of Scientific Research 3 P407 080 04 "Fatigue Fracture Processes and Life Prediction of Metal Alloys in Mechanical and Thermal Conditions".

References

1. Lamber, T., Okrajni, J. and Plaza, M.., Estimation of the structural member low-cycle fatigue durability, *Third International Conference on Low Cycle Fatigue and Elasto-Plastic Behaviour of Materials,* ELSEVIER Berlin, (1992), pp. 823-828.
2. Okrajni, J.,and Plaza, M., Simulation of the fracture process of materials subjected to low-cycle fatigue of mechanical and thermal character, special issue, *Journal of Material Processing Technology,* **53**, (1995), 311-318.
3. Paszkowski, S., *Numerical applications of Czebyszew polynomials and series,* PWN, Warsaw, 1975, in Polish.
4. Perzyna, P., *Theory of viscoplasticity,* PWN, Warsaw, 1966, in Polish.
5. Königer, A.and Liebman, W., (1957) Untersuchungen uber die Haltbarkeit von Schleuder-gusskokillen, aus *Stahl u.Eisen* 79, **23**, **12** (1957), 1730-1742.
6. Ciecela, M., Durability Estimation of Cast Irons for the Metal Moulds, Silesian University. of Technology, Katowice, (1990), in Polish.
7. Ciesla, M., Lamber, T., Okrajni, J., Plaza, M., Effect of service conditions upon the process of metal mould fatigue fracture, *Journal of Theoretical and Applied Mechanics,* **1**, **32** (1994), 59-68.
8. Weroñski, A. and Hejwowski, T., *Thermal Fatigue of Metals,* Marcel and Dekker Inc., New York, 1992.

THERMO-MECHANICAL FATIGUE LIFE PREDICTION METHOD IN TERMS OF ENERGY

K.M. GOLOS
Warsaw University of Technology
Institute of Machine Design Fundamentals
84 Narbutta Str., 02-524 Warsaw, Poland

1. Introduction

Thermo-mechanical fatigue (TMF) problems are encountered in many industries, such as aerospace, power generation, mechanical engineering etc. The last decade has seen increased effort on modelling material behaviour under high operating temperatures and severe TMF environments. One of the current problems is formulation of the method allowing the design and evaluation of engineering systems operating under TMF conditions. In the literature few thermo-mechanical models have been presented. However, none of them has given satisfactory results for all thermo-mechanical loading histories. Life prediction models for TMF generally take the form:

(1) frequency-modified strain-life or stress-life approaches [1-3];
(2) parametric damage approaches [4-7];
(3) continuum damage approaches [8-9];
(4) damage rate approaches [10-12] including microcrack propagation models [13].

In this paper an energy based model is investigated. The total strain energy density equal to the sum of tensile elastic strain energy density and the plastic strain energy is proposed as a damage parameter. The main advantage of the energy based criteria is their capacity to represent the multiaxiality of the stress and strain in a simple and physically correct way [14-18]. They describe at the macroscopic level, the non-linear and hysteretic behaviour of the material which is the result of complex and numerous microstructural changes responsible for thermo-mechanical damage. Recent results show that the cyclic thermo-mechanical deformation behaviour is different when proportional loading is compared with nonproportional loading. To describe thermo-mechanical loading in terms of energy the description of the cyclic stress-strain behaviour is essential. Several constitutive theories have been proposed to describe cyclic deformation behaviour under thermo-mechanical loading [19-21]. In this paper the endochronic constitutive theory (ECT) proposed by Valanis [19] has been used to describe the stress-strain response of the material under thermo-mechanical loading. This constitutive equation has been used to evaluate the damage

J. Bressers and L. Rémy (eds.), Fatique under Thermal and Mechanical Loading, 467–473.

parameter. Based on this parameter a new thermo-mechanical fatigue life prediction method is proposed.

2. Modelling of the Stress-Strain Behaviour

The endochronic constitutive theory has been proposed on the basis of the irreversible thermodynamics. The theory does not have to rely upon the yield surface concept, and the material memory is defined in terms of an intrinsic time scale, a material property at hand. Following Ilyushin's deviatoric vector space the stress vector under the action of axial force and torque may be defined as

$$\overline{\sigma} = \sigma_1 \cdot \overline{n}_1 + \sigma_3 \cdot \overline{n}_3 \tag{1}$$

Where $\overline{n}_1$ and $\overline{n}_3$ are orthonormal base vectors in the subspace, σ_1 is axial stress and $\sigma_3 = \sqrt{3} \cdot \tau$, ($\tau$ is shear stress).
The strain vector can be defined as

$$\overline{\varepsilon} = \varepsilon_1 \cdot \overline{n}_1 + \varepsilon_3 \cdot \overline{n}_3 \tag{2}$$

where is ε_1 axial strain, $\varepsilon_3 = \gamma / \sqrt{3}$ (γ is engineering shear strain).
In the deviatoric vector space, the basic equation of endochronic theory is

$$\overline{\sigma} = \int_0^z \rho \cdot (z - z') \frac{d\varepsilon^p}{dz'} dz' \tag{3}$$

where

$$dz = d\xi / f(\xi, T) \tag{4}$$

$$d\xi = \sqrt{\left(d\varepsilon_1^p\right)^2 + \left(d\varepsilon_3^p\right)^2} \tag{5}$$

For simulating the stable cyclic stress-strain relation, one may assume the function as

$$f(\xi, T) = 1 \tag{6}$$

For thermo-mechanical cyclic loading the intrinsic time scale has to be redefined.
In the proposed model the intrinsic time scale is redefined as

$$dz = \frac{d\xi}{f(\xi, T) g(\xi, T)} \tag{7}$$

where $f(\xi, T)$, $g(\xi, T)$ are a hardening (softening) function and a shape function respectively.
For thermo-mechanical loading one proposes a linear function simply as

$$g(\xi,T)=(1+c_1\phi)(1+c_2(T-T_0)) \tag{8}$$

$$\phi=\sin^2\theta=1-\cos^2\theta \tag{9}$$

$$\cos\theta=\frac{\dot{\varepsilon}^p_{cj}\varepsilon^p_{cj}}{\left|\dot{\varepsilon}^p_{cj}\right|\left|\varepsilon^p_{cj}\right|} \tag{10}$$

where ϕ is the nonproportionality corresponding to the loading path shape: $\phi = 0$ for in-phase loading, and $0 < \phi < 1$ for non-proportional loading. In case where the strain history is prescribed according to Murakami and Read [22] one gets

$$\left[\left(\frac{Q_1}{R+E}\right)^2+\left(\frac{Q_3}{R+3G}\right)^2-f^2(\xi,T)g^2(\xi,T)\right]dz^2+$$
$$+2\left[\frac{d\varepsilon_1 Q_1 E_1}{(R+E)^2}+\frac{3d\varepsilon_3 Q_3 G}{(R+3G)^2}\right]dz+\left[\frac{d\varepsilon_1 E}{R+E}\right]^2+\left[\frac{3d\varepsilon_2 G}{R+3G}\right]^2=0 \tag{11}$$

where

$$d\overline{\sigma}=Rd\overline{\varepsilon}^p-\overline{Q}\,dz \tag{12}$$

$$\overline{Q}=Q_1\overline{n}_1+Q_3\overline{n}_3=\sum_{r=1}^{3}a_r\left(Q_{r1}\overline{n}_1+Q_{r3}\overline{n}_3\right)=\sum_{r=1}^{3}a_r\overline{Q}_r \tag{13}$$

$$d\overline{Q}_r/dz+\alpha_r\overline{Q}_r=R_r d\overline{\varepsilon}^p/dz \tag{14}$$

$$R=\sum_{r=1}^{3}R_r \tag{15}$$

$$d\varepsilon_1^p=d\varepsilon_1-d\sigma_1/E \tag{16}$$

$$d\varepsilon_3^p=d\varepsilon_3-d\sigma_3/E \tag{17}$$

where G and E are the material shear modulus and Young's modulus, respectively. Symbol d(X) stands for the increment of physical quantity (X).The model parameters Q_r and R_r are available from the uniaxial cyclic isothermal stable stress-strain relation.

3. Thermo-Mechanical Damage Parameter

A major research effort is presently made to define a suitable damage parameter which can be used to describe the fatigue process. In addition to the plastic strain which is more often used, a special form of the cyclic strain energy density has been proposed by Golos [14-18] to analyze fatigue damage. This form of the strain energy density combines the plastic strain energy density, ΔW^{p} and the elastic strain energy density, ΔW^{e+} associated with the tensile mode which facilitates crack growth. It is termed the total strain density and for uniaxial thermo mechanical fatigue can be expressed as:

$$\Delta W^{t} = \Delta W^{e+} + \Delta W^{p} \tag{18}$$

The total strain energy density is calculated from the relation

$$\Delta W^{t} = \frac{\Delta\varepsilon^{2}}{8E} + \int_{cycle} (\sigma_{1} - \sigma_{2})\, d\varepsilon^{p} \tag{19}$$

$$\Delta W^{t} = K' N_{f}^{\alpha} \tag{20}$$

where σ_1, σ_2 are functions describing branches of the hysteresis loop.
In Eq. (20) K' and α are material constants, N_f is the number of cycles to fracture in isothermal fatigue. In Fig. 1 the damage parameter is depicted.

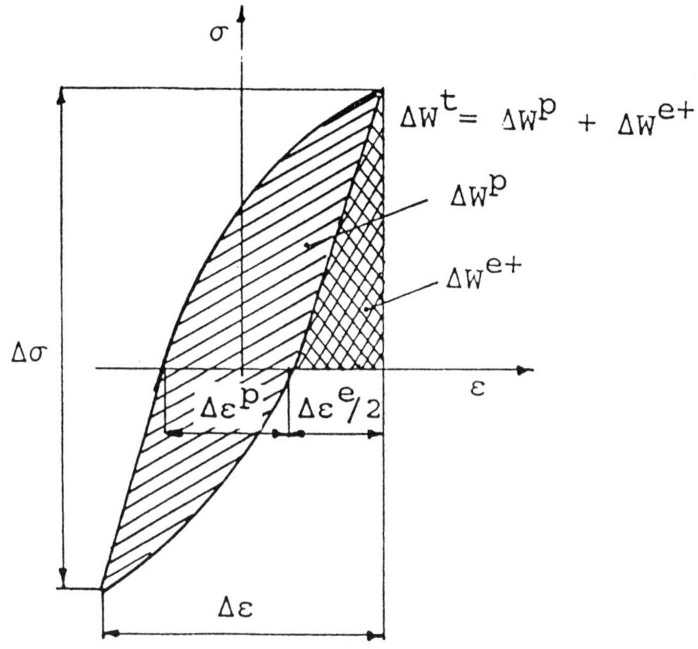

Figure 1. Isothermal stress-strain loop with indication of the damage parameter.

To calculate the total strain energy density under thermomechanical loading the modified endochronic constitutive theory has been used.

In order to calculate the predicted fatigue life of thermal-mechanical fatigue, transforming above equation and supposing a linear damage, we obtain

$$\frac{1}{\Delta D} = N_{f} = K''\left(\Delta W^{t}\right)^{1/a} \tag{21}$$

where ΔD is the damage per cycle.

4. Thermo-Mechanical Fatigue Life Prediction

To describe the influence of the variable temperature on damage, a thermal damage factor $\lambda(T_i)$ was adopted by using isothermal fatigue data [22, 23]

$$\lambda(T_i) = N_f^{Ti} / N_f^{To} \tag{22}$$

where T_o is the reference temperature and T_i is an arbitrary temperature.

When a typical stabilized hysteresis loop of thermo-mechanical cycling is considered and linear damage based on the total strain energy density is supposed, we can get differential equation in the form [23]:

$$\delta\Delta D_{Ti} = \frac{\delta\Delta W^t(T_i)}{\Delta W^t(T_i)}\Delta D_{Ti} = \frac{\delta\Delta W^t(T_i)}{\Delta W^t(T_i)} \cdot \frac{1}{N_f^{To}\lambda(T_i)} \tag{23}$$

where

ΔD_{T_i} - damage per cycle in isothermal fatigue at temperature T_i

$d\Delta D_{Ti}'$- damage element per cycle in thermal-mechanical fatigue when temperature is equal to T_i

The damage for thermo-mechanical fatigue is determined by making the summation of Eq. (23).

5. Comparison with Experimental Data

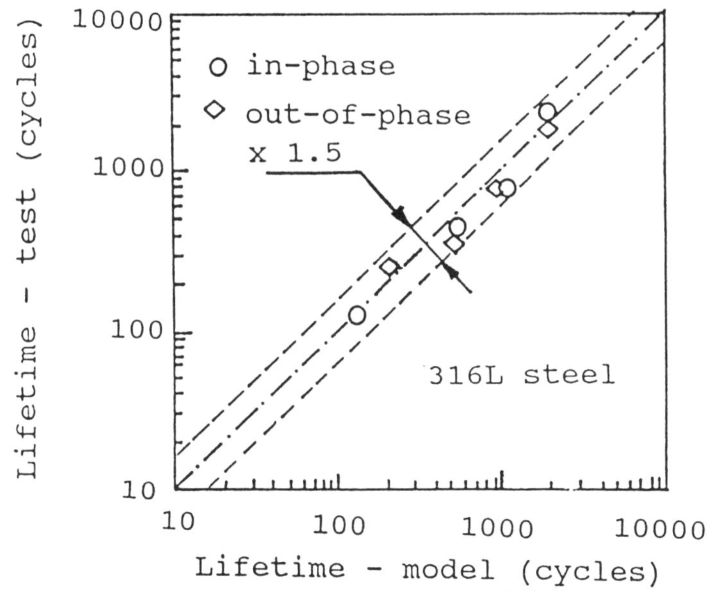

Figure 2. Comparison between experimental and calculated results.

In the analysis experimental data of a 316L type austenic stainless steel have been used [23]. The total strain ranges applied were equal 1%, 1.2%, 1.6%, 2% and 2.4% respectively. The thermo-mechanical tests were made at a temperature range of 250 to 500°C. The comparison of the experimental data and experiments has been made and all results fall in the scatter band of 1.5 (Fig. 2).

It appears that the proposed model for analyzed material gives reasonably good results.

6. Conclusions

The analysis of thermo-mechanical fatigue has resulted in the following conclusions:

(1) A damage parameter - total strain energy density - has been proposed for thermo-mechanical fatigue,
(2) The modified endochronic constitutive theory can describe the cyclic constitutive behaviour of material under thermal mechanical cyclic loading,
(3) With this constitutive theory, the total strain energy density approach can be used to evaluate fatigue life.

It should be noted that the damage model proposed in this paper was only verified by thermo-mechanical fatigue tests in which the maximum temperature was not very elevated. In the case of higher temperatures, this model could possibly become less efficient because of the significant creep damage.

References

1. Coffin, L.F., Fatigue at High Temperatures-Prediction and Interpretation, *Proc. Institution of Mechanical Engineers*, **188**, (1974), London.
2. Coffin, L.F., Overview of Temperature and Environmental Effects on Fatigue of Structural Metals, *Fatigue:Environment and Temperature Effects*, J.J.Burke and V.Weiss, Eds., Plenum Press, New York,(1983), pp. 1-40.
3. Ostergren, W.J., A damage function and associated failure equations for predicting hold time and frequency effects in elevated temperature low cycle fatigue, *Journal of Testing and Evaluation*, **4**, (1976), 327-339.
4. Manson, S.S. et al. *Creep-Fatigue Analysis by Strain Range Partioning*, NACA TM X-67838. 1971.
5. Saltsman, J.F. and Halford, G.R., An update of the total-strain version of strain range partioning, *ASTM STP 942 Low Cycle Fatigue*, Philadelphia, (1988), pp. 329-341.
6. Neu, R.W. and Sehitoglu, H., Thermo-mechanical Fatigue, Oxidation and Creep:Part I. Damage Mechanisms, *Metals Transactions A*, **20A**, (1989).
7. Neu, R.W. and Sehitoglu, H., Thermomechanical Fatigue, Oxidation and Creep:Part I:Life Prediction, *Metals Transactions A*, **20A**, (1989), pp. 1769-1783.
8. Kachanov, L., Fundamentals of Fracture Mechanics, Nauka, Moscow, 1974.
9. Rabotnov,Y.N., *Creep Problems in Structural Members*, Nort Holland Publishing, Amsterdam, 1969.
10. Hult, J., Continuum Damage Mechanics - Capabilities Limitations and Promises, *Mechanisms of Deformation and Fracture*, Pergamon, Oxford, (1979), pp.233-247.
11. Chaboche, J.L., Continuous Damage Mechanics - A Tool to Describe Phenomena Before Crack Initiation, *Nuclear Eng. and Design*, **64**, (1981), 233-247.
12. Lemaitre, J. (1979) *Damage modelling for predicting of plastic or creep-fatigue in structures*, Paper L5-1 SMIRT Conference, Berlin, 1981.

13. Majumdar, S. and Maiya, P.S., A mechanistic model for time dependent fatigue, *ASME Journal of Materials and Technology*, **102**, (1980),159-167.
14. Golos, K., Plastic strain energy under cyclic multiaxial states of stress. *J. of Theoretical and Applied Mechanics*. **26**, (1988), 171-176.
15. Golos, K., Energetistic formulation of fatigue strength criterion, *Archiwum Budowy Maszyn*, **XXXV**, (1988), No1/2, 5-16.
16. Golos, K., Fracture energy criterion for fatigue crack propagation, *Archiwum Budowy Maszyn*, **XXXV**, (1988), No.3.
17. Golos, K., Cumulative Fatigue Damage, *Materials Science and Engineering*, **104A**, (1988), 61-65.
18. Golos, K. and Ellyin, F., A total strain energy density for cumulative fatigue damage, *ASME Journal of Pressure Vessel Technology*, **110**, (1988), 36-41.
19. Valanis, K.C., Fundamental consequences of a new intrinsic time measure: plasticity as a limit of the endochronoc theory, *Arch. Mechanics*, **27**, (1980) 171-191.
20. Ning, J., An endochronic constitutive theory of material under nonproportional cyclic loading, *Mechanics Research Comunications*, **18**, (1991),187-198.
21. Murakami H. and Read, H.E., Endochronic plasticity: some basic properties of plastic flow and failure, *Int. J. Solids Structures*, **23**, (1987), 133-151.
22. Taira, S., Relationship Between Thermal Fatigue and Low-cycle Fatigue at Elevated Temperature, *ASTM STP* **520**, Philadelphia, (1973), pp.80-101.
23. Shi, H.J., Robin, C. and Pluvinage, G.,(1993) *Thermal-mechanical fatigue lifetime prediction of an austenic stainless steel*, *ASTM STP* **1211**, Philadelphia, pp.105-116.

THERMO-MECHANICAL BEHAVIOUR OF SURFACE MOUNT SOLDER JOINT DURING THERMAL CYCLING

V. AUDIGIER[1, 2], S. LASSERRE[2] and J.L. LATAILLADE[2]
[1] *IBM France, IPT Europe*
Bordeaux, France
[2] *Laboratoire Mécanique Endommagement Fiabilité,*
ENSAM, CER de Bordeaux, France

1. Introduction

Surface mount technology (SMT) is a relatively new technology used to attach electronic components to a printed circuit board. SMT provides electronics manufacturers with numerous advantages over conventional through-hole technology (PTH) (Fig. 1), so is widely used in more and more applications. These advantages (Fig. 2) include increased functional density, smaller packages, increased lead count, smaller circuit boards and potential use of both sides of the circuit board. However the smaller packages and the larger lead counts require thinner solder joint between lead and card [1].

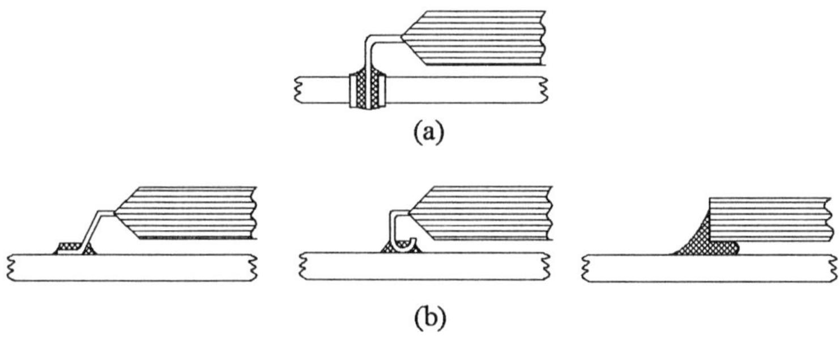

Figure 1. (a) PTH ; (b) SMT (leaded and leadless components).

Eutectic 63Sn/37Pb is used extensively as an interconnection material in electronic packaging. It provides a mechanical, as well as an electrical connection between component and board. So Surface Mount component reliability is directly related with solder joint reliability since 80 to 90% of electronic failures are due to assembly defects : SM reflow process and/or bad lead coplanarity or solderability can create poor, porous solder, which results in cracks which generate assembly failures [2].

J. Bressers and L. Rémy (eds.), Fatique under Thermal and Mechanical Loading, 475–485.

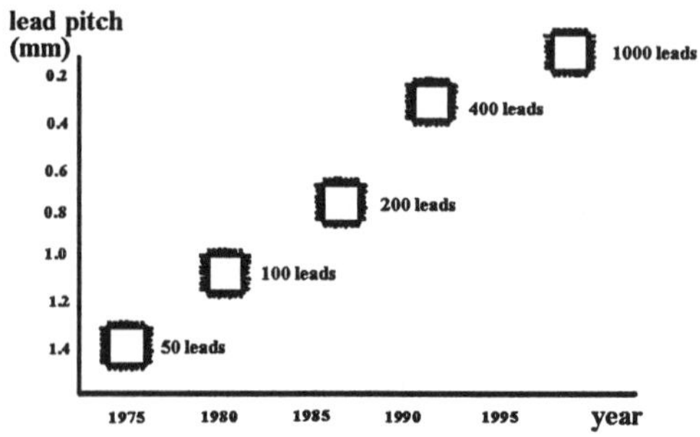

Figure 2. Evolution of lead count and pitch between 1975 and 2000.

The thermal expansion mismatch between board and component is more destructive. As temperature changes from turning power on and off, small differences in thermal expansion properties between component and substrate produce low yet important stresses in solder joints. Residual stresses are generated and irreversible damage is accumulated in the solder which leads to failure. This thermal fatigue cracking has been identified as a major failure mode in electronic packages.

The aim of the study is to understand deformations and stresses that occur in a solder joint during thermal cycling. A two-dimensional finite element analysis has been applied to analyze the thermo-mechanical behaviour of surface mount solder joint during thermal cycling and to determine the main parameters of the thermal cycle. A realistic model, which takes into account the real properties of the solder, has been used to simulate the *elasto-visco-plastic behaviour* of J-lead solder joint. The non-linear numerical analysis is carried out using CAEDS (IDEAS) and ANSYS softwares. The effects of temperature range, hold and ramp times on solder stresses and strains are investigated and then extrapolated on solder fatigue life. Experimental tests are in progress in order to compare these with the numerical simulation predictions. The objective is to find an "efficient accelerated test", faster than the thermal cycles usually used in reliability testing but realistic enough for a good assessment of the assembly reliability level.

The primary long term failure mode of surface mount assemblies is Low Cycle Fatigue, brought about by thermally and mechanically induced cyclic strains. The fatigue damage results from cyclic mechanical strains induced by the mismatch of thermal expansion between the dissimilar components connected together by the solder. This thermal expansion mismatch is caused both by temperature gradients and by mismatch of the coefficient of thermal expansion. Since the solder is a highly visco-plastic material over the typical operating temperature ranges, slow thermal cycling produces primarily time-dependent (anelastic and superplastic) creep strains and relatively negligible amounts of elastic and plastic strains which are dependent on the magnitude of stress but not on time. This phenomenon is described by the low cycle fatigue where simultaneous interaction of creep and fatigue makes damage analysis of a solder joint a complex problem.

2. Theoretical Approach

The Finite Element Method (FEM) became very popular over the past decade as a tool for viscoplastic analysis of the stress and strain histories in solder joints, under the given loads. The solder response is modeled with thermoelastic, creep and rate independent plastic deformation. The concept of conventional creep and plasticity is used, in which creep can be described separately from the rate independent plasticity. Therefore no interaction between the creep and plasticity will be considered in the formulations.

For convenience of computation, the total strain is considered as :

$$\varepsilon = \varepsilon_{el} + \varepsilon_{pl} + \varepsilon_{cr} + \varepsilon_{th} \tag{1}$$

where ε_{el}, ε_{pl}, ε_{cr}, ε_{th} represent the elastic, plastic and creep and thermal components. The total strain is idealized as a partition of time-independent (instantaneous) plastic strain and a time-dependent creep strain.

The elastic response is expressed by the Duhamel-Neumann law [3] :

$$\varepsilon_{ij}^{el} = \frac{1+\nu}{E}\sigma_{ij} - \frac{\nu}{E}\delta_{ij} \cdot \sigma_{kk} \tag{2}$$

The thermal strain caused by thermal expansion is defined as :

$$\varepsilon_{ij}^{th} = \delta_{ij}\,\alpha T \tag{3}$$

where $T = T_{max} - T_{min}$

2.1. PLASTIC CHARACTERISTICS

Solder alloy plasticity is modeled assuming bilinear kinematic hardening. This model takes into account the Bauschinger effect present in the solder and gives good agreement with the real behaviour[4]. The Ramberg-Osgood law [3;6] is used :

$$\varepsilon_{pl} = \left(\frac{\sigma - \sigma_y}{K}\right)^m \tag{4}$$

where σ_y : yield stress
K : Ramberg-Osgood coefficient
m : plastic strain hardening exponent.

In this case, m=1 and $K = E.E_p/(E-E_p)$ where E is the Young modulus and E_p the "plasticity" modulus.

As the solder behaviour is temperature-dependent, the mechanical properties evolution is taken into account during the thermal cycles. Fig. 3 resumes the stress-strain curves used in the FEA for plastic computation.

2.2. CREEP CHARACTERISTICS

To describe the thermally induced creep behaviour of Pb/Sn alloy, a Weertman model [5] is used to relate the creep deformation rate to stress and temperature. The contribution of primary and secondary creep depends on the alloy type : secondary creep controls strain at high temperature for the eutectic alloy, although primary creep is predominant for lead-rich solder alloy. As a first approximation, only steady-state creep is considered for eutectic Pb/Sn and :

$$\varepsilon_{cr} = \varepsilon_o \cdot \sigma^{n_c} \cdot \exp\left(\frac{-\Delta H}{kT}\right) \cdot t \qquad (5)$$

where ε_{cr} : creep equivalent strain,
ε_0 : creep constant,
n_c : creep exponent,
ΔH: activation energy,
k : Boltzmann constant,
T : instantaneous absolute temperature,
t : elapsed time in seconds.

The values used for the creep coefficients are listed in table 1.

3. The Finite Element Analysis

The geometrical modeling is made with CAEDS (IDEAS) v4.2 [7] software while the non-linear calculations are carried out using ANSYS v5.0 A [8]. A 3-dimensional study has been performed, but for computation reasons only a 2-dimensional plane stress study is made here : a non-linear 2D-element-finite simulation with 10 thermal cycles needs approximately 45 hours CP time for 95 hours total elapsed computation time and the 2D main results are globally the same. One typical electronic package with a J-lead is modeled, assembled on a board [9] and is shown in Fig. 4. Only half of the assembly is modeled due to symmetry. The lead is alloy 42 (Fe58/Ni42), the pad copper and the encapsulant glass-epoxy. The board is epoxy FR4. All these materials are considered as linear elastic ; except the board which is orthotropic, all are assumed to be isotropic. The solder alloy (37Pb/63Sn) is treated as a viscoplastic material. All mechanical properties are listed in Table 1 and Fig. 3.

The finite element model uses 4 node quadrilateral thin shell elements with a finer mesh in the critical area (the solder joint). It consists of 4078 elements and 4329 nodes, with 694 elements for the refined solder joint area.

The model is subjected to -20 to +100°C and -40 to +125°C temperature cycling. The thermal cycles have different ramp and hold times listed in table 2.

TABLE 1. Mechanical characteristics [10].

	E *(GPa)*	*α* *(10^{-6}/°C)*	*ν*	*Creep charact. (for σ in kPa)*		
58Fe-42Ni	153	4	0.31			
epoxy	14	13	0.30			
Cu	120	16	0.34			
37Pb-63Sn	*	24	0.40	ε_0= $3.71.10^{-20}$	n_C = 6.28	ΔH/k = 8165.2
FR4	x,y 19 z 9	x,y 13 z 70	xz, yz 0.18 xy 0.39			

* solder plastic characteristics are plotted in Fig. 3.

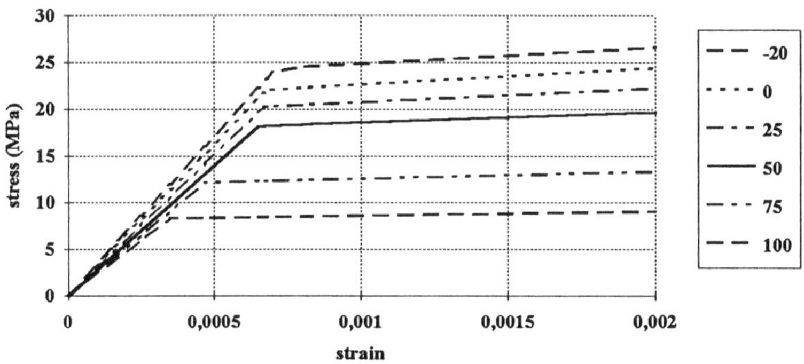

Figure 3. Evolution of solder alloy elasto-plastic characteristics with temperature (in °C).

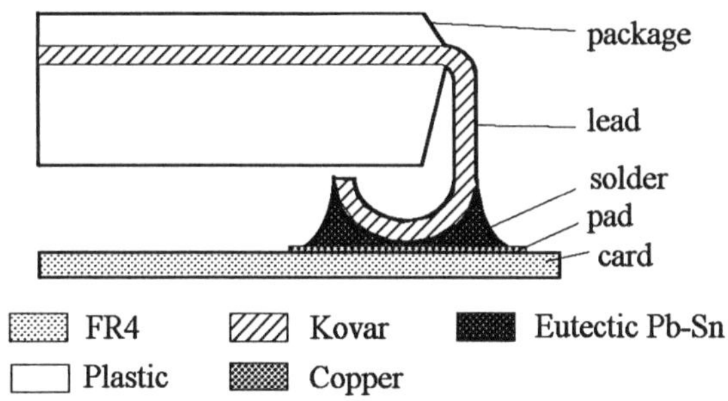

Figure 4. Scheme of the finite element model.

TABLE 2. Thermal profiles parameters.

Temperature range (°C)	*Ramp time (min)*	*Hold time (min)*
-20/+100	3	30
-40/+125	3	15
-40/+125	3	30
-40/+125	20	30

4. Results

4.1. SOLDER JOINT BEHAVIOUR DURING THERMAL CYCLING

The J-lead solder joint is predominantly submitted to shear since it is a thin layer of ductile material between two thicker layers of Fe/Ni alloy and Cu. Fig. 5 shows the shear stress evolution during 2 thermal cycles for an element situated in the critical solder area. Shear stress (Sxy) and the thermal profile (BFETEMP) are plotted as a function of time. The solder joint is submitted to alternate cyclic stresses : tension and compression shear stresses appear. As the solder reaches a homologous temperature ($=T_{use}/T_{melting}$) well beyond 0.5, steady-state creep is significant even under ambient temperatures : a 30 minute dwell time generates stress relaxation from -4 MPa to near 0 MPa at T_{max} and from 13 MPa to 8 MPa at T_{min} (numerical values for a J-lead solder), while irreversible time dependent strains (such as EPCRXY, creep shear strain) are accumulated and increase with every cycle.

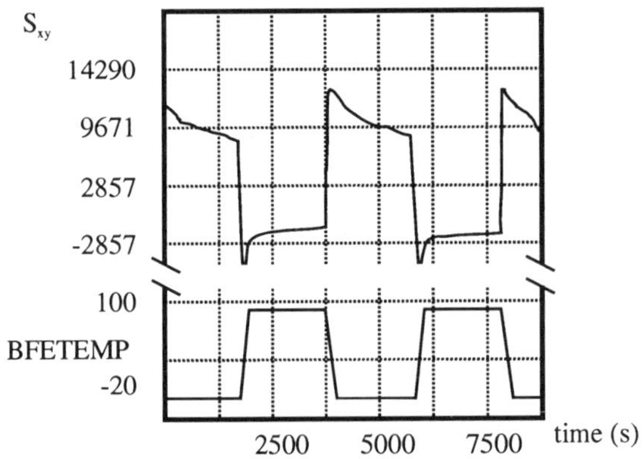

Figure 5. Shear stress (kPa) evolution during 2 thermal cycles for an element situated in a critical solder area; cycles -20/+100°C, 3min ramp and 30min dwell times.

Fig. 6 shows the shear stress distribution in the solder at the beginning of the dwell time at T_{min}. Figs. 7 and 8 show the inelastic (creep and plasticity) shear strains at the end of the dwell time at T_{max}. It can be seen that the critical areas of creep and high level plasticity are situated near the interfaces between lead and solder or between pad and solder. These locations will be the crack initiation locations which correlate well with SEM observations of a solder joint submitted to 414 thermal cycles -20/+100°C, 1 cycle/h, with a crack having initiated at the lead/solder interface (Fig. 9).

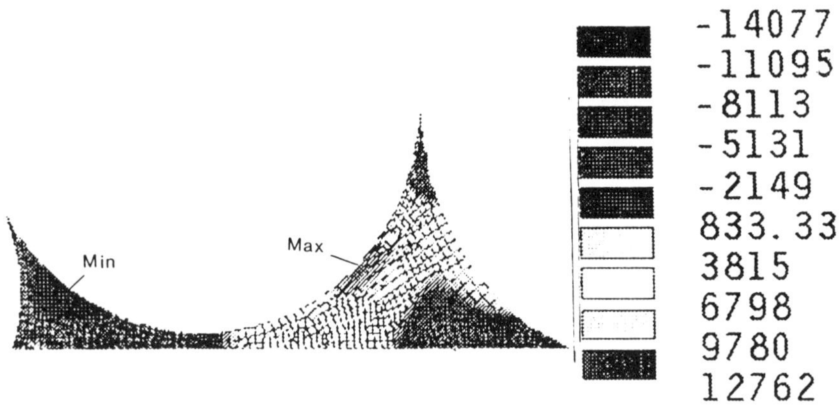

Figure 6. Shear stress (kPa) distribution in the solder at the beginning of the dwell time at -20°C.

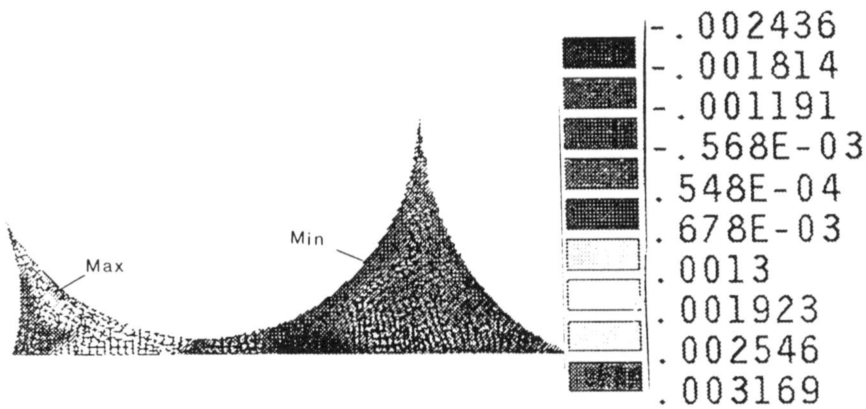

Figure 7. Creep shear strain distribution in the solder at the end of the dwell time at 100°C.

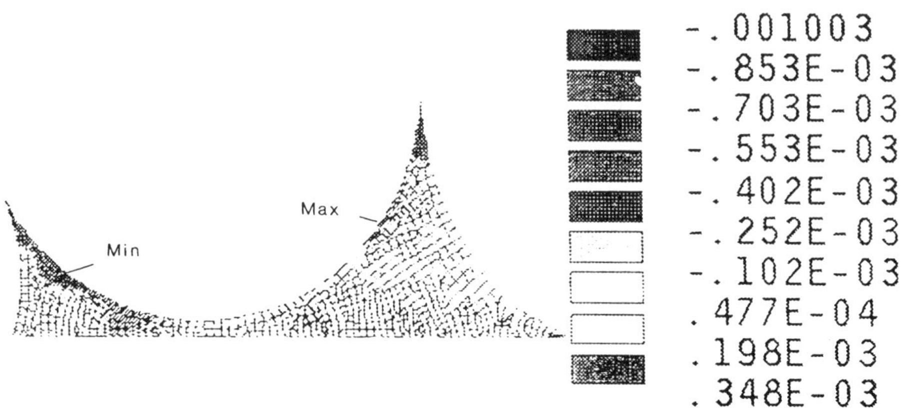

Figure 8. Plastic shear strain distribution in the solder at the end of the dwell time at 100°C.

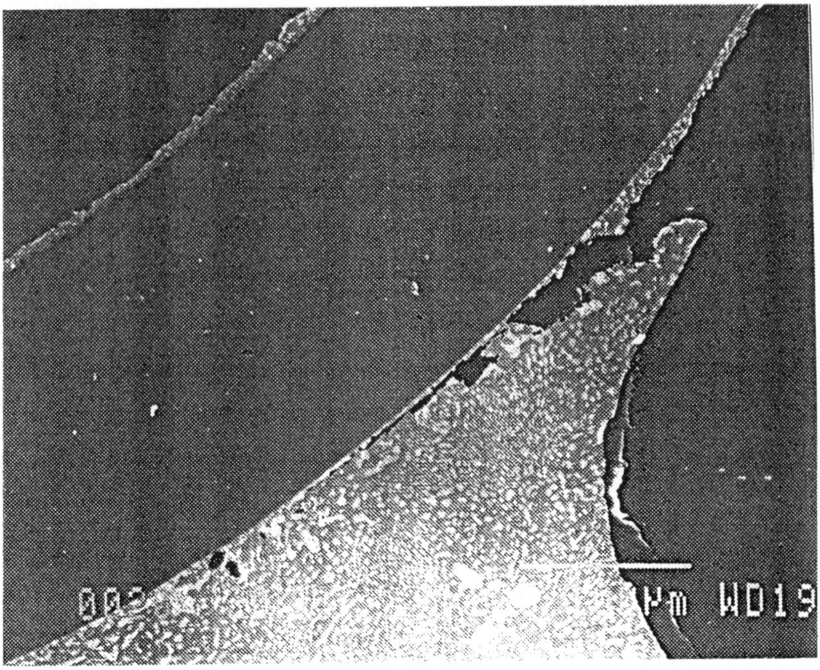

Figure 9. Crack in a J-lead solder joint after 414 cycles -20/+100°C.

It is interesting to study the hysteresis shear stress/strain curve (Fig. 10). During dwell periods, strains are almost constant while stresses are decreasing. Stress relaxation clearly appears during the ramp up. Table 3 summarises the solder shear behaviour with temperature. It must be noted that the thermal mismatch between different materials always exists during the thermal cycles ; this phenomenon is the basis of all others, so it is mentioned in the Table 3 only when it is alone, i.e. when it does not introduce complementary phenomena (creep, ...). It must be mentioned also that the reorientation of principal directions takes place first at the interface lead/solder and then it propagates towards the solder.

TABLE 3. Solder shear behaviour with temperature evolution.

Thermal step		*Shear evolution*	*Main phenomenon*
increase : -20 to 100°C	$-20<T<40$	inversion of the shear at the lead interface	principal stresses reorientation
	$40<T<100$	decrease	creep-relaxation
dwell time at 100°C		decrease	creep-relaxation
decrease :100 to -20°C	$100>T>75$	inversion of the shear at the lead interface	principal stresses reorientation
	$75>T>-20$	increase	thermal mismatch (card/lead)
dwell time at -20°C		decrease	creep-relaxation

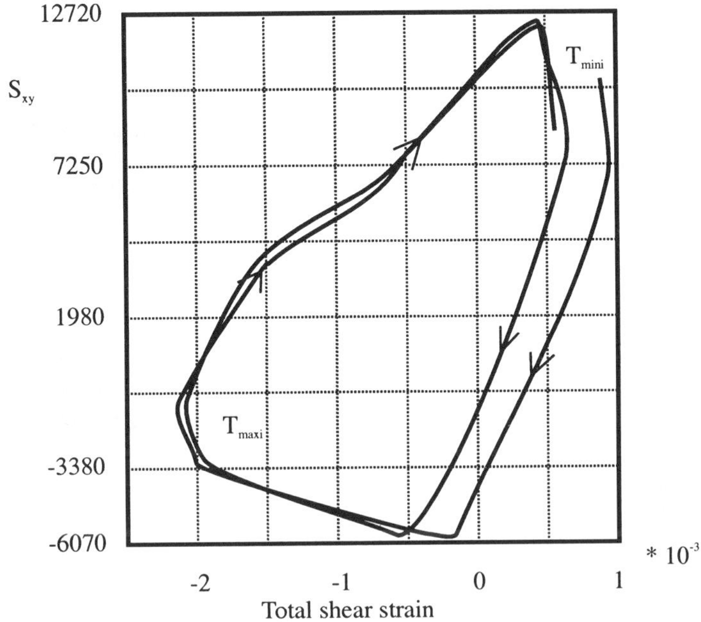

Figure 10. Shear stress/strain curve for -20/+100°C thermal cycle, 3 min ramp and 30 min hold times.

4.2. INFLUENCE OF THE THERMAL CYCLES PARAMETERS

4.2.1. Temperature Range.

Thermal profiles with the same ramp and dwell times but with different temperature ranges are studied : -20/+100°C and -40/+125°C are applied to the model. The global stress evolution is the same and is independent of the temperature range. Table 4 summarises the maxima of the shear stress and strain amplitudes for a critical element in the solder joint. When temperature range increases, shear stress amplitude and total shear strain increase too. The inelastic (plastic and creep) strain increases with the temperature range too.

TABLE 4. Shear stress level for a critical element in J-lead solder joint function of the temperature range.

Thermal cycle (°C)	Temperature range	ΔS_{xy} (MPa) at T_{min}	$\Delta\varepsilon_{xy}$ total	$\Delta\varepsilon_{xy}$ inelastic (pl+cr)
-20/+100	120	26.8	3.1195 e-3	2.6391 e-3
-40/+125	165	31.8	4.2038 e-3	3.4462 e-3

4.2.2. Ramp Time.

Different ramp times are studied : 3 and 20 minutes for a thermal cycle between -40 and +125°C. It can be seen (table 5) that a fast ramp increases the stress level in the solder. The irreversible accumulated total shear strain (elastic + plastic + creep) is sensibly less important for a short ramp time.

TABLE 5. Shear stress and strain level for a critical element in the solder joint as a function of the ramp time.

Ramp time (min)	$\Delta\varepsilon_{xy}$ total (-)	% $\Delta\varepsilon_{xy}$ inelastic/ $\Delta\varepsilon_{xy}$ total	ΔS_{xy} (MPa)
3	4.2038e-3	82.0	19.2
20	4.2108e-3	82.4	16.7

4.2.3. Dwell Time.

For the same temperature range and ramp time, different dwell times have been considered : 15 and 30 minutes. Table 6 shows that the dwell time influences the shear stress level and the amount of inelastic shear strain. In the last case, the influence is greater than that of the ramp time. Shear amplitude increases with dwell duration. For a short dwell time, creep acts partially and stress relaxation is incomplete. The solder fatigue prediction is also wrong because the irreversible strains are not fully accumulated as in the real case. For a long dwell time creep fully appears, stress relaxation is maximum, and irreversible strains are totally accumulated. The creep phenomenon is supposed completely present if the dwell time, measured on the component and not on the thermal equipment, exceeds 15 minutes. So the longer the dwell time, the smaller the solder fatigue life.

TABLE 6. Shear stress level for a critical element in the solder joint as a function of the dwell time.

Dwell time (min)	ΔS_{xy} (MPa)	$\Delta\varepsilon_{xy}$ total (-)	$\Delta\varepsilon_{xy}$ inelastic (pl+cr) (-)	% $\Delta\varepsilon_{xy}$ inelastic/ $\Delta\varepsilon_{xy}$ total
15	14.26	4.1735 e-3	3.2846 e-3	78.7
30	15.98	4.2038 e-3	3.4462 e-3	82.0

It must be noted that the amount of inelastic shear strain is more important at T_{max} (97.9%) than at T_{min} (31.6%).

5. Fatigue Life Prediction

In the automotive and computer industries the challenge consists in designing an accelerated thermal cycling test to adequately simulate the service life in the field. The current fatigue life models don't take into account all key-parameters such as temperature range, hold and ramp time. The traditional strain-based approach, the Coffin-Manson law [11], considers strain from thermal expansion mismatch but doesn't consider creep and stress relaxation during ramp and hold times. Two approaches seem to be closer to reality : a strain-energy based fatigue life model developed by Pan [12] and a modified Coffin-Manson law [13].

Using the finite element results some remarks can be made to determine the most critical areas in the solder joint, which coincide with the first failure location. Typical laws are then used to calculate the fatigue life of the solder and of the electronic assembly. The finite element results and the fatigue prediction correlate and it can be seen that the thermal cycle parameters have a significant influence on the assembly fatigue life :

- if the $\Delta T = T_{max} - T_{min}$ increases, stress level increases too, and fatigue life decreases ;

- if the ramp time increases (i.e. if the ramp rate decreases) , the Von Mises stress level decreases, so the total deformation is reduced and on the opposite the fatigue life increases ;
- if the hold time increases, more stress relaxation occurs and the fatigue life is reduced.

Thus it is important to choose an appropriate thermal cycle to simulate the service conditions in an accelerated test. In particular, the dwell time influence is very important and an accelerated test must have 15 minutes minimum dwell period, otherwise the realistic solder behaviour isn't taken into account.

6. Conclusion

The reliability of surface mount solder joint in electronic assemblies has been studied with a finite element method ; the real mechanical properties (plasticity and creep) of the solder have been taken into account. This study reveals that some of the thermal cycle parameters have a great effect on the thermo-mechanical behaviour of the solder. The accelerated tests used in laboratory in order to simulate a faster ageing of the assembly must also take them into account. The thermo-mechanical fatigue behaviour of the solder joint is predominantly related to the dwell time at T_{min} and T_{max} where inelastic deformation occurs and leads to damage accumulation. So it is important to realize accelerated thermal cycle tests with a dwell duration greater than 15 minutes in order to simulate the realistic solder behaviour.

References

1. Tummala, R., Rymaszewski, E., Van Nostrand, R., *Microelectronics Packaging Handbook*, NY 1989.
2. Lau. J.H., Van Nostrand, R., *Solder joint reliability, Theory and applications,* NY 1991.
3. Lemaitre, P., Chaboche., *Mécanique des matériaux solides,* édition Dunod, 1988.
4. Lau, J., Erasmuss, S., Reliability of fine pitch plastic quad flat pack leads and solder joints under bending, twisting, and thermal conditions, *Journal of Electronic Packaging, Transactions of the ASME,* Vol.**115**, Sep 1993, pp. 322-328.
5. Pao, Y.H., A fracture mechanics approach to thermal fatigue life prediction of solder joints, IEEE *Transactions on Components, Hybrids, and Manufacturing Technology*, Vol **15**, N° 4, August 1992.
6. Hertzberg, R.W., *Deformation and fracture mechanics of engineering materials*, John Wiley, NY, 1989.
7. *CAEDS User's guides*, Structural Dynamics Research Corp.,1991.
8. Kohrke, P., (ed.), *ANSYS User's manuals,* for revision 5.0, Vol **I** to **IV**, Swanson Analysis System, Inc.
9. Audigier, V., *Thermo-mechanical fatigue of surface mount solder joint,* internal paper, 1995.
10. King, J.A., *Material handbook for hybrid microelectronics*, Artech house inc., Boston, London, 1988.
11. Manson, S.S., *Thermal stress and low cycle fatigue*, Mc Graw-Hill, New-York, 1966.
12. Tsung-Yu, P., *Thermal cycling induced plastic deformation in solder joints*, ASME winter annual meeting, NY, 91-WA-EEP-10, pp1-8.
13. Engelmaier, W., Surface mount solder joint long term reliability : design, testing, prediction, *Soldering and surface mount technology*, N°1, pp14-22, Feb 1989.

FATIGUE OF CONTINUOUS FIBER-REINFORCED CERAMIC MATRIX COMPOSITES: REVIEW OF MECHANISMS AND MODELS

B.F. SØRENSEN[1] and J.W. HOLMES[2]
[1] *Materials Department, Risø National Laboratory*
DK-4000 Roskilde, Denmark, EU
[2] *Ceramic Composites Research Laboratory,*
Department of Mechanical Engineering and Applied Mechanics,
The University of Michigan
Ann Arbor, MI 48109-2125, USA

1. Introduction

Ceramics reinforced with aligned continuous fibers have the potential for use in high temperature load carrying structures. However, the desirable damage tolerant behavior observed at room temperature can be lost at high temperature [1] if strong bonding develops along the fiber/matrix interface due to oxidation [2]. Currently, considerable research is being devoted to the development of coatings that remain chemically stable at high temperature and, at the same time, act as weak layers to facilitate fiber/matrix debonding [3], a pre-requisite for damage tolerant behavior. Another important issue is the evolution of damage during thermomechanical cycling. Room temperature fatigue damage has been widely studied, and many aspects are fairly well understood [4] for a detailed discussion. High temperature fatigue and thermomechanical fatigue studies are far more sparse. At high temperatures time-dependent oxidation, diffusion and creep complicate the modelling of fatigue damage. In this paper, we discuss the fatigue behavior of unidirectional ceramic matrix composites (CMC's), and point out similarities and differences between room temperature and high temperature fatigue damage.

2. Damage Mechanisms under Monotonic Tension

Since the microstructural damage evolution under monotonic loading has many similarities to that occurring under cyclic loading, it is instructive to briefly discuss the characteristic damage evolution under monotonic tension and accompanying changes in overall composite behaviour.

Fig. 1a shows a stress-strain curve for a unidirectional CMC tested in uniaxial tension in the axial direction. In order to elucidate the underlying damage mechanisms, it is instructive to divide the stress-strain curve into several sequential

J. Bressers and L. Rémy (eds.), Fatique under Thermal and Mechanical Loading, 487–499.

parts [5]. At low applied stresses the composite behavior is elastic (Stage I). At higher stresses matrix cracks initiate (typically in matrix-rich regions [5, 6]), and evolve into fiber-bridged matrix cracks (Stage II), gradually causing non-linear stress-strain behavior (note, however, that the initial matrix cracking often occurs *without* detectable compliance changes [7]. With increasing stress, the matrix crack spacing reaches a saturated level and remains nearly constant with a further increase in applied stress. Beyond saturation, the deformation is primarily due to elongation of the fibers, and the dominant damage mechanism is interfacial sliding (Stage III). Within Stage III, the stress-strain curve may regain linearity, although with a lower tangential modulus than the initial composite modulus. Distributed fiber failures (Stage IV) may occur before localization (final fracture).

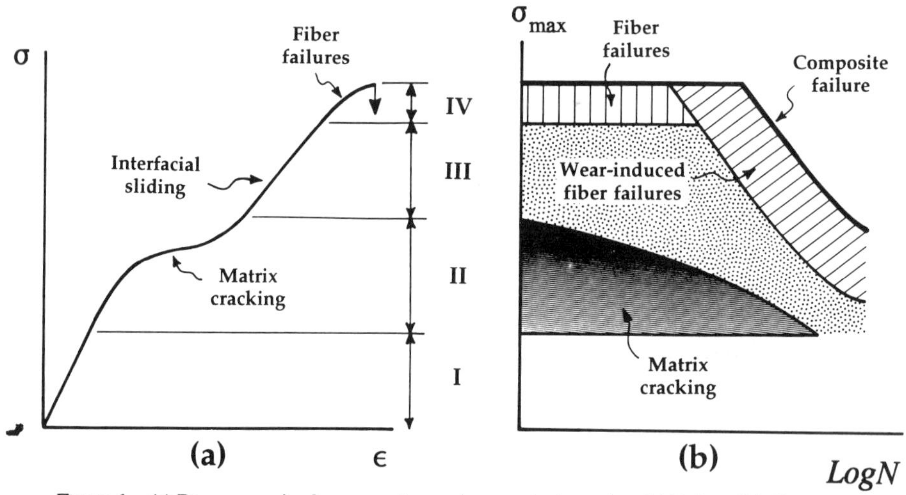

Figure 1. (a) Damage mechanisms operating under monotonic tension. (b) Fatigue life diagram, showing the fatigue mechanisms as a function of maximum applied stress and the number of cycles (modified after Talreja [25]).

The stress-strain behavior may be strongly dependent on the loading rate [8]. The mechanism for this is two-fold: (1) Composites with glass or glass-ceramic matrices may experience slow-crack-growth due to stress-corrosion [9], such that the matrix crack spacing increases with increasing loading rate, and (2) the interfacial frictional shear stress may increase with increasing loading rate [8, 10]. Since damage tolerance is only obtained in composites with relatively low interfacial fracture toughness and low interfacial shear stress, the debonding process will be ignored in the remainder of this paper (*i.e.*, the interface will be treated as purely frictional).

3. Damage Mechanisms under Cyclic Loading (Tension-Tension Fatigue)

During the loading portion of the first fatigue cycle, the evolution of microstructural damage is identical to the damage that develops under monotonic tension to the same load level, *i.e.* below the stress at which matrix cracks initiate, σ_{mc}^{ini} , no damage develops (Stage I). If the maximum applied stress, σ_{max}, exceeds σ_{mc}^{ini} , matrix cracks form (Fig. 1b). Matrix cracking and fiber/matrix debonding cause non-linear stress-strain behavior and hysteresis during cyclic loading (due to frictional sliding), as well as a permanent strain, ε^{*}[11], Fig. 2. The energy dissipation due to repeated forward and reverse frictional sliding results in a temperature rise, ΔT, during cycling [12, 13]. As a result of matrix cracking and interfacial sliding, the composite stiffness (e.g. the average composite modulus, $\overline{E}$ decreases [13-15], as indicated in Fig. 3. When matrix cracking has saturated, the modulus remains nearly constant, although a slight modulus recovery is sometimes observed [13, 15].

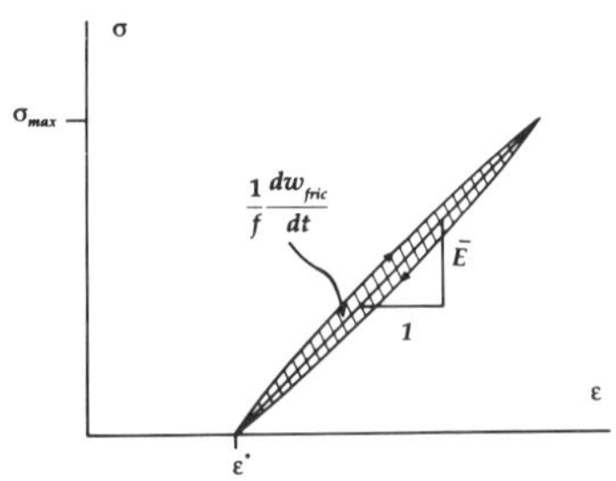

Figure 2. A cyclic stress-strain curve showing hysteresis (the energy dissipation per load cycle $1/f\, dw_{fric}/dt$) and a permanent offset strain, ε^{*}. The average composite modulus is also defined.

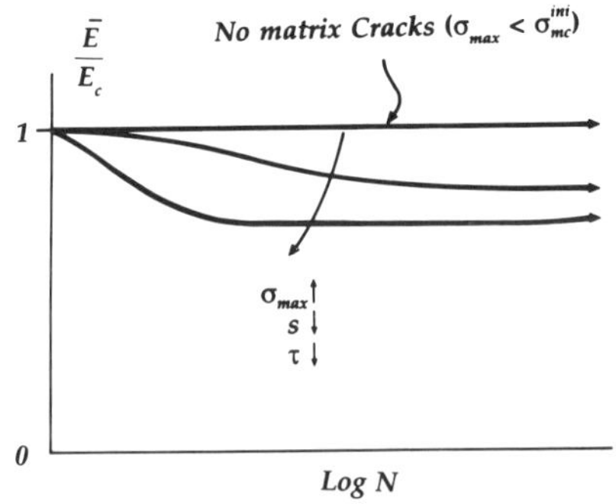

Figure 3. During cycling the average composite modulus decreases as the matrix crack spacing s or the interfacial shear stress τ decreases. The modulus stabilizes when the microstructural damage (matrix cracking, interfacial wear) stabilizes.

Fatigue failure will occur when the residual composite strength, σ_u, has decreased to the value of the maximum applied stress, σ_{max} [16], Fig. 4. Fatigue life is often described by an S-N curve (Fig. 5), *i.e.*, σ_{max} is plotted as a function of the number of cycles to failure N_f. A fatigue limit, σ_{fl}, is defined as the maximum applied stress level that a composite can withstand for a specific number of cycles (typically 10^6 to 10^8) without failure. A fatigue limit should only be claimed to exist if the evolution of fatigue damage stabilizes. For CMC's, the authors suggest the following conditions to be met to specify a fatigue limit: (1) the number of cycles defined as run-out is at least an order of magnitude higher than the highest number of cycles causing fatigue failure, and (2) microstructural damage, as measured by changes in $\overline{E}$ and ΔT remains unchanged during cycling below the fatigue limit. Previous work by the authors suggests that a true fatigue limit may only exist below σ_{mc}^{ini} .

Interfacial sliding and wear play a key role in the behavior of damage tolerant CMCs. For clarity, this mechanism will be described in detail. During loading, fiber/matrix sliding starts from matrix cracks. Sticking friction exists in the middle of the matrix blocks. The fiber slip-length, l_s, increases as the applied stress is increased, and the slip-zone may meet in the middle of the matrix blocks. The latter is termed *partial/full slip* [17]. If unloading occurs before full slip is attained, the composite is said to experience *partial slip*. During unloading the fibers start to slip in the opposite direction (reverse slip). As with loading, the slip-zones start from matrix cracks, while the centre parts of the matrix blocks experience sticking friction. It is the changing slip-length that causes non-linear stress-strain hysteresis loops. Under full slip conditions the stress-strain relationship is linear (Fig. 6).

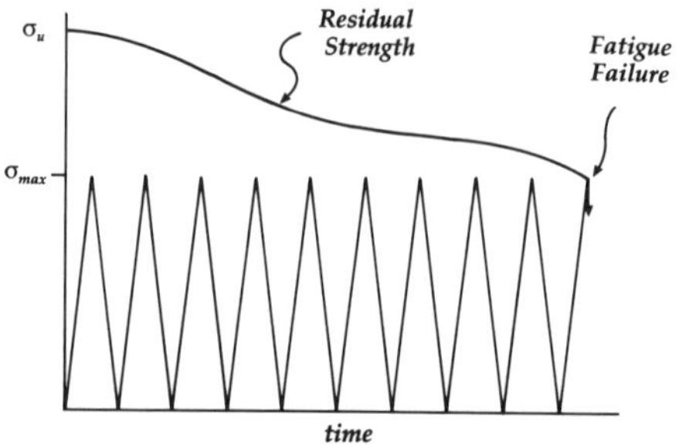

Figure 4. Schematic showing that fatigue failure occurs when the residual composite strength has decreased to the value of the maximum applied stress.

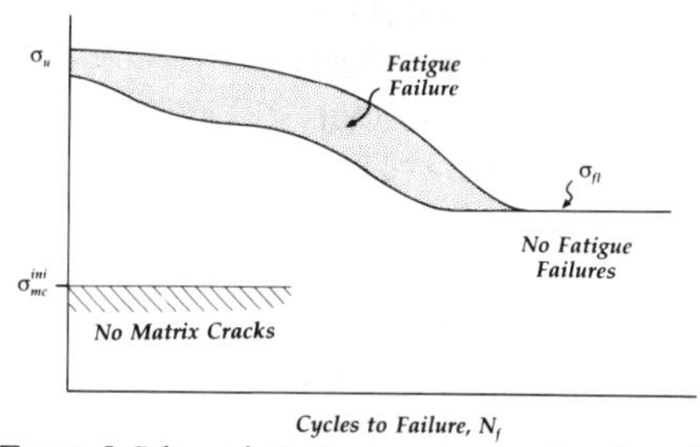

Figure 5. Schematic S-N curve; σ_{max} as a function of the number of cycles to failure.

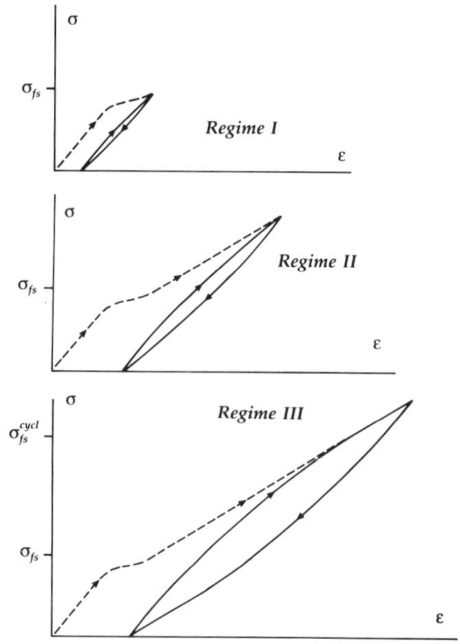

Figure 6. Depending upon the maximum applied stress, the composite may experience partial slip (Regime I), partial/full slip in the first load cycle and partial slip in the remaining (Regime II) or partial/full slip during all cycles (Regime III).

4. Models of Composite Behaviour

The modulus of an *undamaged* composite can be estimated from the rule of mixtures,

$$E_c = (1-v_f)E_m + v_f E_f , \tag{1}$$

where E_f and E_m are the moduli of the fibers and matrix, respectively, and v_f is the fiber volume fraction.

The *cyclic stress-strain behavior* of unidirectional CMC's with a regular matrix crack spacing, *s*, can be described with reasonable accuracy by relatively simple shear-lag models. The basic assumptions of the models are that all fibers are intact, the fiber/matrix interface is treated as purely frictional, and Poisson's effects as well as fiber roughness effects [18] are neglected. For simplicity, interfacial friction is modelled by a constant interfacial shear stress, τ. Recent experiments indicate, however, that τ should not be treated as a material constant. Rather, it is a parameter that is sensitive to wear [13], lubrication [19], velocity [8, 10] and temperature. However, while τ may change during cycling, it is reasonable to assume that τ should remain nearly constant within each load cycle.

We must distinguish between the two stages of interfacial slip. If the maximum stress σ_{max} exceeds a characteristic value, σ_{fs}, given by

$$\sigma_{fs} = \left\{\frac{s}{r}\tau + \sigma_f^{res}\right\}\frac{E_c}{E_m}\frac{\nu_f}{1-\nu_f}, \tag{2}$$

full slip will occur during all stages of the first loading cycle. In Eq. 2, *r* is the fiber radius and σ_f^{res} is the axial residual stress in the fibers (*i.e.*, the axial stress in the fibers in the virgin and unstressed composite). If the cyclic stress range, Δσ, is smaller than a characteristic value

$$\Delta\sigma_{fs}^{cycl} = \frac{2\nu_f}{1-\nu_f}\frac{E_c}{E_m}\frac{s}{r}\tau, \tag{3}$$

then the composite will experience partial slip during cyclic loading [17]. During unloading, the reverse slip-length after complete unloading is (within the approximations of the model) half the initial slip-length during the first load cycle [20]. It is therefore possible that full slip can occur in the first loading cycle, while partial slip will occur in the subsequent unloading and reloading cycles. Therefore, as shown in Fig. 6, a distinction must be made between three situations [21]: Partial slip in the first and subsequent cycles (Regime I), partial/full slip in first cycle, but partial slip in the subsequent cycles (Regime II) and partial/full slip in all cycles (Regime III).

In Regime I the average composite modulus can be estimated from the model of Pryce and Smith [20] to be [22]

$$\overline{E}_{ps} = \frac{\tau}{\dfrac{\tau}{E_c} + \dfrac{r}{4s}\dfrac{\Delta\sigma}{E_f}\left(\dfrac{1-\nu_f}{\nu_f}\dfrac{E_m}{E_c}\right)^2}, \tag{4}$$

where the subscript *ps* indicates partial slip. Since the average composite modulus depends on the applied stress range, it is not a material constant.

The permanent offset strain ε^* (axial strain at $\sigma = 0$) during cycling between zero and σ_{max} is [20]

$$\varepsilon^* = \frac{r}{4s\tau E_f}\left[\sigma_{max}^2\left(\frac{E_m}{E_c}\frac{1-v_f}{v_f}\right)^2 - 4\sigma_{max}\sigma_f^{res}\frac{E_m}{E_c}\frac{1-v_f}{v_f} + 2\sigma_f^{res2}\right], \tag{5}$$

and the frictional energy dissipation per unit time, dw_{fric}/dt, is given by the model of Cho *et al.* [17]

$$\left(\frac{dw_{fric}}{dt}\right)_{ps} = \frac{f}{12}\left[\frac{1-v_f}{v_f}\frac{E_m}{E_f}\right]^2\frac{r}{s}\frac{\Delta\sigma^3}{E_f\tau}, \tag{6}$$

where f is the loading frequency.

In Regime II the average composite modulus and the frictional energy dissipation are still found using (4) and (6), respectively, while the permanent strain becomes [21]

$$\varepsilon^* = -\frac{r}{s}\frac{\sigma_{max}^2}{4E_f\tau}\left(\frac{1-v_f}{v_f}\frac{E_m}{E_c}\right)^2 + \frac{\sigma_{max}}{E_f}\frac{1-v_f}{v_f}\frac{E_m}{E_c} - \frac{s}{2r}\frac{\tau}{E_f} - \frac{\sigma_f^{res}}{E_f}. \tag{7}$$

The average composite modulus in Regime III (partial/full slip) can be found from the model of Sørensen and Talreja [5] (see [22]),

$$\overline{E}_{fs} = \frac{v_f E_f}{1 - v_f \frac{s}{r}\frac{\tau}{\Delta\sigma}}, \tag{8}$$

where the subscript *fs* indicates full slip. For full slip, the rate of frictional energy dissipation can be estimated by [17]

$$\left(\frac{dw_{fric}}{dt}\right)_{fs} = f\frac{s}{r}\frac{\tau}{E_f}\Delta\sigma - \frac{4f}{3}\frac{s^2}{r^2}\left(\frac{v_f}{1-v_f}\frac{E_m}{E_c}\right). \tag{9}$$

The permanent strain in Regime III can be calculated from [5]

$$\varepsilon^* = \frac{s}{2r}\frac{\tau}{E_f} - \frac{\sigma_f^{res}}{E_f}. \tag{10}$$

Note that the equations for ε^* have been derived on the basis of isothermal cycling. As mentioned earlier, frictional energy dissipation by cyclic interfacial slip will cause heating of CMC fatigue specimens. Depending on the stress range and loading frequency the temperature rise can be of the order of 10-100°C [23]; higher temperatures may be present in the vicinity of the fiber/matrix interface.

The above models describe the behavior of multiple-matrix-cracked CMC's. Another important property is the fatigue life, *i.e.*, when the composite strength has decreased to σ_{max}. The composite *strength* can be predicted by models that describe the fiber strength by the two parameter Weibull distribution [16, 24]

$$\sigma_u = \Lambda(m)\ v_f\ \sigma_c , \tag{11}$$

where σ_c is a characteristic strength value given by

$$\sigma_c = \left(\frac{\sigma_o^m L_0 \tau}{r}\right)^{\frac{1}{m+1}} \tag{12}$$

In Eqs. 11 and 12, $\sigma_0(L_0)$ is the stress level where 63.2% of the fibers have failed for a gauge-length L_0, m is the Weibull modulus, and $\Lambda(m)$ is a dimensionless function that depends only on the Weibull modulus. Models predict that the average fiber pull-out length l_p is related to the fiber strength and interfacial sliding shear stress by

$$l_p = \frac{\lambda(m)}{4}\frac{\sigma_c}{\tau} r , \tag{13}$$

where $\lambda(m)$ is a dimensionless function close to unity that depends vaguely on m [24].

5. Fatigue Damage Evolution

In addition to matrix cracking, and as pointed out by Talreja [25], the fiber/matrix interface is the primary site within a brittle matrix composite where room temperature fatigue damage accumulation can evolve. Repeated forward and reverse sliding may cause abrasive wear of roughness along the fiber/matrix interface, and lower the interfacial shear stress by decreasing the interfacial strain mismatch [13]. For ceramic composites with a relatively soft carbon-interphase layer, this wear appears to occur within a relatively short number of cycles. From Eqs. 11 and 12 it follows that a lower value of τ should result in a lower composite residual strength (and longer fiber pull-out length). Thus, a decrease in τ is a likely cause for *low cycle fatigue* failures [16]. The reduction in τ increases the net stress on the fibers and also increases the volume of a fiber subjected to a higher than average stress. Repeated cycling will also induce wear damage along the surface of the fibers (particularly in relatively soft carbon-fibers [26]). It is plausible that for ceramic fibers (which have high wear resistance) it may take a large number of cycles before the wear-induced surface flaws have become larger than the initial flaws in the fibers. Thus, wear-induced damage to the fibers may be a *high cycle fatigue mechanism* (Fig. 7). In models, wear induced fiber damage can be described by a decreasing value of σ_0 and a changing m (or decreasing σ_c). Then, Eqs.[11] and [12] predict a decreasing l_p and σ_u. Sørensen and Holmes [27] have shown that *both* the fatigue life and fatigue limit increase with increasing σ_{min} (for fixed values of σ_{max}). This shows that it is the *amount* of interfacial or fiber wear that controls the fatigue life (Note, that the extent of sliding, and therefore the wear rate, decreases as the stress range decreases). Since the basic damage mechanisms (matrix cracking and interfacial sliding) have been found to be time or velocity-dependent under monotonic tension [8], it is not surprising that the fatigue damage evolution is strongly dependent on the loading frequency. Indeed, it has been shown that the fatigue life decreases sharply with increasing loading frequency [23].

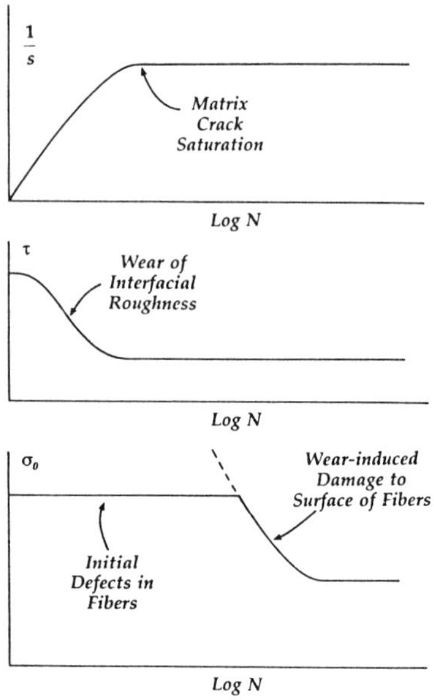

Figure 7. Damage evolution laws for matrix crack spacing, the interfacial shear stress and fiber strength.

It is useful to make a distinction between fatigue damage parameters and damage indicators. *Damage parameters* are defined as the basic microstructural parameters that may change during fatigue, *i.e.* the parameters that cause changes in the behavior of a composite. These are (within the framework of the present models) s, τ and σ_c (or $\sigma_0(L_0)$ and m). We define *damage indicators* as measurable parameters that reflect changes in the underlying damage parameters. In the present study we will consider $\overline{E}$, ε^* and dw/dt as damage indicators (other damage indicators can, however, be chosen, *e.g.*, the opening displacement of matrix cracks). The composite modulus is particularly convenient, since, as noted by Sørensen and Holmes [22], the transition between partial and full slip is predicted to occur at an average modulus given by

$$\overline{E}_{ps\text{-}fs} = \frac{\nu_f E_f}{1 - \dfrac{1-\nu_f}{2}\dfrac{E_m}{E_c}}, \qquad (14)$$

which is *independent of the matrix crack spacing and* τ. Therefore, if the measured value of $\overline{E}$ is lower than $E_{ps\text{-}fs}$ one can conclude that the composite experiences partial/full slip, otherwise partial slip will be present.

6. Fatigue Methodology

In order to determine all relevant (intrinsic) composite parameters, it is useful to choose a methodology that can extract the fatigue damage evolution laws from relatively few experiments which will allow the composite behavior ($\overline{E}$, ε^*, ΔT and fatigue lifetime) to be estimated. According to the models, eight parameters are required to describe the composite behavior: E_f, E_m, r, v_f, σ_f^{res} s, τ and σ_c.

First we note that E_f, E_m, r, v_f and , σ_f^{res} remain constant for a given system, while s, τ and σ_c may change during cycling (Fig. 7). Usually, E_f, and E_m are known. From optical microscopy of a polished cross section of a virgin composite, v_f and r can be determined. The remaining parameters must be found by tensile and fatigue experiments; key experiments are summarized in Fig. 8.

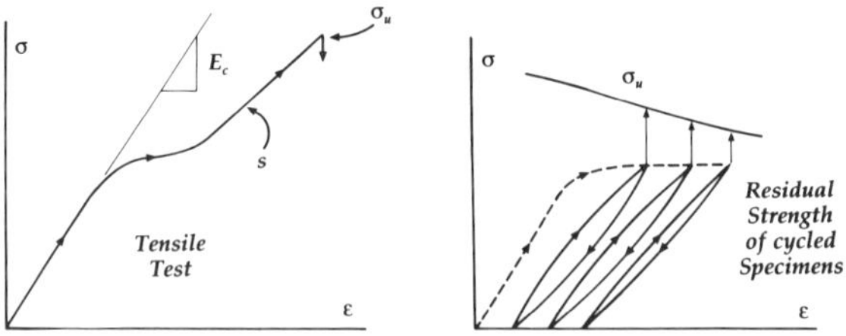

Figure 8. Key experiments required to determine the parameters used to model fatigue damage in a glass-ceramic matrix CMC. Static cracking due to stress-corrosion provides saturated matrix cracking. A tensile test gives the initial composite strength. By measuring the residual strenght of fatigue specimens, the loss of composite strength can be determined.

A *tensile test* conducted to failure can give valuable information. Using the rule of mixtures (Eq. 1), the measured value of E_c provides a check on the fiber volume fraction v_f. The tensile test can also provide σ_u, which can be used to calculate the initial value of σ_c, hereafter denoted $\sigma_c(N=1)$.l_p can be measured by scanning electron microscopy of the fracture surface. Optical microscopy of a polished face of the tensile specimen gives the saturated crack spacing, s.

For matrices that are prone to stress corrosion, matrix cracking can take place under *static loading*, and with time a saturated crack spacing (relatively insensitive to the applied stress level, once it is well above σ_{mc}^{ini} [9]) will be reached [19]. The unloading stress-strain data give $\overline{E}$ and ε^*. Again, s can be measured by optical microscopy. Then $\overline{E}$ and s give τ *(N=1)*, and ε^* provides a means to estimate σ_f^{res}. In this manner all *initial* parameters are determined.

Cyclic tests are needed to determine the fatigue damage evolution laws. The basic assumptions are that during fatigue s, τ and σ_c are functions of N, σ_{max}, σ_{min} and f. In order to study the evolution of fatigue damage, it is preferable to perform cyclic tests on specimens that have been subjected to static loading, such that the matrix crack spacing remains constant during the subsequent fatigue test (the matrix crack spacing appears to be of minor importance for the evolution of strength controlling fatigue damage, see [22] for a detailed discussion). $\tau(N)$ can be calculated from data on $\overline{E}$ as a function of N. The cyclic stress-strain data and the frictional heating data as functions of N can be used as consistency checks on $\tau(N)$. To determine the changes in τ during cyclic loading, the temperature rise of specimens subjected to cyclic loading is measured as a function of the number of cycles [17]. Under quasi-static conditions, there is equilibrium between heat generation *dw/dt* (by frictional energy dissipation) and heat loss *dq/dt* (by radiation, convection and heat conduction) in a test specimen (see Cho *et al.* [17] for a detailed discussion). Hence, from the temperature rise, the frictional energy dissipation, and thus τ, can be estimated [13, 17]. When conducting the fatigue experiments, some specimens should be cycled to failure. Other specimens should be cycled to a preselected number of cycles, N_1, N_2, etc. These specimens are then tested in uniaxial tension to failure. This gives $\sigma_u(N)$, which is used to calculate σ_c at $N = N_1$, $N = N_2$, etc. In this manner, the damage evolution laws for τ and σ_c can be measured.

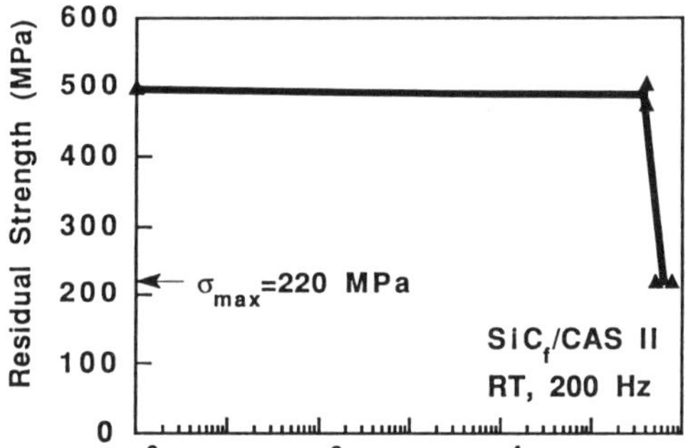

Figure 9. Measured residual strength of $[0]_{16}$ SiC$_f$/ CAS II composites. Two specimens were cycled to failure. Two additional specimens were cycled to 400 000 cycles (approximately 80% of the fatigue life); their strength was almost as high as the virgin material, indicating that fatigue damage remains stable until some event triggers a rapid damage accumulation within a relatively short number of cycles immediately prior to failure.

Results from cyclic experiments on SiC-fiber-reinforced calcium aluminosilicate (Nicalon SiC_f/CAS II) indicate that τ decreases within a short number of cycles and remains nearly constant in subsequent cycles [13] (a slight recovery in τ appears to occur with continued cycling; apparently from wear debris along the interface). Specimens cycled to a preselected number of cycles ($N < N_f$), indicate that the composite strength is retained for up to 90% of the fatigue lifetime [22]. This suggests that the primary fatigue damage mechanism in SiC_f/CAS II is wear damage to the fibers (*i.e.*, decreasing σ_0), and that the decrease in fiber strength occurs within a relatively short number of cycles (Fig. 9) immediately prior to failure.

7. High Temperature Fatigue Damage Mechanisms

An understanding of the damage mechanisms that occur during *high temperature isothermal mechanical cycling* is of major importance. In addition to oxidation damage to the interphase [3], creep of the fibers and matrix may take place [28]. Ideally, future composites will be engineered such that both of these phenomena will occur at minimal rates. In the absence of these phenomena, fatigue damage can still accumulate by the wear of debonded fiber/matrix interfaces and wear of the fibers. High temperatures will affect the thermal mismatch between the fibers and matrix, and, therefore, τ. This will directly affect the composite modulus and residual strength.

Thermal cycling of a multiple-matrix-cracked composite may lead to interfacial sliding due to differential thermal expansion between the fibers and matrix. Thus, precisely the same damage mechanisms that operate during mechanical cycling (decreasing τ and σ_0 due to interfacial sliding) may exist during thermal cycling. This has not yet been studied experimentally. Modelling of the thermal cycling process [29] has revealed two competing effects: (1) Sliding due to the difference in axial thermal expansion between the fibers and matrix, and (2) a change in τ with temperature. This leads to a complicated forward and reverse slipping process as well as slip/stick behavior along the fiber/matrix interface.

An understanding of *in-phase* and *out-of-phase* thermomechanical cycling will also be of major importance for the successful microstructural design and application of CMCs. In addition to thermochemical effects (damage to the interphase material) or creep, purely mechanical fatigue damage can be expected. For multiple-matrix-cracked composites where the thermal expansion coefficient of the matrix is higher than that of the fibers, in-phase cycling will lead to a higher permanent strain, but less interfacial sliding than out-of-phase thermomechanical cycling. This can be understood by noting that interfacial slip occurs due to differences in axial strain between the fibers and matrix. During heating, the thermal strain of the matrix exceeds that of the fibers, whereas during mechanical loading the fiber strain exceeds that of the matrix. Thus, during in-phase cycling, the two strain contributions tend to oppose each other, while the two contributions magnify each other during out-of-phase cycling.

8. Conclusions

This paper has given a broad description of the mechanisms and models that control the thermomechanical fatigue behavior of CMCs, and has highlighted the approaches used to model fatigue life. While the damage mechanisms operating under room temperature mechanical cycling are reasonably well understood (both experimentally and analytically), the behavior under high temperature conditions is still poorly understood. Additional research must be undertaken before reliable design criteria can be stated with any degree of confidence.

Acknowledgements - This work was initiated when B.F.S. was a visiting research investigator at the University of Michigan. B.F.S. was supported by Risø Engineering Science Centre for Structural Characterization and Modelling of Materials. The authors also would like to acknowledge support by the National Science Foundation (Grant No. DMR-9257557) and the Air Force Office of Scientific Research (Grant No. F49620-95-1-0206).

References

1. Prewo, K.M., Johnson, B. and Starrett, S., Silicon carbide fibre-reinforced glass-ceramic composite tensile behavior at elevated temperature, *J. Mater. Sci.*, **24** (1989) 1373-9.
2. Bischoff, E., Rühle, M., Sbaizero, O. and Evans A.G., Microstructural studies of the interface zone of a SiC-fiber-reinforced lithium aluminum silicate glass-ceramic, *J. Am. Ceram. Soc.*, **72** (1989), 741-45.
3. Filipuzzi, L., Camus, G., Thébault, J. and Naslain, R., Effect of high temperature aging treatments on the mechanical behavior of unidirectional SiC/SiC fibrous composites, in *Structural Ceramics, Processing Microstructure and Properties*, J.J. Bentzen, J.B., Bilde-Sørensen, N. Christiansen, A., Horsewell and B. Ralph(eds.), Risø National Laboratory, Roskilde, Denmark, (1990) pp. 283-289.
4. Holmes, J.W. and Sørensen, B.F., Fatigue behavior of continuous fiber-reinforced ceramic matrix composites, *Elevated temperature Behavior of Ceramic Matrix Composites*, S.V. Nair and K. Jakus (eds.), Butterworth Heineman, (1995), 261-326.
5. Sørensen, B.F. and Talreja, R., Analysis of damage in ceramic matrix composites, *Int. J. Damage Mechanics* **2** (1993), 246-71.
6. Karandikar, P. and Chou, T.-W., Characterization and modelling of microcracking and elastic moduli changes in Nicalon/CAS, *Comp. Sci. Techn.*, **46** (1993), 253-63.
7. Kim, R.Y. and Pagano, N. J., Crack initiation in unidirection brittle-matrix composites, *J. Am. Ceram. Soc.*, **74** (1991), 1082-90.
8. Sørensen, B.F. and Holmes, J.W., Effect of loading rate on the monotonic tensile behavior and matrix cracking of a fiber-reinforced ceramic, *J. Am. Ceram. Soc.*, accepted, 1995.
9. Spearing, S.M., Zok, F.M. and Evans, A.G., Stress corrosion cracking in a unidirectional ceramic-matrix composite, *J. Am. Ceram. Soc.*, **77** (1974), 562-70.
10. Goettler, R.W. and Faber K.T., Interfacial shear stress in fiber-reinforced glasses, *Comp. Sci. Techn.*, **37** (1989), 129-47
11. Rousseau, C.Q., Monotonic and cyclic behavior of a silicon carbide/calcium-alumunosilicate ceramic composite, in J.M. Kennedy, H.H. Moeller, W.S. Johnson (eds.), *Thermal and Mechanical Behavior of Metal Matrix and Ceramic Matrix Composites*, ASTM STP 1008, American Society for Testing and Materials, Philadelphia (1990), pp. 136-51.
12. Holmes, J.W. and Shuler, S.F., Temperature rise during fatigue of fibre-reinforced ceramics, *J. Mater. Sci. Lett.*, **9** (1990), 1290-1.
13. Holmes, J.W. and Cho, C., Experimental observations of frictional heating in fiber-reinforced ceramics, *J. Am. Ceram. Soc.* **75** (1992), 929-38.
14. Butkus, L.M., Zawada, L.P. and Hartman, G.A., Room temperature tensile and fatigue properties of silicon-carbide fiber-reinforced ceramic matrix composites, presented at *Aeromat' 90, Advanced Aerospace Materials/Processes Conference*, 21-24 May 1990, Long Beach, California.
15. Zawada, L.P., Butkus, L.M. and Hartman, G.A., Room temperature tensile and fatigue properties of silicon carbide fiber-reinforced aluminosilicate glass, *Ceram. Eng. Sci. Proc.* **11** (1990) 1592-606.

16. Rouby, D. and Reynaud, P., Fatigue behavior related to the interface modification during load cycling in ceramic-matrix composites, *Comp. Sci. Techn.*, **48** (1993), 109-118.
17. Cho, C., Holmes, J.W. and Barber, J.R., Estimation of interfacial shear in ceramic composites from frictional heating measurements, *J. Am. Ceram. Soc.*, **74** (1991), 2802-8.
18. Jero, P.D. and Kerans, R.J., The contribution of interfacial roughness to sliding friction of ceramic fibers in a glass matrix, *Scripta Metall. Mater.*, **24** (1990), 2315-8.
19. Sørensen, B.F. and Holmes, J.W., Improvement in the fatigue life of fiber-reinforced ceramics by use of interface lubrication, *Scripta Metall. Mater.* **32** (1995), 1393-8.
20. Pryce, A.W. and Smith, P.A., Matrix cracking in unidirectional ceramic matrix composites under quasi-static and cyclic loading, *Acta Metall. Mater.*, **41** (1993), 1269-81.
21. Keith, W.P. and Kedward, K.T., The stress-strain behavior of a porous unidirectional ceramic matrix composite, *Composites*, **26** (1995), 163-74.
22. Sørensen, B.F. and Holmes, J.W., Does a true fatigue limit exist for continuous fiber-reinforced ceramic matrix composites?, *J. Am. Ceram. Soc.*, (1995), manuscript under preparation.
23. Holmes, J.W., Wu, X. and Sørensen, B.F., Frequency dependence of fatigue life and internal heating of a fiber-reinforced ceramic matrix composite, *J. Am. Ceram. Soc.*, 77 (1994), 3284-6
24. Curtin, W. A., Theory of mechanical properties of ceramic matrix composites, *J. Am. Ceram. Soc.*, **74** (1991), 2837-45.
25. Talreja, R., Fatigue of fibre-reinforced ceramics, in J.J. Bentzen, J.B. Bilde-Sørensen, N. Christiansen, A. Horsewell and B. Ralph (eds.), *Structural Ceramics, Processing Microstructure and Properties,* Risø National Laboratory, Roskilde, Denmark, (1990), pp. 145-59.
26. Morris, W.L, Cox, B.N. Marshall, D.B., Inman, R.V. and James, M.R., Fatigue mechanisms in graphite/SiC composites at room and high temperature, *J. Am. Ceram. Soc.* 77 (1994), 794-800.
27. Sørensen, B.F. and Holmes, J.W., Influence of stress ratio on the fatigue life of a continuous fiber-reinforced ceramic matrix composite, (1995), manuscript under preparation.
28. Wu, X. and Holmes, J.W., Tensile creep and creep-strain recovery behavior of silicon carbide fiber/calcium aluminosilicate matrix ceramic composites, *J. Am. Ceram. Soc.*, **76** (1993), 2695-700.
29. Cox, B.N., Interfacial sliding near a free surface in a fibrous or layered composite during thermal cycling, *Acta Metall. Mater.*, **38** (1990), 2411-24.

Author Index

Subject Index